HEAT TRANSFER

THERMAL MANAGEMENT OF ELECTRONICS

HEAT TRANSFER
THERMAL MANAGEMENT OF ELECTRONICS

YOUNES SHABANY

CRC Press is an imprint of the
Taylor & Francis Group, an **informa** business

CRC Press
Taylor & Francis Group
6000 Broken Sound Parkway NW, Suite 300
Boca Raton, FL 33487-2742

© 2010 by Taylor and Francis Group, LLC
CRC Press is an imprint of Taylor & Francis Group, an Informa business

No claim to original U.S. Government works

Printed in the United States of America on acid-free paper
10 9 8 7 6 5 4 3 2 1

International Standard Book Number: 978-1-4398-1467-3 (Hardback)

This book contains information obtained from authentic and highly regarded sources. Reasonable efforts have been made to publish reliable data and information, but the author and publisher cannot assume responsibility for the validity of all materials or the consequences of their use. The authors and publishers have attempted to trace the copyright holders of all material reproduced in this publication and apologize to copyright holders if permission to publish in this form has not been obtained. If any copyright material has not been acknowledged please write and let us know so we may rectify in any future reprint.

Except as permitted under U.S. Copyright Law, no part of this book may be reprinted, reproduced, transmitted, or utilized in any form by any electronic, mechanical, or other means, now known or hereafter invented, including photocopying, microfilming, and recording, or in any information storage or retrieval system, without written permission from the publishers.

For permission to photocopy or use material electronically from this work, please access www.copyright.com (http://www.copyright.com/) or contact the Copyright Clearance Center, Inc. (CCC), 222 Rosewood Drive, Danvers, MA 01923, 978-750-8400. CCC is a not-for-profit organization that provides licenses and registration for a variety of users. For organizations that have been granted a photocopy license by the CCC, a separate system of payment has been arranged.

Trademark Notice: Product or corporate names may be trademarks or registered trademarks, and are used only for identification and explanation without intent to infringe.

Library of Congress Cataloging-in-Publication Data

Shabany, Younes.
   Heat transfer : thermal management of electronics / author, Younes Shabany.
     p. cm.
   "A CRC title."
   Includes bibliographical references and index.
   ISBN 978-1-4398-1467-3 (hardcover : alk. paper)
    1. Electronic apparatus and appliances--Thermal properties. 2. Electronic apparatus and appliances--Protection. 3. Heat sinks (Electronics) 4. Electronic packaging. 5. System failures (Engineering)--Prevention. I. Title.

TK7870.25.S53 2010
621.381--dc22
                                                                 2009042415

Visit the Taylor & Francis Web site at
http://www.taylorandfrancis.com

and the CRC Press Web site at
http://www.crcpress.com

To the loves of my life, Saeedeh and Arash

Contents

Preface xiii
About the Author xv

Chapter 1 Introduction 1
        1.1   Semiconductor Technology Trends 3
        1.2   Temperature-Dependent Failures 6
              1.2.1   Temperature-Dependent Mechanical Failures 8
              1.2.2   Temperature-Dependent Corrosion Failures 12
              1.2.3   Temperature-Dependent Electrical Failures 12
        1.3   Importance of Heat Transfer in Electronics 13
        1.4   Thermal Design Process 14
        References 18

Chapter 2 Energy, Energy Transfer, and Heat Transfer 19
        2.1   Energy and Work 19
        2.2   Macroscopic and Microscopic Energies 20
        2.3   Energy Transfer and Heat Transfer 24
        2.4   Equation of State 25
        Problems 28
        References 28

Chapter 3 Principle of Conservation of Energy 29
        3.1   First Law of Thermodynamics 29
        3.2   Energy Balance for a Control Mass 31
        3.3   Energy Balance for a Control Volume 36
        Problems 44
        References 51

Chapter 4 Heat Transfer Mechanisms 53
        4.1   Conduction Heat Transfer 53
        4.2   Convection Heat Transfer 57
              4.2.1   Simplified Correlations for Convection Heat
                         Transfer in Air 58
        4.3   Radiation Heat Transfer 61
        Problems 64
        References 65

Chapter 5	Thermal Resistance Network	67
	5.1 Thermal Resistance Concept	67
	5.2 Series Thermal Layers	72
	5.3 Parallel Thermal Layers	75
	5.4 General Resistance Network	79
	5.5 Thermal Contact Resistance	82
	5.6 Thermal Interface Materials	85
	5.7 Spreading Thermal Resistance	88
	5.8 Thermal Resistance of Printed Circuit Boards (PCBs)	91
	Problems	97
	References	101
Chapter 6	Thermal Specification of Microelectronic Packages	103
	6.1 Importance of Packaging	103
	6.2 Packaging Types	103
	6.3 Thermal Specifications of Microelectronic Packages	110
	6.3.1 Junction-to-Air Thermal Resistance	110
	6.3.2 Junction-to-Case and Junction-to-Board Thermal Resistances	112
	6.3.3 Package Thermal Characterization Parameters	114
	6.4 Package Thermal Resistance Network	115
	6.5 Parameters Affecting Thermal Characteristics of a Package	119
	6.5.1 Package Size	119
	6.5.2 Packaging Material	119
	6.5.3 Die Size	119
	6.5.4 Device Power Dissipation	121
	6.5.5 Air Velocity	121
	6.5.6 Board Size and Thermal Conductivity	122
	Problems	123
	References	125
Chapter 7	Fins and Heat Sinks	127
	7.1 Fin Equation	127
	7.1.1 Infinitely Long Fin	130
	7.1.2 Adiabatic Fin Tip	132
	7.1.3 Convection and Radiation from Fin Tip	133
	7.1.4 Constant Temperature Fin Tip	135
	7.2 Fin Thermal Resistance, Effectiveness, and Efficiency	140
	7.3 Fins with Variable Cross Sections	146
	7.4 Heat Sink Thermal Resistance, Effectiveness, and Efficiency	150
	7.5 Heat Sink Manufacturing Processes	160

Contents

	Problems	164
	References	168

Chapter 8 Heat Conduction Equation ... 169

    8.1 One-Dimensional Heat Conduction Equation for a Plane Wall ... 171
    8.2 General Heat Conduction Equation ... 175
    8.3 Boundary and Initial Conditions ... 177
        8.3.1 Temperature Boundary Condition ... 178
        8.3.2 Heat Flux Boundary Condition ... 179
        8.3.3 Convection Boundary Condition ... 181
        8.3.4 Radiation Boundary Condition ... 183
        8.3.5 General Boundary Condition ... 185
        8.3.6 Interface Boundary Condition ... 185
    8.4 Steady-State Heat Conduction ... 187
        8.4.1 One-Dimensional, Steady-State Heat Conduction ... 187
        8.4.2 Two-Dimensional, Steady-State Heat Conduction ... 191
    8.5 Transient Heat Conduction ... 194
    8.6 Lumped Systems ... 196
        8.6.1 Simple Lumped System Analysis ... 196
        8.6.2 General Lumped System Analysis ... 198
        8.6.3 Validity of Lumped System Analysis ... 202
    Problems ... 204
    References ... 208

Chapter 9 Fundamentals of Convection Heat Transfer ... 209

    9.1 Type of Flows ... 209
        9.1.1 External and Internal Flows ... 209
        9.1.2 Forced and Natural Convection Flows ... 210
        9.1.3 Laminar and Turbulent Flows ... 210
        9.1.4 Steady-State and Transient Flows ... 212
    9.2 Viscous Force, Velocity Boundary Layer, and Friction Coefficient ... 212
    9.3 Temperature Boundary Layer and Convection Heat Transfer Coefficient ... 214
    9.4 Conservation Equations ... 215
    9.5 Boundary Layer Equations ... 217
    References ... 218

Chapter 10 Forced Convection Heat Transfer: External Flows ... 219

    10.1 Normalized Boundary Layer Equations ... 219
    10.2 Reynolds Number, Prandtl Number, Eckert Number, and Nusselt Number ... 221

	10.3	Functional Forms of Friction Coefficient and Convection Heat Transfer Coefficient 223
	10.4	Flow over Flat Plates 227
		10.4.1 Laminar Flow over a Flat Plate with Constant Temperature 227
		10.4.2 Turbulent Flow over a Flat Plate with Uniform Temperature 234
		10.4.3 Flow over a Flat Plate with Uniform Surface Heat Flux 237
	10.5	Flow Across Cylinders 239
	10.6	Cylindrical Pin-Fin Heat Sink 244
	10.7	Procedure for Solving External Forced Convection Problems 246
	Problems 248
	References 253

Chapter 11 Forced Convection Heat Transfer: Internal Flows 255

	11.1	Mean Velocity and Mean Temperature 255
	11.2	Laminar and Turbulent Pipe Flows 257
	11.3	Entry Length and Fully Developed Flow 257
	11.4	Pumping Power and Convection Heat Transfer in Internal Flows 259
	11.5	Velocity Profiles and Friction Factor Correlations 263
	11.6	Temperature Profiles and Convection Heat Transfer Correlations 267
	11.7	Fans and Pumps 271
		11.7.1 Types of Fans 271
		11.7.2 Fan Curve and System Impedance Curve 274
		11.7.3 Fan Selection 276
		11.7.4 Types of Pumps 279
	11.8	Plate-Fin Heat Sinks 281
	Problems 283
	References 285

Chapter 12 Natural Convection Heat Transfer 287

	12.1	Buoyancy Force and Natural Convection Flows 287
	12.2	Natural Convection Velocity and Temperature Boundary Layers 290
	12.3	Normalized Natural Convection Boundary Layer Equations 291
		12.3.1 Grashof and Rayleigh Numbers 293
		12.3.2 Functional Form of the Convection Heat Transfer Coefficient 295

Contents

12.4	Laminar and Turbulent Natural Convection over a Vertical Flat Plate	296
12.5	Natural Convection around Inclined and Horizontal Plates	300
12.6	Natural Convection around Vertical and Horizontal Cylinders	303
12.7	Natural Convection in Enclosures	304
12.8	Natural Convection from Array of Vertical Plates	310
12.9	Mixed Convection	313
Problems		314
References		318

Chapter 13 Radiation Heat Transfer 319

13.1	Radiation Intensity and Emissive Power	320
13.2	Blackbody Radiation	324
13.3	Radiation Properties of Surfaces	326
	13.3.1 Surface Emissivity	326
	13.3.2 Surface Absorptivity	328
	13.3.3 Surface Reflectivity	329
	13.3.4 Surface Transmissivity	330
	13.3.5 Kirchhoff's Law	331
13.4	Solar and Atmospheric Radiations	332
13.5	Radiosity	336
13.6	View Factors	337
13.7	Radiation Heat Transfer between Black Bodies	340
13.8	Radiation Heat Transfer between Nonblack Bodies	341
13.9	Radiation Heat Transfer from a Plate-Fin Heat Sinks	343
Problems		349
References		350

Chapter 14 Computer Simulations and Thermal Design 351

14.1	Heat Transfer and Fluid Flow Equations: A Summary	352
14.2	Fundamentals of Computer Simulation	353
	14.2.1 Steady-State, One-Dimensional Heat Conduction	353
	14.2.2 Steady-State, Two-Dimensional Heat Conduction	356
	14.2.3 Transient Heat Conduction	359
	14.2.4 Fluid Flow and Energy Equations	362
14.3	Turbulent Flows	373
14.4	Solution of Finite-Difference Equations	380
14.5	Commercial Thermal Simulation Tools	381
	14.5.1 Creating the Thermal Model	382

		14.5.2	Creating the Mesh	391
		14.5.3	Solving Flow and Temperature Equations	393
		14.5.4	Review the Results	396
		14.5.5	Presenting the Results	397
	14.6	Importance of Modeling and Simulation in Thermal Design		398
	References			399

Chapter 15 Experimental Techniques and Thermal Design 401

    15.1 Flow Rate Measurement Techniques 401
    15.2 System Impedance Measurement 407
    15.3 Fan and Pump Curve Measurements 409
    15.4 Velocity Measurement Methods 410
    15.5 Temperature Measurement Techniques 414
    15.6 Acoustic Noise Measurements 417
    15.7 Importance of Experimental Measurements in Thermal Design 419
    References 420

Chapter 16 Advanced Cooling Technologies 421

    16.1 Heat Pipes 421
        16.1.1 Capillary Limit 423
        16.1.2 Boiling Limit 424
        16.1.3 Sonic Limit 426
        16.1.4 Entrainment Limit 426
        16.1.5 Other Heat Pipe Performance Limits 427
        16.1.6 Heat Pipe Applications in Electronic Cooling 428
        16.1.7 Heat Pipe Selection and Modeling 430
        16.1.8 Thermosyphons, Loop Heat Pipes, and Vapor Chambers 435
    16.2 Liquid Cooling 439
    16.3 Thermoelectric Coolers 445
    16.4 Electrohydrodynamic Flow 449
    16.5 Synthetic Jet 450
    References 452

Appendix: Tables of Material Properties 455

Index 471

Preface

Design and manufacturing of microelectronic devices and systems are multidisciplinary engineering activities. With the continuous trend toward miniaturization and high power density systems, the dependency between different design disciplines has become even more pronounced. Chip designers, hardware engineers, mechanical engineers, test engineers, reliability engineers, and thermal engineers can no longer work in their own solo environments. They are now expected to interact with each other and ensure that a proposed design from one discipline does not violate the requirements of other disciplines. Some microelectronic companies have gone a step further and looked for multidisciplinary engineers who have knowledge and expertise in multiple design disciplines. These engineers can move the design process through multiple disciplines in parallel, and will reduce the risk of a product being optimized in one aspect while not meeting the requirements of the others.

Thermal engineering is one of the disciplines involved in the design and manufacturing of electronic systems. Appropriate thermal design has become one of the enabling factors for the realization of high power density electronic equipment. In addition to preventing the failures that may result from high temperature, smart and innovative thermal designs will increase the life expectancy of a system; may reduce its emitted acoustic noise, cost, time to market, and energy consumption; and can create significant market differentiations compared to similar products.

I have worked in the field of thermal management of electronics for over 10 years. Most of the mechanical, electrical, hardware, power, and industrial engineers I have worked with have had little to no knowledge of thermal management. I have also been teaching a course on heat transfer in electronics to undergraduate students in electrical and computer engineering majors for over eight years. These students have little or no background in thermodynamics and heat transfer principles. Although there is a general consensus that the engineers who work on the hardware side of the electronic industry will benefit from basic knowledge of thermal engineering, there is no heat transfer book that is written for this group.

The target audience for this book is current and future engineers, with careers in the electronic industry. It educates those engineers on the basics of thermal management of electronics, broadens their engineering knowledge and expertise, and moves them a step closer to being multidisciplinary engineers. It assumes no previous exposure to the sciences of heat transfer and thermodynamics and introduces all the relevant principles accordingly.

This book is structured so that the first seven chapters give readers a basic knowledge of thermal management of electronics such that they will be able to analyze and solve simple electronic cooling problems. Chapters 8 through 13 go into the details of heat transfer fundamentals and may be used by those who are interested in a deeper understanding of the physics of heat transfer. Chapters 14 and 15 discuss computational and experimental methods and tools used in a typical thermal design process. Finally, some advanced cooling techniques are introduced in Chapter 16.

Chapters 1 through 7 and selected parts of Chapters 8 through 13 constitute enough topics for a semester-long undergraduate course in thermal management of electronics, or heat transfer in electronics, for nonmechanical engineering majors.

Chapters 1, 6, 7, 14, 15, 16, and selected parts of the other chapters contain enough topics for an undergraduate course in thermal management of electronics for mechanical engineering students who have already taken courses in thermodynamics and heat transfer.

Chapters 1 through 7 and 14 through 16 can be used as a supplement to a workshop on thermal management of electronics for mechanical, electrical, and power engineers and for project managers who are currently working in the electronics design and manufacturing industry.

I would like to express my sincere thanks to the following reviewers for their many helpful comments and suggestions: Kaveh Azar, Advanced Thermal Solutions; David Copeland, Sun Microsystems; Paul Durbin, Iowa State University; Marc Hodes, Tufts University, Nicole Okamoto, San Jose State University; Sadegh Sadeghipour, Luminant; Yizhang Yang, Advanced Micro Devices; Ross Wilcoxon, Rockwell Collins; and Michael Yovanovich, University of Waterloo.

I am grateful to my wife, Saeedeh, who was my main supporter in this project, and my son, Arash, who gives me the motivation for almost everything I do. I need to thank them for allowing me to spend a lot of time writing this book that I should have spent with them. I cannot make up for those precious times. However, I hope this book will save its readers some time such that they can spend more time with their families.

About the Author

Younes Shabany was born in Rasht in northern Iran, but lived most of his life in Tehran where he went to school until he received his BS in mechanical engineering from Sharif University of Technology in 1991. He then went to Vancouver, Canada where he obtained his MS in mechanical engineering from the University of British Columbia in 1994. He came to the United States and earned a PhD in mechanical engineering with a minor in aeronautics and astronautics from Stanford University in 1999.

Dr. Shabany has over 18 years experience in thermal–fluid engineering. He is currently Director of Thermal Engineering & Design and Thermal Architect in the Advanced Technology Group at Flextronics, Milpitas, California. In this position, he has led thermal design activities in Flextronics' worldwide design centers on a variety of infrastructure, computing, consumer, automotive, medical, and power electronic products. Before Flextronics, he worked for Applied Thermal Technologies, Santa Clara, California, where he was the director for two years. While at Applied Thermal Technologies, he worked with over 60 companies and designed thermal solutions for electronic equipment including telecom and networking equipment, desktop and laptop computers, biomedical equipment, and consumer products.

Dr. Shabany has been a lecturer at San Jose State University, San Jose, California, since the summer of 2001. He has taught undergraduate and graduate courses in heat transfer and advanced mathematical analysis including his favorite course, Heat Transfer in Electronics. He also advises graduate students on their projects and theses.

1 Introduction

We can hardly imagine a world without all the electrical devices and equipment that help us in our daily activities. Consider some of our routine daily activities. We are awakened by the buzzer of our digital clock, use a remote control to turn on the TV or radio to listen to the morning news and traffic reports, turn on the coffeemaker before stepping into the shower or take a few sips of the coffee that was already brewed since we programed it before going to bed the night before, make a piece of toast in the toaster or pick out a couple of eggs from the refrigerator and fix those over the stove, and turn on the computer to surf the internet or to read our e-mails while finishing our breakfast. Then, we pick up our mobile phone, go out of the house, get in our car after we open its door using a remote key, turn the engine on, and (in my case for sure) turn either the radio (in my case again) or CD player on right away. Then, we realize the spouses had used the car last night and had changed the settings on the seat and the mirror, so we use some electromechanical controls to adjust seat location and height and the mirror control buttons to adjust the side mirrors. Alternatively, in most luxury cars, both the driver seat and the mirror adjustments are done by pressing a single button that activates a preprogramed microcontroller. On the way to work, we pass a couple of traffic lights that control the flow of traffic to main highways and a couple more that seem to be smart enough to turn green moments after we reach the respective intersections. At work, we use our magnetized badge to open the employee's entrance door, turn on the computer and, while it is booting up, go to the coffee machine and press a button that makes the coffeemaker brew our favorite kind of coffee. We launch the e-mail program, internet browser, engineering modeling software, or a word processing or slide preparation tool, and make a few phone calls to do the tasks allocated to us at work. After work, we decide to go for exercise before going home. It will be good to listen to some music while working out. So, we pick up our MP3 player and head toward our favorite work out place or park. After the warm-up stretches and possibly some weight lifting exercises, we step on a treadmill or similar cardiovascular equipment. On the way home, we go to a gas station to fill the tank. We insert our credit card into the slot on the gas pump stand, follow the instructions on the LCD monitor, fill up the tank, and pick up the printed receipt. We open the garage door, as we approach the driveway, with the remote control inside our car, turn on the light switches inside the house, listen to the messages on our answering machine, pick up the necessary ingredients from the refrigerator or freezer and fix a dinner in the microwave or on the stove. We usually finish the day by making a couple of phone calls to friends and family, watch our favorite TV shows or news program, and go to bed after setting the buzzer of our digital clock for the next morning.

Almost none of the tasks and activities mentioned above could happen (or would happen as easy as they do now) without some sort of electrical and electronic equipment. What is important to note is that the huge infrastructure that makes all these tasks possible are invisible to us. These include, but not limited to, the telecommunication and networking equipment owned by telephone and internet providers, the data servers and storage equipment in TV stations and movie making studios, the data processing servers and computers in banks and other financial centers, and the complex electronic circuits inside automobiles.

Microelectronic devices such as transistors, capacitors, inductors, transformers, and resistors are the building blocks of all electronic equipment. For such electronics to function, a passage of electrical current is required. A portion of this energy is dissipated as heat due to inefficiencies in a device or the nonzero resistance to electrical current. The power or heat dissipation in resistors is proportional to their electrical resistance and is obtained by Joule's law; $P = I^2R$ where I is the DC current or the root mean square of the AC current and R is the electrical resistance. The power dissipation in inductors and transformers consists of winding loss that is calculated by Joule's law and the core loss, which is due to hysteresis in reversal of magnetic domain and eddy current generated in the core by the changing magnetic field. Although an ideal capacitor does not dissipate any power under a DC current, real capacitors show some resistance and inductance and therefore dissipate some heat. Finally, the heat dissipation in transistors is due to switching, dynamic short circuit and leakage current power dissipations.

Before the invention of transistors, vacuum tubes had been used instead. The efficiency of vacuum tubes was very low. They consumed a lot of electrical energy that was wasted as heat. Their life expectancy was also very short. It is said that the early computer room operators had to continuously move between different stations of the room to replace faulty vacuum tubes. The invention of transistors was a breakthrough in the electronic industry. Transistors replaced vacuum tubes while consuming much less electrical energy and dissipating significantly lower heat.

The early electronic circuits had been built by connecting a few transistors, capacitors, inductors, and resistors on a single-layer printed circuit board. These circuits dissipated very little heat but had limited functionality. Later on, using semiconductor manufacturing processes, more complex electronic circuits were created by building many transistors on a single semiconductor piece called a chip or die. Over a period of about 30 years, integrated circuits (ICs) and large scale integrated circuits (LSICs) with up to a few thousand to a few million transistors were introduced by different manufacturers. For example, Intel's 4004 microprocessor introduced in 1971 had about 2,300 transistors while the Intel Pentium 4 processor introduced in 2000 had about 42 million transistors. During the same time period, transistor size has been reduced and transistor density on the chip has increased. Although a few transistors in early electronic circuits did not generate much heat, a few million of those, on almost the same footprint, generated a significant amount of heat. The heat dissipation concern with vacuum tubes that went away by the invention of transistors came back due to large scale integration of transistors into integrated circuits.

Electronic equipment has continuously grown in functionality and shrunk in size in the last few decades. This has been accompanied by the rise in their power consumptions, and therefore heat dissipations. A review of the trends in semiconductor industry that reveals a steep rise in heat dissipation of electronic equipment will be

Introduction

given in Section 1.1. This will be followed by a review of some of the electronic failures due to overheating in Section 1.2. The importance of *heat transfer* in the design of electronic equipment will be discussed in Section 1.3, and a detailed thermal design process will be given in Section 1.4.

1.1 SEMICONDUCTOR TECHNOLOGY TRENDS

In 1965, Gordon Moore, the cofounder of Intel, predicted that the number of transistors per microprocessor will double every 18 months to 2 years [1]. This trend, known as Moore's law, has been followed very closely by the industry. Figure 1.1 shows that Intel's processors followed Moore's law very closely [2]. Similar trends have been observed for processors manufactured by other vendors.

There are a few organizations that predict and publish roadmaps for semiconductor industry based on either technology developments or customer needs. The Semiconductor Industry Association (SIA) coordinated the first effort of producing what was originally called the National Technology Roadmap for Semiconductors (NTRS). The semiconductor industry became global in the 1990s, as many semiconductor chip manufacturers established facilities in multiple regions of the world. This realization led to the creation of International Technology Roadmap for Semiconductors (ITRS) in the late 1990s. The invitation to cooperate on ITRS was extended by SIA at the World Semiconductor Council in April 1998 to Europe, Japan, Korea, and Taiwan. Since then, full revisions of ITRS were produced every two years from 1999 to 2007 [3].

Figures 1.2 and 1.3 show NTRS/ITRS roadmaps for transistor density in millions per square centimeter (M/cm^2) and feature size in nanometer (nm). Feature size is a measure of the minimum dimension in a semiconductor device and a smaller feature size allows smaller chips with the same number of transistors or higher transistor density as shown in Figure 1.2. It is seen that the semiconductor industry exceeded

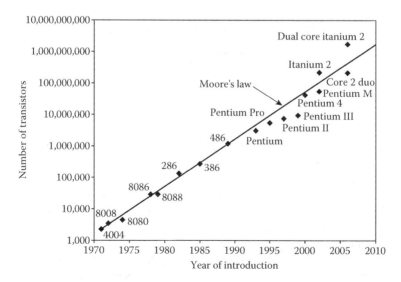

FIGURE 1.1 Intel's processors and Moore's Law.

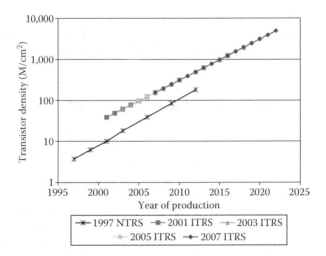

FIGURE 1.2 NTRS/ITRS roadmaps for transistor density. (www.itrs.net)

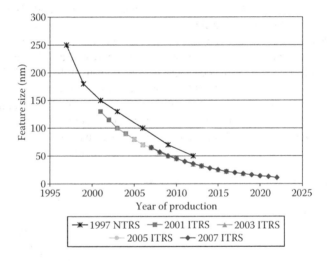

FIGURE 1.3 NTRS/ITRS roadmaps for feature size. (www.itrs.net)

the predictions of the 1997 NTRS roadmap by 2001 and closely followed 2001 through 2005 roadmaps.

The switching power consumption of a chip is proportional to the on-chip clock frequency (f) and the square of the power supply voltage (V); switching power $\propto fV^2$. The NTRS/ITRS roadmaps for on-chip clock frequency are shown in Figure 1.4. It is seen that the 1997 through 2005 roadmaps predicted an exponential increase in on-chip clock frequency. However, during the last few years, the industry pursued ways other than increasing the on-chip clock frequency to improve the performance of microprocessors. These include new architectural designs such as multicore processors and software improvements. As a result of this and to limit the total heat

Introduction

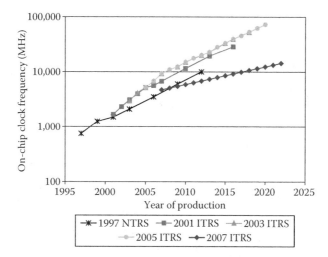

FIGURE 1.4 NTRS/ITRS roadmaps for on-chip clock frequency. (www.itrs.net)

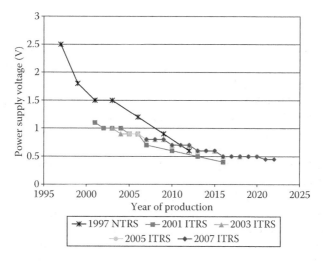

FIGURE 1.5 NTRS/ITRS roadmaps for power supply voltage. (www.itrs.net)

dissipation of the processors, the on-chip clock frequency and its rate of increase has been reduced in recent years. This is reflected in the 2007 roadmap as shown in Figure 1.4. The roadmaps for power supply voltage are shown in Figure 1.5. Although the power supply voltage continues to decrease, its reduction is not enough to compensate for the increase in power consumption due to increase in clock frequency and transistor density.

Total power consumption of a chip is the sum of its switching power consumption, short-circuit power consumption, and the power consumption due to leakage current. A significant portion of the chip power consumption is wasted as heat. For most digital devices it is usually assumed that all the power consumption is

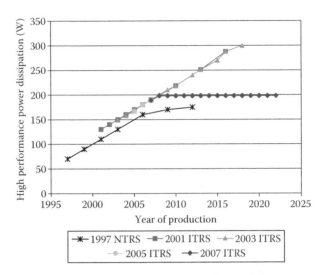

FIGURE 1.6 NTRS/ITRS roadmaps for high-performance processors power dissipation. (www.itrs.net)

dissipated as heat, and for radio frequency components the heat dissipation is equal to the power consumption minus the output radio signal. The NTRS/ITRS roadmaps for power dissipation of high-performance processors are shown in Figure 1.6. Note that (up to 2003) the ITRS roadmap predicted a continuous rise in chip power dissipation while the 2005 and 2007 roadmaps capped the heat dissipation at 200 W due to constraints in the current "system level cooling and test" technologies rather than in the "packaging" [3]. This shows the important role that innovative, cost-effective, and high-performance cooling techniques can play in enabling the use of high-end microelectronic devices.

The semiconductor technology trends indicate that heat dissipation of electronic equipment has been continuously increasing in the last few decades and will continue to be a limiting factor in manufacturing high-end electronic devices in the foreseeable future. If this heat is not properly dissipated, these electronics will overheat and their temperature will rise. This may result in temperature-dependent failures, some of which will be described in Section 1.2.

1.2 TEMPERATURE-DEPENDENT FAILURES

There are maximum occupancy limits for most auditoriums, movie theaters, club rooms, and classrooms. A few factors such as fire safety and constructional rigidity dictate these limits. An additional factor is the air conditioning limit of the room. Many of us have experienced the uncomfortable feeling of being in an overcrowded room. Average heat dissipation from a human body is about 100 watts. Let's consider a classroom with 100 students. The total heat generation, from all those students, is about 10,000 watts. If the air conditioning system in that class is not designed to

Introduction

transfer that heat, the temperature inside the classroom will continue to rise and will exceed the comfortable threshold.

The same thing may happen in an electronic product that is not properly cooled. If the heat generated inside that product is not efficiently transferred out, its temperature will rise. If this temperature rises beyond certain limits, the device may burn or be caught on fire as shown by two examples in Figure 1.7. Even if the temperature rise is not large enough to cause fire or burn the device, failures may still be caused due to high temperature. Some studies suggest that high temperature is the most critical stress factor for microelectronic device failures [4]. Failures

(a)

(b)

FIGURE 1.7 (a) A CPU before and after failing due to overheating and (b) a laptop that is caught in fire due to an overheated battery.

1.2.1 Temperature-Dependent Mechanical Failures

Mechanical failures refer to any kind of excessive deformation, yield, crack, and fracture in a material or the separation of the joint between two pieces. This happens when the force applied to a material creates a stress (force per unit area), which is higher than the yield stress of the material, the joint between two pieces can not tolerate the shear or tensile stress applied to it, or the repetitive application of a small force creates fatigue failure. An example of fatigue failure is the broken paper clip after it is bent back and forth several times.

Materials expand and contract as their temperature increases and decreases, respectively. Consider a beam with length L mounted on two roller joints as shown in Figure 1.8. If we neglect the friction between the roller joints and the surface below, the beam freely expands to length L_e if temperature rises or contracts to length L_c if temperature falls. Therefore, no stress is applied to the beam due to temperature change. However, if the joints on both sides of the beam are fixed, it can not expand or contract freely as temperature changes. In fact, as shown in Figure 1.9, the beam pushes the joints outward, bends slightly, and suffers a compressive stress as temperature increases or pulls the joints inward and suffers a tensile stress as temperature decreases. In either case, if the stress applied to the joints is more than what they can tolerate, the beam will separate from the joints or the joints will separate from the surface. Alternatively, the beam may break due to the excessive tensile stress at low temperatures.

FIGURE 1.8 A metal bar mounted on moving joints (top), can freely expand (middle), and contracts (bottom) due to temperature change.

Introduction

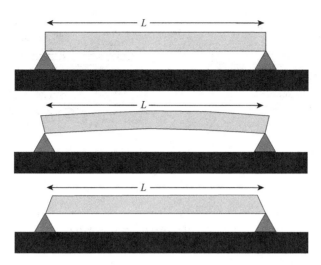

FIGURE 1.9 A metal bar mounted on rigid joints (top), deforms when temperature increases (middle), and goes under tensile stress (bottom), when temperature decreases.

FIGURE 1.10 Two dissimilar materials bonded to each other at room temperature (left) and subjected to different temperatures (middle and right).

Coefficient of thermal expansion (CTE) defines the rate of expansion or contraction of a unit length of a material per unit change in its temperature:

$$\alpha = \frac{1}{L}\left(\frac{\partial L}{\partial T}\right)_P. \tag{1.1}$$

The index P indicates that pressure is kept constant during the measurement of α such that the change in length is due to the change in temperature only. The CTE is measured in ppm/°C where ppm stands for parts per million. For example, a CTE of 20 ppm/°C indicates that a 1 m long beam of this material will expand 20×10^{-6} m or 0.02 mm for every 1°C temperature rise. Different materials have different coefficients of thermal expansion.

Figure 1.10 shows two bonded dissimilar materials with different coefficients of thermal expansion that are attached to each other at room temperature and are subjected to temperatures different from the room temperature. If the CTE of the top material is higher than the bottom one, the middle and the right pictures in Figure 1.10 show the attached pair at a higher and lower than room temperature, respectively. On the other hand, if the CTE of the top material is smaller than the bottom one, the middle and the right pictures in this figure show the attached pair at a lower and higher than room temperature, respectively. In either case, the bond between the

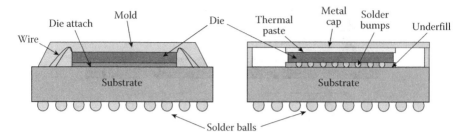

FIGURE 1.11 Schematics of a wire-bonded plastic molded package (left) and a flip-chip metallic package (right).

two materials is subject to shear stress. If this shear stress is higher than what the bond can tolerate, the joint between the two materials will break.

Figure 1.11 shows two examples of microelectronic packaging. The main components of these packages are the die and the substrate. In wire-bonded packages, the electrical connection between the die and the substrate is made by thin gold or aluminum wires while mechanical attachment between the two is through some sort of die attach. The die may be protected from outside ambient by some form of molding compound. In flip-chip packages, both mechanical and electrical connections between the die and the substrate is through solder bumps, and a layer of underfill is sometimes used to increase the mechanical strength of the interface. Flip-chip packages are either bare-die package with an exposed and unprotected die, or plastic molded package in which the die is overmolded, or metallic package in which a metal cap protects the die. A thermal paste between the die and metal cap is used to fill the gap between the die and the metal cap and to improve the heat transfer rate from the die.

Table 1.1 shows that die, substrate, die attach, wire, solder bump, underfill, mold compound, and metal cap materials have different coefficients of thermal expansion [5]. Temperature-dependent mechanical failures arise from a difference in CTE of bonded materials, large time-dependent temperature changes, and large spatial temperature gradients. All of these can cause tensile, compressive, bending, fatigue, and fracture failures. Some of the common temperature-dependent mechanical failures are described below [6].

- Wire Fatigue: The wires that connect die bond pads to substrate and/or leads may fail due to cyclic temperature changes. This failure is a consequence of the cyclic stress applied to the wires, due to the difference between coefficients of thermal expansion of the wire and the package material, as the device is heated and cooled during temperature and power cycling.
- Wire Bond Fatigue: A wire bond subjected to cyclic temperature change experiences cyclic shear stresses between the wire and the bond pad, and between the bond pad and the substrate. Although this shear stress may not be large enough to break the bond the first time it is applied, its cyclic application may result in fatigue failure of the bond.
- Die Fracture: Dies are usually made of silicon, germanium, gallium arsenide, or indium phosphide. Substrate is typically alumina, berylia, or

TABLE 1.1
Coefficients of Thermal Expansion for Typical Packaging Materials

Material	Coefficient of Thermal Expansion (ppm/°C)
Case Materials	
Alumina	4.3–7.4
Copper	16
Molding compound	10–30
Die	
Silicon	2.3–4.7
Germanium	5.7–6.1
Gallium arsenide	5.4–5.7
Die attach	
Silver filled glass	8
Polyimide	40–50
Silicon-based epoxy	60–80
Lead and lead frame	
CDA 194	16.3
OLIN 7025	17.1
Kovar	5.8
Substrate	
Beryllia	6.3–7.5
Silicon carbide	3.5–4.6
Aluminum nitride	4.3–4.7
Alumina	4.3–7.4
Underfill	25–50
Wire bond and pad	
Aluminum	46.4
Copper	16
Gold	14.2
Bumps	
Gold	14.2
95Pb/5Sn	28.7

Source: From Pecht, M., Agarwal, R., McCluskey, F. P., Dishongh, T. J. Javadpour, S., and R. Mahajan, *Electronic Packaging Materials and Their Properties.* Boca Raton, FL: CRC Press, 1999. With permission.

aluminum nitride, having a coefficient of thermal expansion different from the die. As the magnitude of temperature cycle increases during temperature and power cycling, tensile stresses are developed in the center portion of the die and shear stresses are developed at the edge of the die. Ultimate

fracture of the die can occur suddenly, without any plastic deformation, when surface cracks at the center of the die, or at the edge of the die, reach their critical size.
- Die and Substrate Adhesion Fatigue: Since the die, the die attach, and the substrate have a different coefficient of thermal expansion, the bond between the die and the substrate can experience fatigue failure.
- Cracking in Plastic Packages: The CTE of silicon and a typical molding compound are about 3 ppm/°C and 20 ppm/°C, respectively. A typical packaging process involves die attachment with adhesive cure at 270°C (polyimide adhesive) or 170°C (epoxy adhesive), followed by encapsulation and cure of the molding compound at 170°C. At lower operating temperatures, the encapsulation imposes a compressive stress on the die and places the molding compound under tension. Therefore, cracks can be generated and propagated in the molding compound.

1.2.2 Temperature-Dependent Corrosion Failures

Corrosion is defined as the chemical reaction of a material with its surrounding environment. Corrosion can be divided into two types: dry corrosion such as oxidation of aluminum in air and wet corrosion where the reaction occurs in the presence of an electrolyte, a moist environment, or an electromotive force. Two temperature-dependent corrosion failures are described below.

- Corrosion of Metallization and Bond Pads: This is a wet corrosion mechanism and, in the presence of an ionic contaminant such as Cl^- or Na^+ and moisture can provide a conductive path for electrical leakage between adjacent conductors. Corrosion typically begins when the temperature inside the package is below the dew point enabling condensation inside the package. During operation, heat dissipated by the die is often enough to raise the temperature to evaporate the electrolyte and alleviate the humidity problem. Therefore, elevated temperature due to device power acts as a mechanism to slow the corrosion process.
- Stress Corrosion in Packages: This failure results from the acceleration of the fatigue process by corrosion of the advancing fatigue cracks. This failure is typically initiated around 300°C and above, and occurs predominantly in power devices.

1.2.3 Temperature-Dependent Electrical Failures

Electrical failures refer to failures that will inversely affect the performance of the device. These failures may be intermittent or permanent. Some of the common temperature-dependent electrical failures are described below.

- Thermal Runaway: The on-resistance of power transistors increases with temperature. If the power in these transistors is not dissipated properly, their

temperature will increase. This will cause an increase in their on-resistance that, in turn, will result in higher power dissipation and higher temperature. Thermal runaway can destroy the transistor.
- Electrical Overstress: The electrical resistance of silicon reduces as its temperature increases. As the silicon die heats up, the increased temperature and therefore lower electrical resistance, encourages higher current flow. This in turn, further heats up the junction. If the material reaches its melting temperature, permanent damage may result.
- Ionic Contamination: Contamination can happen during packaging and interconnect processing, assembly, testing, stressing, and operation. The mobility of ions is temperature-dependent. Mobile charged ions create uncontrolled current and degrade device performance. A high temperature storage bake and exposure to high temperature during burn-in tests are used to screen out the ionic contamination failures.
- Electromigration: It is the result of momentum transfer from the electrons, which move in the electric field to the ions or atoms of a conductor. This causes the ions and atoms to move from their original positions and to create a void. If a significant number of atoms are moved from their original positions, voids will grow and link together to cause electrical discontinuity or open circuit failure. Electromigration can also cause the atoms of a conductor to pile up and drift toward other nearby conductors, creating a short circuit.

If temperature increases, the atoms vibrate faster and further from their ideal lattice positions and therefore are more likely to be hit by electrons. On the other hand, high current density increases the number of electrons moving in a conductor and, correspondingly, the chance of hitting and displacing an atom. Therefore, electromigration failure usually happens at high current density and high temperature.

1.3 IMPORTANCE OF HEAT TRANSFER IN ELECTRONICS

Electronic device manufacturers specify a maximum allowable operating temperature for their device. They do not guarantee its intended performance and its expected life time if it operates beyond its maximum allowable operating temperature. Therefore, it is important to cool electronic devices such that their operating temperatures remain below the maximum allowable value. As higher performance microelectronic devices with higher power dissipations are introduced to the market, more efficient cooling techniques are required to keep their temperatures below their maximum allowable limits.

There are many benefits in operating electronic devices at lower temperature. For example, it has been known that the clock frequency of a microprocessor increases and its performance improves as its operating temperature decreases. That is the main reason the savvy gamers replace the CPU heat sink in their gaming computer with aftermarket, liquid-cooling modules. On the other hand, the reduction of temperature will reduce the probability of temperature-dependent failures described in the last section. Mean time between failures (MTBF) of an electronic system is defined as the average time between two failures in that system. It has been known that MTBF of a system increases exponentially with the inverse of temperature as temperature decreases. If T is the operating temperature of a system in absolute unit,

$$\text{MTBF} \propto e^{C/T}, \tag{1.2}$$

where C is a system-specific constant [7]. Reliability of a system is inversely related to its MTBF. Therefore, Equation 1.2 shows that MTBF of a system decreases and its reliability increases if it operates at lower temperatures.

Improving performance and increasing reliability are two important reasons for appropriate transfer of heat from electronics. However, other potential consequences of appropriate use of heat transfer techniques in electronic design are reducing or completely eliminating the acoustic noise, reducing energy consumption, and reducing the final cost of the product. As will be described in this book, there are heat transfer mechanisms such as conduction, natural convection, and radiation that do not require airflow generated by a fan. If a product is designed such that it is appropriately cooled by these heat transfer mechanisms, the noise emission, the energy consumption, and the cost associated with the fan will be eliminated. Also, as fans have higher failure rates compared to electrical and nonmoving mechanical parts, the elimination of a fan from a product increases its reliability. Even if complete elimination of a fan is not possible, optimum use of available heat transfer paths in a product allows the use of smaller fans and therefore reduced noise, cost, and energy consumption. Note that less acoustic noise is an important differentiation factor for living room and bedroom electronics, low cost is very critical in consumer electronics, and low energy consumption is important both in handheld electronics and in large telecommunication and networking systems.

1.4 THERMAL DESIGN PROCESS

Possible failures due to overheating of electronics and the importance of efficient heat transfer techniques were explained in Sections 1.2 and 1.3, respectively. The use of appropriate heat transfer techniques, possibly along with some mechanical and electrical design modifications, to sufficiently cool an electronic device or equipment is called thermal design. The heat transfer techniques that may be used to cool electronics will be illustrated in the rest of this book. However, it is important to recognize that thermal design is not a one-shot design task in which the thermal engineer proposes a thermal design once and forever. Instead, similar to other design disciplines such as mechanical, electrical, power, industrial, and so forth, thermal design is a process that goes through multiple phases and levels of details as the design of the product evolves. Thermal engineers need to be involved during the whole product design cycle to propose thermally acceptable mechanical and electrical layouts; to check whether proposed mechanical, electrical, or other design changes can be accommodated while the product can be cooled appropriately; or to provide recommendations to achieve a thermally feasible product after implementing those proposed changes.

Figure 1.12 shows different phases of a thermal design process. Similar flowcharts can be drawn for other design disciplines such as mechanical and electrical designs.

Introduction

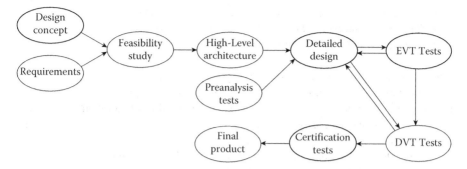

FIGURE 1.12 Different phases of a thermal design process.

An ideal product design cycle aligns the process flows for each of these design disciplines and moves them forward relatively in parallel. This will result in products that meet mechanical, thermal, electrical, and other design requirements simultaneously and will eliminate, or significantly reduce, the risk of making major changes in the product late in the design cycle in order to meet one or more overlooked design requirements. Note that changes are easy to accept and cheap to implement during the early stages of a product design cycle. On the contrary, they are harder or impossible to accept and more expensive to implement as the design cycle moves forward. If all design disciplines work together from the early stages of the product design cycle, they will be able to reach a consensus on the design with possible compromises from each design discipline. Then, having all the design disciplines work together during the rest of the product design cycle guarantees the product continues to meet all design requirements. This may require some changes in the mechanical, thermal, or electrical design of the product at different design phases but those changes will most likely be minor.

The thermal design process starts in a concept phase when, as shown in Figure 1.12, the design concept and requirements it needs to meet are outlined. Examples of design concepts are

- A desktop computer or an LCD television with no fan.
- A fan redundant server that continues to operate without any performance degradation even if one of its fans fails.
- A networking board whose components can be appropriately cooled with an air velocity of 1 m/s or less.
- A high-end phone whose external surface temperature is less than 45°C in all operating conditions.
- An outdoor radio unit that is light enough to be carried by one operator to the top of a pole it would be mounted on.

Design requirements come from

- The design concept itself.
- The market segment the product is intended for.

- The environment in which the product is supposed to operate.
- The safety and certification tests the product has to pass.
- Industry standards the product has to meet.

For example, design requirements for a phone specifies its operating conditions such as ambient temperature and altitude, maximum acceptable external surface temperature, certification tests it has to pass, geometrical form factor, and maximum acceptable dimensions. Design requirements are sometimes divided into those that must be met and those that should be met.

Having identified a design concept and outlined the requirements for it, the next phase in a thermal design process is a feasibility study. The necessary inputs for this phase are design requirements and product specifications. These should include total heat dissipation of the product, its operating temperature and altitude and certification and safety requirements. These design requirements and product specifications will be analyzed, those that have to be modified are identified, and a few thermally feasible designs are proposed in the feasibility study phase. For example, the feasibility study indicates whether a thermal design with no fan is possible or a certain number of fans is needed and whether the product can be cooled with air or liquid cooling is necessary. The analysis tools used during a feasibility study are analytical calculations, very simple computer simulations, and, of course, lessons learned from previous design experiences.

A feasibility study identifies a few thermally feasible design options. The next phase of the thermal design process is a high-level architectural design. A system-level thermal design of the product and cooling design for major power dissipating components are the main objectives of thermal design in this phase. System-level thermal design is usually referred to the bulk thermal design of a product without paying much attention to the fine details inside the product. The inputs necessary for system-level thermal design are

- Total power dissipation of the product.
- Power dissipation per board.
- Power dissipation of the major components.
- Maximum acceptable dimensions for the product.
- Preferred board and major components locations and orientations.
- Thermal and mechanical specifications of major components.

The outputs of architectural design phase are details of one or more high-level thermal designs that include

- Locations and dimensions of inlets and exhausts if the unit is not sealed.
- Internal and external airflow distributions.
- Number of fans, their locations, and dimensions for systems that are cooled with fans.
- Minimum external surface area to dissipate the heat generated inside a sealed product.
- Cooling solutions for major heat dissipating components inside the product.

Recommendations may be made to modify mechanical or electrical layout of the product to make cooling more efficient or cost-effective. Analytical techniques and system-level thermal models, which include major components only, are used in the high-level architectural thermal design phase.

Architectural thermal design results in a few high-level thermal designs. One or a few of these thermal design that are most acceptable to the entire design team will go through detailed thermal design in the next phase of a thermal design process. The additional inputs required for detailed thermal design are a list of all high-power density components including their manufacturer part numbers and power dissipations, components placement on each board or subsystem, components mechanical and thermal specifications, and boards stack up including number and thickness of copper layers. The detailed thermal design shall result in detailed thermal solutions for all high-power density components including recommendations for heat sinks and thermal interface materials, external heat sink dimensions for sealed natural convection-cooled units, inlet and exhaust dimensions and locations for open systems, fan specifications including dimensions and locations, and board thermal conductivity requirements wherever required. If applicable, it should also give mechanical and electrical redesign recommendations that will further improve thermal performance of the product.

The main tools used in the detailed thermal design phase are detailed airflow and thermal models. However, some experimental measurements may also be done in this phase. For example, consider a network server that is cooled by a few fans. Dimensions and locations of these fans have been determined in the architectural phase. Experimental measurements have shown that total airflow rate in these systems will not be significantly affected by minor changes inside the system. Therefore, it is a good design practice that, while detailed thermal design of subsystems and boards are underway, a mechanical chassis with all the fans are built and total airflow rate and air velocity distribution are measured. This helps to either get confidence in the analytical and modeling techniques or to correct those and use more accurate experimentally verified models for the detailed thermal design. The preanalysis tests shown in Figure 1.12 refers to these tests as well as experimental verification of some of the modeling inputs such as fan curve.

So far most of the design was done using analytical and modeling techniques. However, no design process is complete without experimental verification of the product performance. The first set of verification tests is called engineering verification test (EVT). They are performed on early prototypes and are intended to either verify that the prototype meets the design requirements or to identify areas of concern where the design needs to be modified. This means that the design process may go back to the detailed design phase if the prototype fails to pass all the design requirements in the EVT phase. That is the reason for having the arrows between these two phases go in both directions in Figure 1.12. The EVT phase in a thermal design process includes airflow rate and air velocity measurements for fan-based systems, air and component temperature measurements, external surface temperature measurements, and noise level measurements.

Although the EVT phase in Figure 1.12 is shown to start after the detailed design phase, thermal engineers should not wait until the detailed design is complete for all subsystems and boards before starting engineering verification tests for already designed systems and subsystems. In fact, the EVT phase should go in parallel with the detailed design phase as soon as meaningful prototypes of the system or subsystem are available. For example, the airflow rate and air velocity measurements for a network server, classified earlier as preanalysis tests for detailed design of boards, can also be considered as early EVT for the system itself. Another example is the acoustic noise level of a telecommunication chassis, which is mainly due to the fans and the shape of the inlets and exhausts. The acoustic noise level of these chasses must be measured as soon as mechanical chasses with fan trays are available even if the detailed thermal design of the boards is ongoing and no prototype for any board is available.

EVT identify areas that need to be modified, the product will go through necessary design changes, and final prototypes that are expected to meet all the design requirements are built. These prototypes will go through extensive design verification tests (DVT) under their extreme operating conditions. DVT can also be done on units coming off of the production line and in that case it may be called process or production verification tests (PVT). The main objective in the DVT or PVT phase is to make sure the product will pass certification tests, which are usually done by limited certified labs. Most of the DVT tests are done in temperature and altitude chambers where the operating temperature and pressure are set to the extreme values specified through product requirements. External surface temperature of products, air temperature at the exhaust of a chassis, components temperature, and noise level are parameters measured in the DVT phase. If a product goes through extensive EVT, major violations of the design requirements are not expected to be seen during the DVT phase. However, if there are minor issues, design engineers are involved to resolve those issues. That is why there are arrows between the DVT phase and the detailed design phase that go in both directions (Figure 1.12).

Having gone through all these expensive tests during EVT and DVT phases, the product shall go for a final certification test before being released to the market. These tests include safety and market specific standard tests that the product needs to pass to get the necessary labels and to be accepted by customers. For example, all consumer products need to pass UL safety tests and all data center telecommunication units have to meet specific noise level requirements.

REFERENCES

1. Moore, G. E. 1965. Cramming more components onto integrated circuits. *Electronics* 38(8): 114–17.
2. Intel. Moore's Law. www.intel.com/ (accessed June 2009).
3. International Technology Roadmap for Semiconductors. www.itrs.net/ (accessed June 2009).
4. Bar-Cohen, A., and A. D. Kraus. 1983. *Thermal analysis and control of electronic equipment*. New York, NY: McGraw-Hill.
5. Pecht, M., R. Agarwal, F. P. McCluskey, T. J. Dishongh, S. Javadpour, and R. Mahajan. 1999. *Electronic packaging materials and their properties*. Boca Raton, FL: CRC Press.
6. Bar-Cohen, A., and A. D. Kraus. 1990. *Advances in thermal modeling of electronic components and systems*, Vol. 3. New York, NY: ASME Press.
7. Ulrich, R. K., and W. D. Brown. 2006. *Advanced electronic packaging, IEEE press on microelectronic systems*. Hoboken, NJ: John Wiley & Sons, Inc.

2 Energy, Energy Transfer, and Heat Transfer

What is heat? What do we mean by heat transfer? Heat transfer is probably one of the most misunderstood engineering phrases. It is very important to understand that there is no such physical entity as heat that is transferred from one point to another. As we will see in this chapter heat transfer is simply one form of energy transfer. Therefore, it is necessary to first understand what energy and energy transfer are.

2.1 ENERGY AND WORK

After a few hours of hard work we feel exhausted and out of energy. On the other hand, after a good rest we feel energetic and capable of doing work. It seems that there is a direct relationship between the energy of a person and his ability to do any kind of work. In fact, energy and work are two closely related physical quantities. Consider an object which is displaced by an infinitesimal distance dx through the application of a constant force F_x where the index x indicates that the force is in the same direction as dx (see Figure 2.1). The product of the force F_x and the displacement dx is called work;

$$\delta W = F_x \times dx. \tag{2.1}$$

Here δW stands for an infinitesimal amount of work done during the infinitesimal displacement dx. Based on this definition of work, a person who carries a heavy load and walks with it without changing its height with respect to the ground does not do any work even if the load is very heavy and the person feels really tired after carrying it. This is because the force that the person uses to hold the load (i.e., its weight) is normal to the ground while his displacement is parallel to the ground.

If an object is displaced from location x_1 to location x_2 by a force $F_x(x)$, which is always in the x direction and its magnitude is a function of x, the total work done by this force on that object is obtained by the following integral.

$$W = \int_{x_1}^{x_2} F_x(x) \times dx. \tag{2.2}$$

If $F_x(x)$ is constant from x_1 to x_2,

$$W = F_x \times \Delta x, \tag{2.3}$$

where $\Delta x = x_2 - x_1$ is the total displacement of the object.

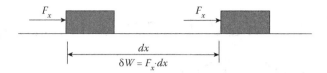

FIGURE 2.1 Work is defined as the product of force and displacement along the force.

The **energy** of an object is defined as its ability to do work or the result of work done on that object. Gasoline that is burned in the engine of a car is the source of energy that moves the car against the road friction and air resistance. This means that chemical energy of the gasoline is released as work done on the car. A load that is lifted by a crane will go through a free fall if it is released. This is because, as a result of the work done by the crane on the load while lifting it, energy has been stored in the load. This energy, called potential energy, will be released as work when the load is released.

The unit of work and energy in the SI system of units is **Joule** and is denoted by J. One Joule is the work done by a force of one Newton to displace an object by one meter; 1 J = 1 N.m.

Example 2.1

A 75 kg person walked the stairs of a 100 m tall building to its roof. How much work did he do to take himself to the roof?

Solution

The person needs to apply a force equal to his weight to lift himself up the stairs. The weight of this person is $F_x = mg$ where $m = 75$ kg is the person's mass and $g = 9.8$ m²/s is the acceleration of gravity. Since the force is constant, the total work during the displacement of 100 m is

$$W = F_x \times \Delta x = mg\Delta x = 75 \text{ kg} \times 9.8 \text{ m}^2/\text{s} \times 100 \text{ m} = 73500 \text{ J}.$$

2.2 MACROSCOPIC AND MICROSCOPIC ENERGIES

Energy of an object can be classified as either macroscopic or microscopic energies [1]. **Macroscopic or observable energy** is related to macroscopic and directly measurable quantities while **microscopic energy** relates to the energy of molecules and atoms.

Macroscopic energies are either kinetic energy or potential energy. **Kinetic energy** of an object with mass m and velocity V (see Figure 2.2) is defined as

$$E_k = \frac{1}{2}mV^2. \tag{2.4}$$

Potential energy is the energy stored in an object because the object is located in a force field. Examples of potential energy are gravitational, electrical, magnetic, and spring potential energies. Gravitational potential energy of an object

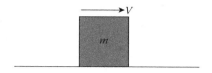

FIGURE 2.2 An object with mass m and velocity V has kinetic energy.

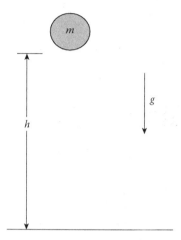

FIGURE 2.3 Gravitational potential energy is defined with respect to a reference level.

with mass m that is located at height h above a reference level (see Figure 2.3), with respect to that reference, is

$$E_g = mgh. \tag{2.5}$$

Note that this is equal to the work done on the object to raise it from the reference level to the height h.

Any spring, when compressed or stretched, applies a push or pull force against the compression or tension, respectively. For many springs, this force is linearly proportional to the amount of compression or expansion; $F_s = kx$, where $x(m)$ is the amount of compression or expansion measured in meter, $k(N/m)$ is the spring constant measure in Newton per meter, and $F_s(N)$ is the spring force measured in Newton. These springs are called linear springs. It can be shown that the potential energy stored in a linear spring that is compressed or stretched by distance x from its uncompressed state, as shown in Figure 2.4, is

$$E_s = \frac{1}{2}kx^2. \tag{2.6}$$

Example 2.2

Consider the cylinder-piston shown in Figure 2.5. The volume of the gas inside the cylinder, $\mathcal{V}$, is compressed by $d\mathcal{V}$ through the application of pressure P. Show that the work done to compress this gas is $\delta W = P d\mathcal{V}$.

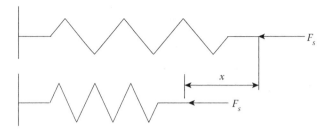

FIGURE 2.4 A compressed spring contains potential energy.

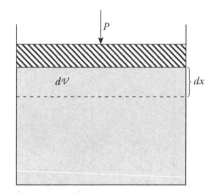

FIGURE 2.5 A gas compressed in a cylinder.

Solution

This can be shown by substituting $F_x = PA$, where A is the piston area, in Equation 2.1 and replacing Adx by $d\mathcal{V}$.

$$\delta W = F_x \times dx = PA \times dx = P d\mathcal{V}. \tag{2.7}$$

Example 2.3

Show that the work done to displace an object from rest is equal to its final kinetic energy if there is no friction against the motion of the object.

Solution

If there is no friction against the motion of an object, the force applied to that object is used to increase its velocity. This force is given by Newton's second law; $F = ma$ where $a = dV/dt$ is the acceleration of the object. Let's assume that the velocity of the object increases from zero to V_2 while it is displaced from x_1 to x_2. The work done on this object can be calculated as follows:

$$W = \int_{x_1}^{x_2} F_x \times dx = \int_{x_1}^{x_2} ma \times dx = \int_{x_1}^{x_2} m \frac{dV}{dt} \times dx$$

$$= \int_0^{V_2} m \frac{dx}{dt} \times dV = \int_0^{V_2} mV \times dV = \frac{1}{2} mV_2^2.$$

Energy, Energy Transfer, and Heat Transfer

Note that the limits of integration changed from x_1 and x_2 to 0 and V_2 as the variable of integration changed from x to V. It is seen that the total work done on this object is equal to its final kinetic energy.

Example 2.4

Show that the potential energy stored in a compressed spring is equal to the work done to compress it.

Solution

Let's assume that the spring is originally in unstretched condition. The work done to compress this spring from $x_1 = 0$ to $x_2 = x$ is

$$W = \int_{x_1}^{x_2} F_x \times dx = \int_0^x F_s \times dx = \int_0^x kx \times dx = \frac{1}{2} kx^2 = E_s.$$

Microscopic energies of material are associated with kinetic and potential energies of its atoms and molecules. Different modes of microscopic energies are briefly explained here.

Translational Energy: Molecules in gases and liquids and electrons in metals move relatively freely. Also, electrons in an atom move around its nucleus. These moving electrons and molecules have kinetic energies that are collectively called translational energy of material. It increases with the temperature of the material.

Rotational Energy: Complex molecules, such as water (H_2O) and carbon dioxide (CO_2), rotate around their center of mass. The kinetic energy associated with this rotation is called rotational energy. Rotational energy also increases with temperature.

Vibrational Energy: Molecules and atoms in a crystal are held together by electronic forces between their electrons and nuclei. The equilibrium distance between the atoms in a molecule or crystal depends on this electronic force. However, this force is very similar to spring force and collision between molecules or excitation of atoms in a crystal may change the distance between the atoms and excite the vibration of atoms in the molecule or crystal. The energy associated with these vibrations is called vibrational energy. This energy is especially important at higher temperatures.

Electronic Bonding Energy: This is the potential energy associated with the electronic bonding force that holds molecules together or binds atoms in a lattice. This is the energy that is released or absorbed in an exothermic or endothermic chemical reaction.

Nuclear Bonding Energy: This is the energy associated with the force that holds the nuclei of atoms together. The nuclear bonding force is much stronger than the electronic bonding force and therefore, the nuclear bonding energy is considerably larger than the electronic bonding energy. Nuclear bonding energy is released in a nuclear fission in which the nucleus of an atom is split into two or more nuclei, and is absorbed or released in nuclear fusion in which two nuclei join together to create a heavier nucleus.

There are other modes of microscopic energies such as kinetic energy due to spin of electrons or nuclei of atoms, potential energy due to external electrical or magnetic forces applied to ionized molecules or free electrons, and so on.

It is impractical to calculate all these different modes of microscopic energies individually. Instead, the **total internal energy**, denoted by U, is defined as the sum of all these microscopic energies.

$$U = E_{\text{translation}} + E_{\text{rotation}} + E_{\text{vibration}} + E_{\text{electronic}} + E_{\text{nuclear}} \quad (2.8)$$

Let's define a **system** as something identified for an engineering analysis. A system may be an object such as the masses shown in Figures 2.2 and 2.3, the spring shown in Figure 2.4, the gas inside the cylinder shown in Figure 2.5, the air inside a computer chassis, an electrical resistor, or an integrated circuit. A system may also refer to a volume in space and some of its content such as the cylinder and piston and gas shown in Figure 2.5, a desktop computer, a cell phone, a car, and so forth. Total internal energy of a system depends on the conditions of that system such as its temperature and pressure as well as its mass. The **internal energy per unit mass**, defined by $u = U/m$, is only a function of the conditions of that system and is independent from its mass.

The sum of total internal energy and macroscopic kinetic and potential energies of a system is called its **total energy** and is denoted by E.

$$E = U + E_k + E_p = U + \frac{1}{2}mV^2 + mgh + \cdots. \quad (2.9)$$

Energy per unit mass of a system is defined as

$$e = \frac{E}{m} = u + \frac{1}{2}V^2 + gh + \cdots. \quad (2.10)$$

2.3 ENERGY TRANSFER AND HEAT TRANSFER

In Section 2.2 we saw that the work done to displace an object from rest is equal to its final kinetic energy if there is no friction, the work done to raise an object is equal to its gravitational potential energy, and the work done to compress a spring is equal to its spring potential energy. On the other hand, the systems that perform these works spend the same amount of energy while performing these works. Therefore, work is a way of energy transfer from a system that does the work to a system that work is done on. In general, **work** is a form of energy transfer between two systems that is accompanied by some form of macroscopically observable motion [1].

There are other ways of energy transfer between two systems that do not involve any macroscopically observable motion. Heating food inside an oven, cooling a can of soda in a refrigerator, and increase in the temperature of a cold object that is in contact with another object at a higher temperature are examples of energy transfer without any macroscopic motion. However, in all of these cases, there is a temperature difference between the two systems that are involved in energy transfer. It can also be seen that, in all these cases, energy is transferred from the high temperature system to the low temperature one. The energy transfer that takes place between two objects or systems, as a result of temperature difference between them, is called **energy transfer as heat** or simply **heat transfer** and is denoted by Q. It is seen that heat transfer is nothing other than a form of energy transfer as a consequence of temperature difference between two systems.

Energy, Energy Transfer, and Heat Transfer

Energy transfer per unit time is called **power** and is denoted by P. Two common units of power are watt (W) and horse power (hp). One watt is equal to 1 Joule of energy transfer in one second and one horse power is equal to 746 W.

Sometimes we are interested in learning how fast heat is transferred between two objects. Heat transfer per unit time is called **heat transfer rate** and is denoted by $\dot{Q}$. Note that heat transfer rate and power are same physical quantities. The relationship between heat transfer rate and heat transfer is given by the following integral.

$$Q = \int_{t_1}^{t_2} \dot{Q} \times dt. \qquad (2.11)$$

If heat transfer rate is constant from time t_1 to time t_2, total heat transfer will be equal to $Q = \dot{Q} \times \Delta t$ where $\Delta t = t_2 - t_1$.

In some cases we are interested in the amount of heat transfer rate per unit area. This is called **heat flux** and is denoted by $\dot{q}$. If heat transfer rate $\dot{Q}$ is uniformly distributed over an area A, the heat flux is $\dot{q} = \dot{Q}/A$. The unit for heat flux is W/m².

Example 2.5

Consider a cylindrical resistor with a diameter of 5 mm and length of 15 mm. If the current through this resistor is 0.1 A and the voltage drop is 12 V, what are the heat transfer rate and the heat flux? Assume that the heat transfer rate is uniformly distributed over the entire area of the resistor.

Solution

The heat transfer rate is equal to the electrical power consumption by this resistor:

$$\dot{Q} = P = VI = 12\,\text{V} \times 0.1\,\text{A} = 1.2\,\text{W}.$$

The heat flux is obtained by dividing the heat transfer rate to the total heat transfer area. If r and L are the radius and length of this resistor, its total surface area is

$$A = 2\pi r^2 + 2\pi r L = 2\pi (0.0025\,\text{m})^2 + 2\pi (0.0025\,\text{m})(0.015\,\text{m}) = 2.75 \times 10^{-4}\,\text{m}^2,$$

and the heat flux is

$$\dot{q} = \frac{\dot{Q}}{A} = \frac{1.2\,\text{W}}{2.75 \times 10^{-4}\,\text{m}^2} = 4363.64\,\text{W/m}^2.$$

2.4 EQUATION OF STATE

The state of a system or its condition is described by values of its properties such as mass, volume, pressure, temperature, internal energy, kinetic or potential energy, polarization, magnetization, and so forth. It has been shown that not all the properties of a system are independent from each other. Consider a **simple compressible substance** that is a substance free from any magnetic or electric force. The state of such a substance is specified by the values of two of its independently variable properties [2,3]. For example, internal energy per unit mass and pressure of such a simple compressible substance can be determined once its temperature and density (mass per unit volume; $\rho = m/V$) are known

$$u = u(T,\rho),$$
$$P = P(T,\rho).$$
(2.12)

These equations are called **equations of state**. They are either presented in graphical or tabular forms or in special cases, explicit algebraic forms. It must be mentioned that not every two properties of a simple compressible substance are always independent from each other. For example, boiling temperature of water is a function of its pressure; water always boils at 100°C if the pressure is 101.42 kPa. Therefore, temperature and pressure of boiling water are not independent from each other although those properties are independent from each other if water is not in a boiling state.

The energy transferred to materials may increase their internal energies and therefore their temperatures. The energy required to increase temperature of unit mass of a material by one degree is called **specific heat**. It is denoted by letter C and measured by J/kg°C. The increase in temperature of a material may happen while it is kept at constant volume or constant pressure. Water that is heated in an open pot over a stove is at constant pressure while a gas that is heated inside a closed container is at constant volume. The energy required to increase temperature of the unit mass of a material by one degree at constant pressure is called **specific heat at constant pressure**, C_p. The energy required to increase temperature of the unit mass of a material by one degree at constant volume is called **specific heat at constant volume**, C_v.

The energy required to increase temperature of a material at constant pressure is always higher than the energy required to increase temperature of the same material by the same amount at constant volume. The reason is that materials at constant pressure can expand once their temperature increases and therefore some of the energy transferred to those materials is spent to do work on their surroundings rather than to increase their own temperatures. This will not happen for a material at constant volume since its volume can not change to cause any movement in its surrounding. This means that more energy is required to change temperature of systems at constant pressure and therefore $C_p > C_v$. Solids and liquids are considered incompressible materials whose volume does not change. Therefore, for solids and liquids $C_p = C_v = C$.

Specific heat at constant volume and specific heat at constant pressure are defined by the following correlations:

$$C_v = \left(\frac{\partial u}{\partial T}\right)_v \quad \text{and} \quad C_p = \left(\frac{\partial h}{\partial T}\right)_p,$$
(2.13)

Energy, Energy Transfer, and Heat Transfer

where h is called **enthalpy** per unit mass and is defined by $h = u + P/\rho$ or $h = u + Pv$ where $v = V/m = 1/\rho$ is specific volume of the system. The importance and physical meaning of enthalpy will be seen in Chapter 3. Note that both C_p and C_v of a simple compressible substance are, in general, functions of two independently variable properties such as temperature and specific volume.

Ideal Gas Law: Pressure, temperature, and volume of many common gases such as air, nitrogen, oxygen, hydrogen, carbon monoxide, and carbon dioxide are correlated by the following equation of state called ideal gas law.

$$PV = mRT. \tag{2.14}$$

Here, m is mass of the gas and R is the gas constant given by

$$R = \frac{8314}{M}, \tag{2.15}$$

where M is the gas molar mass. For example, $M = 28.97$ kg/kmol and $R = 287$ J/kg·K for air. Note that the temperature in ideal gas law has to be expressed in the absolute unit Kelvin, which is related to degree Celsius by

$$T(K) = 273.15 + T(°C). \tag{2.16}$$

Other forms of ideal gas law are $P = \rho RT$ and $Pv = RT$ where $\rho = m/V$ and $v = V/m$ are density and specific volume of the gas, respectively. It can be shown that for an ideal gas:

$$C_p = C_v + R. \tag{2.17}$$

The internal energy per unit mass and the enthalpy per unit mass of an ideal gas are functions of its temperature only. This means that, for an ideal gas, the partial derivatives in Equation 2.13 can be replaced by ordinary derivatives and therefore:

$$du = C_v dT \quad \text{and} \quad dh = C_p dT. \tag{2.18}$$

Both C_p and C_v vary with temperature. If the temperature change is not very large, the change in internal energy per unit mass and enthalpy per unit mass can be estimated by the following equations.

$$\Delta u = C_{v,\text{ave}} \Delta T \quad \text{and} \quad \Delta h = C_{p,\text{ave}} \Delta T, \tag{2.19}$$

where $C_{v,\text{ave}}$ and $C_{p,\text{ave}}$ are average values of C_v and C_p over the temperature range ΔT.

PROBLEMS

2.1 What is internal energy of a material made of?

2.2 When is the energy transfer across a system called work and when is it called heat?

2.3 An 800 kg car is moving with a velocity of 20 m/s. What is its kinetic energy?

2.4 An object with mass m is released at distance h from the ground. If we assume that potential energy is converted to kinetic energy, what is its velocity at the moment that it touches the ground?

2.5 Let's assume that an object with mass m is attached to a spring that is compressed by displacement x. The object is released and the spring is allowed to expand to its nonstretched state. If we assume that the spring potential energy is converted to the kinetic energy of this object, what will the velocity of the object be when the spring is compressed by $x/2$ and when it reaches its uncompressed state?

2.6 An electronic device with surface area of 0.0016 m² dissipates 15 W uniformly over its surface. What is the heat flux on the surface of this device?

2.7 The electrical resistance of a resistor is 5 ohms and the current through it is 10 amps. How much electrical energy is consumed by this resistor during a 24 hour period?

2.8 If all the electrical energy is uniformly dissipated from the 0.01 m² surface area of the resistor in Problem 2.7, what is the heat flux from its surface?

2.9 How does the air density change with temperature if pressure is kept constant? Why?

2.10 What is specific heat of a material?

2.11 Which one is larger? Specific heat at constant volume or specific heat at constant pressure? Why?

2.12 An ideal gas is heated from 50°C to 80°C (a) at constant volume and (b) at constant pressure. For which case will the energy required be greater? Why?

2.13 If the air pressure is doubled at the constant volume, how does its internal energy change?

2.14 If the air volume is doubled while its pressure is halved, how will its internal energy change? Why?

2.15 The volume of an ideal gas is increased at constant temperature from $\mathcal{V}$ to $5\mathcal{V}$. What is the change in internal energy of this gas? Why?

2.16 An ideal gas is compressed at constant temperature from pressure P to pressure 5P. What is the change in internal energy of the gas? Why?

2.17 Pressure and volume of an ideal gas increase 20%. If its original temperature is 25°C, what will be its final temperature?

2.18 Use the definition of enthalpy and show that for an ideal gas $C_p = C_v + R$.

REFERENCES

1. Reynolds, W. C., and H. C. Perkins. 1977. *Engineering thermodynamics*. New York, NY: McGraw-Hill.
2. Moran, M. J., and H. N. Shapiro. 2004. *Fundamentals of engineering thermodynamics*. Hoboken, NJ: John Wiley & Sons, Inc.
3. Sonntag, R. E., and C. Borgnakke. 2007. *Introduction to engineering thermodynamics*. Hoboken, NJ: John Wiley & Sons, Inc.

3 Principle of Conservation of Energy

3.1 FIRST LAW OF THERMODYNAMICS

In Chapter 2 it was shown that energy transfers from one system to another as either work or heat. Energy transfer from one system to another is the only way that a system gains or loses energy. The **first law of thermodynamics** states that energy is not generated or destroyed; it only changes from one form to another or transfers from one system to another. This important physical principle can be expressed by a simple mathematical equation. Consider a system with total initial energy E_i as shown in Figure 3.1. The total energy of this system after some time, during which total energy E_{in} enters the system and total energy E_{out} leaves the system, will be E_f. The first law of thermodynamics requires that the difference between the total energy entering this system and the total energy leaving it be equal to the difference between its final and initial energies [1–3]:

$$E_{in} - E_{out} = E_f - E_i. \tag{3.1}$$

The difference between final and initial energies of a system is called change of energy of that system or energy accumulation in that system; $\Delta E_{system} = E_f - E_i$. Therefore, the first law of thermodynamics can be expressed by the following equation:

$$E_{in} - E_{out} = \Delta E_{system}. \tag{3.2}$$

This is a simple yet powerful equation that is the basis of all the energy and heat transfer analyses. It is also known as **energy balance equation**.

The energy balance equation, Equation 3.2, deals with initial and final states of a system without being concerned with what happens between these two states or how long it takes to reach from an initial state to a final state. However, these are important design parameters in most engineering applications. Let's assume that the state of a system changes during a time interval Δt as shown in Figure 3.2. The energy of the system at times t and $t + \Delta t$ are E_t and $E_{t+\Delta t}$, respectively. If $\dot{E}_{in}$ and $\dot{E}_{out}$ are rates of energy entering and leaving the system per unit time, and are constant during the time interval Δt, the total energy entering and leaving the system are $E_{in} = \dot{E}_{in}\Delta t$ and $E_{out} = \dot{E}_{out}\Delta t$, respectively. Replacing these into Equation 3.2 and dividing the entire equation by Δt gives

$$\dot{E}_{in} - \dot{E}_{out} = \frac{\Delta E_{system}}{\Delta t}. \tag{3.3}$$

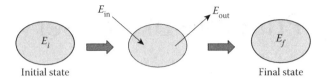

FIGURE 3.1 Energy of a system changes due to energies entering and leaving it.

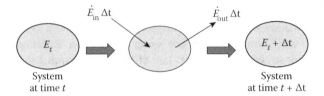

FIGURE 3.2 Energy of a system changes during a time internal Δt.

The limit of this equation as $\Delta t \to 0$ gives the **rate form of the energy balance equation**:

$$\dot{E}_{in} - \dot{E}_{out} = \frac{dE_{system}}{dt}. \tag{3.4}$$

If the energy of a system does not change with time, it is called a **steady-state** system. In this case

$$\dot{E}_{in} = \dot{E}_{out}. \tag{3.5}$$

Engineering systems can be classified into two groups. Some systems do not allow any mass flow in or out of the system. Examples are a can of soda or a bottle of water, gas inside a piston-cylinder assembly, an electric resistor or an integrated circuit, and a light bulb. Such a system is called **closed system**, **fixed mass system**, or **control mass**. Note that energy may enter or exit a control mass. For example, energy transfer as heat occurs from a can of soda in a refrigerator to the air inside the refrigerator and that is why the soda gets cold.

Many engineering systems, on the other hand, involve some form of mass flow in and out. Such a system is called **control volume**. Examples of control volumes are a water heater where cold water flows in and hot water flows out, a car radiator where hot water enters and colder water exits the water tubes while ambient air enters and hot air exits through the external fins, a hair dryer in which room temperature air enters and hot air exits, or an electronic system such as a PC or data server where the ambient air enters to cool the electronics inside the system and leaves at a higher temperature. Appropriate forms of energy balance equations for a control mass and a control volume will be given in the next two sections.

Principle of Conservation of Energy

3.2 ENERGY BALANCE FOR A CONTROL MASS

The energy balance equations for a control mass are those given in Section 3.1 (see Equations 3.2 and 3.4). Note that E_{in} and E_{out} are energy transfer to and from the control mass as either work or heat, and $\dot{E}_{in}$ and $\dot{E}_{out}$ are either work per unit time (power) or heat transfer rate. Since total energy of a system is the sum of its internal, kinetic, and potential energies, the change in its total energy is equal to the sum of the changes in its internal, kinetic, and potential energies. Therefore, the energy balance equations can be written as

$$E_{in} - E_{out} = \Delta U + \Delta E_k + \Delta E_p, \tag{3.6}$$

or in the rate form:

$$\dot{E}_{in} - \dot{E}_{out} = \frac{dU}{dt} + \frac{dE_k}{dt} + \frac{dE_p}{dt}. \tag{3.7}$$

For many systems, including most electronic equipment and devices, **the changes in kinetic and potential energies are very small compared to the change in internal energy**. For these systems:

$$E_{in} - E_{out} = \Delta U, \tag{3.8}$$

or in the rate form:

$$\dot{E}_{in} - \dot{E}_{out} = \frac{dU}{dt}. \tag{3.9}$$

If the system consists of an **ideal gas** or an **incompressible solid** or **liquid**,

$$E_{in} - E_{out} = mC_{v,ave}\Delta T, \tag{3.10}$$

or in the rate form:

$$\dot{E}_{in} - \dot{E}_{out} = mC_v \frac{dT}{dt}. \tag{3.11}$$

Here $\Delta T = T_f - T_i$ is the temperature change of the system and $C_{v,ave}$ is the average specific heat at constant volume over this temperature range.

Example 3.1

Calculate the change in internal energy of 1 kg saturated liquid water as its temperature changes from 25°C to 100°C.

Solution

If we consider saturated liquid water as an incompressible material, the change in its internal energy is related to the change in its temperature as

$$\Delta U = mC_{v,ave}\Delta T,$$

Where $C_{v,ave}$ is the average specific heat at constant volume of water from 25°C to 100°C. The specific heat at constant volume and at constant pressure of saturated liquid water are the same as it is an incompressible material. From table A.11, it is 4182 J/kg°C at 25°C and 4216 J/kg°C at 100°C. Therefore, the average specific heat at constant volume is

$$C_{v,ave} = 0.5\left[C_v(\text{at }25°C) + C_v(\text{at }100°C)\right]$$
$$= 0.5(4182 \text{ J/kg°C} + 4216 \text{ J/kg°C}) = 4199 \text{ J/kg°C},$$

and the change in internal energy of water is

$$\Delta U = mC_{v,ave}\Delta T = 1\text{ kg} \times 4199 \text{ J/kg°C} \times (100 - 25)°C = 314925 \text{ J}.$$

Example 3.2

Calculate the change in kinetic energy of 1 kg of water as its velocity increases from zero to 180 km/h.

Solution

The change in kinetic energy of this mass of water is

$$\Delta E_k = \frac{1}{2}m(V_2^2 - V_1^2) = \frac{1}{2}m(V_2^2 - 0) = \frac{1}{2}mV_2^2.$$

The final velocity of water is $V_2 = 180$ km/h $= 180 \times 1000/3600$ m/s $= 50$ m/s and the change in kinetic energy of water is

$$\Delta E_k = \frac{1}{2}mV_2^2 = \frac{1}{2} \times 1\text{kg} \times (50\text{ m/s})^2 = 1250 \text{ J}.$$

Note that, even with a velocity change from zero to 180 km/h, the change in kinetic energy of water is only 0.4% of the change in its internal energy as calculated in Example 3.1.

Example 3.3

Calculate the change in gravitational potential energy of 1 kg of water as it is raised 1 km above the ground.

Solution

The change in elevation is $\Delta h = 1$ km $= 1000$ m. Therefore, the change in gravitational potential energy is

$$\Delta E_g = mg\,\Delta h = 1\text{ kg} \times 9.8\text{ m/s}^2 \times 1000\text{ m} = 9800 \text{ J}.$$

Principle of Conservation of Energy

Again, even with an elevation change of 1 km, the change in gravitational potential energy of water is only 3% of the change in its internal energy as calculated in Example 3.1, Examples 3.2, and 3.3 confirm our expectation that changes in kinetic and potential energies of many systems are negligible compared to the change in their internal energies.

Energy balance equations are used to calculate work or heat transfer from or to a system, or to obtain an unknown temperature or other relevant parameters of interest. However, correct forms of these equations may not be so obvious in many situations. The following procedure helps to reduce confusion and mistakes when energy balance equations are used.

- Define a system with clear boundaries. There may be different ways to define a system and its boundaries in a specific problem. In these situations, the appropriate system will be the one that leads us toward finding the parameter of interest using the available data. Figure 3.3 shows a closed cylinder that contains two different gases A and B that are separated by a piston that moves upward. Three different systems can be defined for this cylinder; these systems may include gas A, gas B, or both.
- List the relevant assumptions such as ideal gas law, incompressible material, negligible kinetic and potential energy changes, negligible property change with temperature, and so forth.
- Use arrows to indicate flows of energy (work and heat) in or out of the system. Note that it is not important to have the right direction for an unknown work or heat transfer. If we assign a wrong direction to an unknown work or heat transfer, the calculation results in a negative number for those. Figure 3.4 shows work and heat transfer in or out of each system. If we consider gas A as our system, since the piston is compressing this gas, we will assume that work

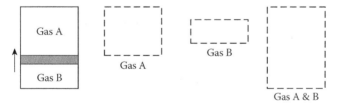

FIGURE 3.3 Three different ways a system can be defined for the piston-cylinder shown on the left.

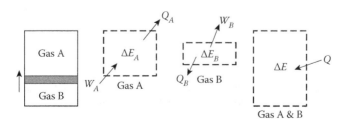

FIGURE 3.4 Flows of energy in and out of systems shown in Figure 3.3.

W_A is done on this system. Since the system is not insulated, heat may enter or exit the system; we assume that heat Q_A is transferring out of the system. Since energy entering and leaving this system, the energy of system may change by ΔE_A. If we choose gas B as our system, we will assume work W_B is done by this system and heat Q_B is transferring out of the system and the energy of the system changes by ΔE_B. In the third choice for a system, gas A and B together, there is no work done on the system as the cylinder does not change volume but heat Q may enter that system and its energy may change by ΔE.

- Write the energy balance equation:

$$E_{in} - E_{out} = \Delta E_{system}$$

For Gas A: $W_A - Q_A = \Delta E_A$

For Gas B: $-W_B - Q_B = \Delta E_B$

For Gas A and B: $Q = \Delta E$.

Example 3.4

Show that the heat transfer to a constant volume control mass, for which the changes in kinetic and potential energies are negligible, is $\delta Q = mC_v \Delta T$.

Solution

The energy balance equation for a control mass with negligible changes in kinetic and potential energies is

$$E_{in} - E_{out} = \Delta U.$$

If the volume of a system is constant, $\delta W = P\delta V = 0$ and no work is done on or by this system. Therefore, the only mode of energy transfer into this system is heat transfer.

$$E_{in} = \delta Q \text{ and } E_{out} = 0 \Rightarrow \delta Q = \Delta U = m\Delta u.$$

Internal energy per unit mass of a system is a function of its temperature and specific volume; $u = u(T, v)$ [2,3]. Therefore, the change in internal energy per unit mass depends on the changes in temperature and volume of the system:

$$\Delta u = \left(\frac{\partial u}{\partial T}\right)_v \Delta T + \left(\frac{\partial u}{\partial v}\right)_T \Delta v.$$

The second term vanishes for a constant volume system and the coefficient of ΔT in the first term is C_v as defined by Equation 2.13. Therefore, **for a constant volume system**, $\Delta u = C_v \Delta T$ and the energy balance equation reduces to a

$$\delta Q = mC_v \Delta T. \qquad (3.12)$$

Principle of Conservation of Energy

Example 3.5

Show that the heat transfer to a control mass of a single-phase material at constant pressure is $\delta Q = mC_p \Delta T$ if the changes in kinetic and potential energies are negligible.

Solution

The energy balance equation for a control mass with negligible changes in kinetic and potential energies is

$$E_{in} - E_{out} = \Delta U.$$

Since the system is at constant pressure, it will expand while heat is transferred to it and does the work $\delta W = P\Delta \mathcal{V}$ on its surrounding. Therefore,

$$E_{in} = \delta Q \text{ and } E_{out} = \delta W = P\Delta \mathcal{V} \Rightarrow \delta Q - P\Delta \mathcal{V} = \Delta U \Rightarrow \delta Q = \Delta U + P\Delta \mathcal{V}.$$

Since pressure is constant and $\Delta P = 0$, we can add $\mathcal{V}\Delta P$ to the right side of this equation:

$$\delta Q = \Delta U + P\Delta \mathcal{V} + \mathcal{V}\Delta P \Rightarrow \delta Q = \Delta U + \Delta(P\mathcal{V})$$

$$\delta Q = \Delta(U + P\mathcal{V}) = m\Delta(u + Pv) = m\Delta h.$$

Enthalpy per unit mass of a system of single-phase material is a function of its temperature and pressure; $h = h(T, P)$ [2,3]. Therefore, the change in enthalpy per unit mass, in general, depends on the change in temperature and pressure of the system:

$$\Delta h = \left(\frac{\partial h}{\partial T}\right)_P \Delta T + \left(\frac{\partial h}{\partial P}\right)_T \Delta P.$$

The second term is zero for a constant pressure system and the coefficient of ΔT in the first term is C_p as defined in Equation 2.13. Therefore, **for a constant pressure system of a single-phase material**, $\Delta h = C_p \Delta T$ and the energy balance equation reduces to

$$\delta Q = mC_p \Delta T. \tag{3.13}$$

Example 3.6

3 kg of air at 80°C temperature and 0.2 MPa pressure is heated in a cylinder until its temperature reaches 150°C at the same pressure; the piston is free to move during the heating. Determine the amount of heat transfer to the air and the work done by the piston.

Solution

Let's define the system to be the air inside the cylinder under the piston and assume that heat transfer is into this system and work is done by this system as shown in Figure 3.5. Let's also assume that air is an ideal gas.

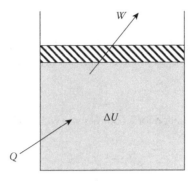

FIGURE 3.5 The air under the piston is the selected system for Example 3.6.

If we neglect the changes in kinetic and potential energies, the energy balance equation for this system is

$$E_{in} - E_{out} = \Delta U,$$

$$Q - W = \Delta U,$$

$$Q - W = mC_{v,ave}(T_2 - T_1).$$

This is a constant pressure system (See Example 2.2). Therefore, the work done by the piston is obtained as follows:

$$W = \int_{\mathcal{V}_1}^{\mathcal{V}_2} P\Delta\mathcal{V} = P(\mathcal{V}_2 - \mathcal{V}_1) = P\mathcal{V}_2 - P\mathcal{V}_1 = mRT_2 - mRT_1 = mR(T_2 - T_1)$$

$$W = 3 \text{ kg} \times 287 \text{ J/kgK} \times (150 - 80)\text{K} = 60270 \text{ J}.$$

From Table A.1, the specific heat at constant pressure of air at 80°C and 150°C are 1010.75 J/kg°C and 1019.30 J/kg°C, respectively. Therefore, specific heats at constant volume $C_v = C_p - R$ at those temperatures are 723.75 J/kg°C and 732.3 J/kg°C, respectively. The average specific heat at constant volume is $C_{v,ave} = 0.5(723.75 + 732.3) = 728.02$ J/kg°C. The heat transfer to the air is

$$Q - 60270 \text{ J} = 3 \text{ kg} \times 728.02 \text{ J/kg°C} \times (150 - 80)°\text{C},$$

$$Q = 213154 \text{ J}.$$

3.3 ENERGY BALANCE FOR A CONTROL VOLUME

Consider a system, such as a water heater, that has one inlet and one exhaust. Let's assume that mass enters and exits this system continuously through its inlet and exhaust. Figure 3.6 shows a fixed amount of mass inside this system at times t and $t + dt$. It is seen that during the time interval dt this mass moves and changes shape. However, it can be considered as a control mass because there is no mass flow through its boundaries. The intersection of these control masses at times t and $t + dt$ is shown

Principle of Conservation of Energy

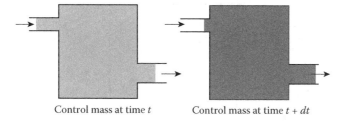

FIGURE 3.6 A fixed amount of mass inside a system at two different times.

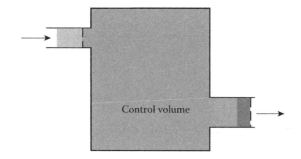

FIGURE 3.7 The intersection of a control mass at two different times represents a control volume.

in Figure 3.7. This intersection represents a control volume. Therefore, the energy balance for a control volume can be obtained by using the energy balance equations for the two control masses at times t and $t + dt$ [1].

First, let's develop a mass balance equation for this control volume. Figure 3.8 shows boundaries of control masses at times t and $t + dt$ and the control volume C as the intersection of these two control masses. Let's assume $d\mathcal{V}_A$ and $d\mathcal{V}_B$ are the volumes of sections A and B that are part of the system at time t and $t + dt$, respectively. Also, let's assume V_i, A_i, ρ_i and V_e, A_e, ρ_e are velocity, flow cross section and density at inlet and exhaust, respectively. These can be assumed to be constant over the small volumes $d\mathcal{V}_A = V_i A_i dt$ and $d\mathcal{V}_B = V_e A_e dt$ because dt and, therefore, these volumes can arbitrarily be chosen to be very small.

The mass of the control mass at times t and $t + dt$ are

$$m(t) = dm_A + m_C \tag{3.14}$$

$$m(t + dt) = dm_B + m_C. \tag{3.15}$$

where dm_A and dm_B represent the mass contained in volumes $d\mathcal{V}_A$ and $d\mathcal{V}_B$, respectively. Since the mass of a control mass does not change,

$$m(t) = m(t + dt) \implies dm_A = dm_B. \tag{3.16}$$

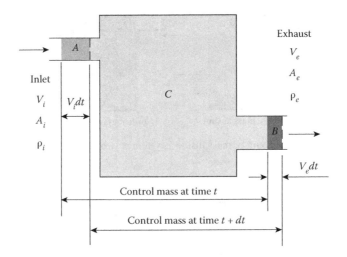

FIGURE 3.8 Mass flow at inlet and exhaust of a control volume.

But
$$dm_A = \rho_i d\mathcal{V}_A = \rho_i A_i V_i dt \quad \text{and} \quad dm_B = \rho_e d\mathcal{V}_B = \rho_e A_e V_e dt. \tag{3.17}$$

Therefore,
$$\rho_i A_i V_i dt = \rho_e A_e V_e dt$$

and
$$\rho_i A_i V_i = \rho_e A_e V_e. \tag{3.18}$$

The product $\rho A V$ is called mass flow rate and is shown by $\dot{m}$ and is meaured by the unit kg/s. Therefore, the **mass balance equation for a control volume with one inlet and one exhaust** can be expressed as:
$$\dot{m}_i = \dot{m}_e. \tag{3.19}$$

If the material entering and leaving the control volume is incompressible, $\rho_i = \rho_0$, the mass balance equation becomes
$$A_i V_i = A_e V_e. \tag{3.20}$$

The product of velocity and area is called **volumetric flow rate** or simply flow rate. It is shown by $\dot{\mathcal{V}}$ and is measured by the unit m³/s.

Although the mass balance equation is derived for a control volume with only one inlet and one exhaust, the result can be generalized for multiple inlets and exhausts. The **mass balance equation for a control volume with n_i inlets and n_e exhausts** can be written as

$$\sum_{i=1}^{n_i} \dot{m}_i = \sum_{e=1}^{n_e} \dot{m}_e \quad \text{or} \quad \sum_{i=1}^{n_i} (\rho A V)_i = \sum_{e=1}^{n_e} (\rho A V)_e. \tag{3.21}$$

Principle of Conservation of Energy

Example 3.7

Write the mass balance equation for the control volume of Figure 3.9.

Solution

The mass balance equation is

$$\dot{m}_2 + \dot{m}_3 = \dot{m}_1 + \dot{m}_4 + \dot{m}_5.$$

or, in terms of density, cross section area, and velocity at every inlet and exhaust

$$(\rho A V)_2 + (\rho A V)_3 = (\rho A V)_1 + (\rho A V)_4 + (\rho A V)_5.$$

A procedure similar to that used to develop the mass balance equation for a control volume is used to derive an appropriate form of the energy balance equation for a control volume. Figure 3.10 shows the same control mass at times t and $t + dt$. In

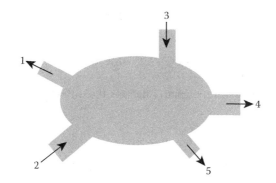

FIGURE 3.9 A control volume with multiple inlets and multiple exhausts.

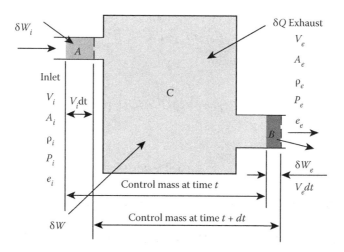

FIGURE 3.10 Energy flows in and out of a control volume.

addition to parameters present in Figure 3.8, P_i and e_i and P_e and e_e represent pressure and total energy per unit mass at inlet and exhaust. Let's also assume that, during time internal dt, a small amount of heat δQ is transferred into the control mass and a small amount of work δW is done on the control mass. In addition, the control mass at time t is compressed by volume $d\mathcal{V}_A$ at inlet and expanded by volume $d\mathcal{V}_B$ at exhaust. These represent work done on the control mass at inlet and by the control mass at exhaust. The work done on the control mass at inlet is

$$\delta W_i = P_i d\mathcal{V}_A = P_i A_i V_i dt = (PAV)_i dt, \qquad (3.22)$$

and the work done by the control mass at exhaust is

$$\delta W_e = P_e d\mathcal{V}_B = P_e A_e V_e dt = (PAV)_e dt. \qquad (3.23)$$

The energy balance for the control mass can be written as

$$E_{in} - E_{out} = dE,$$

$$\delta W + \delta Q + \delta W_i - \delta W_e = dE. \qquad (3.24)$$

The change in energy of the control mass is the difference between its energies at times $t + dt$ and t.

$$dE = E(t + dt) - E(t)$$

$$dE = dE_B + E_C(t + dt) - dE_A - E_C(t)$$

$$dE = dE_B - dE_A + dE_C. \qquad (3.25)$$

But

$$dE_A = e_i dm_A = e_i \rho_i A_i V_i dt = (e\rho AV)_i \qquad (3.26)$$

and

$$dE_B = e_e dm_B = e_e \rho_e A_e V_e dt = (e\rho AV)_e dt. \qquad (3.27)$$

Substituting for δW_i, δW_e and dE in the energy balance equation gives

$$\delta W + \delta Q + (PAV)_i dt - (PAV)_e dt = dE_C + (e\rho AV)_e dt - (e\rho AV)_i dt. \qquad (3.28)$$

Principle of Conservation of Energy

Dividing this by dt gives

$$\frac{\delta W}{dt} + \frac{\delta Q}{dt} + (PAV)_i - (PAV)_e = \frac{dE_C}{dt} + (e\rho AV)_e - (e\rho AV)_i. \tag{3.29}$$

Grouping all terms at inlet and exhaust separately results in

$$\dot{W} + \dot{Q} + \left[(e\rho AV)_i + (PAV)_i\right] - \left[(e\rho AV)_e + (PAV)_e\right] = \frac{dE_C}{dt}, \tag{3.30}$$

$$\dot{W} + \dot{Q} + (\rho AV)_i (e + P/\rho)_i - (\rho AV)_e (e + P/\rho)_e = \frac{dE_C}{dt}. \tag{3.31}$$

Since $\dot{m} = \rho AV$ and $v = 1/\rho$, this equation can be written as

$$\dot{W} + \dot{Q} + \dot{m}_i (e + Pv)_i - \dot{m}_e (e + Pv)_e = \frac{dE_C}{dt}. \tag{3.32}$$

This is the general form of the energy balance equation for a control volume. It must be noted that $e + Pv$ is the sum of the total energy per unit mass and the flow work per unit mass of the material entering or leaving the control volume.

Since $e = u + V^2/2 + E_p/m$, if we neglect the changes in kinetic and potential energies, compared to the change in internal energy, the energy balance equation for a control volume becomes

$$\dot{W} + \dot{Q} + \dot{m}_i (u + Pv)_i - \dot{m}_e (u + Pv)_e = \frac{dE_C}{dt}. \tag{3.33}$$

Since $h = u + Pv$, this equation can be written as

$$\dot{W} + \dot{Q} + \dot{m}_i h_i - \dot{m}_e h_e = \frac{dE_C}{dt}. \tag{3.34}$$

Equation 3.34 can be generalized for a control volume with n_i inlets and n_e exhausts;

$$\dot{W} + \dot{Q} + \sum_{i=1}^{n_i} \dot{m}_i h_i - \sum_{e=1}^{n_e} \dot{m}_e h_e = \frac{dE_C}{dt}. \tag{3.35}$$

This is the general form of the energy balance equation for a control volume with multiple inlets and exhausts and negligible changes in kinetic and potential energies.

For a **steady-state system**

$$\dot{W} + \dot{Q} + \sum_{i=1}^{n_i} \dot{m}_i h_i - \sum_{e=1}^{n_e} \dot{m}_e h_e = 0. \tag{3.36}$$

If there is no work done by or on the control volume, the heat transfer rate to the control volume can be calculated by

$$\dot{Q} = \sum_{e=1}^{n_e} \dot{m}_e h_e - \sum_{i=1}^{n_i} \dot{m}_i h_i + \frac{dE_C}{dt}. \qquad (3.37)$$

Example 3.8

Consider an electronic system with a 400 mm × 35 mm front inlet, a 150 mm × 35 mm left side exhaust, and a 75 mm × 35 mm right side exhaust. The electronic devices will be installed inside this system as shown in Figure 3.11. The inlet air temperature is 40°C and the system operates at an altitude where the air pressure is 90 kPa. The maximum allowable air velocity and air temperature at the exhausts are 2 m/s and 55°C.

  a. What is the maximum allowable mass flow rate and air velocity at the inlet of this system?
  b. What is the maximum allowable heat dissipation by the electronic devices inside this system? Table A.1 shows that $C_{p,ave}$ = 1008 J/kg°C from 40°C To 55°C.

Solution

We chose this electronic system to represent the boundaries of control volume and the air inside it as the content of the control volume. The inlet surface area, A_i, left exhaust surface area, A_{el}, and right exhaust surface area, A_{er}, are

$$A_i = 0.4\,m \times 0.035\,m = 0.014\,m^2$$

$$A_{el} = 0.15\,m \times 0.035\,m = 0.00525\,m^2$$

$$A_{er} = 0.075\,m \times 0.035\,m = 0.002625\,m^2.$$

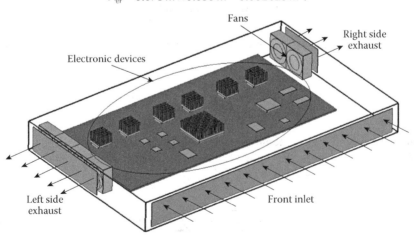

FIGURE 3.11 The system of Example 3.8.

Principle of Conservation of Energy

Mass balance equation for this control volume gives

$$\dot{m}_i = \dot{m}_{el} + \dot{m}_{er}.$$

Ideal gas law is used to calculate air densities at the left and right exhausts. The mass flow rates at these exhausts are

$$\dot{m}_{el} = (\rho A V)_{el} = \left(\frac{P}{RT} A V\right)_{el} = \frac{90{,}000\,\text{Pa}}{287\,\text{J/kgK} \times (273.15 + 55)\text{K}}$$

$$\times 0.00525\,\text{m}^2 \times 2\,\text{m/s} = 0.01\,\text{kg/s}$$

$$\dot{m}_{er} = (\rho A V)_{er} = \left(\frac{P}{RT} A V\right)_{er} = \frac{90{,}000\,\text{Pa}}{287\,\text{J/kgK} \times (273.15 + 55)\text{K}}$$

$$\times 0.002625\,\text{m}^2 \times 2\,\text{m/s} = 0.005\,\text{kg/s}.$$

Substituting into the mass balance equation gives the inlet mass flow rate.

$$\dot{m}_i = 0.01\,\text{kg/s} + 0.005\,\text{kg/s} = 0.015\,\text{kg/s}.$$

The air velocity at inlet is obtained using the ideal gas law to calculate air density and the relationship between mass flow rate, density, velocity, and area.

$$\dot{m}_i = \rho_i A_i V_i = \frac{P_i}{RT_i} A_i V_i$$

$$V_i = \frac{\dot{m}_i RT_i}{P_i A_i} = \frac{0.015\,\text{kg/s} \times 287\,\text{J/kgK} \times (273.15 + 40)\text{K}}{90{,}000\,\text{Pa} \times 0.014\,\text{m}^2} = 1.07\,\text{m/s}.$$

Energy balance Equation 3.37 with steady-state assumption (i.e., $dE_c/dt = 0$) is used to obtain maximum allowable heat dissipation inside this system.

$$\dot{Q} = \sum_{e=1}^{n_e} \dot{m}_e h_e - \sum_{i=1}^{n_i} \dot{m}_i h_i.$$

$$\dot{Q} = \dot{m}_{el} h_{el} + \dot{m}_{er} h_{er} - \dot{m}_i h_i$$

$$\dot{Q} = \dot{m}_{el} C_{p,ave} T_{el} + \dot{m}_{er} C_{p,ave} T_{er} - \dot{m}_i C_{p,ave} T_i$$

$$\dot{Q} = C_{p,ave} (\dot{m}_{el} T_{el} + \dot{m}_{er} T_{er} - \dot{m}_i T_i)$$

$$\dot{Q} = 1008\,\text{J/kg}°\text{C} (0.01\,\text{kg/s} \times 55°\text{C} + 0.005\,\text{kg/s} \times 55°\text{C} - 0.015\,\text{kg/s} \times 40°\text{C})$$

$$\dot{Q} = 226.8\,\text{W}.$$

PROBLEMS

3.1 Using a simple drawing explain the principle of conservation of energy for a closed system.

3.2 What is a control mass? What is a control volume? Give three examples for each.

3.3 What is the appropriate form of energy balance for a surface?

3.4 A 2 kg copper block, initially at 100°C, is brought in contact with a 2 kg plastic block, initially at 20°C. They are kept in contact long enough to reach the same equilibrium temperature. The specific heat of plastic is four times as much as the specific heat of copper. Which block will go through a larger temperature change? Why?

3.5 Consider the electronic enclosure shown below. If $\dot{m}_1 = 0.035$ kg/s, $\dot{m}_2 = 0.015$ kg/s, $\dot{m}_4 = 0.008$ kg/s, $T_1 = 50°C$, $T_2 = T_3 = 70°C$, and $C_p = 1009$ J/kg°C, what are $\dot{m}_3$, $\dot{m}_5$, and total heat dissipation in this system?

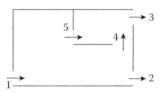

3.6 Consider a desktop computer with one inlet and on exhaust as shown below. If $\dot{m} = 0.02$ kg/s, $T_1 = 25°C$, $T_2 = 40°C$, and $C_p = 1007$ J/kg°C for air, what is total heat dissipation inside this desktop? If pressure and area at inlet of this desktop computer are $P_1 = 70$ kPa and $A_1 = 0.02$ m² respectively, what is the inlet air velocity?

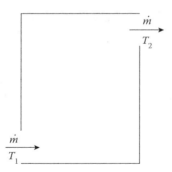

3.7 Consider the electronic system shown below. It consists of three subsystems A, B, and C. Assume the average specific heat of air at constant pressure is $C_{p,\text{ave}} = 1010$ J/kg°C.
   a. If $\dot{m}_1 = 0.02$ kg/s, $\dot{m}_5 = 0.05$ kg/s, what is $\dot{m}_2$?
   b. If $T_1 = 50°C$ and the heat dissipation in subsystem A is $\dot{Q}_A = 400$ W, what is T_3?
   c. If $T_2 = 50°C$, $T_4 = 60°C$, what is total heat dissipated in subsystem B?
   d. If $A_5 = 0.04$ m² and $\rho_5 = 1.05$ kg/m³, what is the air velocity at exhaust 5?

Principle of Conservation of Energy

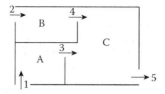

3.8 Consider the system shown below and assume that the average specific heat of air is $C_{p,ave} = 1008$ J/kg°C.
   a. If $\dot{m}_1 = 0.03$ kg/s and $\dot{m}_4 = 0.02$ kg/s, what is $\dot{m}_2$?
   b. If $T_1 = 40°C$, $T_2 = 50°C$, and $T_4 = 55°C$, what is the total heat dissipation inside this system?
   c. If $T_3 = 50°C$, what is the heat dissipation in the bottom portion of this system?
   d. If the area at exhaust 4 is 0.0333 m² and air density is 1.08 kg/m³, what is the air velocity at this exhaust?

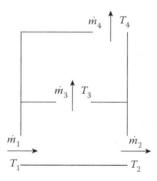

3.9 Consider the electronic enclosure shown below.
   a. If $\dot{m}_1 = \dot{m}_2 = 0.05$ kg/s, $\dot{m}_5 = 0.02$ kg/s, and $\dot{m}_3 = \dot{m}_4$, what is $\dot{m}_3$?
   b. If $T_1 = T_2 = 50°C$, $T_3 = T_4 = 70°C$, and $T_5 = 60°C$ and $C_{p,ave} = 1009$ J/kg°C, what is the total heat dissipated in this enclosure?
   c. If air pressure is 70 kPa and gas constant for air is 287 J/kgK, what is air density at exhaust 5?
   d. If maximum allowable air velocity at exhaust 5 is 1.5 m/s, what is its minimum required area?

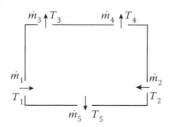

3.10 Consider the piston-cylinder-spring system shown below. The cylinder contains 2 kg of air. The piston diameter is 15 cm. The spring force applied to the piston is $F = kx$ where $k = 2 \times 10^6$ N/m.
   a. Assume air is an ideal gas and find its temperature if $x = 30$ cm.
   b. The cylinder is heated such that the piston moves to $x = 40$ cm. How much work is done against the spring?
   c. Calculate the amount of heat transfer in this process.

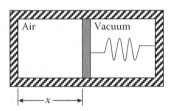

3.11 Consider the insulated vessel shown here, with compartment A of volume 0.03 m³, which is empty, and separated by an insulating membrane from compartment B of volume 0.01 m³, which contains 0.15 kg of air at 25°C. The air is stirred by a fan until the membrane ruptures. The membrane is designed to rupture at a pressure of 2 MPa.
   a. What is the temperature when the membrane ruptures?
   b. Calculate the work done by the fan.
   c. Find the pressure and temperature of the air after the membrane ruptures and the air reaches equilibrium state.

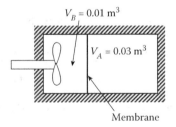

3.12 A car radiator is a heat exchanger that uses air to cool the engine's cooling water. Suppose the hot water from the engine enters the radiator at 80°C and leaves it at 60°C. The water mass flow rate is 0.2 kg/s. If the outside air, at 100 kPa pressure and 25°C temperature, enters the radiator at 10 m/s, what will the air temperature be after passing through the radiator? The cross section of the radiator where air passes through is 40 cm by 60 cm.

Principle of Conservation of Energy

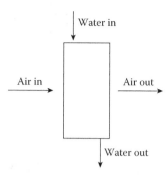

3.13 Consider an electronic system that dissipates 2000 W. Air with a velocity of 2 m/s and a temperature of 50°C enters the system through the 0.01 m² area front inlet and with a volumetric flow rate of $\dot{V} = 0.015$ m³/s and temperature of 25°C through the rear inlet. The system has two exits. The air temperature and the area of the rear exit are 85°C and 0.008 m², respectively, and the area of the front exit is 0.015 m². If the mass flow rate through the front exit is twice as much as the mass flow rate through the rear exit, what are the air temperature at the front exit and the air velocity at the front and rear exits? The air pressure is 1 atm at all inlets and exits and the average constant pressure specific heat of air is $C_{p,ave} = 1009$ J/kg°C over the temperature range of this problem.

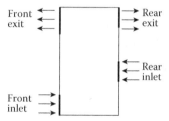

3.14 Consider an electronic system that consists of two main sections. Air, with velocity of 4 m/s and temperature of 55°C, enters the system through a 400 mm × 100 mm front inlet. Part of the inlet air goes through the bottom section of this system and exits through a 400 mm × 75 mm exhaust with velocity of 2.5 m/s and temperature of 72°C. The rest of the air goes through the top section of the system, which dissipates 1500 W, and exits through a 400 mm × 100 mm exhaust area on the top. Determine the total power dissipation in the bottom section and average air velocity and temperature at the exhaust of the top section. The air pressure is 1 atm at the inlet and exhausts and the average constant pressure specific heat of air is $C_{p,ave} = 1009$ J/kg°C over the temperature range of this problem.

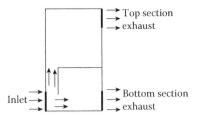

3.15 An electronic enclosure that dissipates 3000 W is cooled by water. The cold water enters the enclosure at 45°C. The warm water that exits the enclosure is cooled down to 45°C in an air-to-water heat exchanger as shown below. Assume that the water pipes are perfectly insulated.
   a. If the maximum allowable warm water temperature is 47°C, what will be the water mass flow rate?
   b. If the outside air, at 85 kPa pressure and 25°C temperature, enters the heat exchanger at 3 m/s speed, what will be the air temperature after passing through the heat exchanger? The cross section of the heat exchanger where air passes through is 35 cm by 55 cm. Assume air properties are constant within the temperature range of this problem.

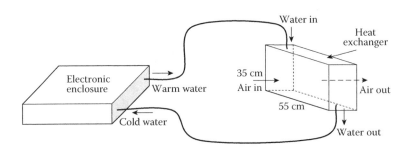

3.16 An electronic enclosure dissipates 300 W and is cooled by air that enters this enclosure through a front inlet and exits it through one rear and two side exhausts. This unit operates at an altitude where the air pressure is 70 kPa. Air enters this unit through the 300 mm wide and 40 mm high inlet with a velocity of 2 m/s and temperature of 50°C. The air velocity and temperature at the 170 mm wide and 40 mm high rear exhaust are 1.5 m/s and 70°C, respectively. If the air temperature and mass flow rate through the 100 mm wide and 40 mm high left and right exhausts are the same, find the air velocity and temperature at these exhausts. Assume $C_{p,ave} = 1009$ J/kg°C for air.

Principle of Conservation of Energy

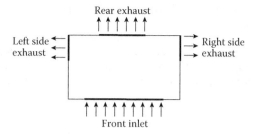

3.17 The heat dissipating components of an electronic enclosure are cooled by four fans that are mounted in front of the enclosure; two on the left side and two on the right side. The air is drawn into the enclosure through a 200 mm × 40 mm inlet in the back and pushed out of the enclosure through two 80 mm × 40 mm exhausts in the front. The total heat dissipation by the electronic components inside the enclosure is 200 W. The air temperature and pressure at inlet are 55°C and 70 kPa. The maximum allowable exhaust air temperature is 70°C and the air pressure at exhaust is also 70 kPa.
   a. What is the minimum required mass flow rate of air? Assume the airflow rates through the left and right side exhausts are equal and they do the same amount of cooling.
   b. What is the air speed at inlet and exhaust?
   c. Now, let's assume that one of the fans on the left side is failed. Therefore, the airflow rate through the left side exhaust is half of the airflow rate through the right side exhaust. What are the air temperatures at the left and right exhausts? Again, assume that air that passes through the left and right exhausts do the same amount of cooling.

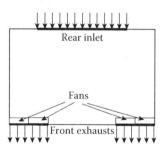

3.18 Consider an electronic system shown below in which the heat dissipating components are located in separate front and rear sections. Air at a temperature of 55°C and pressure of 70 kPa enters the system through the 80 mm × 350 mm front inlet and leaves the system at 70°C and 70 kPa through the 120 mm × 350 mm rear exhaust. The total heat dissipation inside the system in 2800 W.

a. What is the total mass flow rate through inlet and exhaust of this system?
b. What are air velocities at inlet and exhaust?
c. If the heat dissipation in the rear section is 800 W, what is the airflow distribution through the front and rear sections? Assume same exhaust temperature for both front and rear sections.

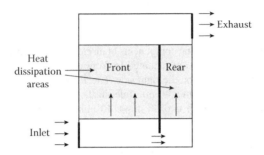

3.19 Consider the electronic system shown below. It consists of an AC to DC power supply that dissipates 100 W and a motherboard that dissipates 400 W. The air temperature and pressure at the inlet of this system are 45°C and 85 kPa, respectively. The power supply has its own fans that provide 0.01 m³/s airflow rate. The main system fans are lined up at the 400 mm × 80 mm exhaust area of the system as shown below. Let's assume that the exhaust air from the power supply mixes well with the rest of the air coming from the 200 mm × 80 mm intake of the system and exits through the exhaust area of the system with a uniform temperature of 60°C.
a. What is the airflow rate at the intake of the system?
b. What are the air velocities at the intake and exhaust of the system?

3.20 Consider the electronic enclosure shown below. Air with velocity of 1 m/s, temperature of 50°C and pressure of 85 kPa enters this enclosure through inlets $A_1 = 0.02$ m², $A_2 = 0.015$ m² and $A_3 = 0.01$ m² and

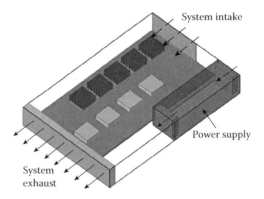

leaves it through exhaust $A_4 = 0.03$ m² at same pressure. Total heat dissipation in this enclosure is 900 W.
a. What are the air velocity and temperature at exhaust?
b. If the heat dissipation is uniformly divided into three separate compartment of this enclosure, what are the air temperatures at internal openings 5 and 6?
Use $C_{p,\text{ave}} = 1009$ J/kg°C in all of your calculations.

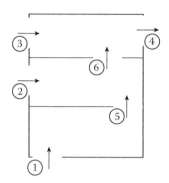

REFERENCES

1. Reynolds, W. C., and H. C. Perkins. 1977. *Engineering thermodynamics*. New York, NY: McGraw-Hill.
2. Moran, M. J., and H. N. Shapiro. 2004. *Fundamentals of engineering thermodynamics*. Hoboken, NJ: John Wiley & Sons, Inc.
3. Sonntag, R. E., and C. Borgnakke. 2007. *Introduction to engineering thermodynamics*. Hoboken, NJ: John Wiley & Sons, Inc.

4 Heat Transfer Mechanisms

In Chapter 2, heat transfer was defined as the form of energy transfer that takes place between two objects that are at different temperatures. The mechanism by which heat transfer occurs depends on whether there is any material medium between the two objects and if that medium is moving or not. Three different mechanisms or modes of heat transfer are conduction, convection, and radiation. An overview of these heat transfer mechanisms are given in this chapter and more detailed discussions will be left for Chapters 8 through 13.

4.1 CONDUCTION HEAT TRANSFER

Heat transfer between two objects, or across a single object, that happens through a material medium and does not involve any fluid motion is called **conduction heat transfer**. Examples of conduction heat transfer are heat transfer from the hot side to the cold side of a solid or a motionless liquid or gas medium, heat transfer from or to our hand when we touch an object, heat transfer through the insulation around a steam pipe, heat transfer from a silicon die to its substrate or the encapsulation covering the die, heat transfer through a printed circuit board, and heat transfer from an electronic device to the heat sink that is mounted on it.

Conduction heat transfer is the energy transfer from more energetic particles of a substance to the adjacent less energetic ones as a result of interaction between these particles. The physical mechanism in which this energy transfer happens is different in different materials. In gases and liquids conduction heat transfer is due to the collision and diffusion of the molecules during their random motion. In metals conduction heat transfer is due to the energy transfer between free electrons. In crystalline solids conduction is due to the energy exchange between vibrating molecules in a lattice.

Consider a plane wall with thickness $\Delta x = x_2 - x_1$ and surface area A whose cross section is shown in Figure 4.1. Temperatures at the left and right sides of this wall are T_1 and T_2, respectively. Experimental observations show that conduction heat transfer rate through this wall is directly proportional to surface area A and temperature difference $\Delta T = T_2 - T_1$, and is inversely proportional to thickness $\Delta x = x_2 - x_1$:

$$\dot{Q}_{\text{cond}} \propto \frac{A \Delta T}{\Delta x}. \tag{4.1}$$

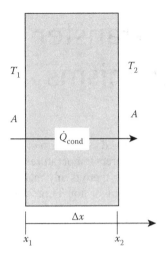

FIGURE 4.1 Conduction heat transfer through a plane wall.

This proportionality correlation can be converted to an equation by introducing a proportionality constant k:

$$\dot{Q}_{cond} = -kA\frac{\Delta T}{\Delta x}. \quad (4.2)$$

The negative sign in this equation is a convention to make the sign of $\dot{Q}_{cond}$ positive if it is along the positive x direction. The sign of $\dot{Q}_{cond}$ will always be opposite the sign of $\Delta T = T_2 - T_1$; that is, $\dot{Q}_{cond}$ will be positive and along the positive x direction if ΔT is negative and vice versa. This is consistent with the fact that heat transfer naturally occurs from high to low temperature sides of an object.

In the limit of $\Delta x \rightarrow 0$, the above equation reduces to

$$\dot{Q}_{cond} = -kA\frac{dT}{dx}. \quad (4.3)$$

This is known as **Fourier's law of heat conduction**.

The k in Equation 4.3 is called **thermal conductivity** of material and its unit in SI system of units is W/m°C. As its role in this equation indicates, it is a measure of how good a material conducts heat. Figure 4.2 shows the spectrum of thermal conductivity values and types of materials that fit in different sections of this spectrum. Thermal conductivities of some materials commonly used in electronic packaging and electronics cooling are shown in Table 4.1 [1]. Crystals such as diamond are the most conductive pure materials. Metals like gold, copper, and aluminum are very good conductive materials and are widely used in different levels of electronic packaging. Gases like air are the least conductive materials.

Thermal conductivity of a material, in general, is not a constant value. Thermal conductivities of gases increase with increasing temperature ($k \propto T^{1/2}$) and decreases with increasing molar mass ($k \propto 1/M^{1/2}$), but it is independent of pressure. Thermal conductivity of most liquids such as saturated ammonia, saturated propane, methane,

Heat Transfer Mechanisms

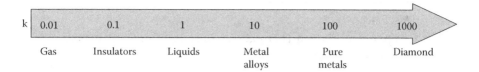

FIGURE 4.2 Spectrum of thermal conductivity values of different materials.

TABLE 4.1
Thermal Conductivities of Some Common Materials

Material	Thermal Conductivity (W/m °C)
Diamond	2000–2300
Copper	380–403
Gold	319
Aluminum	210–237
Silicon	124–148
Germanium	64
Aluminum oxide	15–33
Aluminum nitride	82–320
Silicon carbide	120–270
Solder (95Pb5Sn)	35.5
Solder (37Pb63Sn)	50.6
Epoxy	0.19
Polyimide	0.18
Glass epoxy (FR4)	0.35
Saturated water at 25°C	0.613
Air at 25°C	0.0261

Source: From Pecht, M., Agarwal, R., McCluskey, F. P., Dishongh, T. J., Javadpour, S., and Mahajan, R., *Electronic Packaging Materials and their Properties.* Boca Raton, FL: CRC Press, 1999. With permission.

methanol, and engine oil decreases with increasing temperature. Thermal conductivity of saturated water increases up to about 130°C and then decreases as temperature increases beyond that. Thermal conductivity of most metals decreases slightly as temperature increases above the normal room temperature of 25°C. However, the change in thermal conductivity of metals is usually neglected if the temperature change is not very large.

Table 4.2 compares thermal and electrical conductivities of various materials commonly used in electronic applications. Some materials have similar electrical and thermal behaviors. Metals are good electrical and thermal conductors while polymers are poor electrical and thermal conductors. However, crystalline solids such as diamond and semiconductors such as silicon are good thermal conductors

TABLE 4.2
Comparison between Thermal and Electrical Conductivities of Various Materials

Material	Thermal Conductivity W/m°C	Thermal Conductivity 1/Ω.m
Diamond	2000–2300	3.7×10^{-15}
Copper	380–403	3.4×10^{7}
Aluminum	210–237	5.2×10^{7}
Silicon	124–148	0.0015
Germanium	64	0.112
Alumina	30	10^{-20}
Glass epoxy	0.35	10^{-21}
Polyimide	0.18	10^{-20}

Source: From Pecht, M., Agarwal, R., McCluskey, F. P., Dishongh, T. J., Javadpour, S., and Mahajan, R., *Electronic Packaging Materials and their Properties.* Boca Raton, FL: CRC Press, 1999. With permission.

but poor electrical conductors. This is one reason that makes semiconductors suitable materials for manufacturing integrated circuits.

Pure metals are good thermal conductors but alloys of those metals are usually poor conductors. For example, copper and aluminum are high thermal conductivity metals while commercial bronze, which is 90% copper and 10% aluminum, has a much lower thermal conductivity than any of these metals.

Example 4.1

Consider a 10 mm × 10 mm × 0.7 mm silicon die that dissipates 20 W. The electrical circuitry is on the back side of the die and all the heat is transferred to the front side and is dissipated from there. If thermal conductivity of silicon is 125 W/m°C, what is the temperature difference across the die?

Solution

Equation 4.2 can be used to find temperature difference across the die. Since $\dot{Q}_{cond} = 20$ W, $A = 10$ mm × 10 mm = 100 mm² = 0.0001 m², $\Delta x = 0.7$ mm = 0.0007 m, and $k = 125$ W/m°C,

$$\dot{Q}_{cond} = -kA\frac{\Delta T}{\Delta x} \Rightarrow 20 \text{ W} = -125 \text{ W/m°C} \times 0.0001 \text{m}^2 \frac{\Delta T}{0.0007 \text{m}}$$

$$\Delta T = -1.12°C.$$

Note that the negative sign shows that temperature on the front side of the die is less than that at the back side of the die (i.e., the active side is 1.12°C hotter than the opposite side).

Heat Transfer Mechanisms

Example 4.2

Consider that the die in Example 4.1 is attached to a metal cap through a 0.1 mm thick thermal interface material with thermal conductivity of 5 W/m°C. What is the temperature difference across the thermal interface material?

Solution

For the thermal interface material $\dot{Q}_{cond} = 20$ W, $A = 10$ mm × 10 mm = 100 mm² = 0.0001 m², $\Delta x = 0.1$ mm = 0.0001 m and $k = 5$ W/m°C. Therefore,

$$\dot{Q}_{cond} = -kA\frac{\Delta T}{\Delta x} \Rightarrow 20\text{ W} = -5\text{ W/m°C} \times 0.0001\text{m}^2 \frac{\Delta T}{0.0001\text{m}}$$

$$\Delta T = -4°C.$$

Note that temperature change through the thermal interface material is almost four times as much as temperature change through the die itself. This is usually the case in microelectronic packaging and there have been many research activities on developing high-thermal-conductivity thermal interface materials to minimize this temperature difference.

4.2 CONVECTION HEAT TRANSFER

Heat transfer between an object and the adjacent moving fluid (liquid or gas) is called **convection heat transfer**; see Figure 4.3. If the temperature of the surface is T_s and the velocity and temperature of the moving fluid are U_∞ and T_∞, respectively, the necessary condition for convection heat transfer is that $T_\infty \neq T_s$. Examples of convection heat transfer are heat transfer from a person's face when wind is blowing, blowing into hot food or drink to cool it, cooling electronics inside a PC by air movement generated by a fan, and heat transfer from a boiled egg that is exposed to colder air.

There are two ways that convection heat transfer is created around an object. If the fluid motion is generated by a fan or a pump or wind, it will be called **forced convection**. Examples of electronics that are cooled by forced convection are desktop and notebook CPUs and GPUs, high output power supplies, gaming consoles, printed circuit boards, and components in networking and telecommunication equipment such as servers, routers, and switches.

On the other hand, if the fluid motion is generated by a density difference due to a temperature difference in it, the convection is called **natural** or **free convection**.

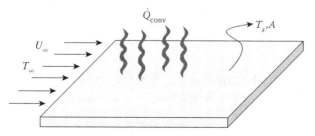

FIGURE 4.3 Convection heat transfer from a surface.

Natural convection is the heat transfer mechanism responsible for heating the whole air inside a room by a heater located at one corner of that room. First, the air near the heater gets hot. The hot air is lighter than the neighboring colder air. Since in a mixture of fluids the lighter fluid moves up, the hot air rises and the neighboring colder air replaces it and goes through the same process. This process will increase the air temperature in the entire room. Heat transfer from a human body is partly due to natural convection. Examples of electronics that are cooled by natural convection are cell phones, handheld computers, most televisions, radios, video, CD and DVD players, and wireless routers.

Convection flows may be laminar or turbulent. Laminar flows are slow, orderly, and streamlined flows while turbulent flows are faster, disordered, and fluctuating flows in which bulks of fluids move from one region to another and mix with each other in a random manner (see Chapter 9 for more details).

Convection heat transfer rate is proportional to the surface area of the object, which is exposed to the moving fluid, A, and the temperature difference between the object and fluid, $T_s - T_\infty$:

$$\dot{Q}_{conv} = hA(T_s - T_\infty). \tag{4.4}$$

This equation is known as **Newton's law of cooling**. The h is called **convection heat transfer coefficient** and its unit in the SI system of units is W/m² °C.

Unlike thermal conductivity, convection heat transfer coefficient is not a material property. It will be shown in Chapters 9 through 12 that convection heat transfer coefficient depends on fluid properties, flow velocity, temperature difference between the surface and the fluid, acceleration of gravity, flow geometry, surface geometry, and type of the flow. Generally, convection heat transfer coefficient in a liquid is few orders of magnitude larger than that in a gas, higher velocity fluid has higher convection heat transfer coefficient, natural convection heat transfer coefficient is larger if either temperature difference between the surface and fluid or the acceleration of gravity is larger, and convection heat transfer coefficient in turbulent flows is larger than in similar smooth laminar flows.

4.2.1 Simplified Correlations for Convection Heat Transfer in Air

Appropriate correlations to calculate convection heat transfer coefficients for some common flows will be given in Chapters 10, 11, and 12. Simplified forms of some of these correlations for convection heat transfer in air are given in this section. Note that these correlations are valid only in the SI system of units where velocity, length, temperature, and heat transfer coefficient are measured in m/s, m, °C and W/m² °C, respectively.

- **Laminar forced convection air flow** with velocity U_∞ over a constant temperature flat plate with length L:

$$h = 3.9 \left(\frac{U_\infty}{L} \right)^{1/2}. \tag{4.5}$$

Heat Transfer Mechanisms

- **Turbulent forced convection air flow** with velocity U_∞ over a constant temperature flat plate with length L:

$$h = 5.5 \left(\frac{U_\infty^4}{L} \right)^{1/5}. \quad (4.6)$$

- **Laminar natural convection from a vertical flat plate** with length L, in the vertical direction, and uniform temperature T_s to air at temperature T_∞:

$$h = 1.4 \left(\frac{T_s - T_\infty}{L} \right)^{1/4}. \quad (4.7)$$

- **Turbulent natural convection from a vertical flat plate** with length L, in the vertical direction, and uniform temperature T_s to air at temperature T_∞:

$$h = 1.1 (T_s - T_\infty)^{1/3}. \quad (4.8)$$

Note that convection heat transfer coefficient in this case is independent from the length of the plate.

- **Natural convection from the top side of a horizontal flat plate** with area A, perimeter P, and uniform temperature T_s to air at temperature T_∞:

$$h = 1.3 \left(\frac{T_s - T_\infty}{A/P} \right)^{1/4} \text{ if } 1.8 \times 10^{-4} < (T_s - T_\infty)(A/P)^3 < 0.18$$

$$h = 1.6 (T_s - T_\infty)^{1/3} \text{ if } 0.18 < (T_s - T_\infty)(A/P)^3 < 1.8 \times 10^3. \quad (4.9)$$

- **Natural convection from the bottom side of a horizontal flat plate** with area A, perimeter P, and uniform temperature T_s to air at temperature T_∞:

$$h = 0.65 \left(\frac{T_s - T_\infty}{A/P} \right)^{1/4}. \quad (4.10)$$

Example 4.3

Consider a 400 mm × 600 mm printed circuit board that is exposed to air with a velocity of 2 m/s and temperature of 25°C. The electronics on this board dissipate 150 W and the board with those electronics will be approximated as a flat plate with uniform temperature. Calculate the board temperature if (a) the airflow is laminar and along the 400 mm side, (b) the airflow is laminar and along the 600 mm side, (c) the air flow is turbulent and along the 400 mm side, and (d) the airflow is turbulent and along the 600 mm side.

Solution

Newton's law of cooling with $\dot{Q}_{conv} = 150$ W, $A = 400$ mm × 600 mm = 240,000 mm² = 0.24 m², $T_\infty = 25°C$, $U_\infty = 2$ m/s and h as calculated below can be used to obtain T_s.

(a) Convection heat transfer coefficient is given by Equation 4.5 with $L = 0.4$ m. Therefore,

$$h = 3.9\left(\frac{U_\infty}{L}\right)^{1/2} = 3.9\left(\frac{2}{0.4}\right)^{1/2} = 8.72 \text{ W/m}^2\text{°C}$$

$\dot{Q}_{conv} = hA(T_s - T_\infty) \Rightarrow 150 \text{ W} = 8.72 \text{ W/m}^2\text{°C} \times 0.24 \text{ m}^2(T_s - 25°C)$

$T_s = 96.67°C$.

(b) Convection heat transfer coefficient is given by Equation 4.5 with $L = 0.6$ m. Therefore,

$$h = 3.9\left(\frac{U_\infty}{L}\right)^{1/2} = 3.9\left(\frac{2}{0.6}\right)^{1/2} = 7.12 \text{ W/m}^2\text{°C}$$

$\dot{Q}_{conv} = hA(T_s - T_\infty) \Rightarrow 150 \text{ W} = 7.12 \text{ W/m}^2\text{°C} \times 0.24 \text{ m}^2(T_s - 25°C)$

$T_s = 112.78°C$.

(c) Convection heat transfer coefficient is given by Equation 4.6 with $L = 0.4$ m. Therefore,

$$h = 5.5\left(\frac{U_\infty^4}{L}\right)^{1/5} = 5.5\left(\frac{2^4}{0.4}\right)^{1/5} = 11.50 \text{ W/m}^2\text{°C}$$

$\dot{Q}_{conv} = hA(T_s - T_\infty) \Rightarrow 150 \text{ W} = 11.50 \text{ W/m}^2\text{°C} \times 0.24 \text{ m}^2(T_s - 25°C)$

$T_s = 79.35°C$.

(d) Convection heat transfer coefficient is given by Equation 4.6 with $L = 0.6$ m. Therefore,

$$h = 5.5\left(\frac{U_\infty^4}{L}\right)^{1/5} = 5.5\left(\frac{2^4}{0.6}\right)^{1/5} = 10.61 \text{ W/m}^2\text{°C}$$

$\dot{Q}_{conv} = hA(T_s - T_\infty) \Rightarrow 150 \text{ W} = 10.61 \text{ W/m}^2\text{°C} \times 0.24 \text{ m}^2(T_s - 25°C)$

$T_s = 83.91°C$.

This example shows that (a) forced convection heat transfer is more efficient if air flows along the shorter side of the plate, and (b) forced convection heat transfer in turbulent flow is more efficient than in a laminar flow with the same velocity.

Heat Transfer Mechanisms

Example 4.4

Consider that the printed circuit board of the Example 4.3 is mounted vertically in air at 25°C. Calculate the board temperature and convection heat transfer coefficient if (a) the 400 mm side is in the vertical direction and natural convection flow is laminar, and (b) the 600 mm side is in the vertical direction and natural convection flow is laminar.

Solution

Newton's law of cooling with $\dot{Q}_{conv} = 150$ W, $A = 400$ mm $\times$ 600 mm $= 240{,}000$ mm² $= 0.24$ m², $T_\infty = 25$°C, and appropriate correlations for h are used to obtain T_s.

(a) Convection heat transfer coefficient is given by Equation 4.7 with $L = 0.4$ m. Therefore,

$$h = 1.4\left(\frac{T_s - T_\infty}{L}\right)^{1/4} = 1.4\left(\frac{T_s - 25}{0.4}\right)^{1/4} = 1.76(T_s - 25)^{1/4} \text{ W/m}^2\text{°C}$$

$$\dot{Q}_{conv} = hA(T_s - T_\infty) \Rightarrow 150 \text{ W} = 1.76(T_s - 25)^{1/4} \text{ W/m}^2\text{°C} \times 0.24 \text{ m}^2 \times (T_s - 25)\text{°C}$$

$$355.11 = (T_s - 25)^{5/4} \Rightarrow T_s = 134.72\text{°C}$$

$$h = 1.76(T_s - 25)^{1/4} \text{ W/m}^2\text{°C} = 1.76(134.72 - 25)^{1/4} \text{ W/m}^2\text{ °C} = 5.70 \text{ W/m}^2\text{°C}.$$

(b) Convection heat transfer coefficient is given by Equation 4.7 with $L = 0.6$ m. Therefore,

$$h = 1.4\left(\frac{T_s - T_\infty}{L}\right)^{1/4} = 1.4\left(\frac{T_s - 25}{0.6}\right)^{1/4} = 1.59(T_s - 25)^{1/4} \text{ W/m}^2\text{°C}$$

$$\dot{Q}_{conv} = hA(T_s - T_\infty) \Rightarrow 150 \text{ W} = 1.59(T_s - 25)^{1/4} \text{ W/m}^2\text{°C} \times 0.24 \text{ m}^2 \times (T_s - 25)\text{°C}$$

$$393.08 = (T_s - 25)^{5/4} \Rightarrow T_s = 144.01\text{°C}$$

$$h = 1.59(T_s - 25)^{1/4} \text{ W/m}^2\text{°C} = 1.59(144.01 - 25)^{1/4} \text{ W/m}^2\text{°C} = 5.25 \text{ W/m}^2\text{°C}.$$

It is seen that the laminar convection heat transfer is more efficient if the shorter side of the plate is along the vertical direction.

4.3 RADIATION HEAT TRANSFER

There are situations in which there is no medium between two objects that are at different temperatures and are somehow facing each other. In these situations heat transfer happens through the exchange of electromagnetic wave or photons and is known as **radiation heat transfer**. The heat transfer from sun to the earth's atmosphere and the heat transfer from the surface of a spaceship to the cold deep sky are examples of radiation heat transfer. It must be mentioned that radiation heat transfer happens

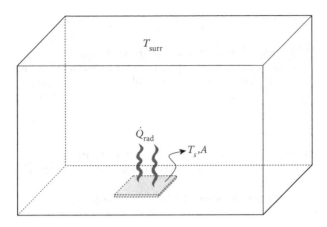

FIGURE 4.4 Radiation heat transfer between an object and its surrounding.

between every two objects with different temperatures, along with other modes of heat transfer, as long as the medium between the two objects is not opaque.

Radiation heat transfer between two objects is a function of their temperatures, surface areas, surface radiation properties, and the relative orientation of those objects with respect to each other. Radiation heat transfer calculations are often complex and cumbersome and will be discussed in detail in Chapter 13. Here, we will consider the simplest case of radiation heat transfer where an object with uniform temperature T_s and surface area A, shown in Figure 4.4, is exposed to a large surrounding with temperature T_{surr}. Note that the word "surrounding" refers to physical walls that make an enclosure that surrounds the object. Three close approximations of this case are a hot object left on a table and exposed to the walls and the roof of a room, an outdoor electronic enclosure such as telecom cabinets exposed to the cold night sky, and a small hot device inside a much larger electronic enclosure. It can be shown that the net radiation heat transfer rate between such an object and its surrounding is

$$\dot{Q}_{rad} = \varepsilon \sigma A (T_s^4 - T_{surr}^4). \tag{4.11}$$

The T_s and T_{surr} are object and surrounding temperatures in an absolute unit such as Kelvin (K) or Rankine (R), which are related to temperature scales Celsius and Fahrenheit and to each other as

$$T(K) = T(°C) + 273.15, \tag{4.12}$$

$$T(R) = T(°F) + 459.67, \tag{4.13}$$

$$T(R) = 1.8\, T(K). \tag{4.14}$$

The $\sigma = 5.67 \times 10^{-8}$ W/m^2.K^4 or 1.714×10^{-9} Btu/h.ft^2.R^4 is **Stefan–Boltzmann constant** and ε is surface emissivity. Surface emissivity, which can take any value

TABLE 4.3
Surface Emissivity of Various Surfaces

Surface	Emissivity
Aluminum foil	0.05
Polished aluminum	0.03
Anodized aluminum	0.84
Polished copper	0.03
Oxidized copper	0.5–0.8
Black paint	0.97
White paint	0.93
White paper	0.9
Snow	0.97

Source: From Cengel, Y. A., *Heat and Mass Transfer: A Practical Approach*. Third Edition. New York, NY: McGraw-Hill, 2007.

between zero and one, is a measure of how good the surface is in emitting radiation. Table 4.3 gives emissivity values for some common surfaces [2].

Equation 4.11 for radiation heat transfer rate between an object and its large surrounding can be written as:

$$\dot{Q}_{rad} = \varepsilon \sigma A (T_s^4 - T_{surr}^4) = \varepsilon \sigma A (T_s^2 + T_{surr}^2)(T_s + T_{surr})(T_s - T_{surr}).$$

If we define

$$h_{rad} = \varepsilon \sigma (T_s^2 + T_{surr}^2)(T_s + T_{surr}), \quad (4.15)$$

then

$$\dot{Q}_{rad} = h_{rad} A (T_s - T_{surr}). \quad (4.16)$$

This equation is similar to Newton's law of cooling for convection heat transfer rate and h_{rad} is called **radiation heat transfer coefficient**.

Although Equation 4.16 is similar to Newton's law of cooling it uses a different reference temperature. The surrounding temperature, T_{surr}, may not be the same as the fluid temperature around an object, T_∞. For example, the air temperature inside an air-conditioned house in summer is usually lower than the temperature of the walls or windows of the house since they may be exposed to the sun, the air temperature inside a glass greenhouse at night is higher than the glass temperature, the air temperature in some regions of a desktop or notebook computer is higher than the computer's casing temperature. In some situations, it is useful to have an equation similar to Equation 4.16 that uses T_∞ as reference temperature. Such an equation will be derived here. First, Equation 4.11 is multiplied and divided by $T_s - T_\infty$;

$$\dot{Q}_{rad} = \varepsilon\sigma A(T_s^4 - T_{surr}^4) = \frac{\varepsilon\sigma(T_s^4 - T_{surr}^4)}{(T_s - T_\infty)} A(T_s - T_\infty). \qquad (4.17)$$

This equation can be written as

$$\dot{Q}_{rad} = h'_{rad} A(T_s - T_\infty), \qquad (4.18)$$

where

$$h'_{rad} = \frac{\varepsilon\sigma(T_s^4 - T_{surr}^4)}{(T_s - T_\infty)} \qquad (4.19)$$

is called **modified radiation heat transfer coefficient**.

Example 4.5

Consider the printed circuit board of Example 4.3. Calculate its average surface temperature if it is surrounded by a large enclosure with temperature of 35°C, its surface emissivity is 0.7, and the only mode of heat transfer is radiation heat transfer.

Solution

In this case $\dot{Q}_{rad} = 150$ W, $A = 0.24$ m², $\varepsilon = 0.7$, $\sigma = 5.67 \times 10^{-8}$ W/m²·K⁴ and $T_{surr} = 35°C = 308.15$ K. Then

$$\dot{Q}_{rad} = \varepsilon\sigma A(T_s^4 - T_{surr}^4) \Rightarrow 150\,W = 0.7 \times 5.67 \times 10^{-8}\,W/m^2 \cdot K^4$$

$$\times 0.24\,m^2 (T_s^4 - 308.15^4)K^4$$

$$T_s = 396.67\,K = 123.54°C.$$

PROBLEMS

4.1 Explain the physical mechanisms of heat conduction in a solid, a liquid, and a gas.

4.2 Explain how the thermal conductivity of a gas changes with temperature.

4.3 Define the three modes of heat transfer and write down equations to calculate heat transfer rate in each mode.

4.4 What are different heat transfer mechanisms from a human body? Which mechanism is more important in windy weather?

4.5 Consider a printed circuit board inside a closed enclosure. What are the mechanisms of heat transfer from this board?

4.6 If $\dot{Q}_{rad} = \varepsilon\sigma A(T_s^4 - T_{surr}^4)$, what is the expression for the radiation heat transfer coefficient h_{rad}?

4.7 Consider two 40 mm × 40 mm × 5 mm thick heat spreaders, one is made of copper and the other is made of aluminum. If heat transfer rates through the thickness of these heat spreaders are the same, which one has less temperature difference across its thickness?

4.8 Consider a 40 mm × 40 mm × 2 mm thick heat spreader with thermal conductivity of 160 W/m°C. If the temperature difference across its thickness is 2°C, what is the heat transfer rate through it?

4.9 If the surface temperature of a device is 90°C and the air temperature and convection heat transfer coefficient are 30°C and 5 W/m² °C, what is the convection heat flux from the surface of this device?

4.10 An electronic device with surface area of 0.01 m² is mounted in an ambient with heat transfer coefficient of 40 W/m²°C and temperature of 45°C. If maximum allowable surface temperature of this device is 100°C, what is its maximum allowable power dissipation?

4.11 A transistor with a surface area of 1 cm² dissipates 0.2 W. Air temperature and heat transfer coefficient around the transistor is 50°C and 40 W/m² °C, respectively. The surrounding temperature is 35°C and the surface emissivity of the transistor is 0.8. Determine the surface temperature of the transistor.

4.12 Consider a 40 mm × 40 mm electronic package that dissipates 5 W. Calculate the average top surface temperature of the package if all the heat is dissipated from the top of the package, the air temperature around the device is 40°C, and
   a. the air flows with a velocity of 3 m/s over the package and flow is laminar,
   b. the package is mounted vertically and natural convection is laminar,
   c. the package is mounted horizontally in air and the top of the package is facing up.

4.13 The emissivity of the package in Problem 4.12 is 0.8 and it is surrounded by an electronic enclosure with temperature 35°C. If the only mode of heat transfer from this package is radiation heat transfer, what is its average surface temperature?

4.14 Multiple modes of heat transfer can happen simultaneously. Let's assume that the package in Problems 4.12 and 4.13 is mounted vertically and dissipates its heat by laminar natural convection as well as radiation heat transfer. Find out its average surface temperature and contributions of natural convection and radiation heat transfer rates.

4.15 Let's assume that the package in Problems 4.12 and 4.13 is located in a laminar airflow with velocity of 3 m/s and temperature of 40°C and simultaneously radiates to the enclosure, which is at 35°C. What is the average surface temperature of this package and contributions of forced convection and radiation heat transfer rates?

REFERENCES

1. Pecht, M., R. Agarwal, F. P. McCluskey, T. J. Dishongh, S. Javadpour, and R. Mahajan. 1999. *Electronic packaging materials and their properties*. Boca Raton, FL: CRC Press.
2. Cengel, Y. A. 2007. *Heat and mass transfer: A practical approach*. Third Edition. New York, NY: McGraw-Hill.
3. Incropera, F. P., D. P. DeWitt, T. L. Bergman, and A. S. Lavine. 2007. Fundamentals of heat and mass *transfer*. Sixth Edition. Hoboken, NJ: John Wiley & Sons, Inc.

5 Thermal Resistance Network

5.1 THERMAL RESISTANCE CONCEPT

Consider an electrical resistor shown in Figure 5.1. If there is a potential difference between the two ends of this resistor, a current will flow through it. The direction of this current is from the high potential to the low potential ends of this resistor, and its magnitude is given by the Ohm's law;

$$I = \frac{V_1 - V_2}{R}. \tag{5.1}$$

Now, consider a layer with thickness L and surface area A as shown in Figure 5.2. If there is a temperature difference across this layer (i.e., $T_1 - T_2 \neq 0$), there will be conduction heat transfer from its high temperature side to its low temperature side. Fourier's law that was introduced in Chapter 4, Section 4.1, gives the magnitude of this conduction heat transfer rate as

$$\dot{Q}_{cond} = kA \frac{T_1 - T_2}{L}, \tag{5.2}$$

or

$$\dot{Q}_{cond} = \frac{T_1 - T_2}{L/kA}. \tag{5.3}$$

It will be shown in Chapter 8 that Equations 5.2 and 5.3 are strictly valid for one-dimensional steady-state heat transfer in a medium with no heat generation and constant thermal conductivity. There are three similarities between Equation 5.3 for conduction heat transfer rate and Ohm's law, Equation 5.1: (1) temperature difference across a layer generates conduction heat transfer just as potential difference across an electrical resistor generates current flow, (2) conduction heat transfer rate is from high temperature to low temperature side of a layer similar to current flow that is from high potential to low potential end of a resistor, and (3) the term L/kA in the denominator of Equation 5.3 plays the same role as R in Equation 5.1. Using these similarities, Equation 5.3 is written similar to Ohm's law:

$$\dot{Q}_{cond} = \frac{T_1 - T_2}{R_{cond}}, \tag{5.4}$$

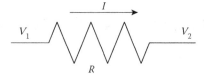

FIGURE 5.1 Potential difference across an electrical resistor creates a current.

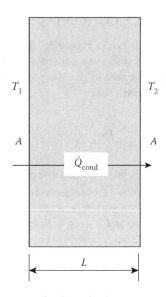

FIGURE 5.2 Conduction heat transfer through a layer.

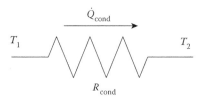

FIGURE 5.3 Equivalent thermal resistance for conduction heat transfer through a layer.

where $R_{cond} = L/kA$ is called **conduction thermal resistance** and is measured with the unit °C/W. This equation indicates that temperature difference across a layer drives the conduction heat transfer. On the other hand, conduction thermal resistance, which is directly proportional to the thickness of the layer and inversely proportional to its conductivity and surface area, opposes the conduction heat transfer. The equivalent thermal resistance for the physical layer shown in Figure 5.2 is shown in Figure 5.3.

It is worthwhile to mention that the electrical resistance of a layer with surface area A and thickness L is $R = L/\sigma A$ where σ is the electrical conductivity of the layer.

Thermal Resistance Network

This shows that heat conduction and electrical conduction have similar characteristics and therefore are governed by similar equations.

Example 5.1

Consider a 40 mm × 40 mm × 1 mm thick copper cap of a flip-chip package that is attached to a same size die dissipating 50 W. If thermal conductivity of copper is 390 W/m°C, what are the conduction thermal resistance of the copper cap and the temperature difference across its thickness?

Solution

With thickness $L = 1$ mm $= 0.001$ m, area $A = 40$ mm × 40 mm $= 1600$ mm² $= 0.0016$ m² and conductivity $k = 390$ W/m°C

$$R_{cond} = \frac{L}{kA} = \frac{0.001 \text{ m}}{390 \text{ W/m°C} \times 0.0016 \text{ m}^2} = 0.0016 \text{°C/W}$$

$$\dot{Q} = \frac{T_1 - T_2}{R_{cond}} \Rightarrow T_1 - T_2 = \dot{Q} \times R_{cond} = 50 \text{ W} \times 0.0016 \text{°C/W} = 0.08 \text{°C}.$$

Thermal resistances can also be defined for convection and radiation heat transfer modes. Newton's law of cooling for convection heat transfer rate, first introduced in Section 4.2:

$$\dot{Q}_{conv} = hA(T_s - T_\infty), \tag{5.5}$$

can be written as

$$\dot{Q}_{conv} = \frac{T_s - T_\infty}{1/hA}. \tag{5.6}$$

If we write this equation in the form of Ohm's law:

$$\dot{Q}_{conv} = \frac{T_s - T_\infty}{R_{conv}}, \tag{5.7}$$

$R_{conv} = 1/hA$ will be called **convection thermal resistance**. It is seen that convection thermal resistance is inversely proportional to the convection heat transfer coefficient and surface area exposed to the fluid. The equivalent resistance for convection heat transfer is shown in Figure 5.4.

It was shown in Chapter 4, Section 4.3, that radiation heat transfer rate between an object at temperature T_s and its large surrounding at temperature T_{surr} can be calculated as

$$\dot{Q}_{rad} = h_{rad} A(T_s - T_{surr}), \tag{5.8}$$

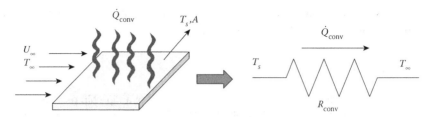

FIGURE 5.4 Equivalent resistance for convection heat transfer.

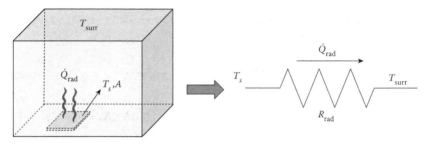

FIGURE 5.5 Equivalent resistance for radiation heat transfer.

where h_{rad} is the radiation heat transfer coefficient defined by Equation 4.15. This equation can also be written as

$$\dot{Q}_{rad} = \frac{T_s - T_{surr}}{1/h_{rad} A}. \tag{5.9}$$

If Equation 5.9 is written as

$$\dot{Q}_{rad} = \frac{T_s - T_{surr}}{R_{rad}}, \tag{5.10}$$

$R_{rad} = 1/h_{rad}A$ will be called **radiation thermal resistance**. The equivalent resistance for radiation heat transfer is shown in Figure 5.5.

Alternatively, following a similar argument given at the end of Section 4.3, the equation for radiation heat transfer rate may be written as

$$\dot{Q}_{rad} = \frac{T_s - T_\infty}{R'_{rad}}, \tag{5.11}$$

where $R'_{rad} = 1/h'_{rad}A$ is called **modified radiation thermal resistance** and h'_{rad} is the modified radiation heat transfer coefficient defined by Equation 4.19. The advantage of Equation 5.11 to Equation 5.10 is that it uses the same reference temperature as Equation 5.7 for convection heat transfer rate. This will be helpful in creating thermal resistance networks that is discussed later in this chapter.

Thermal Resistance Network

Example 5.2

Consider that the copper cap of the flip-chip package in Example 5.1 is exposed to laminar airflow with a velocity of 2 m/s and temperature of 25°C. What are the convection thermal resistance between the copper cap and air and the surface temperature of the copper cap?

Solution

Equation 4.5 can be used to calculate convection heat transfer coefficient in laminar airflow with $U_\infty = 2$ m/s and $L = 40$ mm $= 0.04$ m,

$$h = 3.9 \left(\frac{U_\infty}{L} \right)^{1/2} = 3.9 \left(\frac{2}{0.04} \right)^{1/2} = 27.58 \text{ W/m}^2\text{°C}.$$

The convection thermal resistance is

$$R_{conv} = \frac{1}{hA} = \frac{1}{27.58 \text{ W/m}^2\text{°C} \times (0.04 \text{ m})^2} = 22.66 \text{°C/W}.$$

Note that convection thermal resistance between the cap of the package and air is about four orders of magnitude larger than conduction thermal resistance through the cap itself. The convection thermal resistance and power dissipation of the die are used to obtain surface temperature of the cap.

$$\dot{Q} = \frac{T_s - T_\infty}{R_{conv}} \Rightarrow T_s = T_\infty + \dot{Q} \cdot R_{conv} = 25\text{°C} + 50 \text{ W} \times 22.66 \text{°C/W} = 1158\text{°C} \;!!!!$$

This is definitely not an acceptable temperature for the copper cap. Some design changes or a heat sink is needed to cool the package appropriately for this relatively high power dissipation.

Example 5.3

If the surface emissivity of the package cap in Example 5.2 is 0.8 and the surrounding temperature is 40°C, determine the radiation thermal resistance between it and its surrounding. Neglect convection heat transfer.

Solution

Since radiation heat transfer coefficient is a function of surface temperature, we need to find this temperature first. If we neglect convection heat transfer, all the heat dissipation from the heat spreader will be through radiation heat transfer.

$$\dot{Q}_{rad} = \varepsilon \sigma A (T_s^4 - T_{surr}^4)$$

$$50 \text{ W} = 0.8 \times 5.67 \times 10^{-8} \text{ W/m}^2\text{K}^4 \times (0.04 \text{ m})^2 \left[(T_s + 273.15)^4 - (40 + 273.15)^4 \right] \text{K}^4$$

$$T_s = 641.07\text{°C} = 914.22 \text{ K}.$$

The radiation heat transfer coefficient and radiation thermal resistance are

$$h_{rad} = \varepsilon\sigma(T_s^2 + T_{surr}^2)(T_s + T_{surr})$$

$$h_{rad} = 0.8 \times 5.67 \times 10^{-8} \text{ W/m}^2\text{K}^4 (914.22^2 + 313.15^2)\text{K}^2(914.22 + 313.15)\text{K}$$

$$= 51.99 \text{ W/m}^2\text{°C}$$

$$R_{rad} = \frac{1}{h_{rad}A} = \frac{1}{51.99 \text{ W/m}^2\text{°C} \times (0.04 \text{ m})^2} = 12.02\text{°C/W},$$

Note that radiation thermal resistance is less than convection thermal resistance in this case. This is mainly due to the high temperature of the metal cap.

Example 5.4

What are the modified radiation heat transfer coefficient and modified radiation thermal resistance for the package cap in Example 5.3?

Solution

The modified radiation heat transfer coefficient is

$$h'_{rad} = \frac{\dot{Q}_{rad}}{A(T_s - T_\infty)} = \frac{50 \text{ W}}{0.04^2 \text{ m}^2 \times (641.07 - 25)\text{°C}} = 50.72 \text{ W/m}^2\text{°C},$$

and the modified radiation thermal resistance is

$$R'_{rad} = \frac{1}{h'_{rad}A} = \frac{1}{50.72 \text{ W/m}^2\text{°C} \times (0.04 \text{ m})^2} = 12.32\text{°C/W}.$$

5.2 SERIES THERMAL LAYERS

Consider a composite layer that is made of three layers with different thicknesses and thermal conductivities as shown in Figure 5.6. Let's assume that this composite layer is exposed to a fluid with temperature $T_{\infty,1}$ and heat transfer coefficient h_1 on

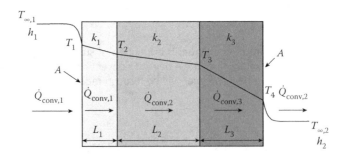

FIGURE 5.6 Heat transfer through a composite layer.

Thermal Resistance Network

the left side and another fluid with temperature $T_{\infty,2}$ and heat transfer coefficient h_2 on the right side. If $T_{\infty,1} > T_{\infty,2}$, heat will be transferred from the fluid on the left side into this composite layer and, through these layers, to the fluid on the right side. A typical qualitative temperature profile is shown in Figure 5.6 in which T_1 through T_4 represent uniform temperatures at the fluid-surface and solid-solid interfaces. If the layers are insulated on top and bottom, the heat transfer is one dimensional. If there is no heat generation inside these layers and heat transfer is in steady-state condition, the heat transfer rate entering each interface will be the same as the heat transfer rate leaving it:

$$\dot{Q}_{conv,1} = \dot{Q}_{cond,1} = \dot{Q}_{cond,2} = \dot{Q}_{cond,3} = \dot{Q}_{conv,2} = \dot{Q}. \tag{5.12}$$

Here, $\dot{Q}$ is the common notation used to represent the heat transfer rate through each layer. If we define thermal resistance of each layer, including the convection layers on the left and side sides, as

$$R_{conv,1} = \frac{1}{h_1 A}, \; R_{cond,1} = \frac{L_1}{k_1 A}, \; R_{cond,2} = \frac{L_2}{k_2 A}, \; R_{cond,3} = \frac{L_3}{k_3 A}, \; R_{conv,2} = \frac{1}{h_2 A},$$

Equation 5.12 can be written as

$$\dot{Q} = \frac{T_{\infty,1} - T_1}{R_{conv,1}} = \frac{T_1 - T_2}{R_{cond,1}} = \frac{T_2 - T_3}{R_{cond,2}} = \frac{T_3 - T_4}{R_{cond,3}} = \frac{T_4 - T_{\infty,2}}{R_{conv,2}}. \tag{5.13}$$

There is a mathematical identity that states if

$$\frac{a_1}{b_1} = \frac{a_2}{b_2} = \frac{a_3}{b_3} = \cdots = c, \tag{5.14}$$

then

$$\frac{a_1 + a_2 + a_3 + \cdots}{b_1 + b_2 + b_3 + \cdots} = c. \tag{5.15}$$

Applying this mathematical identity to Equation 5.13 gives

$$\dot{Q} = \frac{T_{\infty,1} - T_1 + T_1 - T_2 + T_2 - T_3 + T_3 - T_4 + T_4 - T_{\infty,2}}{R_{conv,1} + R_{cond,1} + R_{cond,2} + R_{cond,3} + R_{conv,2}}, \tag{5.16}$$

or

$$\dot{Q} = \frac{T_{\infty,1} - T_{\infty,2}}{R_{total}}, \tag{5.17}$$

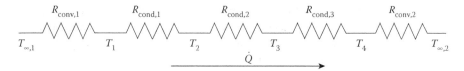

FIGURE 5.7 Thermal resistance network for the composite layer shown in Figure 5.6.

where $R_{total} = R_{conv,1} + R_{cond,1} + R_{cond,2} + R_{cond,3} + R_{conv,2}$ is total thermal resistance between fluids on the left and right sides of the composite layer. This equation implies that if multiple layers of different materials are attached to each other in a direction normal to the direction of heat transfer, those layers will act like a series set of thermal resistances. The equivalent thermal resistance network for the composite layer shown in Figure 5.6 is shown in Figure 5.7.

Similar equations for $\dot{Q}$ can be written using different pairs of points on the resistance network shown in Figure 5.7. Examples are

$$\dot{Q} = \frac{T_{\infty,1} - T_1}{R_{conv,1}} = \frac{T_2 - T_4}{R_{cond,2} + R_{cond,3}} = \frac{T_3 - T_{\infty,2}}{R_{cond,3} + R_{conv,2}} = \cdots.$$

In general

$$\dot{Q} = \frac{T_i - T_j}{R_{total,i-j}}, \qquad (5.18)$$

where T_i and T_j are temperatures at two points and $R_{total,i-j}$ is the total thermal resistance between those two points.

Example 5.5

Consider a 25 mm × 25 mm × 1 mm thick silicon die attached to a same size 2 mm-thick copper cap through a 0.1 mm thick thermal interface material (TIM) as shown in Figure 5.8. Convection heat transfer coefficient on the top side of the copper cap is 2500 W/m²°C. If thermal conductivity of silicon, copper, and thermal interface material are 125, 390, and 5 W/m°C, respectively, what is the total thermal resistance from the active (bottom) side of the silicon die to outside ambient?

Solution

Figure 5.8 shows the copper cap and the silicon die attached to each other with a thermal interface material and their equivalent thermal resistance network.

Thermal Resistance Network

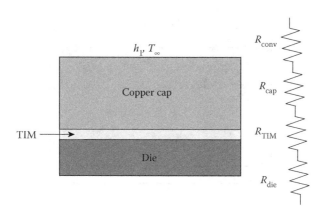

FIGURE 5.8 Physical construction and equivalent thermal resistance network for Example 5.5.

The individual thermal resistances are calculated as follows:

$$R_{die} = \left(\frac{L}{kA}\right)_{die} = \frac{0.001\,m}{125\,W/m°C \times 0.025^2\,m^2} = 0.013°C/W$$

$$R_{TIM} = \left(\frac{L}{kA}\right)_{TIM} = \frac{0.0001\,m}{5\,W/m°C \times 0.025^2\,m^2} = 0.032°\,C/W$$

$$R_{cap} = \left(\frac{L}{kA}\right)_{cap} = \frac{0.002\,m}{390\,W/m°C \times 0.025^2\,m^2} = 0.008°C/W$$

$$R_{conv} = \frac{1}{hA} = \frac{1}{2500\,W/m^2°C \times 0.025^2\,m^2} = 0.64°C/W.$$

Total thermal resistance from the active side of the die to outside ambient is the sum of all these thermal resistances.

$$R_{die\text{-}ambient} = R_{die} + R_{TIM} + R_{cap} + R_{conv}$$

$$R_{die\text{-}ambient} = 0.013°C/W + 0.032°C/W + 0.008°C/W + 0.64°C/W = 0.693°C/W.$$

5.3 PARALLEL THERMAL LAYERS

Consider the two-layer composite layer shown in Figure 5.9. Surface area, thickness, and thermal conductivity for the top layer are A_1, L, and k_1 and for the bottom layer are A_2, L, and k_2, respectively. Let's assume that the entire left and right sides of

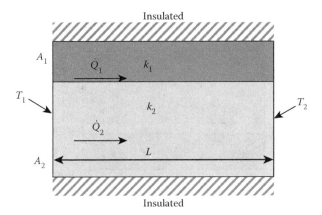

FIGURE 5.9 A composite layer with heat transfer along the line of contact between the layers.

this composite layer are at uniform temperatures T_1 and T_2, respectively, the top and bottom sides are insulated, there is no heat generation in these layers and heat conduction is in steady-state condition. If $T_1 > T_2$, there will be conduction heat transfer through each layer. If $\dot{Q}_1$ and $\dot{Q}_2$ are conduction heat transfer rates through the top and bottom layers,

$$\dot{Q}_1 = \frac{T_1 - T_2}{R_1} \quad \text{and} \quad \dot{Q}_2 = \frac{T_1 - T_2}{R_2}, \tag{5.19}$$

where $R_1 = L/k_1 A_1$ and $R_2 = L/k_2 A_2$ are conduction thermal resistances of the top and bottom layers, respectively. Total heat transfer rate through this composite layer is the sum of $\dot{Q}_1$ and $\dot{Q}_2$;

$$\dot{Q} = \dot{Q}_1 + \dot{Q}_2 = \frac{T_1 - T_2}{R_1} + \frac{T_1 - T_2}{R_2} = (T_1 - T_2)\left(\frac{1}{R_1} + \frac{1}{R_2}\right). \tag{5.20}$$

On the other hand, the equation for $\dot{Q}$ in the form of Ohm's law is

$$\dot{Q} = \frac{T_1 - T_2}{R_{\text{total}}}. \tag{5.21}$$

An expression for R_{total} is obtained by comparing Equations 5.20 and 5.21,

$$\frac{1}{R_{\text{total}}} = \frac{1}{R_1} + \frac{1}{R_2}. \tag{5.22}$$

This shows that a composite layer with heat transfer rate along the contact lines between the layers acts like a set of parallel thermal resistances. The equivalent thermal resistance network for the composite layer of Figure 5.9 is shown in Figure 5.10.

Thermal Resistance Network

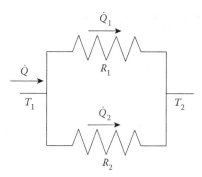

FIGURE 5.10 Equivalent thermal resistance network for the composite layer shown in Figure 5.9.

Example 5.6

Consider the copper cap of the package in Examples 5.2 to 5.4 and let's assume both convection and radiation heat transfers happen simultaneously from its surface. Calculate equivalent thermal resistance from its surface to air around it and the temperature of the heat spreader.

Solution

Since convection and radiation heat transfers happen simultaneously, the cap to ambient thermal resistance is the parallel equivalent of convection and modified radiation thermal resistances. Example 5.2 gave $R_{conv} = 22.66°C/W$ and Example 5.4 showed that $R'_{rad} = 12.32°C/W$ if radiation was the only heat transfer mode. However, R'_{rad} depends on surface temperature, which is unknown here. Therefore, the solution is obtained by iteration. Let's start with R'_{rad} calculated in Example 5.4 that was based on $T_s = 641.07°C$. The total thermal resistance between the cap and ambient is

$$\frac{1}{R_{cap\text{-}ambient}} = \frac{1}{R_{conv}} + \frac{1}{R'_{rad}} \Rightarrow \frac{1}{R_{cap\text{-}ambient}} = \frac{1}{22.66} + \frac{1}{12.32}$$

$$R_{cap\text{-}ambient} = 7.98°C/W.$$

The surface temperature of the cap is

$$\dot{Q} = \frac{T_s - T_\infty}{R_{cap\text{-}ambient}} \Rightarrow T_s = T_\infty + R_{cap\text{-}ambient} \times \dot{Q}$$

$$T_s = 25°C + 7.98°C/W \times 50\ W$$

$$T_s = 424°C.$$

This calculated value for T_s is very different from the value used to calculate R'_{rad}. Therefore, the calculation has to be repeated using a new value for T_s. Let's choose the average value of the originally assumed and newly calculated surface temperatures as a new guess for T_s.

$$T_s = 0.5(641.07 + 424) = 532.54\,°C$$

$$h'_{rad} = \frac{\dot{Q}_{rad}}{A(T_s - T_\infty)} = \frac{\varepsilon\sigma A(T_s^4 - T_{surr}^4)}{A(T_s - T_\infty)} = \frac{\varepsilon\sigma(T_s^4 - T_{surr}^4)}{(T_s - T_\infty)}$$

$$h'_{rad} = \frac{0.8 \times 5.67 \times 10^{-8} W/m^2 K^4 ((273.15+532.54)^4 - (273.15+40)^4) K^4}{(532.54-25)K}$$

$$= 36.8 W/m^2\,°C$$

$$R'_{rad} = \frac{1}{h'_{rad} A} = \frac{1}{36.8 W/m^2\,°C \times (0.04m)^2} = 16.98\,°C/W$$

$$\frac{1}{R_{cap\text{-}ambient}} = \frac{1}{R_{conv}} + \frac{1}{R'_{rad}} \Rightarrow \frac{1}{R_{cap\text{-}ambient}} = \frac{1}{22.66} + \frac{1}{16.98}$$

$$R_{cap\text{-}ambient} = 9.71\,°C/W$$

$$T_s = 25\,°C + 9.71\,°C/W \times 50\,W = 510.39\,°C.$$

Again, the calculated value of T_s is different from the assumed value used to calculate R'_{rad}. Two more iterations, as summarized in the table below, result in a converged solution of $T_s = 520.65\,°C$.

Step	Assumed T_s, °C	h'_{rad}, W/m²°C	R'_{rad}, °C/W	$R_{cap\text{-}ambient}$, °C/W	Calculated T_s, °C
1	641.07	50.72	12.32	7.98	424
2	532.54	36.8	16.98	9.71	510.39
3	521.46	35.55	17.58	9.90	520.02
4	520.74	35.47	17.62	9.91	520.65

The convection and radiation heat transfer rates are

$$\dot{Q}_{conv} = \frac{T_s - T_\infty}{R_{conv}} = \frac{(520.65 - 25)\,°C}{22.66\,°C/W} = 21.87\,W,$$

$$\dot{Q}_{rad} = \frac{T_s - T_\infty}{R'_{rad}} = \frac{(520.65 - 25)\,°C}{17.62\,°C/W} = 28.13\,W.$$

Note that, even with simultaneous convection and radiation heat transfers, the cap temperature is too high. In order to reduce this temperature to acceptable levels, convection and/or radiation heat transfer rates need to be enhanced by either increasing the heat transfer coefficient or increasing the surface area. The latter option will be discussed in Chapter 7.

Thermal Resistance Network

5.4 GENERAL RESISTANCE NETWORK

Equivalent thermal resistance networks for more complex composite layers can be obtained by dividing those into sets of series and parallel layers. Consider the composite layer shown in Figure 5.11, which is subject to uniform temperatures T_1 and T_2 on the left and right sides and is insulated on the top and bottom. The surface area, thickness and thermal conductivity for layer 1 are A_1, L_1, and k_1, for layer 2 are A_2, L_2, and k_2, and for layer 3 are $A_1 + A_2$, L_3, and k_3, respectively. The construction of this composite layer suggests that layers 1 and 2 can be treated as parallel layers that are in series with layer 3. An equivalent resistance network for this composite layer is shown in Figure 5.12 where $R_1 = L_1 / k_1 A_1$, $R_2 = L_2 / k_2 A_2$, and $R_3 = L_3 / k_3 (A_1 + A_2)$.

If R_{12} is the equivalent thermal resistance of layers 1 and 2,

$$\frac{1}{R_{12}} = \frac{1}{R_1} + \frac{1}{R_2} \Rightarrow R_{12} = \frac{R_1 R_2}{R_1 + R_2}, \qquad (5.23)$$

and the total thermal resistance of this composite layer is

$$R_{\text{total}} = R_{12} + R_3. \qquad (5.24)$$

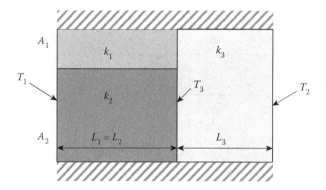

FIGURE 5.11 A general composite layer.

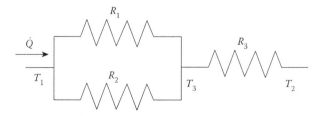

FIGURE 5.12 An equivalent thermal resistance for the composite layer shown in Figure 5.11.

Total heat transfer rate through this composite layer can be calculated using any of the following three equations;

$$\dot{Q} = \frac{T_1 - T_2}{R_{total}} = \frac{T_1 - T_3}{R_{12}} = \frac{T_3 - T_2}{R_3}. \tag{5.25}$$

The heat transfer rates through layers 1 and 2 are

$$\dot{Q}_1 = \frac{T_1 - T_3}{R_1} \quad \text{and} \quad \dot{Q}_2 = \frac{T_1 - T_3}{R_2}. \tag{5.26}$$

There are other ways to come up with an equivalent thermal resistance network for the composite layer of Figure 5.11. For example, this composite layer can be treated as a four-layer composite as shown in Figure 5.13. In this case, layer 3 is divided into two layers along the contact line of layers 1 and 2. Let's name the layers on the top and bottom of the dashed line in Figure 5.13 layer 3′ and 3″, respectively. Layers 1 and 3′ are in series and so are layers 2 and 3″. The equivalent thermal resistance of layers 1 and 3′ is in parallel with the equivalent thermal resistance of layers 2 and 3″. The thermal resistance network for this case is shown in Figure 5.14. If $R_{13'}$ and $R_{23''}$ are equivalent thermal resistances of layers 1 and 3′ and 2 and 3″, respectively,

$$R_{13'} = R_1 + R_{3'} = \frac{L_1}{k_1 A_1} + \frac{L_3}{k_3 A_1} \tag{5.27}$$

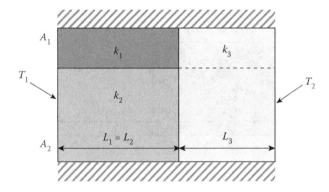

FIGURE 5.13 Composite layer of Figure 5.11 with layer 3 divided into two layers.

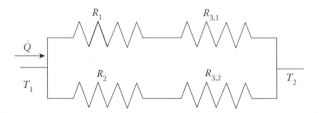

FIGURE 5.14 Equivalent resistance network of the composite layer shown in Figure 5.13.

Thermal Resistance Network

and

$$R_{23''} = R_2 + R_{3''} = \frac{L_1}{k_2 A_2} + \frac{L_3}{k_3 A_2}. \tag{5.28}$$

The total thermal resistance of this composite layer is found as follows:

$$\frac{1}{R_{\text{total}}} = \frac{1}{R_{13'}} + \frac{1}{R_{23''}} \Rightarrow R_{\text{total}} = \frac{R_{13'} R_{23''}}{R_{13'} + R_{23''}} = \frac{(R_1 + R_{3'})(R_2 + R_{3''})}{R_1 + R_{3'} + R_2 + R_{3''}}. \tag{5.29}$$

Total heat transfer rate through this layer is

$$\dot{Q} = \frac{T_1 - T_2}{R_{\text{total}}}. \tag{5.30}$$

It is expected that the heat transfer rate or the temperature difference calculated by assuming the resistance network of Figures 5.12 and 5.14 will be different. Each of these resistance networks are based on a different assumption. The resistance network of Figure 5.12 is obtained by assuming every plane normal to the direction of heat transfer is isothermal. This method underestimates the total thermal resistance and therefore overpredicts the total heat transfer rate. On the other hand, the resistance network of Figure 5.14 is based on the assumption that every plane parallel to the direction of heat transfer is insulated. This assumption overestimates the total thermal resistance and therefore underpredicts the total heat transfer rate. The true results lie somewhere between these two limits. The difference between the predictions of these two methods is small if the difference between thermal resistances of parallel layers is small and the heat transfer is predominantly one dimensional.

Example 5.7

Let's assume that the silicon die of Example 5.5 is mounted on a same size 1.2 mm-thick substrate through a 0.1 mm-thick die-attach material. If the thermal conductivity of substrate and die-attach are 0.2 and 2 W/m°C, respectively, and the convection heat transfer coefficient on the bottom side of the substrate is 20 W/m²°C, what is the thermal resistance from the active side of the die to ambient? The ambient temperature on the top side of the cap and bottom side of the substrate is the same.

Solution

There are two paths for heat transfer from the active side of the die to ambient; one path is through the die, TIM, and cap to the ambient over the top of the cap, and the other path is through the die attach and substrate to the ambient on the bottom side of the substrate. As shown in Figure 5.15, these two paths are in parallel and the total thermal resistance from the active side of the die to the ambient is the parallel equivalent of the thermal resistances of these two paths.

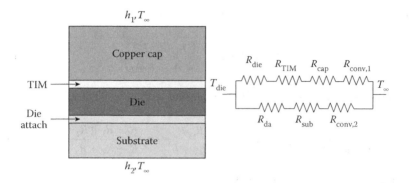

FIGURE 5.15 Physical layout and thermal resistance network for Example 5.7.

From Example 5-5, $R_{die} = 0.013°C/W$, $R_{TIM} = 0.032°C/W$, $R_{cap} = 0.008°C/W$ and $R_{conv,1} = 0.64°C/W$. The conduction thermal resistances of the die attach, the substrate, and the convection thermal resistance on the bottom side of the substrate are

$$R_{da} = \left(\frac{L}{kA}\right)_{da} = \frac{0.0001\,m}{2\,W/m°C \times 0.025^2\,m^2} = 0.08°C/W,$$

$$R_{sub} = \left(\frac{L}{kA}\right)_{sub} = \frac{0.0012\,m}{0.2\,W/m°C \times 0.025^2\,m^2} = 9.6°C/W,$$

$$R_{conv,2} = \frac{1}{h_2 A} = \frac{1}{20\,W/m^2\,C \times 0.025^2\,m^2} = 80°C/W.$$

Total thermal resistance form the active side of the die to ambient is

$$\frac{1}{R_{die\text{-}ambient}} = \frac{1}{R_{die} + R_{TIM} + R_{cap} + R_{conv,1}} + \frac{1}{R_{da} + R_{sub} + R_{conv,2}}$$

$$\frac{1}{R_{die\text{-}ambient}} = \frac{1}{0.013 + 0.032 + 0.008 + 0.64} + \frac{1}{0.08 + 9.6 + 80}$$

$$R_{die\text{-}ambient} = 0.688°C/W.$$

Note that die to ambient thermal resistance is only slightly less than the sum of the die, TIM, copper cap, and top convection thermal resistances. This shows that, in this case, the heat transfer path through the die attach and substrate is not an effective heat transfer path.

5.5 THERMAL CONTACT RESISTANCE

A perfect contact or attachment between two layers requires every point on the surface of a first layer to be in contact with a corresponding point on the surface of a second layer. This is possible only if both surfaces are perfectly smooth or if there is some form of chemical bond that creates molecular bonds between the particles

Thermal Resistance Network

of those two layers. However, no matter how smooth two contacting surfaces may appear to the naked eye, they are microscopically rough, and this prevents having a perfect contact between those two surfaces. Figure 5.16 shows how rough the contact area between surfaces may be in microscopic scale. The actual contacts between the two surfaces are the point contacts with air gaps in between those. Since air is a very poor heat conductor, the air gaps between the two contacting surfaces create a thermal resistance against the heat transfer. This resistance per unit area of contact is called **thermal contact resistance**, R_c, and is measured by the unit $(°C/W)/m^2$ or $°C·m^2/W$.

If the two contacting surfaces become smoother, the size of the air gaps and, consequently, the thermal contact resistance will be less. Also, if the two layers are pressed harder against each other, the actual contact area between the surfaces will increase, the air gaps will become smaller, and the thermal contact resistance will decrease.

Figure 5.17 shows temperature variation in a two-layer composite with either a perfect or a real contact interface. The picture on the left shows the two layers in perfect contact with each other. In this case the temperature variation is continuous

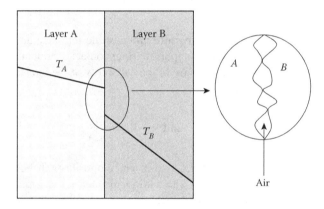

FIGURE 5.16 The contact area between two surfaces is rough in a microscopic scale.

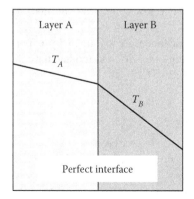 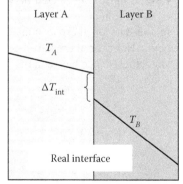

FIGURE 5.17 Temperature profile in two layers with perfect and real interface.

at the contact interface. The temperature of layer A at this interface is the same as the temperature of layer B there. The same two layers with a real contact are shown on the right side. Here, there is a discontinuity in temperature profile across the interface; the temperature of layer A at this interface is not the same as the temperature of layer B there. If R_c and A are thermal contact resistance and surface area of the interface, **thermal interface resistance** is defined as

$$R_{int} = R_c / A, \tag{5.31}$$

and temperature change across the interface is

$$\Delta T_{int} = \dot{Q} \cdot R_{int} = \dot{Q} \cdot R_c / A, \tag{5.32}$$

where $\dot{Q}$ is the heat transfer rate through the interface.

Heat transfer rate through the interface can be expressed by an equation similar to Newton's law of cooling,

$$\dot{Q} = h_c A \Delta T_{int}, \tag{5.33}$$

where h_c is called **thermal contact conductance** of the interface and is measured with the unit W/m²°C. Some sources report thermal contact conductance, instead of thermal contact resistance, as a measure of the capability of the interface to transfer heat. However, it is clear that

$$h_c = \frac{1}{R_c} \quad \text{and} \quad R_{int} = \frac{1}{h_c A}. \tag{5.34}$$

Thermal resistance networks for a two-layer composite with perfect and real interfaces are shown in Figure 5.18. The additional interface thermal resistance is seen in the case of a real interface. The interface thermal resistance must be included in resistance networks created for engineering applications such as the resistance networks between a heat source and a heat sink in an electronic cooling application. However, it may be neglected between two low conductive thermal insulating materials.

Most of the data on thermal contact resistance are obtained by experimental measurements. However, there have been many theories and models for calculating thermal contact resistance between two surfaces [1,2].

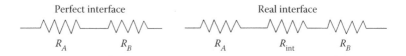

FIGURE 5.18 Resistance network for a two-layer composite with perfect and real interface.

5.6 THERMAL INTERFACE MATERIALS

Thermal interface resistance will significantly reduce the heat transfer rate or increase the temperature difference between the two layers and is an undesirable phenomenon in electronic cooling applications. One way to reduce this undesirable thermal resistance is to insert soft, compliant, and high thermal conductivity materials between the two contacting surfaces. Such a material, called **thermal interface material (TIM)**, replaces most of the air gaps between the two surfaces. Since the thermal conductivity of interface materials is one or two orders of magnitude higher than the thermal conductivity of air, this will reduce the interface thermal resistance.

Thermal interface materials are made of soft base materials that are filled with thermally conductive solid particles. Typical base materials are silicon, hydrocarbon oils, polymeric rubbers, epoxy, and wax. Common fillers are aluminum oxide, magnesium oxide, aluminum nitride, boron nitride, and diamond powder. Examples of common interface materials are thermal greases and pastes, elastomeric pads, thermally conductive tapes, phase-change materials, gels, thermally conductive adhesives, solder, and low-melt alloys.

Thermal interface materials reduce the air gap between the two surfaces. However, they do not completely eliminate those air gaps. In fact, as shown in Figure 5.19, there will be interface resistances on two sides of a TIM due to air gaps between that TIM and the surfaces it is attached to. Therefore, effective thermal resistance of a thermal interface material inserted between layers A and B is the sum of the bulk resistance of the TIM, R_{bulk}, which is a conduction thermal resistance, and the interface resistances between the TIM and layer A, $R_{int,1}$, and the interface resistance between the TIM and layer B, $R_{int,2}$:

$$R_{TIM} = R_{bulk} + R_{int,1} + R_{int,2}, \quad (5.35)$$

$$R_{TIM} = \frac{t}{kA} + R_{int,1} + R_{int,2}. \quad (5.36)$$

t, k, and A are compressed thickness (sometimes called bond-line thickness), thermal conductivity and area of the thermal interface material. Note that pressure reduces both

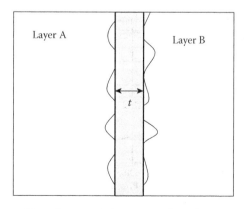

FIGURE 5.19 A thermal interface material compressed to thickness t between two surfaces.

bond-line thickness and two interface thermal resistances. Therefore, as shown in Figure 5.20, thermal resistance of a thermal interface material decreases as pressure increases.

Important properties of interface materials are

- Thermal conductivity: Higher thermal conductivity interface materials are desired to reduce the interface thermal resistance. Note that bulk thermal resistance of a TIM is inversely proportional to thermal conductivity. Thermal conductivity of most common interface materials is about 1–10 W/m°C with elastomer pads and thermal tapes having the lowest and gels and greases having the highest thermal conductivities.
- Compliance or conformability: Softer and more compliant interface materials conform better with the roughness of a surface, leave less air gap, and result in lower $R_{int,1}$ and $R_{int,2}$.
- Thickness: Conduction or bulk thermal resistance of a TIM is proportional to its thickness. Therefore, thinner interface materials are desired to reduce their thermal resistances. Figure 5.21 shows measured thermal resistances of two interface materials at different thicknesses. It is seen that thermal resistance variation with thickness is fairly linear.
- Modulus of elasticity: This is a measure of mechanical strength of the interface material and the amount of pressure required to compress it. If the modulus of elasticity of a TIM is low, it will require less pressure to fill the gaps between the contacting surfaces. Lower applied pressure reduces the risk of damaging the electronic package. Note that, as shown in Figure 5.22, the pressure required to compress a thermal interface material increases exponentially as the compression rate increases. It is also seen that the pressure required to compress an initially thicker material is less than that for a thinner material.

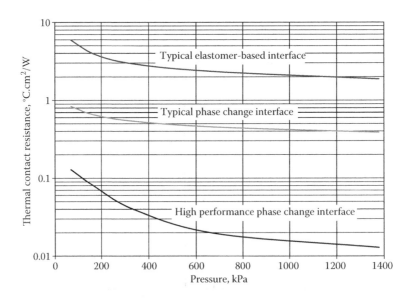

FIGURE 5.20 Variation of thermal resistance of a TIM with pressure.

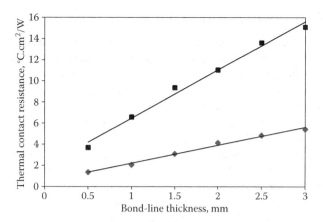

FIGURE 5.21 Variation of thermal resistance of two TIMs with bond-line thickness.

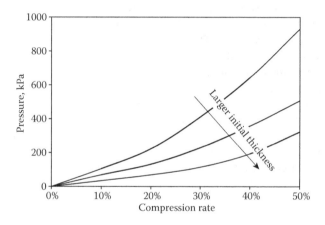

FIGURE 5.22 Pressure versus compression ratio of a TIM.

- Dielectric strength: Thermal interface materials may or may not be electrically insulating. Dielectric strength is a measure of electrical insulation of an interface material. If one or both of the contacting surfaces are electrically charged, they have to be electrically insulated from each other and a high dielectric strength TIM will be required in these situations.
- Long-term stability and reliability: Thermal interface materials will go through temperature cycles and have to survive other harsh operating conditions. They should keep their properties such as thermal conductivity, modulus of elasticity, and dielectric strength within acceptable limits over their life.
- Ease of use and reworkability: Thermal interface materials should be easy to use and apply. It is sometimes desirable that a thermal interface material can be easily removed from a surface and put back on there during a repair process that requires access to the surface.

5.7 SPREADING THERMAL RESISTANCE

The concept of thermal resistance and thermal resistance network can be used in situations where heat transfer is one dimensional. The one-dimensional assumption is a good approximation as long as the dimension in the direction of heat transfer is small compared to dimensions in the other two directions, and the heat source (hot side) and heat sink (cold side) have the same areas. On the other hand, in most situations, heat is transferred from a small heat source to a much larger heat sink. Examples are heat transfer from the silicon die of a flip-chip package to the copper heat spreader and the heat transfer from a device to the base of a larger heat sink. Heat flow path in these situations is strongly three dimensional, and a one-dimensional analysis will introduce significant errors. However, even in these situations, it will be useful if an equivalent one-dimensional thermal resistance is defined. This thermal resistance will be larger than the one-dimensional conduction thermal resistance based on the thickness, thermal conductivity, and surface area of the heat sink. The additional thermal resistance is called **spreading** or **constriction thermal resistance**. Spreading is used to describe the situation where the heat source is smaller than the heat sink and constriction is used for the case where heat is transferred from a large heat source to a small heat sink. They are assumed to be equal in value for the same geometry and boundary conditions.

Consider the case shown in Figure 5.23 where a circular heat source with radius r_1 is attached to a circular plate, called heat sink, with radius r_2, thickness t and thermal conductivity k. Heat leaves the heat sink from the top surface that is exposed to a fluid with temperature T_∞ and uniform convection heat transfer coefficient h. All the other exposed surfaces are insulated. It is obvious that the maximum heat source temperature will happen at its center. If T_{max} and T_∞ are maximum source and ambient temperatures respectively and $\dot{Q}$ is the source heat dissipation, thermal resistance between the source and ambient can be defined as

$$R_{tot} = \frac{T_{max} - T_\infty}{\dot{Q}}. \qquad (5.37)$$

Lee et al. [3] obtained expressions for T_{max} and R_{tot} by solving a three-dimensional partial differential equation for temperature. The result is in the form of an infinite

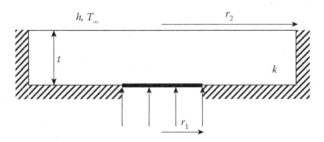

FIGURE 5.23 A circular heat source with radius r_1 and a circular heat sink with thickness t, thermal conductivity k, and radius r_2.

Thermal Resistance Network

series and will not be presented. Instead, a simpler approximate method based on the concept of spreading thermal resistance is used here. The total thermal resistance is considered to be the sum of the one-dimensional conduction thermal resistance of the heat sink, the spreading resistance in the heat sink, and the convection thermal resistance,

$$R_{tot} = R_{cond} + R_{sp} + R_{conv}, \qquad (5.38)$$

$$R_{tot} = \frac{t}{\pi k r_2^2} + R_{sp} + \frac{1}{\pi h r_2^2}. \qquad (5.39)$$

Note that πr_2^2 is the surface area of the heat sink. Lee et al. [3] showed that the following approximate expression for the spreading thermal resistance provides a good approximation to their exact solution.

$$R_{sp} = \frac{(1-\varepsilon)\phi}{\pi k r_1}. \qquad (5.40)$$

Here

$$\phi = \frac{\tanh(\lambda\tau) + \dfrac{\lambda}{Bi}}{1 + \dfrac{\lambda}{Bi}\tanh(\lambda\tau)}, \quad \lambda = \pi + \frac{1}{\varepsilon\sqrt{\pi}}, \quad Bi = \frac{hr_2}{k}, \quad \tau = \frac{t}{r_2}, \quad \varepsilon = \frac{r_1}{r_2}. \qquad (5.41)$$

Although Equation 5.40 was derived for a circular heat source and sink, it can be used to calculate spreading resistance between a square-shape heat source with surface area A_c and a square-shape heat sink with surface area A_s if r_1 and r_2 are defined as

$$r_1 = \sqrt{\frac{A_c}{\pi}} \quad \text{and} \quad r_2 = \sqrt{\frac{A_s}{\pi}}. \qquad (5.42)$$

Figure 5.24 compares spreading resistances calculated with Equations 5.40 through 5.42 with exact values obtained by computational solutions of appropriate partial differential equations for temperature. Good agreements between the analytical and computer simulation results are seen for both circular and square shaped heat source and sinks.

It must be emphasized that the above expression for spreading thermal resistance is valid only for the geometry and boundary conditions shown in Figure 5.23. Similar expressions have been derived for other boundary conditions, but the case considered here is the most common case in electronic thermal management.

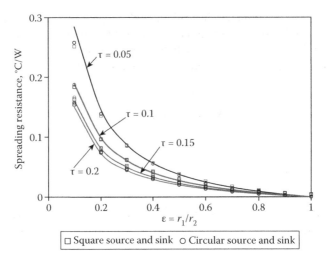

FIGURE 5.24 A comparison between predictions of Equation 5.40 (solid lines) and exact computer simulation results (square and circles) when $Bi = 1$.

Example 5.8

Consider a 10 mm × 10 mm × 1 mm thick silicon die attached to a 25 mm × 25 mm × 2 mm-thick copper heat spreader through a 0.1 mm thick thermal interface material (TIM). The convection heat transfer coefficient on the other side of the heat spreader is 2500 W/m²°C. If the thermal conductivities of silicon, copper, and the thermal interface material are 125, 390, and 5 W/m°C, respectively, what is the total thermal resistance from the active side of the silicon die to outside ambient?

Solution

This is similar to Example 5.5 except that the die is smaller than the heat spreader. Therefore, there will be additional spreading thermal resistance through the copper heat spreader. The conduction thermal resistance through the die, interface material, and the heat spreader, as well as the convection thermal resistance from the top side of the heat spreader to ambient are

$$R_{die} = \left(\frac{L}{kA}\right)_{die} = \frac{0.001\,m}{125\,W/m°C \times 0.01^2\,m^2} = 0.08°C/W,$$

$$R_{TIM} = \left(\frac{L}{kA}\right)_{TIM} = \frac{0.0001\,m}{5\,W/m°C \times 0.01^2\,m^2} = 0.2°C/W,$$

$$R_{hs} = \left(\frac{L}{kA}\right)_{hs} = \frac{0.002\,m}{390\,W/m°C \times 0.025^2\,m^2} = 0.008°C/W,$$

$$R_{conv} = \frac{1}{hA} = \frac{1}{2500\,W/m^2°C \times 0.025^2\,m^2} = 0.64°C/W.$$

If we neglect heat transfer from all sides of the heat spreader except the top side, the spreading thermal resistance can be obtained using the correlations given above. Note that the heat source area is $A_c = 0.01^2$ m² and the heat sink area is $A_s = 0.025^2$ m².

$$r_1 = \sqrt{\frac{A_c}{\pi}} = \sqrt{\frac{0.01^2 \text{ m}^2}{\pi}} = 0.0056 \text{ m}$$

$$r_2 = \sqrt{\frac{A_s}{\pi}} = \sqrt{\frac{0.025^2 \text{ m}^2}{\pi}} = 0.0141 \text{ m}$$

$$\varepsilon = \frac{r_1}{r_2} = \frac{0.0056 \text{ m}}{0.0141 \text{ m}} = 0.397$$

$$\tau = \frac{t}{r_2} = \frac{0.002 \text{ m}}{0.0141 \text{ m}} = 0.142$$

$$Bi = \frac{hr_2}{k} = \frac{2500 \text{ W/m}^2{}^\circ\text{C} \times 0.0141 \text{ m}}{390 \text{ W/m}^\circ\text{C}} = 0.09$$

$$\lambda = \pi + \frac{1}{\varepsilon\sqrt{\pi}} = \pi + \frac{1}{0.397\sqrt{\pi}} = 4.563$$

$$\phi = \frac{\tanh(\lambda\tau) + \frac{\lambda}{Bi}}{1 + \frac{\lambda}{Bi}\tanh(\lambda\tau)} = \frac{\tanh(4.563 \times 0.142) + \frac{4.563}{0.09}}{1 + \frac{4.563}{0.09}\tanh(4.563 \times 0.142)} = 1.714$$

$$R_{sp} = \frac{(1-\varepsilon)\phi}{\pi k r_1} = \frac{(1-0.397) \times 1.714}{\pi \times 390 \text{ W/m}^\circ\text{C} \times 0.0056 \text{ m}} = 0.151^\circ\text{C/W}.$$

Total thermal resistance from the active side of the silicon die to outside ambient is

$$R_{\text{die-ambient}} = R_{\text{die}} + R_{\text{TIM}} + R_{\text{hs}} + R_{\text{sp}} + R_{\text{conv}},$$

$$R_{\text{die-ambient}} = 0.08^\circ\text{C/W} + 0.2^\circ\text{C/W} + 0.008^\circ\text{C/W} + 0.151^\circ\text{C/W} + 0.64^\circ\text{C/W}$$

$$= 1.079^\circ\text{C/W}.$$

5.8 THERMAL RESISTANCE OF PRINTED CIRCUIT BOARDS (PCBs)

Microelectronic devices are mounted on printed circuit boards (PCBs) and connected to each other through trace layers in the printed circuit board (PCB). This is called second-level interconnect; the connection between an integrated circuit inside a device and the leads of the device is called first-level interconnect. PCBs are complex multilayer structures in which layers of high thermal conductivity copper (trace, ground, and power layers) are sandwiched between layers of dielectric and low thermal conductivity glass epoxy (see Figure 5.25).

The thermal conductivity of copper (about 400 W/m°C) is more than 1000 times larger than the thermal conductivity of glass epoxy (about 0.35 W/m°C). Due to of

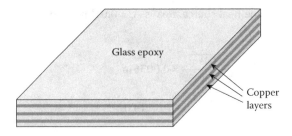

FIGURE 5.25 Schematic of a printed circuit board with three internal copper layers.

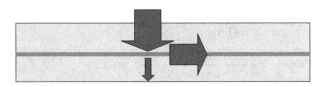

FIGURE 5.26 A qualitative picture of heat transfer path in a PCB.

this large difference between the thermal conductivity of copper and glass epoxy, the thermal conductivity of PCB, as a whole, is strongly anisotropic. This is qualitatively shown in Figure 5.26 where most of the heat that arrives to an area of a copper layer is transferred in the lateral (or planar) direction into the copper layer itself rather than directly transferring through the copper and the glass epoxy layers. This shows that PCBs have orthotropic heat transfer properties, and their thermal conductivities in planar and normal directions are different.

PCBs are commonly modeled as a single orthotropic object with effective planar and normal thermal conductivities, k_p and k_n, which are calculated as shown below. Consider a PCB with thicknesses t and surface area A as shown in Figure 5.27. Let's assume copper layers have thicknesses $t_{c,1}$, $t_{c,2}$, and $t_{c,3}$, epoxy layers have thicknesses $t_{e,1}$, $t_{e,2}$, $t_{e,3}$ and $t_{e,4}$, and k_c and k_e are thermal conductivities of copper and glass epoxy, respectively. If we assume one-dimensional heat transfer in the direction normal to surface A, this PCB is a series layers of thermal resistances similar to that discussed in Section 5.2. Total thermal resistance of this PCB in the normal direction, R_n, is

$$R_n = R_{e,1} + R_{c,1} + R_{e,2} + R_{c,2} + R_{e,3} + R_{c,3} + R_{e,4}, \quad (5.43)$$

$$R_n = R_{e,1} + R_{e,2} + R_{e,3} + R_{e,4} + R_{c,1} + R_{c,2} + R_{c,3}, \quad (5.44)$$

$$R_n = \frac{t_{e,1}}{k_e A} + \frac{t_{e,2}}{k_e A} + \frac{t_{e,2}}{k_e A} + \frac{t_{e,2}}{k_e A} + \frac{t_{c,1}}{k_c A} + \frac{t_{c,2}}{k_c A} + \frac{t_{c,3}}{k_c A}. \quad (5.45)$$

If $t_e = t_{e,1} + t_{e,2} + t_{e,3} + t_{e,4}$ and $t_c = t_{c,1} + t_{c,2} + t_{c,3}$ are total glass epoxy and copper layer thicknesses, respectively,

Thermal Resistance Network

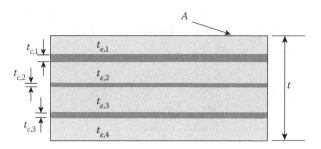

FIGURE 5.27 A typical PCB with three inner copper layers.

$$R_n = \frac{t_e}{k_e A} + \frac{t_c}{k_c A}. \tag{5.46}$$

Thermal resistance of an object with the same total thickness t, surface area A, and thermal conductivity k_n is $t/k_n A$. Equating this with the normal thermal resistance of PCB given by Equation 5.46 gives an expression for the **effective normal thermal conductivity** of a PCB:

$$k_n = \frac{t}{t_e/k_e + t_c/k_c}. \tag{5.47}$$

Similarly, effective planar thermal resistance of a PCB is obtained by considering one-dimensional heat transfer in a direction parallel to the interface between copper and glass epoxy layers. Consider the PCB shown in Figure 5.28 with thickness t, length L, and width W. Let's assume heat transfer is one dimensional and parallel to the length L. Total thermal resistance of this PCB in planar direction, R_p, is obtained by considering copper and glass epoxy layers as a set of parallel layers:

$$\frac{1}{R_p} = \frac{1}{R_{e,1}} + \frac{1}{R_{c,1}} + \frac{1}{R_{e,2}} + \frac{1}{R_{c,2}} + \frac{1}{R_{e,3}} + \frac{1}{R_{c,3}} + \frac{1}{R_{e,4}}, \tag{5.48}$$

$$\frac{1}{R_p} = \frac{1}{R_{e,1}} + \frac{1}{R_{e,2}} + \frac{1}{R_{e,3}} + \frac{1}{R_{e,4}} + \frac{1}{R_{c,1}} + \frac{1}{R_{c,2}} + \frac{1}{R_{c,3}}. \tag{5.49}$$

Thermal resistance of the layer i of glass epoxy and copper are $R_{e,i} = L/(k_e t_{e,i} W)$ and $R_{c,i} = L/(k_c t_{c,i} W)$, respectively, Therefore,

$$\frac{1}{R_p} = \frac{k_e t_{e,1} W}{L} + \frac{k_e t_{e,2} W}{L} + \frac{k_e t_{e,3} W}{L} + \frac{k_e t_{e,4} W}{L} + \frac{k_c t_{c,1} W}{L} + \frac{k_c t_{c,2} W}{L} + \frac{k_c t_{c,3} W}{L}, \tag{5.50}$$

$$\frac{1}{R_p} = \frac{k_e t_e W + k_c t_c W}{L}, \tag{5.51}$$

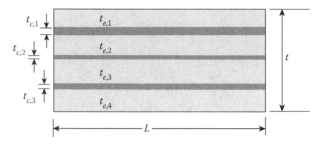

FIGURE 5.28 Cross section of a PCB with thickness t, length L and width W (normal to the page).

$$R_p = \frac{L}{(k_e t_e + k_c t_c)W}. \qquad (5.52)$$

Thermal resistance of an object with the same total thickness L, surface area tW and thermal conductivity k_p is $L/k_p tW$. Equating this with Equation 5.52 gives an expression for the **effective planar thermal conductivity** of a PCB:

$$k_p = \frac{k_e t_e + k_c t_c}{t}. \qquad (5.53)$$

Although these equations were derived for a PCB with three inner copper layers, they are valid for any PCB with any number of inner or outer copper layers. However, it is important to remember the assumption behind these expressions, which is one-dimensional heat transfer in normal or planar directions. They will provide reasonable accuracy if either of these assumptions are valid. However, there are many situations in which this one-dimensional heat transfer assumption is not valid. Examples are when small heat sources, such as surface mount gallium arsenide field effect transistors (GaAs FET) and radio frequency (RF) transistors and miniature voltage regulators, are mounted on a much larger PCB. Spreading thermal resistance becomes very important in these situations and simple one-dimensional models do not work. It has been shown that the true effective parallel and normal thermal conductivities of a PCB are functions of heat transfer coefficient on top and bottom sides of the PCB, source to PCB size ratio, as well as the arrangement and location of the copper layers [4,5].

There have also been studies suggesting the modeling of a PCB as a homogenous medium. Different suggestions for the value of an effective thermal conductivity for a homogeneous model of a PCB included k_n or k_p as defined above, the geometric mean of these two, $\sqrt{k_n \times k_p}$, or the harmonic mean of those, $2k_n \times k_p / (k_n + k_p)$. Each of these models provide reasonable approximations in specific situations. However, it has been shown that the placement of various layers, source size and placement, and convective boundary conditions have significant effects on the value of the effective thermal conductivity for a homogeneous model of a PCB [6]. A new homogenous model of PCBs treats every copper layer as a fin (see Chapter 7) and therefore accounts for the increase in the effective heat transfer area as heat penetrates further into the PCB [7].

PCBs may include thermal vias, as shown in Figure 5.29, to improve heat transfer through the thickness of the boards as the power and ground layers spread the heat mainly in the plane of the boards. Thermal vias are created by drilling holes in PCBs and copper plating the inner sides of those holes. Therefore, thermal vias are basically thin copper shells that go through the thickness of a PCB. Thermal vias are normally connected to copper layers at both ends of the PCB as shown in Figure 5.29. Thermal vias improve thermal conductivity of the board in the normal direction and have to be considered when calculating effective normal thermal conductivity of a PCB.

The simplest model for a PCB with thermal vias is a one-dimensional model. Thermal resistance of a single thermal via in the direction normal to the PCB is

$$R_{via} = \frac{t}{k_c \pi (r_{via,o}^2 - r_{via,i}^2) + k_f \pi r_{via,i}^2}. \tag{5.54}$$

Where t is the thickness of the PCB and the length of the via, k_c is thermal conductivity of copper, k_f is the thermal conductivity of a filler material such as air or solder that fills the via, and $r_{via,o}$ and $r_{via,i}$ are outer and inner radii of the via, respectively ($r_{via,o} = r_{via,i}$ + copper plating thickness). Note that the effect of the term including k_f is normally small and this term can be neglected without major effect on the accuracy of the results. The heat path through the vias' copper plating is in parallel with the heat path through the rest of the PCB area. Consider a PCB with area A, thickness t, and N thermal vias. If k_n is the normal thermal conductivity of this PCB with

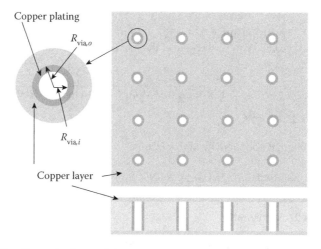

FIGURE 5.29 (See color insert following page 240.) A printed circuit board with thermal vias.

no thermal vias, calculated with Equation 5.47, the normal thermal resistance of the portion of the PCB without thermal vias is

$$R_n = \frac{t}{k_n(A - \pi N r_{via,o}^2)}. \tag{5.55}$$

The normal thermal resistance of a PCB with thermal via, $R_{n,\text{with-via}}$, is the parallel equivalent of thermal resistances of N vias and normal thermal resistance of PCB in the areas with no thermal vias,

$$\frac{1}{R_{n,\text{with-via}}} = \frac{N}{R_{via}} + \frac{k_n(A - \pi N r_{via,o}^2)}{t}, \tag{5.56}$$

And the effective normal thermal conductivity of a PCB with thermal vias, $k_{n,\text{with-via}}$, is

$$k_{n,\text{with-via}} = \frac{t}{AR_{n,\text{with-via}}}. \tag{5.57}$$

A more accurate model for calculating effective normal thermal conductivity of a PCB with thermal vias will be given in Section 7.3.

Example 5.9

Consider a 2.1 mm thick printed circuit board with four inner copper layers. If the thickness of each copper layer is 35 μm and thermal conductivities of copper and glass epoxy are 390 W/m°C and 0.35 W/m°C, respectively, determine effective normal and planar thermal conductivities of this PCB.

Solution

Since $t_c = 4 \times 35$ μm $= 140$ μm $= 0.14$ mm, $t_e = 2.1$ mm $- 0.14$ mm $= 1.96$ mm, $k_c = 390$ W/m°C and $k_e = 0.35$ W/m°C, the normal thermal conductivity of this PCB is

$$k_n = \frac{t}{t_e/k_e + t_c/k_c} = \frac{2.1\,\text{mm}}{1.96\,\text{mm}/0.35\,\text{W/m°C} + 0.14\,\text{mm}/390\,\text{W/m°C}}$$

$$= 0.375\,\text{W/m°C}.$$

The planar thermal conductivity of this PCB is

$$k_p = \frac{k_e t_e + k_c t_c}{t} = \frac{0.35\,\text{W/m°C} \times 1.96\,\text{mm} + 390\,\text{W/m°C} \times 0.14\,\text{mm}}{2.1\,\text{mm}}$$

$$= 26.327\,\text{W/m°C}.$$

Note that the planar thermal conductivity of this PCB is two orders of magnitude larger than its normal thermal conductivity. It is also seen that its normal thermal conductivity has not been improved significantly by the copper layers, and is almost the same as the thermal conductivity of glass epoxy.

Thermal Resistance Network

PROBLEMS

5.1 Consider a double-pane window with two glass layers separated by air. Explain what happens to total thermal resistance of this double-pane window if temperature increases. Assume thermal conductivity of glass does not change with temperature and there is no convection and radiation heat transfer through the air.

5.2 Write down expressions for conduction thermal resistance, convection thermal resistance, and radiation thermal resistance.

5.3 What is thermal contact resistance?

5.4 How does thermal contact resistance between two surfaces change with temperature? Why?

5.5 How does thermal contact resistance between two surfaces change with pressure? Why?

5.6 Why is it necessary to have a thermal interface material between a heat sink and an electronic device, and how do they improve the heat transfer rate from the device?

5.7 What is the equivalent thermal resistance of the composite layer shown below?

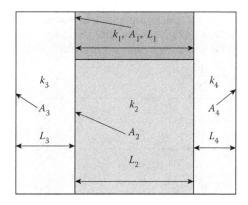

5.8 What is the equivalent thermal resistance of the resistance network shown below if $R_1 = 10°C/W$, $R_2 = 15°C/W$, and $R_3 = 35°C/W$?

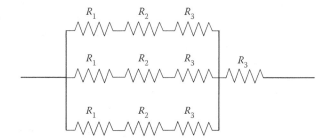

5.9 What is the temperature at the perfect interface of the two layers shown below if we assume the left and right sides are insulated?

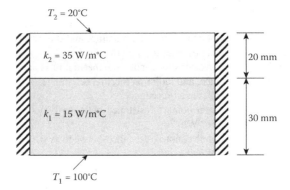

5.10 Let's assume the interface between the two layers shown below is not perfect. Draw the equivalent thermal resistance network for this composite layer and identify each resistance in that network. If the heat transfer rate per unit area from the bottom to the top of this layer is 6000 W/m², what is the thermal resistance across the interface? Assume the left and right sides are insulated.

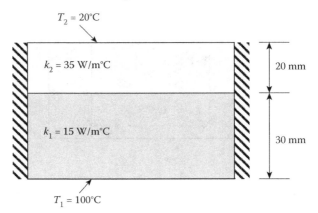

5.11 A 2 mm-thick PCB includes two 0.035 mm-thick copper layers with thermal conductivity of 400 W/m°C. If the thermal conductivity of the PCB material is 0.35 W/m°C, what are the effective normal and planar thermal conductivities of this PCB?

5.12 A 4 m high and 6 m long wall is constructed of two large 2 cm thick steel plates ($k = 15$ W/m°C) separated by 1 cm thick and 20 cm wide steel bars placed 99 cm apart. The remaining space between the steel plates is filled with fiberglass insulation ($k = 0.035$ W/m°C). If the air temperature and heat transfer coefficient at one side of the wall are 40°C and 10 W/m²°C and at the other side of the wall are 10°C and 15 W/m²°C, determine the total heat transfer rate through the wall and the wall temperature at both sides.

Thermal Resistance Network

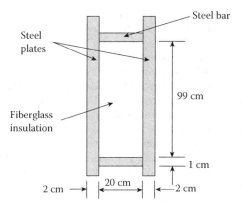

5.13 The composite wall of an oven consists of three materials, two of which are of known thermal conductivity, $k_A = 20$ W/m°C and $k_C = 50$ W/m°C, and known thickness, $L_A = 0.3$ m and $L_C = 0.15$ m. The third material, B, which is sandwiched between materials A and C, is of known thickness, $L_B = 0.15$ m, but unknown thermal conductivity k_B. Under steady-state operating conditions, measurements reveal an outer surface temperature of $T_{s,o} = 20°C$, an inner surface temperature of $T_{s,i} = 600°C$, and an oven air temperature of $T_\infty = 800°C$. The inside convection coefficient is $h = 25$ W/m²°C. What is the thermal conductivity of material B?

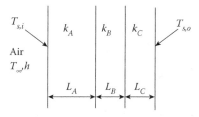

5.14 Sketch the thermal resistance network for the composite wall shown below and obtain its total thermal resistance, along x or y, per unit depth normal to the paper.

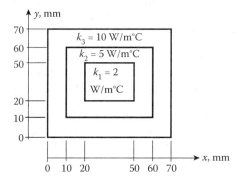

5.15 Consider a 50 mm × 50 mm × 5 mm silicon block sandwiched between two different composite layers as shown in a representative cross section below. The composite layer on the right consists of a 0.5 mm thick thermal paste with thermal conductivity of 10 W/m°C and a 1 mm-thick copper spreader with thermal conductivity of 395 W/m°C. The composite layer on the left is made of 25 pairs of 1 mm × 1 mm × 50 mm solder with thermal conductivity of 40 W/m°C and 1 mm × 1 mm × 50 mm underfill with thermal conductivity of 1 W/m°C that are connected to a 2 mm-thick PCB. The PCB is made of glass epoxy with thermal conductivity of 0.35 W/m°C and four copper layers each 0.035 mm thick. The convection heat transfer coefficient on the left and right sides of this composite layer are 30 W/m²°C and 5 W/m²°C, respectively.
  a. What is the normal thermal resistance of the PCB?
  b. Calculate thermal resistance between the left side of the silicon layer and the fluid at temperature $T_{\infty,1}$, and the right sides of the silicon layer and the fluid at temperature $T_{\infty,2}$.

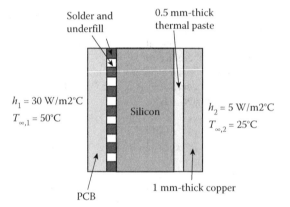

5.16 Let's assume that the dimensions of the silicon die in Problem 5.15 are 10 mm × 10 mm × 1 mm and there are five pairs of 1 mm × 1 mm × 10 mm solder and 1 mm × 1 mm × 10 mm underfill between the silicon and the PCB. If all the other conditions are the same, calculate thermal resistance between the left side of the silicon layer and the fluid at temperature $T_{\infty,1}$, and the right sides of the silicon layer and the fluid at temperature $T_{\infty,2}$. Assume the PCB can be treated as a homogenous medium with normal thermal conductivity k_n.

5.17 Consider a 400 mm × 400 mm printed circuit board (PCB) with small microelectronic devices on its top side as shown in the picture below. The microelectronic devices dissipate 250 W total. This PCB is 2 mm thick and has three 0.07 mm-thick copper layers sandwiched between layers of glass epoxy. Its top and bottom sides can be approximated as flat plates. The air velocities on the top and bottom are 5 m/s and 3 m/s, respectively. The airflow is turbulent on the top and laminar on the bottom. Thermal conductivity of copper and glass epoxy are 390 W/m°C and 0.35 W/m°C.

Thermal Resistance Network

a. Calculate normal and planar thermal conductivities of this PCB.
b. Sketch the thermal resistance network between the top side of this PCB and the air above and below it, and calculate the total thermal resistance between the top side of this PCB and the air above and below it. Note that the heat generated by the electronics on the top of this PCB is transferred partially to the air on the top and partially, through the PCB, to the air on the bottom.
c. If the air temperature is 25°C, what are the temperatures of the top and bottom of the PCB?

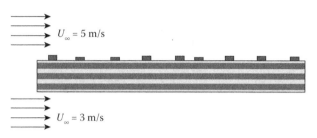

REFERENCES

1. Yovanovich, M. M. 1981. New contact and gap conductance correlations for conforming rough surfaces. AIAA 16th Thermophysics Conference, Palo Alto, California.
2. Bahrami, M., J. R. Culham, and M. M. Yovanovich. 2004. Modeling thermal contact resistance: A scale analysis approach. *Journal of Heat Transfer* 126:896–905.
3. Lee, S., S. Song, V. Au, and K. P. Moran. 1995. Constriction/spreading resistance model for electronics packaging. Proceedings of ASME/JSME Thermal Engineering Conference, Maui, Hawaii, Vol. 4, pp. 199–206.
4. Shabany, Y. 2002. Component size and effective thermal conductivity of printed circuit boards. Proceedings of the 8th Intersociety Conference on Thermal and Thermomechanical Phenomena in Electronic Systems (ITherm 2002), San Diego, California, pp. 489–94.
5. Shabany, Y. 2003. Effects of boundary conditions and source dimensions on the effective thermal conductivity of a printed circuit board. Proceedings of IPACK03, Maui, Hawaii.
6. Culham, J. R., and M. M. Yovanovich. 1998. Factors affecting the calculation of effective conductivity in printed circuit boards. Proceedings of the 6th Intersociety Conference on Thermal and Thermomechanical Phenomena in Electronic Systems (ITherm'98), Seattle, Washington, pp. 460–67.
7. Liu, W., M. Lee, Y. Shabany, and M. Asheghi. 2005. A novel scheme in thermal modeling of printed circuit boards. Proceedings of IPACK05, San Francisco, California.

6 Thermal Specification of Microelectronic Packages

6.1 IMPORTANCE OF PACKAGING

Consider a desktop computer and a DVD player as two examples of electronic systems. Both of these require power to function. That is usually provided by connecting them to power outlets through appropriate cables and wall plugs. They also require some form of input to provide the desired outputs. A computer receives input from a mouse, keyboard, or voice and displays the output on a monitor. A DVD player receives its input from a DVD and its output signal is transferred to a TV and is displayed on a TV monitor. In both cases there are input and output connectors; USB, serial, parallel, microphone, headphone, and other connectors for computer and coaxial, video, and audio connectors for DVD player. These input–output (I/O) connectors for power and signal are usually located on enclosures that contain the electronics of these systems. These enclosures or packages also protect the electronic devices against undesired contacts with other objects and exposure to dust, condensate, and so forth. Another role for these enclosures is that they usually come with some form of cooling method. Almost all desktop computers have one or more cooling fans that blow or draw air through the enclosure. This air is used to cool hot components such as a CPU, graphics module, hard disk drives, CD and DVD players, PCI cards, and power supply. The DVD players, on the other hand, may not come with a cooling fan. However, there are air vents on their enclosures that allow fresh air to enter and hot air to leave. This natural motion of air cools the electronic components inside the DVD player. These two examples show that packaging of electronics provides I/O interface for power and signal, protects electronic devices against a possibly harmful environment, and provides necessary cooling for heat dissipating electronics.

An integrated circuit is a miniaturized electronic system. It requires power to operate and needs ways to receive and send a signal to other electronic devices. It is a very fragile and expensive piece that has to be protected against possible harsh outside environments. It generates a certain amount of heat that has to be properly dissipated to avoid overheating and the consequential failures and damages. These are achieved by appropriate packaging of the integrated circuit. Although many different packaging techniques have been developed, the main objectives of all those are providing I/O connections to the device, protecting it, and dissipating the heat generated by it.

6.2 PACKAGING TYPES

Integrated circuits are produced in large varieties including those with very low to very high power dissipations and those that require only a few, to those that need more than 1000, I/O connections. A broad range of packaging techniques have been developed to serve various integrated circuits. Some of the common techniques will be explained here.

Figure 6.1 shows two examples of a relatively simple package called Small Outline Integrated Circuit (SOIC) or Small Outline Package (SOP). These are low-profile rectangular surface-mount components. A cross section of this package is shown in Figure 6.2. The die is bonded to an inner lead frame. External terminals are gull wing formed leads and exit parallel to the seating plane on two opposite sides of the molded package. The mold material is usually plastic but ceramic packages are also available. The electrical connection between the die and the leads are provided by gold wire bonds. Other packages that belong to this family are Plastic Small Outline Package (PSOP), Ceramic Small Outline Package (CSOP), Shrink Small Outline Package (SSOP; which is a smaller version of SOP), Thin Small Outline Package (TSOP; with a thickness of about 1.0 mm, which is thinner than regular SOP), and Thin Shrink Small Outline Package (TSSOP; which is thinner than 1.0 mm and offers smaller lead pitch). Lead counts of these packages range from 8 to 80 and lead pitches are about 0.4 to 1.27 mm.

An increase in the number of I/O connections can be achieved if they are located on all four sides of the package. Early versions of Quad Flat Pack (QFP) packages were similar to SOPs as far as electrical and mechanical connections and the gull wing shaped leads were concerned. However, as shown in Figure 6.3, they offered leads on all four sides. Different members of this family are Plastic Quad Flat Pack

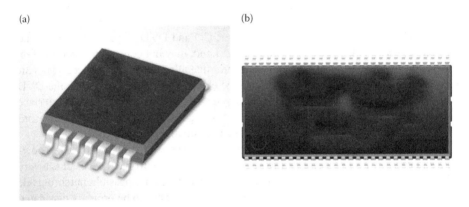

FIGURE 6.1 A 14-lead and a 56-lead SOP. (Courtesy of Practical Components.)

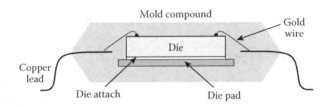

FIGURE 6.2 Cross section of a small outline package.

Thermal Specification of Microelectronic Packages 105

(PQFP), Low Profile Quad Flat Pack (LQFP), and Thin Quad Flat Pack (TQFP; which are thinner than regular PQFPs). Any of these variations may come with the die facing up or down, with the die attach pad exposed or covered inside the mold, and with or without an internal heat slug. Schematics of the internal construction of some of these packages are shown in Figures 6.4 and 6.5. The copper heat slug provides a lower thermal resistance path for the heat generated by the die. If the die is faced up and the heat slug is attached to the bottom of the die (as shown in Figure 6.4a) or the die pad is exposed (as shown in Figure 6.5a), the heat is mainly dissipated through the bottom of the package. The heat slug or the die pad in some of these packages may be soldered to the corresponding printed circuit boards. On the other hand, if the die is faced down and the heat slug is attached to the top of the die (as shown in Figure 6.4b) or the die pad is exposed (as shown in Figure 6.5b),

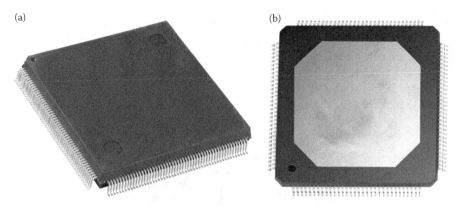

FIGURE 6.3 A 240-lead QFP (a) and a 128-lead LQFP with exposed top (b). (Courtesy of Practical Components.)

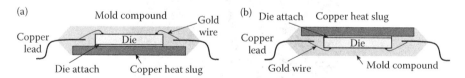

FIGURE 6.4 Cross section of a thermally enhanced LQFP with die facing up (a) and die facing down (b).

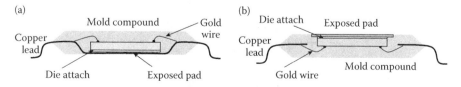

FIGURE 6.5 Cross section of an exposed pad TQFP with die facing up (a) and die facing down (b).

the heat is mainly dissipated through the top of the package. These packages may be used with a heat sink (see Chapter 7) mounted on them to improve heat transfer rate from the package.

Another package type with leads on all four sides is the Plastic Leaded Chip Carrier (PLCC) package. They have J-shaped leads on all four sides and the lead counts range from 20 to 84. The PLCC packages can be square or rectangle. These packages require less board space compared to the equivalent QFP packages. An example of a PLCC package is shown in Figure 6.6 and the internal construction of this package is shown in Figure 6.7. It is seen that, except for the shape of the leads, the mechanical and electrical construction of a PLCC package is similar to SOP and QFP packages.

The I/O connections are located on two sides of the SOP packages and on four sides of the QFP and PLCC packages. This limits the maximum number of I/O connections to the available perimeter of the package and minimum allowable lead pitch. A significant increase in the number of I/O connections is obtained if the package bottom area is used for I/O connections. Pin Grid Array (PGA) packages are an example of these packages. Figure 6.8 shows the top and bottom of a PGA package. The pins of the package are either inserted through holes in a printed circuit board and then soldered to it or inserted into an appropriate socket that is already soldered to the board itself.

Figure 6.9a shows the schematic cross section of a Plastic Pin Grid Array (PPGA) package. The die is located at the center of the package and the substrate

FIGURE 6.6 An 84-lead PLCC. (Courtesy of Practical Components.)

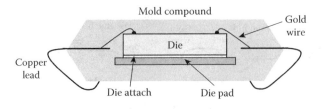

FIGURE 6.7 Cross section of a PLCC package.

Thermal Specification of Microelectronic Packages 107

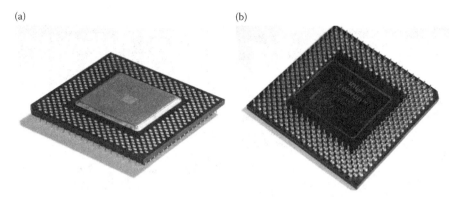

FIGURE 6.8 (**See color insert following page 240.**) Top (a) and bottom (b) of a Pin Grid Array package. (http://en.wikipedia.org/wiki/Pin_grid_array.)

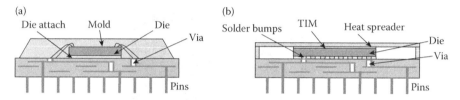

FIGURE 6.9 Schematic cross section of (a) a PPGA and (b) a FCPGA.

is a laminated structure similar to a printed circuit board. The pins go through this laminated structure and are connected to a trace, ground, or power layer. The connection between these pins and different locations of the die are through wire bonds between the die and traces on top of the substrate, and the vias in the substrate. Vias are microscale copper shells that are generated by drilling small holes in the substrate and copper plating the holes. A version of the PGA package is Flip-Chip Pin Grid Array (FCPGA or FC-PGA) shown in Figure 6.9b. The die is facing down and its back is either exposed or attached to a copper heat spreader. The connection between the pins and the circuit on the die are through solder bumps between the die and traces on top of the substrate, and the vias in the substrate. These packages show better thermal performance compared to the similar plastic packages.

The PGA and SOP packages had been produced with larger number of pins and leads and therefore smaller spacing between those pins or leads. This had increased the challenges in the soldering process and the risk of an electrical short between adjacent pins or leads. A solution to this problem was a Ball Grid Array (BGA) package. Ball Grid Array packages are similar to their ancestor PGA packages but the pins had been replaced with solder balls. Figure 6.10 shows the bottom sides of three BGA packages. This package are placed on a PCB such that the solder balls are in contact with appropriate solder pads on the PCB. The PCB and the packages will be heated until the solder balls are melted and the package are attached to the PCB. The amount and material properties of solder and the soldering temperature are selected such that the solder balls are not completely melted. They are only semiliquefied and will stay separate from the neighboring solder balls.

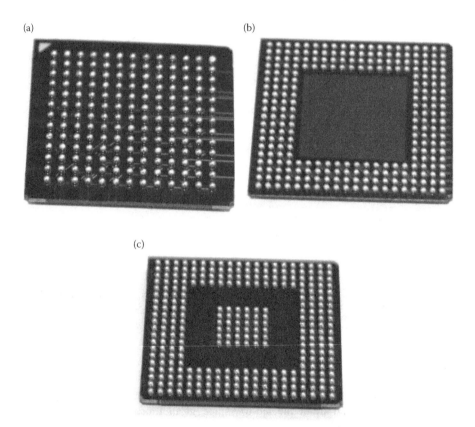

FIGURE 6.10 (See color insert following page 240.) Bottom sides of a 23 mm × 23 mm 169 ball BGA (a), a 27 mm × 27 mm 256 ball BGA (b), and a 27 mm × 27 mm 292 ball BGA (c).

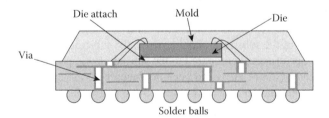

FIGURE 6.11 Schematic cross section of a plastic ball grid array.

Figure 6.11 shows the schematic cross section of a plastic ball grid array (PBGA) package. The die is mounted on a laminated substrate similar to a printed circuit board. The connection between the integrated circuit on the die and the solder balls at the bottom of the substrate is through gold wire bonds and vias in the substrate. The die and the gold wires are covered by a plastic mold and that is why the package is called a PBGA.

Thermal Specification of Microelectronic Packages 109

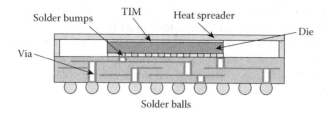

FIGURE 6.12 Schematic cross section of a flip chip ball grid array.

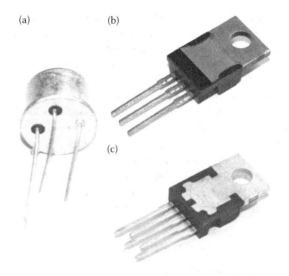

FIGURE 6.13 (See color insert following page 240.) Three different TO packages: (a) TO-92, (b) three-lead TO-220, and (c) five-lead TO-220.

A common version of the BGA packages is Flip-Chip Ball Grid Array (FCBGA or Flip-Chip BGA). Figure 6.12 shows a schematic cross section of a flip-chip BGA package. It is seen that there is no wire bonding between the die and the substrate. Instead, electrical connection between the die and substrate is through solder bumps, or flip-chip bumps that are located on the active side of the die. The die is flipped upside down and these solder bumps contact the solder pads on the substrate. The gaps between the solder bumps and the substrate are filled by an underfill material to increase the strength of the attachment between the die and the substrate and to protect solder bumps against excessive pressure. The package is molded or, as shown in Figure 6.12, covered by a metallic (usually copper) lid that contacts the back side of the die and improves heat transfer from the package.

Another common packaging technique that is used for small but high power transistors and integrated circuits is a Transistor Outline (TO) package. Figure 6.13 shows three different forms of this package; other forms are also available. The TO-92 is a small, cheap package and is used for low power applications. The TO-220 and

TO-263 (not shown in this figure) are plastic packages that have a metal tab that may be in electrical contact with the integrated circuit inside the package. The TO-220 is a through-hole package that is mounted vertically on a printed circuit board, and may be attached to a heat sink for improved thermal performance. The TO-263 is a surface-mount package and its leads and metal tab are soldered to the board for electrical and heat conduction.

6.3 THERMAL SPECIFICATIONS OF MICROELECTRONIC PACKAGES

All integrated circuits contain a small region in which most of the electronic activity is concentrated and therefore most of the heat is generated. This region is the hottest part of the die and is called **junction**. Maximum allowable junction temperature of a device is limited by certain requirements on its performance and reliability and by material properties of the die and the package. The main objective of thermal management of electronic systems is to design appropriate cooling methods that keep the junctions of those devices below their respective maximum allowable temperatures. Microelectronic packages provide heat transfer paths from their dies to their external surfaces where the above mentioned cooling techniques can be applied. Therefore, an important characteristic of a microelectronic package is its efficiency in transferring heat from its die. Different metrics have been used to characterize thermal efficiency of a microelectronics package. These metrics, commonly called package thermal specifications or thermal characteristics, will be discussed in this section.

6.3.1 Junction-to-Air Thermal Resistance

Consider one of the packages shown in Figures 6.1 through 6.13 is mounted on a board. The heat generated by the circuitry on the die in these packages will go out of the package through different paths. However, in all cases, the final sink to dump the heat is the surrounding air. Therefore, a good measure of cooling efficiency of a package is its junction-to-air thermal resistance defined as

$$R_{ja} = \frac{T_j - T_a}{\dot{Q}}, \qquad (6.1)$$

where T_j and T_a are junction and ambient air temperatures and $\dot{Q}$ is the total power (heat) dissipated by the chip. The JEDEC Solid State Technology Association [1] (or JEDEC) that issues standards and publications for measuring and reporting thermal specifications of microelectronic packages uses the letter θ instead of R for thermal resistance. JEDEC also distinguishes between thermal resistances in natural and forced convection by using the notation $θ_{ja}$ for junction-to-ambient air thermal resistance in natural convection and $θ_{jma}$ for junction-to-moving air thermal resistance in forced convection. However, in order to be consistent with the rest of this book, we will use R for thermal resistance and R_{ja} for both natural and forced convection junction-to-air thermal resistances. At zero forced air velocity, R_{ja} is the junction-to-air

Thermal Specification of Microelectronic Packages

thermal resistance in natural convection and R_{ja} at a nonzero air velocity refers to forced convection junction-to-air thermal resistance at that velocity.

The procedures to measure natural convection junction-to-air thermal resistance of a package are outlined in detail in JEDEC standard JESD51-2 [2]. The package with a test die is mounted on a 76.2 mm × 114.3 mm test board if its maximum dimension is less that 27 mm or on a 101.6 mm × 114.3 mm board if its maximum dimension is from 27 mm to 45 mm. The test board is 1.6 mm thick. It is either a low thermal conductivity 1s0p (one signal layer and no ground or power plane) board or a high thermal conductivity 2s2p (two signal layers, one power plane and one ground plane) board. The package is centered in the frontal 76.2 mm × 76.2 mm or 101.6 mm × 101.6 mm areas of the test board. The test board is mounted horizontally to a vertical wall, through an edge connector, and the entire assembly is covered by a 0.0283 m³ cubic enclosure as shown in Figure 6.14. The enclosure shall be made of a low thermal conductivity material such as polycarbonate, cardboard, wood or any other material with an equivalent thermal conductivity. The ambient air temperature is measured by a thermocouple mounted 25.4 mm below the test board and 25.4 mm away from a side wall. The junction temperature is measured by the change in a temperature sensitive variable such as the voltage drop across a forward-biased diode located on the test die. The power (heat) dissipation is calculated by measuring voltage and current applied to the device under test. The measured temperatures and power dissipation are used in Equation 6.1 to find junction-to-air thermal resistance of the package.

The procedures to measure forced convection junction-to-air thermal resistance are given in JEDEC standard JESD51-6 [3]. The test board and the location of the

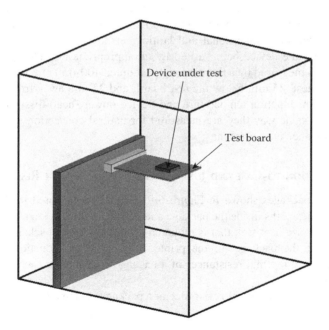

FIGURE 6.14 JEDEC enclosure for measuring natural convection junction-to-air thermal resistance.

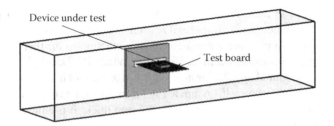

FIGURE 6.15 A JEDEC test section for measuring forced convection junction-to-air thermal resistance.

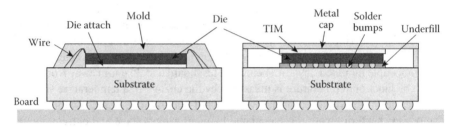

FIGURE 6.16 Schematics of a plastic ball grid array (left) and a metallic flip-chip ball grid array (right).

package on the test board are similar to those for measuring natural convection junction-to-air thermal resistance. The test board will be placed in the test section of a wind tunnel, as shown in Figure 6.15, which is able to generate relatively steady, uniform, one-dimensional and laminar airflow. Air velocity is measured upstream from the device being tested by an appropriate air velocity probe. Air temperature is measured by a thermocouple mounted 100–150 mm upstream of the device under test, 25 mm below the test board, and 25 mm away from the wall of the test section. Junction temperature and device power (heat) dissipation will be measured the same way they are measured for natural convection junction-to-air thermal resistance measurement.

6.3.2 Junction-to-Case and Junction-to-Board Thermal Resistances

Consider the packages shown in Figure 6.16. The heat generated in the die goes through complex paths inside the package and gets out of the package either through the top (to air or a heat sink that is mounted on the top of the package) or through the bottom of the package (to the printed circuit board). Junction-to-case and junction-to-board thermal resistances of a package characterize each of these heat transfer paths.

Junction-to-case thermal resistance of a package is defined as

$$R_{jc} = \frac{T_j - T_c}{\dot{Q}_{jc}}, \tag{6.2}$$

Thermal Specification of Microelectronic Packages

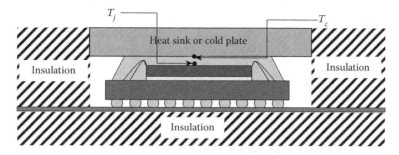

FIGURE 6.17 Test setup for junction-to-case thermal resistance measurement.

where T_j and T_c are junction and case temperatures and $\dot{Q}_{jc}$ is part of the heat that flows from the die to the case of the package. The case is typically referred to the external surface in the primary heat transfer path of the package. It is the top surface for SOP, PGA, BGA, PLCC, and QFP packages with the die facing down and the bottom surface for metal tabbed packages such as TO-220 and TO-263, and QFP packages with the die facing up. The JEDEC standards use the notation θ_{jcx} for junction-to-case thermal resistance where x is replaced by *top* or *bot* depending on whether the case is the top or the bottom surface of the package. The JEDEC standard for the measurement of junction-to-case thermal resistance is under development. However, it is commonly measured in situations where almost all of the heat flows through the case of the package and $\dot{Q}_{jc} \approx \dot{Q}$. This can be achieved with a test setup similar to that shown in Figure 6.17. A very efficient heat sink or cold plate is attached to the case of the package and the entire bottom of the test board and the top around the package are insulated. The case temperature is measured at the center and the junction temperature and power dissipation are measured as explained earlier.

Junction-to-board thermal resistance of a package is defined as

$$R_{jb} = \frac{T_j - T_b}{\dot{Q}_{jb}}, \quad (6.3)$$

where T_j and T_b are junction and board temperatures and $\dot{Q}_{jb}$ is part of the heat that flows from the junction to the board. The JEDEC standards use the notation θ_{jb} for junction-to-board thermal resistance. The procedure to measure junction-to-board thermal resistance is described in JEDEC standard JESD51-8 [4]. The package is mounted on a 2s2p board described in Section 6.3.1. The test setup is shown in Figure 6.18. A ring-style, liquid-cooled cold plate clamps both sides of the test board such that most of the heat generated in the die flows from the package into the test board and towards the sides of the package. An insulation material with thermal conductivity less than 0.1 W/m°C is used to minimize heat transfer through the case of the package and through the board right under the package. These conditions create a situation in which almost all the heat generated in the die flows laterally into the test board as shown in Figure 6.18 and $\dot{Q}_{jb} \approx \dot{Q}$. Junction temperature and power (heat) dissipation are measured as described earlier. The thermocouple that

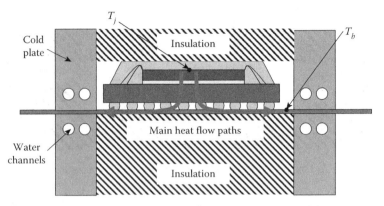

FIGURE 6.18 Test setup for junction-to-board thermal resistance measurement.

measures the board temperature for a leaded package is attached to the foot of a lead that is located halfway along one side of the package. For an area array package, the thermocouple is attached to a test board trace located halfway along one side of the package and within 1 mm distance from the package.

6.3.3 PACKAGE THERMAL CHARACTERIZATION PARAMETERS

Junction-to-case and junction-to-board thermal resistances are measured in situations where almost all of the heat generated in the die flows to the case or to the board, respectively. This is a good assumption if a heat sink is attached to the case of the package or to the board. It is also a good approximation if there is no heat sink but the package is mounted on a high thermal conductivity board with relatively low or no forced airflow over the package such that most of the heat flows into the board. In these situations, junction-to-case or junction-to-board thermal resistance, and case or board temperature and power dissipation of the package, can be used to estimate junction temperature of the device.

However, in most applications only a portion of the heat generated in the die is dissipated from the case and the rest is transferred to the board. Since it is not possible to measure individual power dissipations to the case and to the board, $\dot{Q}_{jc}$ and $\dot{Q}_{jb}$, Equations 6.2 and 6.3 can not be used to calculate junction temperature of the device. Thermal characterization parameters are defined to overcome this problem.

Junction-to-top thermal characterization parameter, Ψ_{jt}, is defined as

$$\Psi_{jt} = \frac{T_j - T_t}{\dot{Q}}, \qquad (6.4)$$

where T_j and T_t are junction and package top temperatures and $\dot{Q}$ is the total power dissipation of the die. T_t is the temperature at the top center of the package. Similarly, junction-to-board thermal characterization parameter, Ψ_{jb}, is defined as

Thermal Specification of Microelectronic Packages

$$\Psi_{jb} = \frac{T_j - T_b}{\dot{Q}}, \tag{6.5}$$

where T_j and T_b are junction and board temperatures and $\dot{Q}$ is the total power dissipation of the die. The board temperature is measured at the same location it is measured for the junction-to-board thermal resistance measurement.

Both junction-to-top and junction-to-board thermal characterization parameters are measured at the same time and environment as junction-to-air thermal resistance is measured. It is necessary to emphasize that thermal characterization parameters are not thermal resistances. This is because the heat transfer rate $\dot{Q}$ used in calculating those is not the actual heat transfer rate between the two points. Junction-to-top thermal characterization parameter is often significantly lower than junction-to-case thermal resistance and junction-to-board thermal characterization parameter is lower than or approximately equal to junction-to-board thermal resistance. Also, junction-to-top thermal characterization parameter is usually lower than junction-to-board thermal characterization parameter.

6.4 PACKAGE THERMAL RESISTANCE NETWORK

Let's consider the packages shown in Figure 6.19. The heat generated in the die is dissipated through two separate paths; one toward the top case of the package and one through the bottom of the package or through the leads to the board. Junction-to-case and junction-to-board thermal resistances account for total thermal resistances from the junction of the die to the case and to the board, respectively. The heat that reaches the case of the package faces a convection thermal resistance between the case and the ambient air that is called case-to-air thermal resistance, R_{ca}. The rest

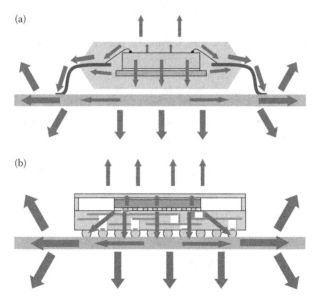

FIGURE 6.19 Heat flow paths in a QFP (a) and a FCBGA (b) package.

of the heat goes to the board and is dumped into the ambient, both from the top and from the bottom of the board. Thermal resistance between the component side of the board (the point where T_b is measured) and the air is called board-to-air thermal resistance and is denoted by R_{ba}.

Figure 6.20 shows the equivalent thermal resistance network of a typical package. This resistance network represents total thermal resistance between the junction and the ambient air around this package and therefore is equivalent to its junction-to-air thermal resistance. It shows the following relationship between different thermal characteristics of a package:

$$\frac{1}{R_{ja}} = \frac{1}{R_{jc} + R_{ca}} + \frac{1}{R_{jb} + R_{ba}}, \quad (6.6)$$

$$R_{ja} = \frac{(R_{jc} + R_{ca})(R_{jb} + R_{ba})}{R_{jc} + R_{ca} + R_{jb} + R_{ba}}. \quad (6.7)$$

This equation shows that if $R_{jb} + R_{ba}$ is much larger than $R_{jc} + R_{ca}$,

$$R_{ja} \approx R_{jc} + R_{ca}. \quad (6.8)$$

This means that almost all the heat generated in the die leaves the package through its case. Similarly, if $R_{jc} + R_{ca}$ are much larger than $R_{jb} + R_{ba}$,

$$R_{ja} \approx R_{jb} + R_{ba}, \quad (6.9)$$

which means that almost all the heat leaves the package through the bottom of the package.

Junction-to-case and junction-to-board thermal resistances of a package are conduction thermal resistances. They depend on package geometry and its material properties. Therefore, they are properties of the package itself and do not change from one application to another where environmental conditions such as air velocity, board dimensions, and board material may change. On the other hand, case-to-air thermal resistance is a convection and radiation type thermal resistance and depends on package geometry, surface property, power dissipation, as well as ambient conditions

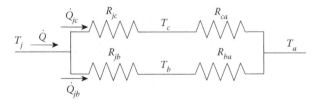

FIGURE 6.20 Equivalent thermal resistance network for a typical package.

Thermal Specification of Microelectronic Packages

such as air velocity and temperature, and surrounding temperature. Similarly, board-to-air thermal resistance of a package depends on the conduction thermal resistance of the board and convection and radiation thermal resistances from the board to air. Therefore, board-to-air thermal resistance of a package depends on board geometry, thermal conductivity, and surface property as well as package power dissipation and ambient conditions. Since junction-to-air thermal resistance of a package depends on the thermal resistances mentioned above, it is a function of package, board, and ambient conditions. The same thing can be said for the package thermal characterization parameters. This is an important observation and shows that junction-to-air thermal resistance and thermal characterization parameters are valid only for the exact conditions they were measured at and shall not be used to predict junction temperature of that package in other conditions. On the other hand, junction-to-case and junction-to-board thermal resistances of a package are functions of the package itself and do not change if conditions in which the package operates change.

Example 6.1

Consider an electronic package with junction-to-case, case-to-air, junction-to-board, and board-to-air thermal resistances of 2.5°C/W, 15°C/W, 8°C/W, and 25°C/W, respectively. What is its junction-to-air thermal resistance? If this package dissipates 5 W and the air temperature is 25°C, what are its case and junction temperatures?

Solution

Junction-to-air thermal resistance of this package is the parallel equivalent of junction-to-case plus case-to-air and junction-to-board plus board-to-air thermal resistances as given in Equation 6.6:

$$\frac{1}{R_{ja}} = \frac{1}{R_{jc}+R_{ca}} + \frac{1}{R_{jb}+R_{ba}} \Rightarrow \frac{1}{R_{ja}} = \frac{1}{2.5°C/W+15°C/W} + \frac{1}{8°C/W+25°C/W}$$

$$R_{ja} = 11.4°C/W.$$

Junction temperature of this package is

$$T_j = T_a + \dot{Q} \cdot R_{ja} \Rightarrow T_j = 25°C + 5\,W \times 11.4°C/W \Rightarrow T_j = 82°C.$$

Heat transfer rate through the case of the package is

$$\dot{Q}_{jc} = \frac{T_j - T_a}{R_{jc}+R_{ca}} \Rightarrow \dot{Q}_{jc} = \frac{82°C - 25°C}{2.5°C/W+15°C/W} \Rightarrow \dot{Q}_{jc} = 3.26\,W,$$

and case temperature of the package is

$$T_c = T_j - \dot{Q}_{jc} \cdot R_{jc} \Rightarrow T_c = 82°C - 3.26\,W \times 2.5°C/W \Rightarrow T_c = 73.85°C.$$

Example 6.2

Consider a 40 mm × 40 mm × 2 mm package that is exposed to laminar forced convection airflow on its case. The junction-to-case, junction-to-board, and board-to-air thermal resistances of this package are 4°C/W, 10°C/W, and 25°C/W, respectively, and it dissipates 10 W. What is the minimum air velocity to keep the junction temperature of this package below 105°C if the ambient air temperature is 25°C? Neglect radiation heat transfer rate.

Solution

The air velocity affects case-to-air and therefore junction-to-air thermal resistances of this package. To keep its junction temperature below 105°C, the maximum acceptable junction-to-air thermal resistance is

$$R_{ja} = \frac{T_j - T_a}{\dot{Q}} \Rightarrow R_{ja} = \frac{105°C - 25°C}{10\,W} = 8°C/W.$$

Then, its maximum acceptable case-to-air thermal resistance is obtained as follows:

$$\frac{1}{R_{ja}} = \frac{1}{R_{jc} + R_{ca}} + \frac{1}{R_{jb} + R_{ba}} \Rightarrow \frac{1}{8°C/W} = \frac{1}{4°C/W + R_{ca}} + \frac{1}{10°C/W + 25°C/W}$$

$$R_{ca} = 6.37°C/W.$$

Since radiation heat transfer is neglected, case-to-air thermal resistance is a convection thermal resistance. Therefore, minimum convection heat transfer coefficient that results in this case-to-air thermal resistance is obtained as shown below:

$$R_{ca} = \frac{1}{hA} \Rightarrow h = \frac{1}{A \cdot R_{ca}} = \frac{1}{0.04^2\,m^2 \times 6.37°C/W} \Rightarrow h = 98.12\,W/m^2°C.$$

The minimum velocity to achieve this convection heat transfer coefficient is obtained by using the correlation given in Chapter 4 for laminar flow of air over a flat plate:

$$h = 3.9\left(\frac{U_\infty}{L}\right)^{1/2} \Rightarrow 98.12\,W/m^2°C = 3.9\left(\frac{U_\infty}{0.04\,m}\right)^{1/2} \Rightarrow U_\infty = 25.3\,m/s.$$

This is a very high air velocity for electronics cooling applications. Therefore, other provisions are necessary to achieve 105°C junction temperature. One such provision is using a heat sink that will be discussed in Chapter 7.

Thermal Specification of Microelectronic Packages

6.5 PARAMETERS AFFECTING THERMAL CHARACTERISTICS OF A PACKAGE

6.5.1 PACKAGE SIZE

Both internal conduction thermal resistances (junction-to-case and junction-to-board) and external convection and radiation thermal resistances (case-to-air and board-to-air) are inversely proportional to the package surface area. Therefore, larger packages have lower thermal resistances. Figure 6.21 shows that natural convection junction-to-air thermal resistances of three different packages reduce as the number of leads or pins, and therefore their sizes, increase.

6.5.2 PACKAGING MATERIAL

Higher conductivity packaging materials such as ceramic and metallic covers and high conductivity substrates such as ceramic will reduce junction-to-case and junction-to-board and therefore junction-to-air thermal resistance of a package. For example, since thermal conductivity of ceramic is much higher than that of plastic, ceramic packages have lower thermal resistances compared to a similar plastic package. This is shown in Figure 6.22 where natural convection junction-to-air thermal resistances of a ceramic quad flat pack (CQFP) and a similar PQFP are compared with each other.

6.5.3 DIE SIZE

A die is usually much smaller than the substrate that it mounts on or the lid that it is attached to. The spreading thermal resistance between the die and substrate or the die and lid, and therefore thermal resistance of the package, increases as die

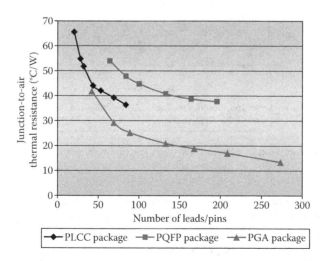

FIGURE 6.21 Variation of natural convection junction-to-air thermal resistances of three different packages with the number of leads/pins. (http://www.intel.com/)

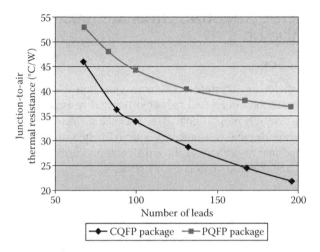

FIGURE 6.22 Variation of natural convection junction-to-air thermal resistance of QFP package with the number of leads and material. (http://www.intel.com/)

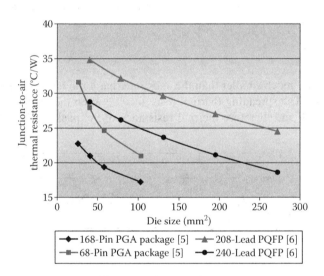

FIGURE 6.23 Variation of natural convection junction-to-air thermal resistance with die size.

size decreases. Figure 6.23 shows how natural convection junction-to-air thermal resistance of two PGA and two PQFP packages increase as the size of their dies decrease. This becomes more pronounced for a die with nonuniform power dissipation where the circuitry that generates most of the heat is concentrated in a small region of the die; the size of the heat source in this case is much smaller than the actual size of the die.

Thermal Specification of Microelectronic Packages

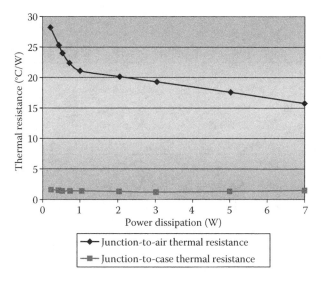

FIGURE 6.24 Variation of junction-to-air and junction-to-case thermal resistances with power dissipation. (http://www.intel.com/)

6.5.4 Device Power Dissipation

An increase in power dissipation of a device increases its temperature. The hotter an object, the larger the natural convection and radiation heat transfer rates from that object will be (see Chapter 4, Sections 4.2 and 4.3). Therefore, if the power dissipation of a device increases, its case-to-air and board-to-air thermal resistances decrease. This will result in a reduction in junction-to-air thermal resistance of the package. Figure 6.24 shows variation of junction-to-air and junction-to-case thermal resistances of a typical package with its power dissipation. It is seen that its junction-to-air thermal resistance decreases while its junction-to-case thermal resistance stays the same as its power dissipation increases. The insensitivity of junction-to-case thermal resistance to power dissipation is expected since it is a conduction-type thermal resistance and depends only on package material and dimensions.

6.5.5 Air Velocity

Higher air velocity results in more efficient convection heat transfer and therefore smaller case-to-air and board-to-air thermal resistances (see Chapter 4, Section 4.2). Therefore, junction-to-air thermal resistance of a package decreases with an increase in air velocity. Figures 6.25 and 6.26 show how junction-to-air thermal resistances of a few different packages change with air velocity. Note that the decrease in junction-to-air thermal resistance slows down as air velocity increases. This indicates that increasing the air velocity beyond a certain value may not result in a meaningful decrease in thermal resistance and therefore will not be a good design recommendation. Figure 6.25 shows also that, as expected, air velocity does not affect junction-to-case thermal resistance of a package.

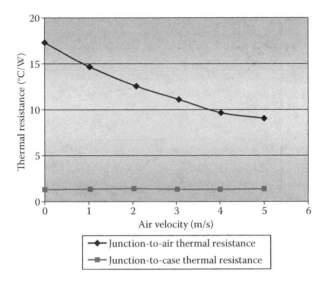

FIGURE 6.25 Variations of junction-to-air and junction-to-case thermal resistances of a 168-pin PGA package with air velocity. (http://www.intel.com/)

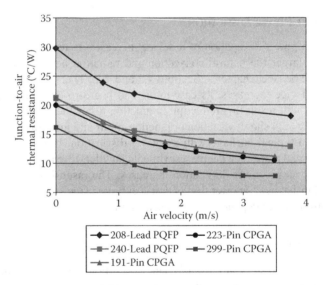

FIGURE 6.26 Variations of junction-to-air thermal resistance with air velocity. (http://www.xilinx.com/)

6.5.6 BOARD SIZE AND THERMAL CONDUCTIVITY

A higher thermal conductivity board reduces the conduction part of the board-to-air thermal resistance and larger board size reduces the convection and radiation part of the board-to-air thermal resistance. Therefore, larger and/or higher thermal conductivity boards reduce board-to-air and therefore junction-to-air thermal

Thermal Specification of Microelectronic Packages

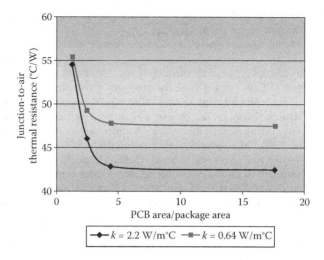

FIGURE 6.27 Variation of junction-to-air thermal resistance of a 132-lead PQFP package with PCB size and thermal conductivity. (http://www.intel.com/)

resistance of a package. Figure 6.27 shows that junction-to-air thermal resistance of a 132-lead PQFP package is less if it is mounted on a board with higher thermal conductivity. It also shows that the thermal resistance decreases as the PCB area increases. However, there is no reduction in thermal resistance for PCB to package area ratios larger than five. The reason is that the board temperature reduces at larger distances from the package, the heat flux which is proportional to temperature difference between the board and air reduces, and those areas do not help in heat dissipation from the package.

PROBLEMS

6.1 How does junction to air thermal resistance of an electronic device change with its size? Why?

6.2 How does the junction-to-air thermal resistance of a package change with air velocity? Why?

6.3 In the square duct shown below $T_1 = 120°C$ and $T_2 = 40°C$. If the conduction heat transfer rate through the walls of the duct is 200 W, what is the thermal resistance of this duct?

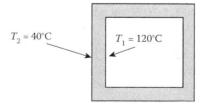

6.4 The junction-to-case and case-to-air thermal resistances of a package are $R_{jc} = 5°C/W$ and $R_{ca} = 18°C/W$, respectively. If this package dissipates 6 W and the air temperature is 25°C, what is the junction temperature of this package? Assume that all the heat is dissipated through the case of this package.

6.5 The maximum allowable junction temperature for an electronic device is 105°C. If air temperature around this device is 25°C and the junction-to-case and case-to-air thermal resistances of this device are 5°C/W and 35°C/W, what will be the maximum power this device can dissipate before reaching its maximum allowable junction temperature? Assume that all the heat is dissipated through the case of this package.

6.6 The maximum allowable junction temperature of an electronic device is 105°C. If the power dissipation of this device is 2 W and its junction-to-case and case-to-air thermal resistances are 5°C/W and 35°C/W, what will be the maximum allowable air temperature for this device? Assume that all the heat is dissipated through the case of this package.

6.7 The junction-to-case, case-to-air, junction-to-board, and board-to-air thermal resistances of a package are 3°C/W, 15°C/W, 8°C/W, and 20°C/W, respectively. If this packages dissipates 7 W and the air temperature is 35°C, determine its junction, case, and board temperatures.

6.8 Junction-to-case, junction-to-board, case-to-air, and board-to-air thermal resistances of a package are 2°C/W, 10°C/W, 20°C/W, and 30°C/W, respectively. What percent of the power generated in the die is dissipated from the case of this package?

6.9 An electronic device dissipates all its heat through its leads. This device dissipates 5 W and its junction to lead thermal resistance is 15°C/W. If the lead temperature is 60°C, what will its junction temperature be?

6.10 Junction-to-case and case-to-air thermal resistances of a package are 5°C/W and 25°C/W, respectively. If this package dissipates 2 W and 80% of this heat is dissipated through its case and the air temperature is 25°C, what will its junction temperature be?

6.11 Junction-to-case, case-to-air, junction-to-board, and board-to-air thermal resistance of a device are 2°C/W, 15°C/W, 10°C/W, and 20°C/W, respectively. What is its junction-to-air thermal resistance? If junction and case temperatures of this device are 120°C and 100°C, respectively, what is its total power dissipation?

6.12 Consider a 35 mm × 35 mm × 2 mm package whose case is exposed to laminar forced convection air flow. Plot the junction-to-air thermal resistance of this package versus air velocity if its junction-to-case, junction-to-board, and board-to-air thermal resistances are 1.5°C/W, 12°C/W, and 23°C/W, respectively.

6.13 Consider a package with a square case that is exposed to laminar forced convection air flow with a velocity of 2 m/s. Plot junction-to-air thermal resistance of this package versus its length if its junction-to-case, junction-to-board, and board-to-air thermal resistances are 1.5°C/W, 12°C/W, and 23°C/W, respectively. Note that the junction-to-case thermal resistance of this package will also change with its size. However, this change was not considered here.

6.14 Consider an electronic package in which a 12 mm × 12 mm × 0.8 mm silicon die is attached to a 35 mm × 35 mm × 1 mm copper cap through a 0.25 mm thick thermal interface material (TIM) as shown below. This package is mounted on a PCB and air with a laminar velocity of $U_\infty = 2$ m/s is flowing over the package. If thermal conductivity of silicon, TIM, and copper are 125 W/m°C, 10 W/m°C, and 390 W/m°C, respectively, calculate junction-to-case and case-to-air thermal resistances of this package.

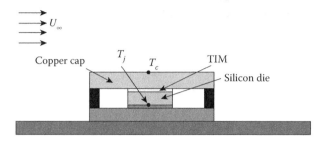

6.15 Consider a 50 mm × 50 mm × 2 mm electronic device with junction-to-case thermal resistance of 5°C/W and junction-to-board thermal resistance of 15°C/W. This device is mounted on a 50 mm × 50 mm printed circuit board (PCB) with thickness of 2 mm. The PCB is made of glass epoxy with thermal conductivity of 0.35 W/m°C. However, 20 copper fillings with a diameter of 0.5 mm are added to the PCB to increase its normal thermal conductivity. Convection heat transfer coefficient over the top of the device and the bottom of the PCB are 25 W/m²°C and 5 W/m²°C, respectively, and thermal conductivity of copper is 390 W/m°C.
   a. What is the case-to-air thermal resistance of this device?
   b. What is the normal thermal resistance of the PCB?
   c. What is the thermal resistance between the bottom of the PCB and air?
   d. Sketch the total resistance network between junction of this device and air and obtain junction-to-air thermal resistance of this device.

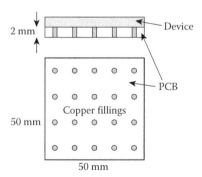

REFERENCES

1. JEDEC Solid State Technology Association. http://www.jedec.org/.
2. EIA/JEDEC Standard. 1995, December. *Integrated circuits thermal test method environment conditions: Natural convection (still air)*. Electronic Industries Association (EIA)/JESD51-2.

3. EIA/JEDEC Standard. 1999, March. *Integrated circuits thermal test method environmental conditions: Forced convection (moving air).* Electronic Industries Association (EIA)/JESD51-6.
4. JEDEC Standard. 1999, October. *Integrated circuits thermal test method environment conditions: Junction-to-board.* JESD51-8.
5. Intel. http://www.intel.com/.
6. Xilinx. http://www.xilinx.com/.

7 Fins and Heat Sinks

Consider the surface area A shown in Figure 7.1a. This can be the top surface of an electronic package, the surface of a heat spreader that is attached to multiple electronic components of a board, or the external surface of an electronic system such as a handheld computer or a pole-mount wireless radio unit. The total heat transfer rate from this surface is equal to the sum of convection and radiation heat transfer rates given by Equations 4.4 and 4.18 in Chapter 4:

$$\dot{Q} = \dot{Q}_{conv} + \dot{Q}_{rad} = h_{conv} A(T_s - T_\infty) + h'_{rad} A(T_s - T_\infty), \qquad (7.1)$$

or

$$\dot{Q} = h_{comb} A(T_s - T_\infty), \qquad (7.2)$$

where

$$h_{comb} = h_{conv} + h'_{rad} \qquad (7.3)$$

is called the combined heat transfer coefficient and includes the effects of both convection and radiation heat transfers.

In most electronic cooling applications the objective is to increase heat transfer rate for a specific surface temperature or to reduce surface temperature for a given heat dissipation. Equation 7.2 shows that this is possible by either increasing convection and/or radiation heat transfer coefficients or increasing surface area. Fins and heat sinks primarily do the latter, but can also affect the local heat transfer coefficient. They are used to increase the heat transfer rate by increasing the heat transfer area. Figures 7.1b and 7.1c shows two examples of common fin geometries, straight or plate fin, and pin or cylindrical fin.

7.1 FIN EQUATION

Consider a typical fin with length L, cross section area A_c, cross section perimeter P, and thermal conductivity k as shown in Figure 7.2. Let's assume $h = h_{comb}$ is the combined radiation and convection heat transfer coefficient from the fin surface, T_∞ is ambient temperature, and T is fin temperature. Both h and T_∞ are assumed to be uniform over the entire fin surface but T, A_c, P, and k may vary in the x direction only. Heat enters the fin at its base where $x = 0$ and transfers through it by conduction. Part of the heat that is conducted through the fin is transferred out of the fin by convection and radiation from its surface as shown in Figure 7.3. Conduction heat transfer through the fin implies that the fin temperature drops continuously from $x = 0$ to $x = L$. A small element of this fin at location x, and with thickness Δx, is shown in Figure 7.2.

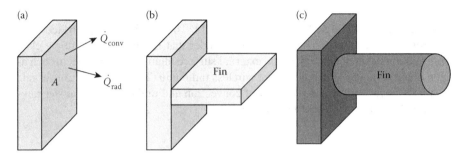

FIGURE 7.1 A surface with no fin (a), with straight fin (b) and with pin fin (c).

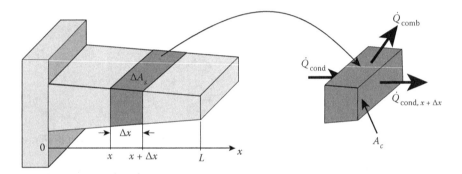

FIGURE 7.2 Schematic of a typical fin showing heat transfer rates in and out of a small element of the fin.

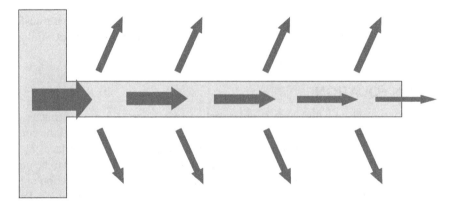

FIGURE 7.3 Heat flow path is a typical fin.

Fins and Heat Sinks

If we assume steady-state condition, the energy balance equation for this element can be written as

$$\dot{E}_{in} = \dot{E}_{out}. \quad (7.4)$$

If there is no heat generation in the fin and heat conduction through the fin is one-dimensional, energy will enter this element at location x by conduction heat transfer, $\dot{Q}_{cond,x}$, and will leave it at location $x + \Delta x$ by conduction heat transfer, $\dot{Q}_{cond,x+\Delta x}$, and from its exposed surface by combined convection and radiation heat transfer, $\dot{Q}_{comb}$. Therefore, the energy balance equation becomes

$$\dot{Q}_{cond,x} = \dot{Q}_{cond,x+\Delta x} + \dot{Q}_{comb}. \quad (7.5)$$

If A is the total exposed area of this fin and ΔA is the exposed area of the small element with length Δx, $\dot{Q}_{comb} = h\Delta A(T - T_\infty)$ and the energy balance equation becomes

$$\dot{Q}_{cond,x+\Delta x} - \dot{Q}_{cond,x} + h\Delta A(T - T_\infty) = 0. \quad (7.6)$$

Dividing this equation by Δx gives

$$\frac{\dot{Q}_{cond,x+\Delta x} - \dot{Q}_{cond,x}}{\Delta x} + h\frac{\Delta A}{\Delta x}(T - T_\infty) = 0. \quad (7.7)$$

Taking the limit of Equation 7.7 as $\Delta x \to 0$ gives

$$\frac{d\dot{Q}_{cond}}{dx} + h\frac{dA}{dx}(T - T_\infty) = 0. \quad (7.8)$$

Since $\dot{Q}_{cond} = -kA_c \, dT/dx$,

$$\frac{d}{dx}\left(kA_c \frac{dT}{dx}\right) - h\frac{dA}{dx}(T - T_\infty) = 0. \quad (7.9)$$

Equation 7.9 is the general form of the fin equation. The only assumptions behind its derivation were steady-state condition, no heat generation in the fin, and one-dimensional heat conduction inside the fin.

Let's consider a simplified case where fin thermal conductivity, its cross section and its perimeter are constant. The constant perimeter assumption gives $A = PL$ and $dA/dx = P$, and Equation 7.9 becomes

$$\frac{d^2T}{dx^2} - \frac{hP}{kA_c}(T - T_\infty) = 0. \tag{7.10}$$

Defining a new dependent variable $\theta = T - T_\infty$ and introducing a new parameter $a = \sqrt{hP/kA_c}$ reduces Equation 7.10 to the following simpler form:

$$\frac{d^2\theta}{dx^2} - a^2\theta = 0. \tag{7.11}$$

Equation 7.11 is the fin equation for a fin with constant thermal conductivity, cross section, and perimeter. It is a linear, homogenous, second-order ordinary differential equation with constant coefficients. The general solution to this equation is

$$\theta = C_1 e^{ax} + C_2 e^{-ax}, \tag{7.12}$$

where coefficients C_1 and C_2 are determined by two boundary conditions at fin base, $x = 0$, and the fin tip, $x = L$. A common boundary condition at $x = 0$ is fixed temperature boundary condition:

$$T(x=0) = T_b \quad \text{or} \quad \theta(x=0) = \theta_b = T_b - T_\infty. \tag{7.13}$$

Here T_b is the fin base temperature and $\theta_b = T_b - T_\infty$ is the fin base temperature difference with ambient. Four possible boundary conditions at $x = L$ will be discussed in the following sections.

7.1.1 INFINITELY LONG FIN

Fin temperature drops from its value at the base, T_b, to a value between the base and ambient temperatures at the tip. If the fin is sufficiently long, the fin temperature at the tip will approach the ambient temperature as shown in Figure 7.4. Therefore, two appropriate boundary conditions for an infinitely long fin are

$$\begin{aligned} T(x=0) &= T_b \quad \text{or} \quad \theta(x=0) = \theta_b, \\ T(x=L) &= T_\infty \quad \text{or} \quad \theta(x=L) = 0 \quad \text{as} \quad L \to \infty. \end{aligned} \tag{7.14}$$

Fins and Heat Sinks

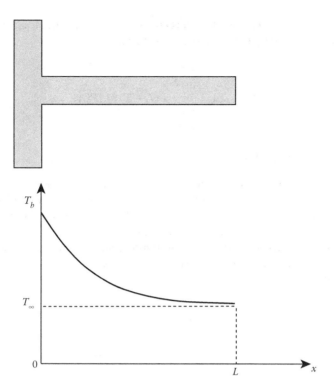

FIGURE 7.4 Temperature variation in a very long fin.

The general form of the solution to the fin equation, Equation 7.11, is

$$\theta = C_1 e^{ax} + C_2 e^{-ax} \quad \text{where} \quad a = \sqrt{\frac{hP}{kA_c}}.$$

The second boundary condition gives

$$0 = C_1 e^{aL} + C_2 e^{-aL} \quad \text{as} \quad L \to \infty.$$

This requires $C_1 = 0$. Then, the first boundary condition gives

$$\theta_b = C_2 e^{-a \times 0} \Rightarrow C_2 = \theta_b.$$

Therefore, the solution to the fin equation for **an infinitely long fin** with constant thermal conductivity, cross section, and perimeter that is exposed to an ambient with uniform heat transfer coefficient and temperature is

$$\theta = \theta_b e^{-ax} \quad \text{or} \quad T - T_\infty = (T_b - T_\infty)e^{-ax} \quad \text{where} \quad a = \sqrt{\frac{hP}{kA_c}}. \quad (7.15)$$

Total heat transfer rate from this fin is equal to conduction heat transfer rate to its base and is calculated by Fourier's law of heat conduction:

$$\dot{Q}_f = -kA_c \left.\frac{dT}{dx}\right|_{x=0} = (T_b - T_\infty)\sqrt{hPkA_c}. \tag{7.16}$$

7.1.2 Adiabatic Fin Tip

The combined convection and radiation heat transfer rates from a fin is proportional to its surface area. The surface area of the tip of a fin is usually a small fraction of the total surface area of that fin. Therefore, the heat transfer rate from the tip of a fin may be neglected compared to the heat transfer rate from the rest of the fin. The adiabatic fin tip, which assumes zero heat transfer rate from the fin tip and therefore zero temperature gradient there, represents this approximation. The boundary conditions for this case are

$$T(x=0) = T_b \quad \text{or} \quad \theta(x=0) = \theta_b,$$

$$\left.\frac{dT}{dx}\right|_{x=L} = 0 \quad \text{or} \quad \left.\frac{d\theta}{dx}\right|_{x=L} = 0. \tag{7.17}$$

Applying these boundary conditions to the general solution of the fin equation, Equation 7.12, gives

$$\theta(x=0) = \theta_b \quad \Rightarrow \quad C_1 + C_2 = \theta_b$$

$$\left.\frac{d\theta}{dx}\right|_{x=L} = 0 \quad \Rightarrow \quad C_1 e^{aL} - C_2 e^{-aL} = 0.$$

Solving these two equations for C_1 and C_2 gives

$$C_1 = \frac{\theta_b e^{-aL}}{e^{aL} + e^{-aL}} \quad \text{and} \quad C_2 = \frac{\theta_b e^{aL}}{e^{aL} + e^{-aL}},$$

and therefore the solution to the fin equation for a **fin with adiabatic fin tip** is

$$\theta = \theta_b \frac{\cosh a(L-x)}{\cosh aL} \quad \text{or} \quad T - T_\infty = (T_b - T_\infty)\frac{\cosh a(L-x)}{\cosh aL}. \tag{7.18}$$

Fins and Heat Sinks

The total heat transfer rate from this fin is

$$\dot{Q}_f = -kA_c \left. \frac{dT}{dx} \right|_{x=0} = \sqrt{hPkA_c}\,(T_b - T_\infty)\tanh aL. \tag{7.19}$$

Note that as L becomes very large, $\tanh aL \to 1$ and Equation 7.19 for the heat transfer rate from a fin with an adiabatic tip becomes the same as Equation 7.16 for the heat transfer rate from an infinitely long fin. Also note that if $aL = 1, 2,$ and 3, $\tanh aL = 0.762, 0.964,$ and 0.995, respectively, and the heat transfer rate from a fin with an adiabatic tip is 76.2%, 96.4%, and 99.5% of the heat transfer rate from an infinitely long fin with similar geometries and ambient conditions. This shows that a fin with length $L = 3/a$ is practically an infinitely long fin and any further increase in its length does not increase heat transfer rate from that fin.

7.1.3 Convection and Radiation from Fin Tip

Since the tip of a fin is exposed to ambient, heat will be transferred from it by convection and radiation heat transfer modes. Therefore, the most accurate boundary condition at the tip of a fin is the combined convection and radiation heat transfer rates. Conduction heat transfer rate to the tip of a fin is $\dot{Q}_{\text{cond,tip}} = -kA_c\,(dT/dx)_{x=L}$ and combined convection and radiation heat transfer rates from the fin tip is $\dot{Q}_{\text{conv,tip}} = hA_c\,(T(x=L) - T_\infty)$. Equating these two gives the boundary condition at the fin tip. The two boundary conditions for this fin are

$$T(x=0) = T_b \quad \text{or} \quad \theta(x=0) = \theta_b,$$

$$-k \left. \frac{dT}{dx} \right|_{x=L} = h(T(x=L) - T_\infty) \quad \text{or} \quad -k \left. \frac{d\theta}{dx} \right|_{x=L} = h\theta(x=L). \tag{7.20}$$

Applying these boundary conditions to the general solution of the fin equation, Equation 7.12, gives

$$\theta(x=0) = \theta_b \quad \Rightarrow \quad C_1 + C_2 = \theta_b,$$

$$-k \left. \frac{d\theta}{dx} \right|_{x=L} = h\theta(x=L) \quad \Rightarrow \quad C_1 a e^{aL} - C_2 a e^{-aL} = -\frac{h}{k}\left(C_1 e^{aL} + C_2 e^{-aL}\right).$$

Solving these two equations for C_1 and C_2 gives

$$C_1 = \frac{\theta_b \left[e^{-aL} - \dfrac{h}{ka} e^{-aL} \right]}{(e^{aL} + e^{-aL}) + \dfrac{h}{ka}(e^{aL} - e^{-aL})} \quad \text{and} \quad C_2 = \frac{\theta_b \left[e^{aL} + \dfrac{h}{ka} e^{aL} \right]}{(e^{aL} + e^{-aL}) + \dfrac{h}{ka}(e^{aL} - e^{-aL})}.$$

This results in the following solution for the fin equation for a **fin with convection and radiation from fin tip**.

$$\theta = \theta_b \frac{\cosh a(L-x) + \dfrac{h}{ak}\sinh a(L-x)}{\cosh aL + \dfrac{h}{ak}\sinh aL} \qquad (7.21)$$

or

$$T - T_\infty = (T_b - T_\infty)\frac{\cosh a(L-x) + \dfrac{h}{ak}\sinh a(L-x)}{\cosh aL + \dfrac{h}{ak}\sinh aL}. \qquad (7.22)$$

The total heat transfer rate from this fin is

$$\dot{Q}_f = -kA_c\left.\frac{dT}{dx}\right|_{x=0} = \sqrt{hPkA_c}\,(T_b - T_\infty)\frac{\sinh aL + \dfrac{h}{ak}\cosh aL}{\cosh aL + \dfrac{h}{ak}\sinh aL}. \qquad (7.23)$$

These expressions for temperature distribution and heat transfer rate are relatively cumbersome. However, it has been shown that a fin with combined convection and radiation heat transfers from its tip can be approximated by a slightly longer fin with an adiabatic fin tip as shown in Figure 7.5. The additional fin length is chosen such that the added fin surface area, due to the increase in its length, is equal to the fin tip area. The total heat transfer area of a fin with constant cross section A_c, perimeter P, and length L is $PL + A_c$. On the other hand, the heat transfer area of the longer fin with adiabatic tip is PL_c where L_c, called the corrected length, is

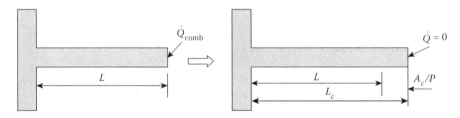

FIGURE 7.5 A fin with length L and heat transfer from its tip and its equivalent fin with length L_c and adiabatic tip.

Fins and Heat Sinks

the length of the longer fin. Equating the heat transfer surface areas of these two fins gives

$$L_c = L + \frac{A_c}{P}. \quad (7.24)$$

The corrected length of a pin or cylindrical fin with pin diameter D is $L_c = L + D/4$ and for a thin plate or rectangular fin with thickness t and width W is $L_c = L + 0.5\, tW/(t + W)$ and approaches $L_c = L + t/2$ if $W \gg t$.

The expressions for temperature distribution and total heat transfer rate for **a fin with corrected length L_c** are obtained by replacing L with L_c in the corresponding expressions derived for the fin with adiabatic tip:

$$\theta = \theta_b \frac{\cosh a(L_c - x)}{\cosh aL_c} \quad \text{or} \quad T - T_\infty = (T_b - T_\infty)\frac{\cosh a(L_c - x)}{\cosh aL_c}, \quad (7.25)$$

$$\dot{Q}_f = -kA_c \left.\frac{dT}{dx}\right|_{x=0} = \sqrt{hPkA_c}\,(T_b - T_\infty)\tanh aL_c. \quad (7.26)$$

7.1.4 Constant Temperature Fin Tip

Another possible condition at the tip of a fin is a constant temperature. This condition may happen if the fin is attached to two bases at $x = 0$ and $x = L$ at different temperatures, as shown in Figure 7.6, or if the tip of the fin is exposed to convection and radiation heat transfers with very large heat transfer coefficients. The latter can happen if the tip of the fin is exposed to a material that is going through a constant temperature phase change process such as melting, evaporation, or condensation. If T_L is the temperature at the fin tip, the boundary conditions are

$$\begin{aligned} T(x = 0) = T_b \quad &\text{or} \quad \theta(x = 0) = \theta_b, \\ T(x = L) = T_L \quad &\text{or} \quad \theta(x = L) = \theta_L, \end{aligned} \quad (7.27)$$

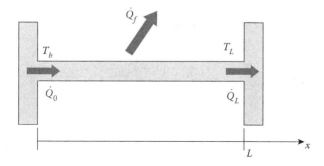

FIGURE 7.6 A fin with constant temperature at both ends.

where $\theta_b = T_b - T_\infty$ and $\theta_L = T_L - T_\infty$. Applying these boundary conditions to the general solution of the fin equation, Equation 7.12, gives two equations for constants C_1 and C_2:

$$\theta(x=0) = \theta_b \quad \Rightarrow \quad C_1 + C_2 = \theta_b$$

$$\theta(x=L) = \theta_L \quad \Rightarrow \quad C_1 e^{aL} + C_2 e^{-aL} = \theta_L.$$

The solution of these two equations gives

$$C_1 = \frac{\theta_L - \theta_b e^{-aL}}{e^{aL} - e^{-aL}} \quad \text{and} \quad C_2 = \frac{\theta_b e^{aL} - \theta_L}{e^{aL} - e^{-aL}},$$

and the solution of the fin equation for a **fin with the constant tip temperature** becomes

$$\theta = \frac{\theta_b \sinh a(L-x) + \theta_L \sinh ax}{\sinh aL} \tag{7.28}$$

or

$$T - T_\infty = \frac{(T_b - T_\infty)\sinh a(L-x) + (T_L - T_\infty)\sinh ax}{\sinh aL}. \tag{7.29}$$

The heat transfer rate to the base of this fin at $x = 0$ is

$$\dot{Q}_0 = -kA_c \left.\frac{dT}{dx}\right|_{x=0} = \sqrt{hPkA_c}\, \frac{(T_b - T_\infty)\cosh aL - (T_L - T_\infty)}{\sinh aL}, \tag{7.30}$$

and the heat transfer rate from the tip of the fin is

$$\dot{Q}_L = -kA_c \left.\frac{dT}{dx}\right|_{x=L} = \sqrt{hPkA_c}\, \frac{(T_b - T_\infty) - (T_L - T_\infty)\cosh aL}{\sinh aL}. \tag{7.31}$$

Note that $\dot{Q}_0$ and $\dot{Q}_L$ calculated by Equations 7.30 and 7.31 are positive if they are along the positive x direction. Therefore, heat transfer rates from the base at $x = 0$ to

Fins and Heat Sinks

the fin is $\dot{Q}_0$ and from the base at $x = L$ to the fin is $-\dot{Q}_L$ and total heat transfer rate from the exposed surface of the fin is

$$\dot{Q}_f = \dot{Q}_0 - \dot{Q}_L.$$

Example 7.1

Consider an aluminum rectangular fin with base area of 2 mm × 40 mm and length of 30 mm. Let's assume that the combined convection and radiation heat transfer coefficient is 30 W/m²°C, ambient temperature is 25°C, fin base temperature is 90°C, and thermal conductivity of aluminum is 210 W/m°C. Compute temperature distribution in this fin and heat transfer rate from it assuming (a) the fin is infinitely long, (b) the fin tip is adiabatic, (c) combined convection and radiation heat transfer coefficient from the fin tip is 30 W/m²°C, and (d) the temperature at the fin tip is 50°C. Use both exact and approximate methods based on the corrected fin length for part c.

Solution

The fin perimeter and cross sectional area are

$$P = 2(2\text{ mm} + 40\text{ mm}) = 2(0.002\text{ m} + 0.04\text{ m}) = 0.084\text{ m},$$

$$A_c = 2\text{ mm} \times 40\text{ mm} = 0.002\text{ m} \times 0.04\text{ m} = 8 \times 10^{-5}\text{ m}^2.$$

Some of the parameters that appear in all correlations for fin temperature and heat transfer rate are

$$a = \sqrt{\frac{hP}{kA_c}} = \sqrt{\frac{30\text{ W/m}^2\text{°C} \times 0.084\text{ m}}{210\text{ W/m°C} \times 8 \times 10^{-5}\text{ m}^2}} = 12.25\text{ m}^{-1},$$

$$aL = 12.25\text{ m}^{-1} \times 0.03\text{ m} = 0.368,$$

$$\sqrt{hPkA_c} = \sqrt{30\text{ W/m}^2\text{°C} \times 0.084\text{ m} \times 210\text{ W/m°C} \times 8 \times 10^{-5}\text{ m}^2} = 0.206\text{ W/°C}.$$

a. If we consider the fin as an infinitely long fin,

$$T - T_\infty = (T_b - T_\infty)e^{-ax} \Rightarrow T = 25 + (90 - 25)e^{-12.25x} \Rightarrow T = 25 + 65e^{-12.25x}$$

$$\dot{Q}_f = (T_b - T_\infty)\sqrt{hPkA_c} = (90 - 25)\text{°C} \times 0.206\text{ W/°C} = 13.39\text{ W}.$$

b. If the fin tip is adiabatic,

$$T - T_\infty = (T_b - T_\infty)\frac{\cosh a(L-x)}{\cosh aL} \Rightarrow T = 25 + (90-25)\frac{\cosh[12.25(0.03-x)]}{\cosh(12.25 \times 0.03)}$$

$$T = 25 + 60.84\cosh[12.25(0.03-x)]$$

$$\dot{Q}_f = (T_b - T_\infty)\sqrt{hPkA_c}\tanh aL = (90-25)°C \times 0.206 \text{ W/°C} \times \tanh(12.25 \times 0.03)$$

$$= 4.71 \text{ W}.$$

c. If the fin tip is exposed to convection and radiation with combined heat transfer coefficient of 30 W/m²°C, the exact solution gives

$$T - T_\infty = (T_b - T_\infty)\frac{\cosh a(L-x) + \dfrac{h}{ak}\sinh a(L-x)}{\cosh aL + \dfrac{h}{ak}\sinh aL}$$

$$T = 25 + (90-25)\frac{\cosh[12.25(0.03-x)] + \dfrac{30}{12.25 \times 210}\sinh[12.25(0.03-x)]}{\cosh(12.25 \times 0.03) + \dfrac{30}{12.25 \times 210}\sinh(12.25 \times 0.03)}$$

$$T = 25 + 60.6\cosh[12.25(0.03-x)] + 0.707\sinh[12.25(0.03-x)]$$

$$\dot{Q}_f = \sqrt{hPkA_c}(T_b - T_\infty)\frac{\sinh aL + \dfrac{h}{ak}\cosh aL}{\cosh aL + \dfrac{h}{ak}\sinh aL}$$

$$\dot{Q}_f = 0.206 \text{ W/°C} \times (90-25)°C \frac{\sinh(12.25 \times 0.03) + \dfrac{30}{12.25 \times 210}\cosh(12.25 \times 0.03)}{\cosh(12.25 \times 0.03) + \dfrac{30}{12.25 \times 210}\sinh(12.25 \times 0.03)}$$

$$\dot{Q}_f = 4.85 \text{ W}.$$

The approximate solution can be found by replacing

$$L_c = L + A_c/P = 0.03 + 8 \times 10^{-5}/0.084 = 0.031 \text{ m}$$

into the correlations used in part b:

$$T - T_\infty = (T_b - T_\infty)\frac{\cosh a(L_c - x)}{\cosh aL_c} \Rightarrow T = 25 + (90-25)\frac{\cosh[12.25(0.031-x)]}{\cosh(12.25 \times 0.031)}$$

$$T = 25 + 60.78\cosh[12.25(0.031-x)]$$

Fins and Heat Sinks

$$\dot{Q}_f = (T_b - T_\infty)\sqrt{hPkA_c}\tanh aL = (90-25)°C \times 0.206 \text{ W/°C} \times \tanh(12.25 \times 0.031)$$

$$= 4.85 \text{ W}.$$

This is the same as the fin heat transfer rate calculated by the exact method.

d. If the fin tip is at 50°C,

$$T - T_\infty = \frac{(T_b - T_\infty)\sinh a(L-x) + (T_L - T_\infty)\sinh ax}{\sinh aL}$$

$$T = 25 + \frac{(90-25)\sinh[12.25(0.03-x)] + (50-25)\sinh 12.25x}{\sinh(12.25 \times 0.03)}$$

$$T = 25 + 172.95\sinh[12.25(0.03-x)] + 66.52\sinh 12.25x$$

$$\dot{Q}_0 = \sqrt{hPkA_c}\frac{(T_b - T_\infty)\cosh aL - (T_L - T_\infty)}{\sinh aL}$$

$$= 0.206 \text{ W/°C}\frac{(90-25)\cosh(12.25 \times 0.03) - (50-25)}{\sinh(12.25 \times 0.03)}$$

$$\dot{Q}_0 = 24.36 \text{ W}.$$

The heat transfer rate through the tip of the fin is

$$\dot{Q}_L = \sqrt{hPkA_c}\frac{(T_b - T_\infty) - (T_L - T_\infty)\cosh aL}{\sinh aL}$$

$$= 0.206 \text{ W/°C}\frac{(90-25) - (50-25)\cosh(12.25 \times 0.03)}{\sinh(12.25 \times 0.03)}$$

$$\dot{Q}_L = 20.99 \text{ W}$$

$$\dot{Q}_f = \dot{Q}_0 - \dot{Q}_L = 24.36 \text{ W} - 20.99 \text{ W} = 3.37 \text{ W}.$$

Figure 7.7 shows temperature distribution in this fin with different boundary conditions at its tip. It is seen that the temperature distributions in the fin with an adiabatic tip and in the fin with convection from the tip (obtained by both exact and approximate methods) are the same. These methods also result in approximately the same heat transfer rate from the fin. On the other hand, the temperature drop through the length of the fin and the heat transfer rate from the fin, if we assume the fin is infinitely long, is too large. This shows that the assumption of an infinitely long fin is not an appropriate assumption for this fin. This should have been expected since $aL = 0.368$ and $\tanh aL = 0.352 < 1$. It will be shown in Example 7.2 that this assumption is physically impossible for this fin.

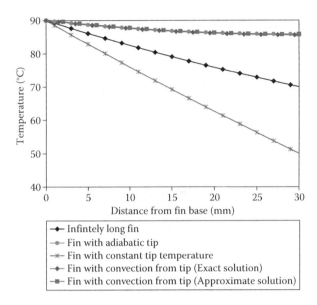

FIGURE 7.7 Temperature distribution in the fin of Example 7.1 with different conditions at its tip.

7.2 FIN THERMAL RESISTANCE, EFFECTIVENESS, AND EFFICIENCY

Fin thermal resistance is defined as the ratio of temperature difference between fin base and its ambient to the heat transfer rate from the fin

$$R_f = \frac{T_b - T_\infty}{\dot{Q}_f}. \tag{7.32}$$

Using Equations 7.16, 7.19, 7.23, and 7.26 for fin heat transfer rate, the following expressions for fin thermal resistance are obtained.

- Infinitely long fin

$$R_f = \frac{1}{\sqrt{hPkA_c}} \tag{7.33}$$

- Fin with adiabatic tip

$$R_f = \frac{1}{\sqrt{hPkA_c}\tanh aL} \tag{7.34}$$

Fins and Heat Sinks

- Fin with convection and radiation from tip

$$R_f = \frac{\cosh aL + \dfrac{h}{ak}\sinh aL}{\sqrt{hPkA_c}\left[\sinh aL + \dfrac{h}{ak}\cosh aL\right]} \quad \text{or} \quad R_f = \frac{1}{\sqrt{hPkA_c}\,\tanh aL_c} \quad (7.35)$$

Fins are used to improve heat transfer by increasing convection and radiation surface areas and therefore reducing convection and radiation thermal resistances. On the other hand, a fin creates an additional conduction thermal resistance between its base and the outside ambient. Therefore, the addition of a fin to a surface may or may not increase total heat transfer rate from that surface. A measure of fin performance is **fin effectiveness** and is defined as the ratio of heat transfer rate from the fin to heat transfer rate from the base of the fin if the fin was not there. The addition of a fin to a surface and the associated cost and weight increase can be justified only if fin effectiveness is reasonably high. Let's assume that the heat transfer rate from the fin shown in Figure 7.8 is $\dot{Q}_f$ and the heat transfer rate from the base of the same fin, A_b, if the fin is not there is $\dot{Q}_{\text{nofin}} = hA_b(T_b - T_\infty)$. Then, fin effectiveness is defined as

$$\varepsilon_f = \frac{\dot{Q}_f}{\dot{Q}_{\text{nofin}}} = \frac{\dot{Q}_f}{hA_b(T_b - T_\infty)}. \quad (7.36)$$

Equation 7.36 can be written as

$$\varepsilon_f = \frac{1/hA_b}{(T_b - T_\infty)/\dot{Q}_f}, \quad (7.37)$$

where $1/hA_b = R_b$ is the combined convection and radiation thermal resistance of the fin base if there is no fin and $(T_b - T_\infty)/\dot{Q}_f = R_f$ is the thermal resistance of the fin.

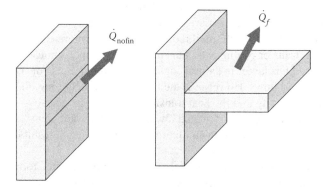

FIGURE 7.8 Heat transfer rate from a fin compared to heat transfer rate from its base if the fin was not present.

Therefore, fin effectiveness can be defined as the ratio of the fin base thermal resistance to the fin thermal resistance:

$$\varepsilon_f = \frac{R_b}{R_f}. \tag{7.38}$$

Using Equations 7.33 through 7.35 for R_f, and considering that for a fin with constant cross section $A_b = A_c$ and therefore $R_b = 1/hA_c$, the following expressions are obtained for fin effectiveness.

- Infinitely long fin

$$\varepsilon_f = \sqrt{\frac{kP}{hA_c}} \tag{7.39}$$

- Fin with adiabatic tip

$$\varepsilon_f = \sqrt{\frac{kP}{hA_c}} \tanh aL \tag{7.40}$$

- Fin with convection and radiation from its tip

$$\varepsilon_f = \sqrt{\frac{kP}{hA_c}} \frac{\sinh aL + \frac{h}{ak}\cosh aL}{\cosh aL + \frac{h}{ak}\sinh aL} \quad \text{or} \quad \varepsilon_f = \sqrt{\frac{kP}{hA_c}} \tanh aL_c \tag{7.41}$$

Several conclusions can be drawn from Equation 7.39 regarding the effectiveness of an infinitely long fin. First, fin effectiveness is proportional to thermal conductivity of fin material. That is why metals such as aluminum and copper are commonly used as fin materials. Second, the more effective fin geometry is the one with a higher perimeter to cross section area ratio. This is usually achieved by using thin, wide plate fins or pin fins with small radii. Third, a fin is more effective in mediums with lower heat transfer coefficients. This means that the percentage increase in heat transfer due to a fin is larger in natural convection than in forced convection or in gas flows than in liquid flows. However, if the fin is not infinitely long, the effects of the fin thermal conductivity and the heat transfer coefficient in a have to be considered as well. For example, for a short fin and low heat transfer coefficient (i.e. small L and h), changing the fin material from aluminum to copper may not result in a significant improvement in the fin effectiveness.

Fins are manufactured from materials such as aluminum and copper with finite thermal conductivities. Figure 7.9 shows that the temperature drop in a fin will be steeper for lower thermal conductivity fins and more gradual for higher thermal conductivity fins. If the thermal conductivity of a fin approaches infinity, the entire

Fins and Heat Sinks

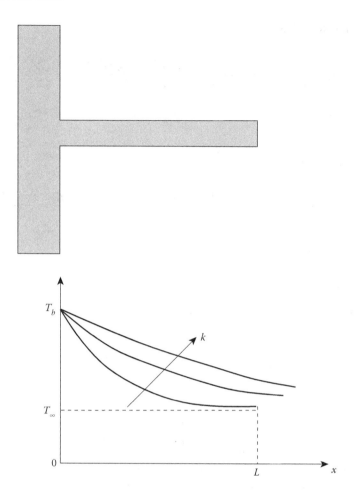

FIGURE 7.9 Fin temperature drops from its base to its tip.

fin will be at the base temperature. This will result in maximum heat transfer rate from that fin,

$$\dot{Q}_{f,\max} = hA_f(T_b - T_\infty), \tag{7.42}$$

where A_f is the fin heat transfer area. **Fin efficiency** is defined as the ratio of actual heat transfer rate to maximum heat transfer rate from a fin

$$\eta_f = \frac{\dot{Q}_f}{\dot{Q}_{f,\max}} = \frac{\dot{Q}_f}{hA_f(T_b - T_\infty)}. \tag{7.43}$$

Using Equations 7.16, 7.19, 7.23, and 7.26 for fin heat transfer rate, the following correlations are obtained for fin efficiency.

- Infinitely long fin:

$$\eta_f = \frac{\dot{Q}_f}{hA_f(T_b - T_\infty)} = \frac{(T_b - T_\infty)\sqrt{hPkA_c}}{hA_f(T_b - T_\infty)} = \frac{\sqrt{hPkA_c}}{hPL} = \frac{1}{L}\sqrt{\frac{kA_c}{hP}}$$

$$\eta_f = \frac{1}{aL} \tag{7.44}$$

- Fin with adiabatic tip:

$$\eta_f = \frac{\dot{Q}_f}{hA_f(T_b - T_\infty)} = \frac{(T_b - T_\infty)\sqrt{hPkA_c}\tanh aL}{hA_f(T_b - T_\infty)} = \frac{\sqrt{hPkA_c}\tanh aL}{hPL} = \frac{\tanh aL}{L}\sqrt{\frac{kA_c}{hP}}$$

$$\eta_f = \frac{\tanh aL}{aL}$$

$$\tag{7.45}$$

- Fin with convection and radiation from its tip:

$$\eta_f = \frac{1}{aL_c}\frac{\sinh aL + \dfrac{h}{ak}\cosh aL}{\cosh aL + \dfrac{h}{ak}\sinh aL} \quad \text{or} \quad \eta_f = \frac{\tanh aL_c}{aL_c} \tag{7.46}$$

Note that for this case the fin tip contributes to heat transfer rate and therefore $A_f = PL + A_c = PL_c$ and that is why aL_c appears in the denominators of both exact and approximate expressions for fin efficiency.

The expressions that define fin thermal resistance, fin effectiveness, and fin efficiency give three equations to calculate heat transfer rate from a fin:

$$\dot{Q}_f = \frac{T_b - T_\infty}{R_f}, \tag{7.47}$$

$$\dot{Q}_f = \varepsilon_f hA_b(T_b - T_\infty), \tag{7.48}$$

$$\dot{Q}_f = \eta_f hA_f(T_b - T_\infty). \tag{7.49}$$

These equations show the following relationship between fin thermal resistance, effectiveness, and efficiency:

$$R_f = \frac{1}{\varepsilon_f hA_b} = \frac{1}{\eta_f hA_f} \quad \text{and} \quad \varepsilon_f = \frac{A_f}{A_b}\eta_f. \tag{7.50}$$

Fins and Heat Sinks

Example 7.2

Calculate efficiency, effectiveness, and thermal resistance of the fin of Example 7.1 assuming (a) infinitely long fin, (b) adiabatic fin tip, and (c) convection and radiation from the fin tip.

Solution

The fin base area, A_b, and aL for this fin are

$$A_b = 0.002\,\text{m} \times 0.04\,\text{m} = 8 \times 10^{-5}\,\text{m}^2$$

$$a = \sqrt{\frac{hP}{kA_c}} = \sqrt{\frac{30\,\text{W/m}^2{}^\circ\text{C} \times 0.084\,\text{m}}{210\,\text{W/m}{}^\circ\text{C} \times 8 \times 10^{-5}\,\text{m}^2}} = 12.25\,m^{-1}$$

$$aL = 12.25\,m^{-1} \times 0.03\,\text{m} = 0.368.$$

The three assumptions for the condition at the tip of the fin give the following results.

a. If the fin is infinitely long,

$$A_f = PL = 2(0.002 + 0.04)\,\text{m} \times 0.03\,\text{m} = 0.00252\,\text{m}^2$$

$$\eta_f = \frac{1}{aL} = \frac{1}{0.368} = 2.717$$

$$\varepsilon_f = \frac{A_f}{A_b}\eta_f = \frac{0.00252\,\text{m}^2}{8 \times 10^{-5}\,\text{m}^2} \times 2.717 = 85.6$$

$$R_f = \frac{1}{\eta_f h A_f} = \frac{1}{2.717 \times 30\,\text{W/m}^2{}^\circ\text{C} \times 0.00252\,\text{m}^2} = 4.87\,{}^\circ\text{C/W}.$$

Note that the calculated efficiency is larger than one that is physically impossible. This shows that this fin does not qualify as an infinitely long fin and can not be approximated as such.

b. If the fin tip is adiabatic, it does not contribute to the heat transfer rate from the fin and therefore the tip area does not count as a fin surface area.

$$A_f = PL = 2(0.002 + 0.04)\,\text{m} \times 0.03\,\text{m} = 0.00252\,\text{m}^2$$

$$\eta_f = \frac{\tanh aL}{aL} = \frac{\tanh(0.368)}{0.368} = 0.957$$

$$\varepsilon_f = \frac{A_f}{A_b}\eta_f = \frac{0.00252\,\text{m}^2}{8 \times 10^{-5}\,\text{m}^2} \times 0.957 = 30.1$$

$$R_f = \frac{1}{\eta_f h A_f} = \frac{1}{0.957 \times 30\,\text{W/m}^2{}^\circ\text{C} \times 0.00252\,\text{m}^2} = 13.82\,{}^\circ\text{C/W}.$$

c. If the fin tip transfers heat through convection and radiation, similar to the rest of the fin, the fin tip area accounts as the fin surface area, and therefore

$$A_f = PL + A_c = 2(0.002 + 0.04)\,\text{m} \times 0.03\,\text{m} + 0.002\,\text{m} \times 0.04\,\text{m} = 0.0026\,\text{m}^2.$$

The exact method gives

$$L_c = L + \frac{A_c}{P} = 0.03 + 8 \times 10^{-5} / 0.084 = 0.031\,\text{m}$$

$$aL_c = 12.25\,\text{m}^{-1} \times 0.031\,\text{m} = 0.38$$

$$\eta_f = \frac{1}{aL_c} \frac{\sinh aL + \dfrac{h}{ak}\cosh aL}{\cosh aL + \dfrac{h}{ak}\sinh aL} = \frac{1}{0.38} \frac{\sinh(0.368) + \dfrac{30}{12.25 \times 210}\cosh(0.368)}{\cosh(0.368) + \dfrac{30}{12.25 \times 210}\sinh(0.368)} = 0.954$$

$$\varepsilon_f = \frac{A_f}{A_b}\eta_f = \frac{0.0026\,\text{m}^2}{8 \times 10^{-5}\,\text{m}^2} \times 0.954 = 31.0$$

$$R_f = \frac{1}{\eta_f h A_f} = \frac{1}{0.954 \times 30\,\text{W/m}^2{}^\circ\text{C} \times 0.0026\,\text{m}^2} = 13.44\,^\circ\text{C/W},$$

and the approximate solution gives

$$\eta_f = \frac{\tanh aL_c}{aL_c} = \frac{\tanh(12.25 \times 0.031)}{12.25 \times 0.031} = 0.954$$

$$\varepsilon_f = \frac{A_f}{A_b}\eta_f = \frac{0.0026\,\text{m}^2}{8 \times 10^{-5}\,\text{m}^2} \times 0.954 = 31.0$$

$$R_f = \frac{1}{\eta_f h A_f} = \frac{1}{0.954 \times 30\,\text{W/m}^2{}^\circ\text{C} \times 0.0026\,\text{m}^2} = 13.44\,^\circ\text{C/W}.$$

Note that both exact and approximate methods predict the same results.

7.3 FINS WITH VARIABLE CROSS SECTIONS

The correlations given in Section 7.1 and 7.2 are valid for fins with constant cross sections. However, most manufacturing techniques produce fins with variable cross sections. The equation for a fin with variable cross section and constant thermal conductivity is

$$\frac{d}{dx}\left(A_c \frac{d\theta}{dx}\right) - \frac{hP}{k}\theta = 0, \qquad (7.51)$$

where $\theta = T - T_b$ and A_c and P are functions of x.

Fins and Heat Sinks

Two common examples of fins with nonuniform cross sections are triangular and circular fins shown in Figure 7.10. The solutions of Equation 7.51 can be used to show that **efficiency of a triangular fin** is

$$\eta_f = \frac{1}{aL}\frac{I_1(2aL)}{I_0(2aL)} \quad \text{with} \quad a = \sqrt{\frac{2h}{kt}}, \tag{7.52}$$

where t is the fin thickness at the base, and the **efficiency of a circular fin** is

$$\eta_f = \frac{2r_i}{a(r_{oc}^2 - r_i^2)}\frac{K_1(ar_i)I_1(ar_{oc}) - I_1(ar_i)K_1(ar_{oc})}{K_0(ar_i)I_1(ar_{oc}) + I_0(ar_i)K_1(ar_{oc})} \quad \text{with} \quad a = \sqrt{\frac{2h}{kt}}, \tag{7.53}$$

where t is the fin thickness and $r_{oc} = r_o + t/2$ is the correct outer radius of the fin [1,2]. Here I_0 and K_0 are modified, zero-order Bessel functions of the first and second kinds, respectively, and I_1 and K_1 are modified, first-order Bessel functions of the first and second kinds, respectively. Note that the heat transfer surface area of a triangular fin is $A_f = 2W[L^2 + t^2/4]$ and for a circular fin is $A_f = 2\pi(r_{oc}^2 - r_i^2)$.

Figure 7.11 shows efficiency of triangular fin and the efficiency of rectangular fin with the same base thickness as the triangular fin. Note that the efficiency curve for the rectangular fin is representative of fin efficiency for any fin with constant cross section. It must be noted that the triangular fin uses half as much material as the rectangular fin with the same base thickness and same aL_c. Therefore, although the efficiency curve for the triangular fin is below the one for the rectangular fin, it is easily shown that the triangular fin is more efficient than the rectangular fin for the

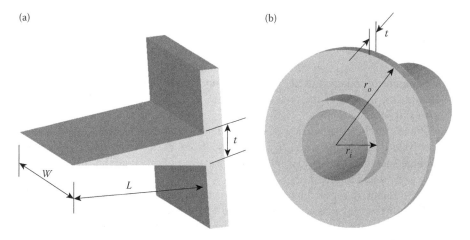

FIGURE 7.10 Two examples of fins with nonuniform cross sections; (a) triangular fin. (b) circular fin.

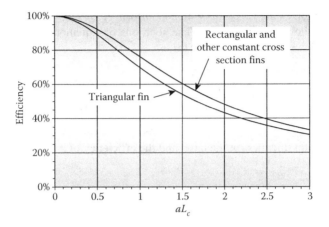

FIGURE 7.11 Efficiency of triangular fins and fins with constant cross section.

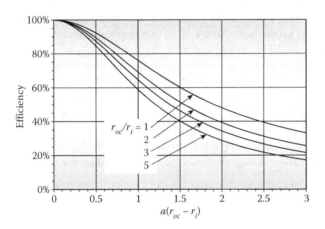

FIGURE 7.12 Efficiency of circular fins.

same amount of material used. This is an important observation as the use of the triangular fin instead of the rectangular fin results in better thermal performance while using less material.

Figure 7.12 shows efficiency of circular fins with various outer to inner radius ratios. Note that the case for $r_{oc}/r_i = 1$ represents the case where both r_{oc} and r_i are very large such that the circular fin approaches a straight rectangular fin whose efficiency is shown in Figure 7.12.

The correlations for circular fin can be used to obtain a more accurate model for a PCB with thermal vias, compared to the one-dimensional model introduced in Chapter 5, Section 5.8. Consider a PCB with area A, thickness t, and N thermal vias as shown in Figure 7.13. If k_n is normal thermal conductivity of this PCB with no

Fins and Heat Sinks

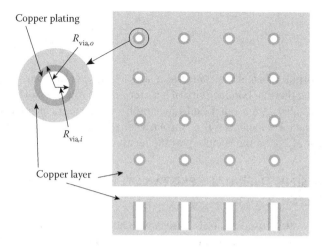

FIGURE 7.13 A printed circuit board with thermal vias.

thermal vias, h_{top} and h_{bot} are heat transfer coefficients on the top and bottom of PCB, $t_{c,top}$ and $t_{c,bot}$ are thickness of copper layers on the top and bottom of PCB, normal thermal resistance, and thermal conductivity of PCB with thermal vias, $R_{n,\text{with-via}}$ and $k_{n,\text{with-via}}$, are calculated as follows

$$\alpha = r_{\text{via},o} \qquad \beta = \sqrt{\frac{A}{N\pi}} \tag{7.54}$$

$$a_{top} = \sqrt{\frac{h_{top}}{k_c t_{c,top}}} \qquad a_{bot} = \sqrt{\frac{h_{bot}}{k_c t_{c,bot}}} \tag{7.55}$$

$$R_{\text{fin,top}} = \frac{1}{2\pi\alpha k_c t_{c,top} a_{top}} \left(\frac{K_0(a_{top}\alpha)I_1(a_{top}\beta) + I_0(a_{top}\alpha)K_1(a_{top}\beta)}{K_1(a_{top}\alpha)I_1(a_{top}\beta) - I_1(a_{top}\alpha)K_1(a_{top}\beta)} \right) - \frac{N}{Ah_{top}} \tag{7.56}$$

$$R_{\text{fin,bot}} = \frac{1}{2\pi\alpha k_c t_{c,bot} a_{bot}} \left(\frac{K_0(a_{bot}\alpha)I_1(a_{bot}\beta) + I_0(a_{bot}\alpha)K_1(a_{bot}\beta)}{K_1(a_{bot}\alpha)I_1(a_{bot}\beta) - I_1(a_{bot}\alpha)K_1(a_{bot}\beta)} \right) - \frac{N}{Ah_{bot}} \tag{7.57}$$

$$R_{\text{via}} = \frac{t}{k_c \pi (r_{\text{via},o}^2 - r_{\text{via},i}^2) + k_f \pi r_{\text{via},i}^2} \tag{7.58}$$

$$\frac{1}{R_{n,\text{with-via}}} = \frac{N}{R_{\text{fin,top}} + R_{\text{via}} + R_{\text{fin,bot}}} + \frac{k_n(A - \pi N r_{\text{via},o}^2)}{t} \tag{7.59}$$

$$k_{n,\text{with-via}} = \frac{t}{AR_{n,\text{with-via}}}. \qquad (7.60)$$

The R_{via} is normal thermal resistance of a single via, k_c is thermal conductivity of copper, k_f is the thermal conductivity of a material such as air or solder that fills the via, and $r_{\text{via},o}$ and $r_{\text{via},i}$ are outer and inner radii of the via, respectively. Note that the effect of the term including k_f is normally small and this term can be neglected without major effect on the accuracy of the results.

7.4 HEAT SINK THERMAL RESISTANCE, EFFECTIVENESS, AND EFFICIENCY

A single fin is rarely used by itself to dissipate heat from a heat source. Instead, a series of fins mounted on a common base called a **heat sink** is used. Figure 7.14 shows examples of heat sinks used for dissipating heat generated by electronic components.

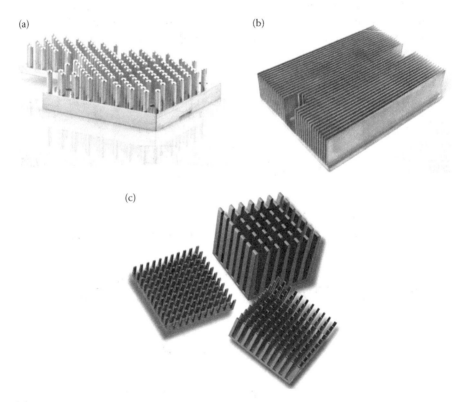

FIGURE 7.14 (See color insert following page 240.) Examples of (a) cylindrical pin fin heat sink. (Courtesy of Cool Innovations.) (b) rectangular plate fin heat sink. (Courtesy of Auras Technology Co. Ltd.) (c) rectangular pin fin heat sink. (Courtesy of Aavid Thermalloy.)

Fins and Heat Sinks

Consider the schematic of a straight fin heat sink shown in Figure 7.15. Let's assume that the heat sink base temperature, on the side where the fins are mounted, is T_b and the ambient temperature is T_∞. If the total heat transfer rate through the heat sink is $\dot{Q}_{hs}$, the heat sink convection-radiation thermal resistance is defined as

$$R_{hs,cr} = \frac{T_b - T_\infty}{\dot{Q}_{hs}}. \tag{7.61}$$

Heat is transferred from the heat sink base through a parallel path that consists of multiple fins as well as the unfinned surfaces of the base. Therefore, the convection-radiation thermal resistance of the heat sink is equal to the parallel equivalent of thermal resistances of all the fins and the thermal resistance of the unfinned area of the heat sink base:

$$\frac{1}{R_{hs,cr}} = \frac{N_f}{R_f} + \frac{1}{R_{\text{unfinned}}}. \tag{7.62}$$

Here, N_f is the number of fins, $R_f = 1/\eta_f h A_f$, $R_{\text{unfinned}} = 1/h_{\text{unfinned}}$ where A_{unfinned} and h_{unfinned} are the surface area and the combined heat transfer coefficient of the base that is not occupied by fins. If $h = h_{\text{unfinned}}$ and the heat sink is made of N_f fins, the unfinned area will be $A_{hsb} - N_f A_b$ where A_{hsb} is the surface area of the heat sink base. Therefore, the **heat sink convection-radiation thermal resistance** is

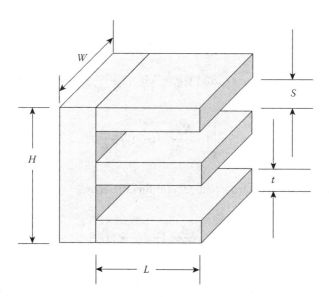

FIGURE 7.15 Schematic of a straight fin heat sink.

$$R_{hs,cr} = \frac{1}{N_f \eta_f h A_f + (A_{hsb} - N_f A_b)h} = \frac{1}{[N_f \eta_f A_f + (A_{hsb} - N_f A_b)]h}. \quad (7.63)$$

The total fin area for this heat sink is $N_f A_f$ and the total heat sink heat transfer area is $A_t = N_f A_f + A_{\text{unfinned}} = N_f A_f + A_{hsb} - N_f A_b$. Substituting $A_{hsb} - N_f A_b = A_t - N_f A_f$ in the above equation gives

$$R_{hs,cr} = \frac{1}{N_f \eta_f h A_f + (A_t - N_f A_f)h} = \frac{1}{[A_t - N_f A_f(1 - \eta_f)]h}$$

$$= \frac{1}{\left[1 - \dfrac{N_f A_f}{A_t}(1 - \eta_f)\right] A_t h}. \quad (7.64)$$

The total heat transfer rate from heat sink is

$$\dot{Q}_{hs} = \frac{T_b - T_\infty}{R_{hs,cr}} = \left[1 - \frac{N_f A_f}{A_t}(1 - \eta_f)\right] A_t h (T_b - T_\infty). \quad (7.65)$$

Fin effectiveness was defined as the ratio of heat transfer rate from a fin to the heat transfer rate from the base of the fin if there was no fin. Similarly, **heat sink effectiveness** is defined as the ratio of heat transfer rate from a heat sink to the heat transfer rate from the base of the heat sink if there were no fins. If $\dot{Q}_{hsb}$ is the heat transfer rate from the heat sink base if there were no fins, heat sink effectiveness is defined as

$$\varepsilon_t = \frac{\dot{Q}_{hs}}{\dot{Q}_{hsb}}. \quad (7.66)$$

Using Equation 7.65 for $\dot{Q}_{hs}$ and $\dot{Q}_{hsb} = h A_{hsb}(T_b - T_\infty)$, heat sink effectiveness is given by

$$\varepsilon_t = \left[1 - \frac{N_f A_f}{A_t}(1 - \eta_f)\right] \frac{A_t}{A_{hsb}} \quad (7.67)$$

or, after substituting from Equation 7.50 for fin efficiency,

$$\varepsilon_t = \left[1 - \frac{N_f A_f}{A_t}\left(1 - \varepsilon_f \frac{A_b}{A_f}\right)\right] \frac{A_t}{A_{hsb}}. \quad (7.68)$$

Maximum heat transfer rate from a heat sink is obtained if the entire heat sink is at the base temperature;

Fins and Heat Sinks

$$\dot{Q}_{hs,\max} = A_t h(T_b - T_\infty) \tag{7.69}$$

Heat sink efficiency is defined as the ratio of actual heat transfer rate from a heat sink to the maximum possible heat transfer rate from the same heat sink:

$$\eta_t = \frac{\dot{Q}_{hs}}{\dot{Q}_{hs,\max}} \tag{7.70}$$

Using Equation 7.69 for $\dot{Q}_{hs,\max}$ and Equation 7.65 for $\dot{Q}_{hs}$ it can be shown that

$$\eta_t = \left[1 - \frac{N_f A_f}{A_t}(1 - \eta_f)\right]. \tag{7.71}$$

This is an important correlation that relates heat sink efficiency to fin efficiency, fin area, number of fins, and total heat sink area.

It can easily be shown that heat sink thermal resistance, effectiveness and efficiency are related by the following correlations:

$$R_{hs,cr} = \frac{1}{\eta_t h A_t} = \frac{1}{\varepsilon_t h A_{hsb}} \quad \text{and} \quad \varepsilon_t = \frac{A_t}{A_{hsb}} \eta_t. \tag{7.72}$$

The convection-radiation thermal resistance of a heat sink is the thermal resistance between the surface of the heat sink where the fins are mounted and the ambient around the heat sink. However, a heat sink is attached to a heat source such as an electronic component on the other side of the base. Since the heat sink base has a nonzero thickness, there is an additional conduction thermal resistance between the heat source side and the fin side of that base. If the heat sink base thermal conductivity is k, its base thickness is t_b and its width and length are W and H, the heat sink base conduction thermal resistance is

$$R_{hs,b} = \frac{t_b}{kWH}. \tag{7.73}$$

Another contribution to thermal resistance of a heat sink is the spreading resistance. In most practical situations, the heat source area is smaller than the heat sink base area. This will introduce a spreading thermal resistance, $R_{hs,s}$, that can be calculated using equations given in Chapter 5, Section 5.7 and an effective heat transfer coefficient calculated by

$$h_{\text{eff}} = \frac{1}{R_{hs,cr} A_{hsb}}. \tag{7.74}$$

The last contribution to heat sink thermal resistance comes from the fact that one of the assumptions made in the fin analysis may not be valid for a heat sink. It has been assumed that the local ambient temperature, T_∞, is constant around a fin. This is a reasonable assumption for a single fin. However, if multiple fins are closely packed together to create a heat sink, the fluid temperature will increase in the flow direction as it absorbs energy. This will introduce an additional thermal resistance called caloric thermal resistance. If $\dot{m}$ is the fluid mass flow rate through the heat sink and C_p is the fluid specific heat at constant pressure, the heat sink caloric thermal resistance is given by

$$R_{hs,c} = \frac{1}{2\dot{m}C_p}. \tag{7.75}$$

It is seen that the caloric thermal resistance is larger for a small mass flow rate or fluids with low specific heats as the fluid temperature rise can be significant in these cases. Note that if T_∞ is the fluid temperature before entering the heat sink, and h is the average heat transfer coefficient with respect to this temperature, there is no need to account for the caloric thermal resistance.

Total heat sink thermal resistance is the sum of the convection-radiation thermal resistance, base conduction thermal resistance, spreading thermal resistance, and caloric thermal resistance,

$$R_{hs} = R_{hs,cr} + R_{hs,b} + R_{hs,s} + R_{hs,c}. \tag{7.76}$$

Example 7.3

Consider an aluminum heat sink with base dimensions of 80 mm × 80 mm × 5 mm and 20 aluminum fins that are 2 mm thick and 40 mm long as shown in Figure 7.16. If thermal conductivity of aluminum is 210 W/m°C and heat transfer coefficient is 15 W/m²°C, compute convection-radiation thermal resistance and base conduction thermal resistance for this heat sink.

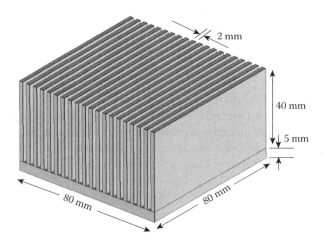

FIGURE 7.16 Plate fin heat sink of Example 7.3.

Fins and Heat Sinks

Solution

The convection-radiation thermal resistance of the heat sink is a function of the number of fins, fin efficiency, fin area, and total heat transfer area of the heat sink. The heat transfer rate from the fin tips will be taken into account and the approximate method, based on the fin corrected length, will be used to calculate fin efficiency

$$L_c = L + \frac{A_c}{P} = 0.04\,\text{m} + \frac{0.002\,\text{m} \times 0.08\,\text{m}}{2(0.002\,\text{m} + 0.08\,\text{m})} = 0.041\,\text{m}$$

$$a = \sqrt{\frac{hP}{kA_c}} = \sqrt{\frac{15\,\text{W/m}^2{}^\circ\text{C} \times 2(0.002 + 0.08)\,\text{m}}{210\,\text{W/m}^\circ\text{C} \times (0.002 \times 0.08)\,\text{m}^2}} = 8.56\,\text{m}^{-1}$$

$$aL_c = 8.56\,\text{m}^{-1} \times 0.041\,\text{m} = 0.35$$

$$\eta_f = \frac{\tanh aL_c}{aL_c} = \frac{\tanh(0.35)}{0.35} = 0.961.$$

The fin area and total heat transfer area of the heat sink are

$$A_f = PL + A_c = PL_c = 2(0.002\,\text{m} \times 0.08\,\text{m}) \times 0.041\,\text{m} = 0.0067\,\text{m}^2,$$

$$A_t = N_f A_f + (A_{hsb} - n_f A_b),$$

$$A_t = 20 \times 0.0067\,\text{m}^2 + (0.08\,\text{m} \times 0.08\,\text{m} - 20 \times 0.002\,\text{m} \times 0.08\,\text{m}) = 0.1372\,\text{m}^2.$$

The heat sink efficiency is

$$\eta_t = 1 - \frac{N_f A_f}{A_t}(1 - \eta_f) = 1 - \frac{20 \times 0.0067\,\text{m}^2}{0.1372\,\text{m}^2}(1 - 0.961) = 0.962,$$

and the heat sink convection-radiation thermal resistance is

$$R_{hs,cr} = \frac{1}{\eta_t h A_t} = \frac{1}{0.962 \times 15\,\text{W/m}^2{}^\circ\text{C} \times 0.1372\,\text{m}^2} = 0.505\,^\circ\text{C/W}.$$

The heat sink base conduction thermal resistance is

$$R_{hs,b} = \frac{t_b}{kWH} = \frac{0.005\,\text{m}}{210\,\text{W/m}^\circ\text{C} \times 0.08\,\text{m} \times 0.08\,\text{m}} = 0.004\,^\circ\text{C/W}.$$

Example 7.4

Consider a pin fin copper heat sink with base dimensions of 60 mm × 60 mm × 4 mm and 64 cylindrical fins with 3 mm diameter and 30 mm length as shown in Figure 7.17. The heat source area is 400 mm², thermal conductivity of copper is 390 W/m°C, heat transfer coefficient is 25 W/m²°C, mass flow rate of air approaching

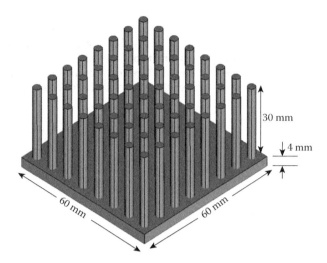

FIGURE 7.17 Pin fin heat sink of Example 7.4.

the heat sink is 0.004 kg/s and the specific heat of air is 1005 J/kg°C. Calculate total thermal resistance of this heat sink. Compare the results with those for the same heat sink made of aluminum with thermal conductivity of 210 W/m°C.

Solution

Total heat sink thermal resistance is the sum of the convection-radiation thermal resistance, base conduction thermal resistance, spreading thermal resistance, and caloric thermal resistance

$$R_{hs} = R_{hs,cr} + R_{hs,b} + R_{hs,s} + R_{hs,c}.$$

The convection-radiation thermal resistance is obtained as follows:

$$L_c = L + \frac{A_c}{P} = 0.03\,m + \frac{(\pi \times 0.003^2 / 4)\,m^2}{\pi \times 0.003\,m} = 0.03075\,m$$

$$a = \sqrt{\frac{hP}{kA_c}} = \sqrt{\frac{25\,W/m^2°C \times \pi \times 0.003\,m}{390\,W/m°C \times (\pi \times 0.003^2 / 4)\,m^2}} = 9.245\,m^{-1}$$

$$aL_c = 9.245\,m^{-1} \times 0.03075\,m = 0.284$$

$$\eta_f = \frac{\tanh aL_c}{aL_c} = \frac{\tanh(0.284)}{0.284} = 0.974.$$

The fin area and total heat transfer area of the heat sink are

$$A_f = PL + A_c = PL_c = \pi \times 0.003\,m \times 0.03075\,m = 0.00029\,m^2,$$

$$A_t = N_f A_f + (A_{hsb} - n_f A_b),$$

$$A_t = 64 \times 0.00029\,m^2 + (0.06\,m \times 0.06\,m - 64 \times \pi \times 0.003^2 / 4\,m^2) = 0.0217\,m^2.$$

Fins and Heat Sinks

The heat sink efficiency is

$$\eta_t = 1 - \frac{N_f A_f}{A_t}(1-\eta_f) = 1 - \frac{64 \times 0.00029 \, m^2}{0.0217 \, m^2}(1-0.974) = 0.978,$$

and the heat sink convection-radiation thermal resistance is

$$R_{hs,cr} = \frac{1}{\eta_t h A_t} = \frac{1}{0.978 \times 25 \, W/m^2 °C \times 0.0217 \, m^2} = 1.88 °C/W.$$

The conduction thermal resistance through the heat sink base is

$$R_{hs,b} = \frac{t_b}{kWH} = \frac{0.004 \, m}{390 \, W/m°C \times 0.06 \, m \times 0.06 \, m} = 0.0028 °C/W.$$

The spreading thermal resistance is calculated using the correlation given in Section 5.7 and the effective heat transfer coefficient:

$$h_{eff} = \frac{1}{R_{hs,cr} A_{hsb}} = \frac{1}{1.88 °C/W \times 0.06 \, m \times 0.06 \, m} = 147.8 \, W/m^2 °C.$$

$$r_1 = \sqrt{\frac{A_c}{\pi}} = \sqrt{\frac{0.0004 \, m^2}{\pi}} = 0.0113 \, m$$

$$r_2 = \sqrt{\frac{A_s}{\pi}} = \sqrt{\frac{0.06 \times 0.06 \, m^2}{\pi}} = 0.0338 \, m$$

$$\varepsilon = \frac{r_1}{r_2} = \frac{0.0113 \, m}{0.0338 \, m} = 0.334$$

$$\tau = \frac{t_b}{r_2} = \frac{0.004 \, m}{0.0338 \, m} = 0.118$$

$$Bi = \frac{hr_2}{k} = \frac{147.8 \, W/m^2 °C \times 0.0338 \, m}{390 \, W/m°C} = 0.0128$$

$$\lambda = \pi + \frac{1}{\varepsilon\sqrt{\pi}} = \pi + \frac{1}{0.334\sqrt{\pi}} = 4.831$$

$$\phi = \frac{\tanh(\lambda\tau) + \frac{\lambda}{Bi}}{1 + \frac{\lambda}{Bi}\tanh(\lambda\tau)} = \frac{\tanh(4.831 \times 0.118) + \frac{4.831}{0.0128}}{1 + \frac{4.831}{0.0128}\tanh(4.831 \times 0.118)} = 1.933$$

$$R_{hs,s} = \frac{(1-\varepsilon)\phi}{\pi k r_1} = \frac{(1-0.334) \times 1.933}{\pi \times 390 \, W/m°C \times 0.0113 \, m} = 0.093 °C/W.$$

Finally, the caloric thermal resistance is

$$R_{hs,c} = \frac{1}{2\dot{m}C_p} = \frac{1}{2\times 0.004 \text{ kg/s} \times 1005 \text{ J/kg°C}} = 0.124°C/W,$$

and total thermal resistance of this heat sink is

$$R_{hs} = R_{hs,cr} + R_{hs,b} + R_{hs,s} + R_{hs,c} = 1.88°C/W + 0.0028°C/W + 0.093°C/W$$
$$+ 0.124°C/W$$
$$R_{hs} = 2.1°C/W.$$

If the same heat sink is made of aluminum with thermal conductivity of 210 W/m°C, the heat sink convection-radiation thermal resistance, base conduction thermal resistance, and spreading thermal resistance will be different while the caloric thermal resistance remains the same. Similar calculations as above shows that, for an aluminum heat sink, $R_{hs,cr} = 1.92°C/W$, $R_{hs,b} = 0.0053°C/W$, and $R_{hs,s} = 0.173°C/W$ and therefore $R_{hs} = 2.22°C/W$.

Example 7.5

Consider a 20 mm × 20 mm electronic package whose junction-to-case, case-to-air, junction-to-board, and board-to-air thermal resistances are 2°C/W, 25°C/W, 10°C/W, and 15°C/W, respectively, and dissipates 10 W. Calculate junction temperature of this package if (a) it operates with no heat sink and (b) the copper heat sink of Example 7.4 is mounted on this package through a 0.1 mm-thick interface material with equivalent thermal conductivity of 1 W/m°C. The air temperature is 25°C for both cases.

Solution

The ambient air temperature is $T_a = 25°C$ and the power dissipation of the package is $\dot{Q} = 10W$.

a. The thermal resistance network of the package with no heat sink is shown in Figure 7.18. Junction-to-air thermal resistance of this package is calculated as shown below

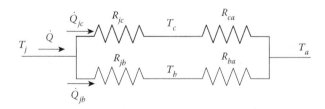

FIGURE 7.18 Resistance network for the package in Example 7.5.

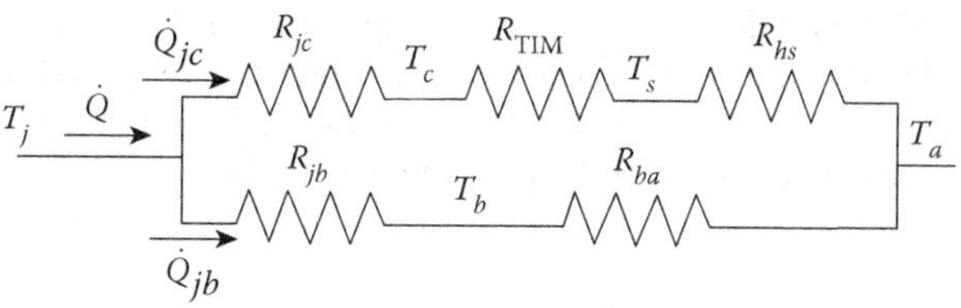

Fins and Heat Sinks

FIGURE 7.19 Resistance network for the package in Example 7.5 with a heat sink.

$$\frac{1}{R_{ja}} = \frac{1}{R_{jc}+R_{ca}} + \frac{1}{R_{jb}+R_{ba}} \Rightarrow \frac{1}{R_{ja}} = \frac{1}{2°C/W + 25°C/W} + \frac{1}{10°C/W + 15°C/W}$$

$R_{ja} = 12.98°C/W$.

The junction temperature of this package is

$$T_j = T_a + \dot{Q} \times R_{ja} \Rightarrow T_j = 25°C + 10\,W \times 12.98°C/W \Rightarrow T_j = 154.8°C.$$

It can be shown that heat transfer rate through the case and board are $\dot{Q}_{jc} = 4.81\,W$ and $\dot{Q}_{jb} = 5.19\,W$, respectively.

b. The thermal resistance network for the package with heat sink is obtained by replacing R_{ca} with the sum of the interface thermal resistance, R_{TIM}, and heat sink thermal resistance, R_{hs}, as shown in Figure 7.19.

Thermal resistance of the interface material is calculated using the correlation for conduction thermal resistance

$$R_{TIM} = \frac{t_{TIM}}{k_{TIM} A_{TIM}} = \frac{0.1 \times 10^{-3}\,m}{1\,W/m°C \times (0.02\,m \times 0.02\,m)} = 0.25°C/W.$$

Junction-to-air thermal resistance of the package with heat sink is given by

$$\frac{1}{R_{ja}} = \frac{1}{R_{jc}+R_{TIM}+R_{hs}} + \frac{1}{R_{jb}+R_{ba}},$$

$$\frac{1}{R_{ja}} = \frac{1}{2°C/W + 0.25°C/W + 2.1°C/W} + \frac{1}{10°C/W + 15°C/W},$$

$R_{ja} = 3.71°C/W$.

The junction temperature of this package is

$$T_j = T_a + \dot{Q} \cdot R_{ja} \Rightarrow T_j = 25°C + 10\,W \times 3.71°C/W \Rightarrow T_j = 62.1°C.$$

Note that significant reduction in junction temperature of the package is obtained by using the heat sink. It can be shown that heat transfer rate through the case and board are $\dot{Q}_{jc} = 8.53 \text{ W}$ and $\dot{Q}_{jb} = 1.47 \text{ W}$, respectively. It is seen that the heat transfer rate through the case of the package has significantly increased by using a heat sink.

7.5 HEAT SINK MANUFACTURING PROCESSES

Heat sinks are usually classified by their manufacturing processes. Common classifications include extrusion, die-cast, bonded fin, folded fin, swaged fin, skived, machined, and forged heat sinks. The choice of a manufacturing process for a heat sink depends on its mechanical specifications such as fin dimensions and spacing, heat sink material, production volume, and target cost. A recent review of these manufacturing techniques was given by Iyengar and Bar-Cohen [3]. Some of these processes and their respective advantages and disadvantages are discussed in this section.

Extrusion is the most commonly used process in heat sink manufacturing. This is a process by which a metal bar, usually aluminum or one of its alloys, is shaped as a heat sink by pushing it through an extrusion die. The output of the extrusion process is a long heat sink bar that is then cut to desired lengths. Some examples of heat sinks produced by extrusion are shown in Figure 7.20. Extruded heat sinks are plate-fin heat sinks and therefore are unidirectional flow heat sinks. However, multiple cuts can be made through the fins to generate cross-cut extrusion heat sinks similar to the ones shown in Figure 7.20b. These cross-cut heat sinks are multidirectional flow heat sinks. The biggest advantage of the extrusion process is its low cost. However, the extrusion process can not be used to manufacture heat sinks with very high aspect ratio (fin height to spacing ratio, L/s) and thin fins. Most extrusion heat sinks are built with aspect ratios less than 10 and fins thicker than 1 mm.

Another commonly used low cost and high volume heat sink manufacturing technique is die casting. Die casting is a manufacturing process in which molten metal is injected under high pressure into reusable metal molds or dies to form a product. An advantage of die casting compared to extrusion is that most intricate details and special features of a heat sink can be included in the die-casting process. This makes die casting the favorite choice for producing complex-shaped electronic enclosures that act as a heat sink as well. Figure 7.21 shows an example of a die-cast heat sink. Die-cast parts have lower thermal conductivity compared to similar extruded parts due to higher porosity in the die-cast metal alloy. Also, the die-cast process is limited to heat sinks with aspect ratios less than 6 and fin thicknesses of more than 2–3 mm. This is due to the fact that the molten metal alloy may not flow well into small crevices of the die to produce thinner fins.

In a bonded-fin heat sink, shown in Figure 7.22, the base is extruded with channels in which individual fins are inserted. The fins are bonded to the base by thermal epoxy, brazing, or soldering. The advantage of bonded-fin heat sink technology is that it allows the production of heat sinks with aspect ratios as high as 60 and fins as thin as 0.7–0.8 mm. There is also no limitation on materials and the fins and the base

Fins and Heat Sinks 161

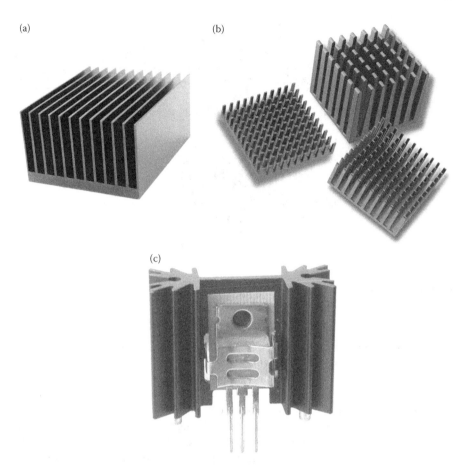

FIGURE 7.20 (**See color insert following page 240.**) Examples of extrusion heat sinks. (Courtesy of Aavid Thermalloy, LLC.)

FIGURE 7.21 Example of a die-cast heat sink.

FIGURE 7.22 A bonded-fin heat sink. (Courtesy of Aavid Thermalloy, LLC.)

can be made from different materials. The drawbacks of this process are high cost and unidirectional flow in the heat sink.

Another technology with the ability to produce high aspect ratio heat sinks is folded fin heat sink technology shown in Figure 7.23. A continuous thin sheet of fin material is cut to width and folded to a desired fin pitch. Then, the fin stock is bonded to a base with thermal epoxy, brazing, or a soldering process. Folded fin heat sinks can be built with fins as thin as 0.25 mm and aspect ratios as high as 50. Also, fin stock and base can be made from different materials. Folded fin heat sinks with a copper base and aluminum fins as well as folded fin heat sinks with a copper base and partially copper and aluminum fins are very common. Disadvantages of folded-fin heat sinks are their high cost, unidirectional flow, and fragile fins.

Swaged-fin heat sinks are bonded-fin heat sinks that use a swaging process to bond the fins to the base. The swaging process, as shown in Figure 7.24, can be described as a cold forming process and is used to fabricate high fin density heat sinks. This process involves the placement of fins with tapered bases into grooves of a base plate and application of pressure, by rolling wheels, on the opposite sides of each fin. The vertical and lateral pressures on the base unit material pushes the fin toward the bottom of the groove in the base. This secure connection provides good thermal contact between the fins and base and prevents air and moisture from entering the grooves, thereby preventing corrosion [4]. Advantages and disadvantages of swaged-fin heat sinks are similar to bonded-fin heat sinks. However, swaged-fin heat sinks have better thermal contact between the fins and the base.

Another manufacturing process for producing heat sinks with thin and dense (high aspect ratio) fins is skiving. Skiving is defined as cutting in thin layers. In this process, a reciprocating cutter shaves thin slices of fins from a metal block. This is a very high throughput and therefore low cost process that is able to generate heat

Fins and Heat Sinks

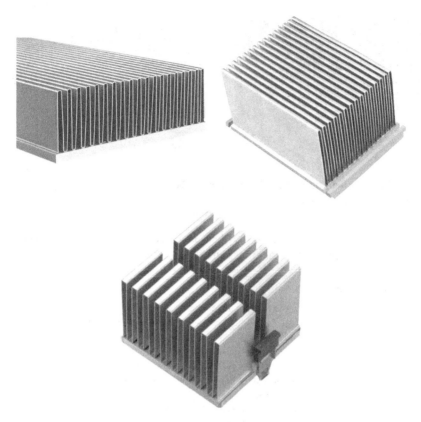

FIGURE 7.23 (**See color insert following page 240.**) Examples of folded-fin heat sinks. (Courtesy of Aavid Thermalloy, LLC.)

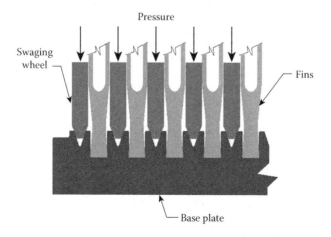

FIGURE 7.24 Schematic of the swaging process. (Courtesy of R-Theta Thermal Solutions, Inc.)

FIGURE 7.25 (See color insert following page 240.) Examples of skived fin heat sinks. (Courtesy of Auras Technology Co. Ltd.)

sinks with high aspect ratios. Skived heat sinks can be manufactured from both aluminum and copper blocks as shown in Figure 7.25.

PROBLEMS

7.1 Explain how the fins enhance heat transfer from a surface.

7.2 If the length of a fin is doubled, will the heat transfer rate from the fin double as well? Why?

7.3 Explain why very low thermal conductivity materials are not suitable for making heat sinks. Also, explain when using higher thermal conductivity materials may not significantly improve the thermal performance of heat sinks.

7.4 The junction-to-case and case-to-air thermal resistances of a package are $R_{jc} = 5°C/W$ and $R_{ca} = 18°C/W$, respectively. If this package dissipates 6 W and the air temperature is 25°C, what is the junction temperature of this package? Assume all the heat is dissipated through the case of this package.

7.5 If the maximum allowable junction temperature for the package in Problem 7.4 is 105°C, what is the total thermal resistance of a heat sink and thermal interface material that are required to cool it?

7.6 The junction temperature of a 5 W package should not exceed 105°C. If the junction-to-case thermal resistance of this package is $R_{jc} = 7°C/W$ and the air temperature around it is 55°C, determine the maximum thermal resistance of a heat sink plus thermal interface material that are required for this package. Assume all the heat is dissipated from the case of this package.

7.7 Consider the aluminum pin fin heat sink shown below. The heat sink base is 50 mm × 50 mm × 2 mm. There are 16 fins mounted on the heat sink base. The fins are 30 mm long and their diameters are 5 mm. The heat transfer coefficient around the heat sink is 20 W/m²°C. Thermal conductivity of aluminum is 237 W/m°C.
   a. What is the fin efficiency?
   b. What is the efficiency of the heat sink?
   c. What is convection-radiation thermal resistance of the heat sink?
   d. This heat sink is used to cool a 50 mm × 50 mm, 10 W chip with junction-to-case thermal resistance of 2°C/W. If the air temperature

is 25°C, what will the junction temperature of the chip be? Assume all the heat is dissipated from the case, and the thermal resistance of the interface material and caloric thermal resistance are negligible.

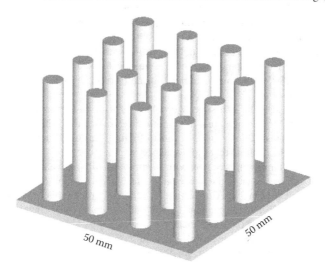

7.8 Consider two similar shape heat sinks; one made of pure aluminum with $k = 230$ W/m°C and one made of pure copper with $k = 400$ W/m°C. The heat sinks are 40 mm wide, 60 mm long and have 8 plate fins each 30 mm long and 2 mm thick. Air with a velocity of 3 m/s and temperature of 25°C is flowing between the fins of these heat sinks.
   a. Assume that airflow over the fins can be treated as laminar forced convection flow over a flat plate and find convection heat transfer coefficient.
   b. What are the aluminum and copper fin efficiencies?
   c. What are the aluminum and copper heat sinks convection thermal resistances? Neglect any radiation heat transfer rate.
   d. Considering that copper is heavier and more expensive than aluminum, is it worthwhile to choose copper heat sink rather than aluminum heat sink in this case?

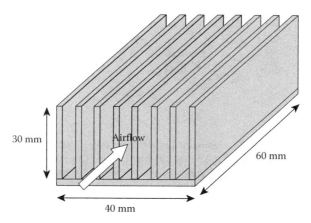

7.9 Consider an aluminum heat sink with base dimensions of 45 mm × 45 mm × 3 mm and 20 plate fins that are 1 mm thick. If thermal conductivity of aluminum is 210 W/m°C and heat transfer coefficient is 20 W/m^2°C, plot the sum of the convection-radiation and base conduction thermal resistances of this heat sink versus fin lengths of 10 mm to 100 mm and comment on the results.

7.10 Consider the heat sink of Problem 7.9 with 40 mm long fins, plot total thermal resistance of the heat sink versus source areas of 25 mm^2 to 2025 mm^2, and comment on the results. Neglect caloric thermal resistance.

7.11 Consider the heat sink of Problem 7.9 with 40 mm long fins and 25 mm^2 source area, plot total thermal resistance of the heat sink versus base plate thicknesses of 1 mm to 10 mm, and comment on the results. Neglect caloric thermal resistance.

7.12 An aluminum heat sink with 47 mm × 47 mm base has 10 triangular fins as shown in the figure below. The fins are 2 mm thick at the base and 40 mm long. Convection heat transfer around this heat sink is 30 W/m^2°C and thermal conductivity of aluminum is 170 W/m°C.
   a. What is the fin efficiency?
   b. What is the heat sink efficiency?
   c. What is the convection thermal resistance of this heat sink?

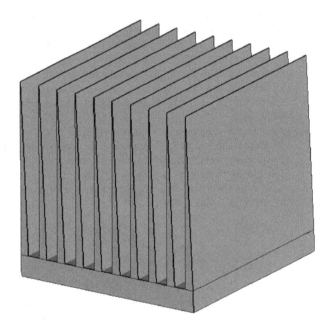

7.13 Consider the aluminum pin fin heat sink shown below. The base dimensions are 100 mm × 100 mm × 5 mm. There are 80 fins with diamond-shape cross sections that are 50 mm long. Thermal conductivity of aluminum is 170 W/m°C and convection heat transfer coefficient

Fins and Heat Sinks 167

is 30 W/m²°C. This heat sink will be used to cool a 40 mm × 40 mm package. Calculate total thermal resistance of this heat sink for this package. Neglect caloric thermal resistance.

Note: Cross section area and perimeter of a diamond with diagonals D_1 and D_2 are $A_c = D_1 D_2 / 2$ and $P = 2\sqrt{D_1^2 + D_2^2}$, respectively.

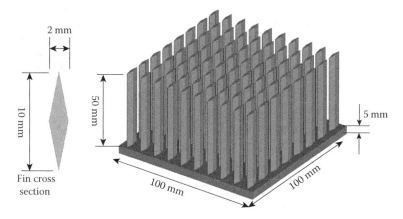

7.14 Consider the copper pin fin heat sink shown below. Its base dimensions are 100 mm × 100 mm × 5 mm and has 100 fins with square cross sections. The fins are 50 mm long and 2 mm × 2 mm cross section. Air with a velocity of 2 m/s blows into this heat sink. Thermal conductivity of copper is 390 W/m°C and the convection heat transfer coefficient is 30 W/m²°C. Calculate total thermal resistance of this heat sink if it is mounted on a 100 mm × 100 mm package.

Note: Use air properties at 25°C in all your calculations.

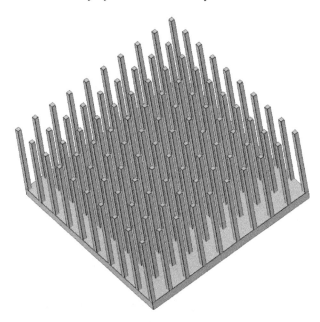

REFERENCES

1. Cengel, Y. A. 2007. *Heat and mass transfer: A practical approach.* Third Edition. New York, NY: McGraw-Hill.
2. Incropera, F. P., D. P. DeWitt, T. L. Bergman, and A. S. Lavine. 2006. *Fundamentals of heat and mass transfer.* Sixth Edition. Hoboken, NJ: John Wiley & Sons, Inc.
3. Iyengar, M., and A. Bar-Cohen. 2007. Design for manufacturability of forced convection air cooled fully ducted heat sinks. *Electronics Cooling* 13(3). This article can be seen at http://www.electronics-cooling.com/html/2007_aug_a1.php (accessed June 2009).
4. Zaghlol, A. M., W. Leonard, and R. Culham. 2003. Characterization of mixed metals swaged heat sinks for concentrated heat source. Proceedings of IPACK03, Maui, Hawaii.

8 Heat Conduction Equation

So far the Fourier's law of heat conduction, first introduced in Chapter 4, Section 4.1, has been used to calculate conduction heat transfer rate in a medium. This approach is reasonably accurate, and provides good first-order engineering approximations, in cases where heat conduction is mainly one-dimensional and there is no heat generation in the medium. Conduction heat transfer through the length of a thin fin, in the base of a heat sink that is the same size as the electronic device it is mounted on, and through the interface material inserted between a heat sink and an electronic package are examples in which this approach is reasonably accurate. However, there are other situations that either one-dimensional heat conduction assumption is not valid or there is heat generation in the medium. Conduction heat transfer in a heat spreader attached to a much smaller die, inside a foil heater with electrical joule heating, and in a printed circuit board are examples in which this simple approach will give erroneous results.

The location of a point P in a Cartesian coordinate system is specified by its three coordinates x, y, and z as shown in Figure 8.1. Temperature at that location and time t can be expressed as $T(x,y,z,t)$, and the conduction heat transfer rate in each direction is obtained using Fourier's law of conduction in that direction;

$$\dot{Q}_x = -k_x A_x \frac{\partial T}{\partial x}, \quad \dot{Q}_y = -k_y A_y \frac{\partial T}{\partial y}, \quad \dot{Q}_z = -k_z A_z \frac{\partial T}{\partial z}. \tag{8.1}$$

Several points have to be clarified regarding these equations:

1. Since temperature can be a function of three spatial coordinates and time, partial derivatives were used. Conduction heat transfer rate in any direction is proportional to temperature gradient in that direction only. Temperature gradient in one direction, say x, does not cause conduction heat transfer in other directions, say y and z.
2. Thermal conductivity of a material may be different in various directions. That is why k_x, k_y, and k_z were used to indicate thermal conductivities in x, y, and z directions, respectively. Examples of such materials are printed circuit boards and graphite. Thermal conductivity of a material may also change from one point to another and over time. An example of material with variable thermal conductivity is a gas medium. Thermal conductivity of gases change with temperature. Therefore, if the temperature of a gas

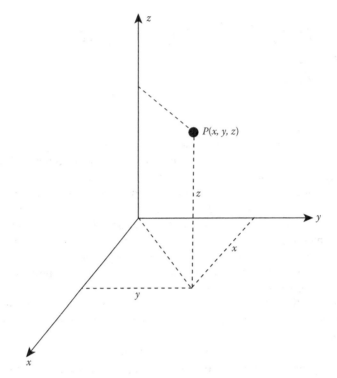

FIGURE 8.1 A typical point in a Cartesian coordinate system.

medium changes with location and time, its thermal conductivity will change correspondingly.

3. A_x, A_y, and A_z refer to surface areas normal to x, y, and z directions, respectively, over which the corresponding temperature gradients are constant.
4. Temperature, thermal conductivity, and surface area are in general functions of three space coordinates and time. Therefore, conduction heat transfer rate in each direction is a function of three space coordinates and time; $\dot{Q}_x(x,y,z,t)$, $\dot{Q}_y(x,y,z,t)$, and $\dot{Q}_z(x,y,z,t)$.
5. Since conduction heat transfer rate has three components in three space directions, it is a vector quantity as shown in Figure 8.2. If $\hat{i}$, $\hat{j}$, and $\hat{k}$ are unit vectors in x, y, and z directions, conduction heat transfer rate vector is expressed as

$$\vec{\dot{Q}}(x,y,z,t) = \dot{Q}_x(x,y,z,t)\hat{i} + \dot{Q}_y(x,y,z,t)\hat{j} + \dot{Q}_z(x,y,z,t)\hat{k}. \qquad (8.2)$$

If temperature and therefore conduction heat transfer rate are functions of time, the heat transfer is called **transient heat transfer**. Examples of transient conduction heat transfer are temperature rise in a die after it is powered on, temperature

Heat Conduction Equation

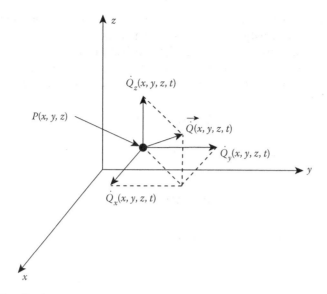

FIGURE 8.2 Conduction heat transfer vector at a typical point in a Cartesian coordinate system.

change inside a potato put in an oven, cooling a can of soda inside a refrigerator, and air temperature rise inside a closed electronic enclosure after the electronics are powered on. On the other hand, if temperature and conduction heat transfer rate do not change with time, heat transfer is called **steady-state heat transfer**. All transient cases mentioned above will stabilize after some time and become examples of steady-state heat conduction if their environment and power inputs do not change with time.

Consider a medium in which conduction heat transfer is the only heat transfer mechanism. The principle of conservation of energy can be used to derive a differential equation that describes the variation of temperature inside this medium. This equation is called **heat conduction equation** and its solution will give temperature distribution in the medium as a function of location and time. Derivation of heat conduction equation and its solutions for a few simplified cases will be given in this chapter.

8.1 ONE-DIMENSIONAL HEAT CONDUCTION EQUATION FOR A PLANE WALL

In was shown in Chapter 3, Section 3.2 that heat transfer to a constant pressure control mass of a single-phase material, for which the change in kinetic and potential energies could be neglected, was

$$\delta Q = mC_p \Delta T. \tag{8.3}$$

Most mediums in which conduction heat transfer is the only or the dominant heat transfer mechanism satisfy these conditions. Therefore, Equation 8.3 will be used to derive heat conduction equations. Since δQ is the net amount of heat transfer into the medium, $\delta Q = \delta Q_{in} - \delta Q_{out}$, where δQ_{in} and δQ_{out} are total heat transfers in and out of the medium, respectively. Therefore, Equation 8.3 can be written as

$$\delta Q_{in} - \delta Q_{out} = mC_p \Delta T. \tag{8.4}$$

If $\dot{Q}_{in}$ and $\dot{Q}_{out}$ are heat transfer rates in and out of the medium during the time interval Δt,

$$\delta Q_{in} = \dot{Q}_{in} \Delta t \quad \text{and} \quad \delta Q_{out} = \dot{Q}_{out} \Delta t, \tag{8.5}$$

and Equation 8.4 will take the rate form:

$$\dot{Q}_{in} - \dot{Q}_{out} = mC_p \frac{\Delta T}{\Delta t}. \tag{8.6}$$

This equation will be called **heat balance equation** and will be used to derive heat conduction equations. Let's remember that this equation is actually the energy balance equation for a constant pressure closed system of a single-phase material in which the changes in kinetic and potential energies are negligible.

Let's start with deriving the one-dimensional heat conduction equation for a thin plane wall. A wall can be considered a **thin wall** if its thickness is much smaller than its other dimensions. A pane of glass, a printed circuit board, a thin interface material, a thin foil heater, and a heat spreader or a metal lid on top of a package are examples of a thin wall. Consider a **thin plane wall** whose dimensions are parallel to the three Cartesian coordinates as shown in Figure 8.3. If conditions on the sides of this wall that are parallel to the yz plane are uniform, heat transfer through the wall will be one-dimensional and in the thickness, or x, direction.

Let's consider a slice of the wall shown in Figure 8.3 that goes from points x to $x + \Delta x$. Conduction heat transfer rates at the left and right sides of this slice are $\dot{Q}_x$ and $\dot{Q}_{x+\Delta x}$, respectively. In addition, energy can be transferred as heat to this slice. For example, if current I passes through a slice with electrical resistance R, the electric power $I^2 R$ will be converted into internal energy. We will consider this energy transfer as heat in the volume of this slice as a **heat generation** inside this slice. Another example is a glass window exposed to sun. The radiation energy from the sun passes through the window and is partially absorbed by it. The absorbed solar radiation can also be considered energy transfer as heat, or heat generation inside the glass. Heat generation is a volumetric phenomenon. Therefore, if heat generation per unit volume of a medium is $\dot{g}$, total heat generation inside this medium, $\dot{G}$, is given by

$$\dot{G} = \int_V \dot{g}\, dV, \tag{8.7}$$

Heat Conduction Equation

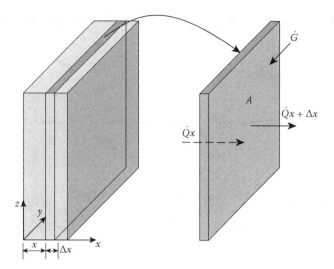

FIGURE 8.3 Schematic of a thin plane wall showing the heat flow in and out of one of its typical slices.

where $\mathcal{V}$ is the total volume of the medium. If heat generation inside the volume is constant, $\dot{G} = \dot{g}\mathcal{V}$.

The heat balance equation for the slice shown in Figure 8.3 can be written as

$$\dot{Q}_x + \dot{G} - \dot{Q}_{x+\Delta x} = mC_p \frac{\Delta T}{\Delta t}. \tag{8.8}$$

The thickness of this slice can be selected small enough such that density, ρ, and heat generation per unit volume, $\dot{g}$, are constant throughout the volume of the slice, $\mathcal{V}$. Therefore, $m = \rho \mathcal{V}$ and $\dot{G} = \dot{g}\mathcal{V}$. Since $\mathcal{V} = A\Delta x$, where A is the wall surface area parallel to yz plane, $m = \rho A \Delta x$ and $\dot{G} = \dot{g} A \Delta x$. Substituting these into Equation 8.8 results in

$$\dot{Q}_x + \dot{g} A \Delta x - \dot{Q}_{x+\Delta x} = \rho A \Delta x C_p \frac{\Delta T}{\Delta t}, \tag{8.9}$$

which can be rearranged as

$$-\frac{1}{A} \frac{\dot{Q}_{x+\Delta x} - \dot{Q}_x}{\Delta x} + \dot{g} = \rho C_p \frac{\Delta T}{\Delta t}. \tag{8.10}$$

This is the finite-difference form of the heat conduction equation for a thin plane wall. A differential equation governing heat conduction in this wall can be obtained by taking the limit of Equation 8.10 as both Δx and Δt approach zero. Note that

$$\lim_{\Delta x \to 0} \frac{\dot{Q}_{x+\Delta x} - \dot{Q}_x}{\Delta x} = \frac{\partial \dot{Q}_x}{\partial x} \quad \text{and} \quad \lim_{\Delta t \to 0} \frac{\Delta T}{\Delta t} = \frac{\partial T}{\partial t}. \tag{8.11}$$

Therefore, if both Δx and Δt approach zero, Equation 8.10 becomes

$$-\frac{1}{A}\frac{\partial \dot{Q}_x}{\partial x} + \dot{g} = \rho C_p \frac{\partial T}{\partial t}. \tag{8.12}$$

Finally, substituting Fourier's law of heat conduction for $\dot{Q}_x$ gives

$$-\frac{1}{A}\frac{\partial}{\partial x}\left(-k_x A_x \frac{\partial T}{\partial x}\right) + \dot{g} = \rho C_p \frac{\partial T}{\partial t}. \tag{8.13}$$

Note that $A_x = A$ and is constant in this one-dimensional analysis. Also, since heat conduction is in one direction only, the index x is removed from thermal conductivity and k will be used to represent thermal conductivity in the direction of heat conduction. This results in the general from of the **one-dimensional heat conduction equation in a plane wall**:

$$\frac{\partial}{\partial x}\left(k\frac{\partial T}{\partial x}\right) + \dot{g} = \rho C_p \frac{\partial T}{\partial t}. \tag{8.14}$$

Several simplified forms of the one-dimensional heat conduction equation are

1. Transient heat conduction with constant thermal conductivity:

$$\frac{\partial^2 T}{\partial x^2} + \frac{\dot{g}}{k} = \frac{\rho C_p}{k}\frac{\partial T}{\partial t} \tag{8.15}$$

2. Transient heat conduction, constant thermal conductivity, and no heat generation:

$$\frac{\partial^2 T}{\partial x^2} = \frac{\rho C_p}{k}\frac{\partial T}{\partial t} \tag{8.16}$$

3. Steady-state heat conduction:

$$\frac{d}{dx}\left(k\frac{dT}{dx}\right) + \dot{g} = 0 \tag{8.17}$$

4. Steady-state heat conduction with constant thermal conductivity:

$$\frac{d^2 T}{dx^2} + \frac{\dot{g}}{k} = 0 \tag{8.18}$$

Heat Conduction Equation

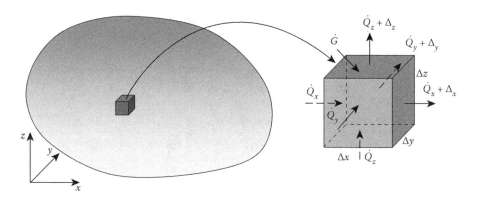

FIGURE 8.4 A general three-dimensional domain with a small cubical element.

5. Steady-state heat conduction, constant thermal conductivity, and no heat generation:

$$\frac{d^2T}{dx^2} = 0 \qquad (8.19)$$

8.2 GENERAL HEAT CONDUCTION EQUATION

Consider a medium in which conduction heat transfer can happen in three dimensions. Let's also assume that conduction is the only heat transfer mechanism in this medium. A small cubical volume of this medium with dimensions Δx, Δy, and Δz and volume $\mathcal{V} = \Delta x \Delta y \Delta z$ is shown in Figure 8.4.

Let's assume conduction heat transfer rates through the left and right sides of this volume are $\dot{Q}_x$ and $\dot{Q}_{x+\Delta x}$, through the front and rear sides are $\dot{Q}_y$ and $\dot{Q}_{y+\Delta y}$, and through the bottom and top are $\dot{Q}_z$ and $\dot{Q}_{z+\Delta z}$, and $\dot{G}$ is the total heat generation inside this volume. The heat balance equation, Equation 8.6, for this volume becomes

$$\dot{Q}_x + \dot{Q}_y + \dot{Q}_z + \dot{G} - \dot{Q}_{x+\Delta x} - \dot{Q}_{y+\Delta y} - \dot{Q}_{z+\Delta z} = mC_p \frac{\Delta T}{\Delta t}. \qquad (8.20)$$

This volume can be selected small enough such that material density and heat generation per unit volume are constant inside it. This means that

$$m = \rho \mathcal{V} = \rho \Delta x \Delta y \Delta z \quad \text{and} \quad \dot{G} = \dot{g}\mathcal{V} = \dot{g}\Delta x \Delta y \Delta z. \qquad (8.21)$$

Substituting these into Equation 8.20, dividing it by $\Delta x \Delta y \Delta z$ and rearranging gives

$$-\frac{1}{\Delta y \Delta z}\frac{\dot{Q}_{x+\Delta x} - \dot{Q}_x}{\Delta x} - \frac{1}{\Delta x \Delta z}\frac{\dot{Q}_{y+\Delta y} - \dot{Q}_y}{\Delta y} - \frac{1}{\Delta x \Delta y}\frac{\dot{Q}_{z+\Delta z} - \dot{Q}_z}{\Delta z} + \dot{g} = \rho C_p \frac{\Delta T}{\Delta t}. \qquad (8.22)$$

This is the finite-difference form of the three-dimensional heat conduction equation. The limit of this equation as Δx, Δy, Δz and Δt approach zero is

$$-\frac{1}{\Delta y \Delta z}\frac{\partial \dot{Q}_x}{\partial x} - \frac{1}{\Delta x \Delta z}\frac{\partial \dot{Q}_y}{\partial y} - \frac{1}{\Delta x \Delta y}\frac{\partial \dot{Q}_z}{\partial z} + \dot{g} = \rho C_p \frac{\partial T}{\partial t}. \quad (8.23)$$

Fourier's law of heat conduction gives

$$\dot{Q}_x = -k_x A_x \frac{\partial T}{\partial x} = -k_x \Delta y \Delta z \frac{\partial T}{\partial x},$$

$$\dot{Q}_y = -k_y A_y \frac{\partial T}{\partial y} = -k_y \Delta x \Delta z \frac{\partial T}{\partial y}, \quad (8.24)$$

$$\dot{Q}_z = -k_z A_z \frac{\partial T}{\partial z} = -k_z \Delta x \Delta y \frac{\partial T}{\partial z}.$$

The **general form of heat conduction equation** is obtained by substituting these into Equation 8.23;

$$\frac{\partial}{\partial x}\left(k_x \frac{\partial T}{\partial x}\right) + \frac{\partial}{\partial y}\left(k_y \frac{\partial T}{\partial y}\right) + \frac{\partial}{\partial z}\left(k_z \frac{\partial T}{\partial z}\right) + \dot{g} = \rho C_p \frac{\partial T}{\partial t}. \quad (8.25)$$

This is the heat conduction equation for an anisotropic medium for which thermal conductivities in three directions may be different. It is the appropriate form of heat conduction equation that is used in general purpose thermal simulation tools. However, most engineering applications involve isotropic materials and all analytical solutions presented in this chapter are for those cases. The **general form of heat conduction equation for an isotropic medium** where $k_x = k_y = k_z = k$ is

$$\frac{\partial}{\partial x}\left(k \frac{\partial T}{\partial x}\right) + \frac{\partial}{\partial y}\left(k \frac{\partial T}{\partial y}\right) + \frac{\partial}{\partial z}\left(k \frac{\partial T}{\partial z}\right) + \dot{g} = \rho C_p \frac{\partial T}{\partial t}. \quad (8.26)$$

Several simplified forms of this equation are

1. Transient heat conduction with constant thermal conductivity;

$$\frac{\partial^2 T}{\partial x^2} + \frac{\partial^2 T}{\partial y^2} + \frac{\partial^2 T}{\partial z^2} + \frac{\dot{g}}{k} = \frac{1}{\alpha}\frac{\partial T}{\partial t} \quad (8.27)$$

Here $\alpha = k/\rho C_p$ is called thermal diffusivity of material.

2. Transient heat conduction, constant thermal conductivity, and no heat generation;

$$\frac{\partial^2 T}{\partial x^2} + \frac{\partial^2 T}{\partial y^2} + \frac{\partial^2 T}{\partial z^2} = \frac{1}{\alpha}\frac{\partial T}{\partial t} \quad (8.28)$$

Heat Conduction Equation

This form of the equation is called the **diffusion equation**.

3. Steady-state heat conduction;

$$\frac{\partial}{\partial x}\left(k\frac{\partial T}{\partial x}\right)+\frac{\partial}{\partial y}\left(k\frac{\partial T}{\partial y}\right)+\frac{\partial}{\partial z}\left(k\frac{\partial T}{\partial z}\right)+\dot{g}=0 \qquad (8.29)$$

4. Steady-state heat conduction with constant thermal conductivity:

$$\frac{\partial^2 T}{\partial x^2}+\frac{\partial^2 T}{\partial y^2}+\frac{\partial^2 T}{\partial z^2}+\frac{\dot{g}}{k}=0 \qquad (8.30)$$

This equation is called the **Poisson equation**.

5. Steady-state heat conduction, constant thermal conductivity, and no heat generation:

$$\frac{\partial^2 T}{\partial x^2}+\frac{\partial^2 T}{\partial y^2}+\frac{\partial^2 T}{\partial z^2}=0 \qquad (8.31)$$

This is the simplest form of the three-dimensional heat conduction equation and is called the **Laplace equation**.

8.3 BOUNDARY AND INITIAL CONDITIONS

Consider steady-state, one-dimensional heat conduction in a medium with constant thermal conductivity and no heat generation. The heat conduction equation for this medium is Equation 8.19. This is a second-order, linear, and homogeneous differential equation whose solution is obtained by integrating twice. However, each integration introduces a new constant and the final solution will have two undetermined constants;

$$T(x) = C_1 x + C_2. \qquad (8.32)$$

The constants C_1 and C_2 can be determined if, for example, we know temperatures at two different x locations, or temperature at one location and conduction heat flux at the same or another location. These are called **boundary conditions**. The complete solution of every second-order differential equation, no matter how the solution is obtained, requires two boundary conditions. Similarly, it can be shown that every first-order differential equation requires one boundary condition. If the independent variable in the differential equation is time, the information that is used to find the undetermined constant is called **initial condition**.

The general form of the heat conduction equation is Equation 8.25 and is second-order in space and first-order in time. Therefore, it requires two boundary conditions for each space direction (a total of six) and one initial condition. Different possible boundary conditions are explained in the following sections.

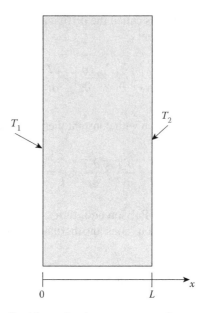

FIGURE 8.5 A plane wall with two fixed temperature surfaces.

8.3.1 Temperature Boundary Condition

In many situations a surface is kept at a fixed temperature or can be reasonably approximated as a uniform temperature surface. An example is a surface covered by a material that is going through some form of phase change such as evaporation, condensation, or freezing. If a surface at coordinate $x = x_0$ is at temperature T_0, the boundary condition can be expressed as

$$T(x_0, y, z, t) = T_0. \tag{8.33}$$

This boundary condition specifies the temperature at the plane $x = x_0$ at any time t. Figure 8.5 shows a plane wall with both surfaces at fixed temperatures. The boundary conditions on the left and right sides of this wall can be expressed as

$$T(0,t) = T_1 \quad \text{and} \quad T(L,t) = T_2. \tag{8.34}$$

Example 8.1

Consider steady-state heat conduction in a thin plane wall with constant thermal conductivity and no heat generation. If the wall thickness is 5 mm and temperatures at the left and right sides of this wall are 105°C and 70°C, respectively, obtain an expression for the temperature as a function of position in this wall.

Solution

Let's select a coordinate system such that the left and right sides of this wall are at coordinates $x = 0$ mm and $x = 5$ mm, respectively. Temperature of this wall at steady-state is a function of x only; $T = T(x)$. The differential equation governing heat conduction inside this wall and the associated boundary conditions are

$$\frac{d^2T}{dx^2} = 0, \quad \text{Boundary Conditions:} \begin{cases} T(0) = 105, \\ T(5) = 70. \end{cases}$$

The general solution of this equation is

$$T(x) = C_1 x + C_2.$$

The application of the boundary conditions gives

$T(0 \text{ mm}) = 105°C \quad \Rightarrow \quad C_2 = 105°C,$

$T(5 \text{ mm}) = 70°C \quad \Rightarrow \quad 70°C = C_1 \cdot 5 \text{ mm} + 105°C \quad \Rightarrow \quad C_1 = -7°C/\text{mm}.$

Substituting for C_1 and C_2 gives the expression for the temperature distribution

$$T(x) = -7x + 105$$

Note that this equation is valid only if the units for x and T are mm and °C, respectively.

8.3.2 Heat Flux Boundary Condition

Consider a surface that is covered by electric foil heaters. These heaters apply a relatively uniform heat flux on this surface. Another case of a uniform heat flux surface is an opaque surface exposed to the sun as solar heat flux is absorbed fairly uniformly over the surface. These cases are situations in which heat flux boundary condition can be used. If a surface at coordinate x_0 is exposed to a specified heat flux $\dot{q}_0$, in the positive x direction, the boundary condition is expressed as

$$-k \frac{\partial T}{\partial x} \bigg|_{(x_0, y, z, t)} = \dot{q}_0. \tag{8.35}$$

Figure 8.6 shows a thin plane wall with specified heat fluxes on both surfaces. These boundary conditions are expressed as

$$-k \frac{\partial T}{\partial x} \bigg|_{(0,t)} = \dot{q}_1 \quad \text{and} \quad -k \frac{\partial T}{\partial x} \bigg|_{(L,t)} = -\dot{q}_2. \tag{8.36}$$

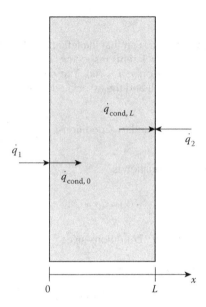

FIGURE 8.6 A plane wall with two specified heat flux surfaces.

Note that conduction heat flux is assumed to be in the positive x direction and therefore the standard form of the Fourier's law of heat conduction with the negative sign is used above. On the other hand, depending on whether the applied heat flux is in the positive or negative x direction, it will be given a positive or negative sign, respectively.

Two special cases of specified heat flux boundary are symmetry line or plane and adiabatic or insulated surface. In both cases the heat flux is zero and, since the heat flux is proportional to temperature gradient, the boundary conditions are expressed as $\partial T/\partial n = 0$ where n is the coordinate direction normal to the symmetric or adiabatic surface.

Example 8.2

Consider steady-state heat conduction in a 10 mm × 10 mm × 0.7 mm thick silicon die. The circuitry on the active side of the die dissipates 2 W and the temperature on the back side of the die is 100°C. If thermal conductivity of silicon is 148 W/m°C, obtain an expression for temperature distribution inside the die.

Solution

Let's select a coordinate system such that the active and the back sides of the die are at $x = 0$ mm and $x = 0.7$ mm, respectively. The heat flux at the active side of the die is

$$\dot{q} = \frac{2\,\text{W}}{0.01\,\text{m} \times 0.01\,\text{m}} = 20000 \text{ W/m}^2.$$

Heat Conduction Equation

The heat conduction equation and the boundary conditions are

$$\frac{d^2T}{dx^2} = 0, \quad \text{Boundary Conditions:} \begin{cases} -k\dfrac{dT}{dx}\bigg|_{(x=0)} = 20000 \text{ W/m}^2, \\ T(0.0007\text{ m}) = 100°\text{C}. \end{cases}$$

The general solution of this equation is

$$T(x) = C_1 x + C_2,$$

and the application of the boundary conditions gives

$$-k\frac{dT}{dx}\bigg|_{(x=0\ m)} = 20000 \text{ W/m}^2 \Rightarrow -148 \text{ W/m°C} \times C_1 = 20000 \text{ W/m}^2$$

$$\Rightarrow C_1 = -135.14°\text{C/m},$$

$T(0.0007\text{ m}) = 100°\text{C} \Rightarrow 100°\text{C} = -135.14 \times 0.0007 \text{ m} + C_2 \Rightarrow C_2 = 100.1°\text{C}.$

Substituting for C_1 and C_2 gives the following expression for the temperature distribution inside this die:

$$T(x) = -135.14x + 100.1.$$

Note that the units for x and T in this equation are m and °C, respectively. It is also seen that the temperature variation across the die is very small and the entire die can be approximated as a uniform temperature object. However, this is the result of the high thermal conductivity of the die and uniform power dissipation assumption on the active side of the die. The latter assumption is not true in many practical situations where the power dissipation in the die is fairly nonuniform.

8.3.3 CONVECTION BOUNDARY CONDITION

Consider a surface that is exposed to a fluid with a different temperature such that convection heat transfer happens between that surface and fluid. The appropriate boundary condition for this situation is

conduction heat flux to (from) surface = convection heat flux from (to) surface.

If the temperature at the surface $x = x_0$ is $T(x_0,y,z,t)$ and it is exposed to a fluid with temperature T_∞ and convection heat transfer coefficient h, convection boundary condition will be expressed as

$$-k\frac{\partial T}{\partial x}\bigg|_{(x_0,y,z,t)} = h\big[T(x_0,y,z,t) - T_\infty\big], \tag{8.37}$$

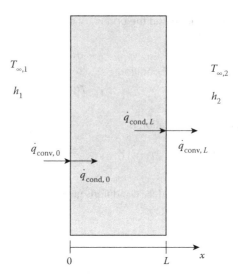

FIGURE 8.7 A thin plane wall exposed to convection on both sides.

where both conduction and convection are assumed to be in the positive x direction.

Figure 8.7 shows a thin plane wall exposed to convection heat transfer on both sides. Convection boundary conditions are expressed as follows

$$-k\frac{\partial T}{\partial x}\bigg|_{(0,t)} = h_1[T_{\infty,1} - T(0,t)],$$
$$-k\frac{\partial T}{\partial x}\bigg|_{(L,t)} = h_2[T(L,t) - T_{\infty,2}].$$
(8.38)

Example 8.3

Let's consider the silicon die of Example 8.2 and assume that the back side of the die is exposed to an ambient of 25°C and convection heat transfer coefficient of 2000 W/m²°C. Obtain an expression for temperature distribution inside this die.

Solution

The differential equation and the boundary conditions are

$$\frac{d^2T}{dx^2} = 0, \quad \text{Boundary Conditions:} \begin{cases} -k\dfrac{dT}{dx}\bigg|_{(x=0)} = 20000 \text{ W/m}^2, \\ -k\dfrac{dT}{dx}\bigg|_{(x=0.0007\text{m})} = h(T(0.0007\text{ m}) - 25). \end{cases}$$

Heat Conduction Equation

The general solution of this equation is

$$T(x) = C_1 x + C_2,$$

and the application of the boundary conditions gives

$$-k \frac{dT}{dx}\bigg|_{(x=0 \text{ m})} = 20000 \text{ W/m}^2 \Rightarrow$$

$$-148 \text{ W/m°C} \times C_1 = 20000 \text{ W/m}^2 \Rightarrow C_1 = -135.14 \text{°C/m},$$

$$-k \frac{dT}{dx}\bigg|_{(x=0.0007\text{m})} = h(T(0.0007 \text{ m}) - 25) \Rightarrow$$

$$(-148 \text{ W/m°C})(-135.14) = 2000 \text{ W/m}^{2°}C(-135.14 \times 0.0007 + C_2 - 25)\text{°C}$$

$$\Rightarrow C_2 = 35.09\text{°C}.$$

Substituting for C_1 and C_2 gives the following expression for temperature distribution

$$T(x) = -135.14x + 35.09.$$

Note that temperature at the active side of the die (i.e., at $x = 0$) is 35.09°C.

8.3.4 Radiation Boundary Condition

A surface that is exposed to a surrounding with a different temperature may send (receive) net radiation heat to (from) its surrounding. The boundary condition in this case can be expressed as

conduction heat flux to (from) surface = radiation heat flux from (to) surface.

If the temperature at the surface $x = x_0$ is $T(x_0,y,z,t)$, its emissivity is ε, and it is exposed to a surrounding with temperature T_{surr}, radiation boundary condition will be expressed as

$$-k \frac{\partial T}{\partial x}\bigg|_{(x_0,y,z,t)} = \varepsilon \sigma \left[T(x_0,y,z,t)^4 - T_{surr}^4 \right], \tag{8.39}$$

where both conduction and radiation are assumed to be in the positive x direction. Note also that temperatures have to be expressed in an absolute unit such as Kelvin.

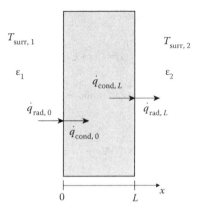

FIGURE 8.8 A thin plane wall exposed to radiation on both sides.

Radiation boundary conditions for both sides of the thin plane wall shown in Figure 8.8 can be written as

$$-k\frac{\partial T}{\partial x}\bigg|_{(0,t)} = \varepsilon_1 \sigma \left[T_{surr,1}^4 - T(0,t)^4\right],$$

$$-k\frac{\partial T}{\partial x}\bigg|_{(L,t)} = \varepsilon_2 \sigma \left[T(L,t)^4 - T_{surr,2}^4\right].$$

(8.40)

Example 8.4

Let's consider the silicon die of Example 8.2 and assume that the back side of this die has an emissivity of 0.7 and is exposed to a surrounding at 25°C. Obtain an expression for temperature distribution inside this die.

Solution

The differential equation and the boundary conditions are

$$\frac{d^2T}{dx^2} = 0, \quad \text{Boundary Conditions:} \begin{cases} -k\dfrac{dT}{dx}\bigg|_{(x=0)} = 20000 \text{ W/m}^2. \\ -k\dfrac{dT}{dx}\bigg|_{(x=0.0007\text{m})} = \varepsilon \sigma (T(0.0007 \text{ m})^4 - 298^4). \end{cases}$$

The general solution of this equation is

$$T(x) = C_1 x + C_2,$$

Heat Conduction Equation

and the application of the boundary conditions gives

$$-k\frac{dT}{dx}\bigg|_{(x=0\text{ m})} = 20000 \text{ W/m}^2 \Rightarrow$$

$$-148 \text{ W/m°C} \times C_1 = 20000 \text{ W/m}^2 \Rightarrow C_1 = -135.14\text{°C/m},$$

$$-k\frac{dT}{dx}\bigg|_{(x=0.0007\text{m})} = \varepsilon\sigma(T(0.0007 \text{ m})^4 - 298^4) \Rightarrow$$

$$(-148 \text{ W/m°C})(-135.14\text{°C/m}) = 0.7 \times 5.67 \times 10^{-8} \text{ W/m}^2\text{K}^4$$

$$\left[(-135.14 \times 0.0007 + C_2 + 273)^4 - 298^4\right]\text{K}^4 \Rightarrow C_2 = 572.91\text{°C}.$$

Substituting for C_1 and C_2 gives the following expression for the temperature distribution:

$$T(x) = -135.14x + 572.91.$$

It is seen that the die temperature is significantly high and that shows the radiation heat transfer alone is not able to cool it appropriately.

8.3.5 GENERAL BOUNDARY CONDITION

A surface may be exposed to a specified uniform heat flux as well as convection and radiation heat transfers to its ambient and surrounding. An example is the surface of an outdoor electronic cabinet that is exposed to solar heat flux, and exchanges natural convection and radiation heat transfers with its ambient and surrounding. The boundary condition in this case can be expressed as

conduction heat flux to (from) surface = total heat flux from (to) surface.

Figure 8.9 shows a thin plane wall that is exposed to three boundary conditions at the same time. Boundary conditions at the left and right sides of this wall are

$$-k\frac{\partial T}{\partial x}\bigg|_{(0,t)} = \dot{q}_1 + h(T_{\infty,1} - T(0,t)) + \varepsilon_1\sigma(T_{\text{surr},1}^4 - T(0,t)^4),$$

$$-k\frac{\partial T}{\partial x}\bigg|_{(L,t)} = -\dot{q}_2 + h(T(L,t) - T_{\infty,2}) + \varepsilon_2\sigma(T(L,t)^4 - T_{\text{surr},2}^4).$$
(8.41)

8.3.6 INTERFACE BOUNDARY CONDITION

Consider two layers with different materials A and B in perfect contact with each other as shown in Figure 8.10. Any point on the contact surface has a unique temperature

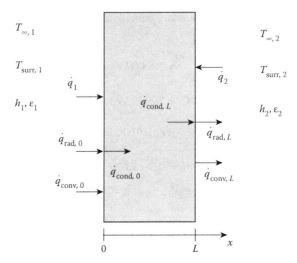

FIGURE 8.9 A thin plane wall with general boundary conditions on both sides.

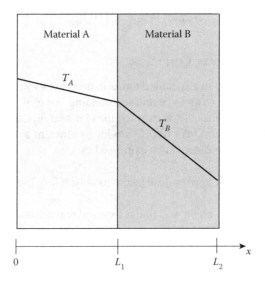

FIGURE 8.10 Interface of two materials with perfect contact.

no matter if that point is on material A or material B. On the other hand, since a surface can not accumulate energy, the heat flux to the interface surface from side A is equal to the heat flux from the interface surface on side B. Figure 8.10 shows two thin plane walls in perfect contact with each other at $x = L_1$. If $T_A(x)$ and $T_B(x)$ are temperatures in materials A and B, respectively, the interface boundary conditions will be expressed in the following forms:

Heat Conduction Equation

$$T_A(L_1) = T_B(L_1)$$

$$-k_A \left. \frac{\partial T_A}{\partial x} \right|_{x=L_1} = -k_B \left. \frac{\partial T_B}{\partial x} \right|_{x=L_1}. \tag{8.42}$$

8.4 STEADY-STATE HEAT CONDUCTION

8.4.1 ONE-DIMENSIONAL, STEADY-STATE HEAT CONDUCTION

Solutions of heat conduction equation for a thin plane wall with constant thermal conductivity, no heat generation, and various boundary conditions were obtained in Section 8.3. It was shown that temperature distribution inside such a wall was linear. For example, if temperatures at the left and right sides of a wall with thickness L are T_1 and T_2, respectively, temperature distribution will be given by

$$T(x) = \frac{T_2 - T_1}{L} x + T_1. \tag{8.43}$$

Conduction heat transfer rate through this wall is obtained using Fourier's law of heat conduction:

$$\dot{Q}_{\text{cond}} = -kA \frac{dT}{dx} = -kA \frac{T_2 - T_1}{L}. \tag{8.44}$$

It is just shown that Equation 8.44, which was first introduced in Chapter 4 as an approximation to Fourier's law of conduction, is strictly valid for one-dimensional, steady-state heat conduction with constant thermal conductivity and no heat generation.

Now, let's consider a plane thin wall with thickness L and uniform heat generation per unit volume $\dot{g}$, whose left and right sides are at fixed temperatures T_1 and T_2 as shown in Figure 8.11. The steady-state heat conduction equation and the boundary conditions are

$$\frac{d^2T}{dx^2} + \frac{\dot{g}}{k} = 0, \quad \text{Boundary Conditions:} \begin{cases} T(0) = T_1, \\ T(L) = T_2. \end{cases} \tag{8.45}$$

The general solution to this differential equation is obtained by integrating twice,

$$T(x) = -\frac{\dot{g}}{2k} x^2 + C_1 x + C_2. \tag{8.46}$$

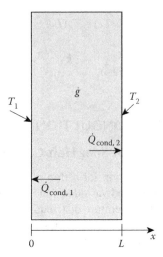

FIGURE 8.11 A thin plane wall with uniform internal heat generation and fixed temperature surfaces.

The application of the boundary conditions gives the values for constants C_1 and C_2.

$$T(0) = T_1 \Rightarrow C_2 = T_1$$

$$T(L) = T_2 \Rightarrow -\frac{\dot{g}}{2k}L^2 + C_1 L + T_1 = T_2 \Rightarrow C_1 = \frac{T_2 - T_1}{L} + \frac{\dot{g}}{2k}L.$$

The temperature distribution inside the wall will have the form:

$$T(x) = -\frac{\dot{g}}{2k}x^2 + \left(\frac{T_2 - T_1}{L} + \frac{\dot{g}}{2k}L\right)x + T_1. \tag{8.47}$$

Maximum temperature happens at the point in which temperature derivative is zero.

$$\frac{dT}{dx} = 0 \Rightarrow -\frac{\dot{g}}{k}x_{\max} + \frac{T_2 - T_1}{L} + \frac{\dot{g}}{2k}L = 0 \Rightarrow x_{\max} = \frac{k}{\dot{g}}\frac{T_2 - T_1}{L} + \frac{L}{2}.$$

Note that $x_{\max} = L/2$ if $T_1 = T_2$. Substituting $x_{\max}$ into Equation 8.47 gives the maximum temperature inside the wall:

$$T_{\max} = \frac{k}{2\dot{g}}\left(\frac{T_2 - T_1}{L}\right)^2 + \frac{\dot{g}}{8k}L^2 + T_1. \tag{8.48}$$

Heat Conduction Equation

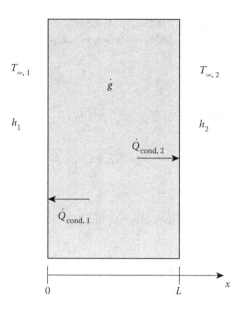

FIGURE 8.12 A thin plane wall with uniform internal heat generation and convection on both sides.

Conduction heat transfer rate to the left and right sides of this wall are

$$\dot{Q}_{cond,1} = \dot{Q}_{cond}(0) = kA\frac{dT}{dx}\bigg|_{x=0} = kA\left(\frac{T_2 - T_1}{L} + \frac{\dot{g}}{2k}L\right),$$

$$\dot{Q}_{cond,2} = \dot{Q}_{cond}(L) = -kA\frac{dT}{dx}\bigg|_{x=L} = -kA\left(\frac{T_2 - T_1}{L} - \frac{\dot{g}}{2k}L\right). \quad (8.49)$$

Note that

$$\dot{Q}_{cond,1} + \dot{Q}_{cond,2} = \dot{g}AL = \dot{G}$$

is the total heat generated inside this wall.

Next, let's consider the same thin plane wall with convection boundary conditions on both sides as shown in Figure 8.12. The steady-state heat conduction equation and the boundary conditions for this case are

$$\frac{d^2T}{dx^2} + \frac{\dot{g}}{k} = 0, \quad \text{Boundary Conditions:} \begin{cases} k\dfrac{dT}{dx}\bigg|_{x=0} = h_1(T(0) - T_{\infty,1}) \\ -k\dfrac{dT}{dx}\bigg|_{x=L} = h_2(T(L) - T_{\infty,2}) \end{cases} \quad (8.50)$$

The general solution to this differential equation is given by Equation 8.46. The application of boundary conditions gives two equations for two constants C_1 and C_2.

$$k\frac{dT}{dx}\bigg|_{x=0} = h_1(T(0) - T_{\infty,1}) \quad \Rightarrow \quad kC_1 = h_1(C_2 - T_{\infty,1})$$

$$-k\frac{dT}{dx}\bigg|_{x=L} = h_2(T(L) - T_{\infty,2}) \quad \Rightarrow \quad -k\left(-\frac{\dot{g}}{k}L + C_1\right) = h_2\left(-\frac{\dot{g}}{2k}L^2 + C_1 L + C_2 - T_{\infty,2}\right).$$

The solution of these two equations gives C_1 and C_2:

$$C_1 = \frac{2kh_1\dot{g}L + h_1 h_2 \dot{g}L^2 - 2kh_1 h_2(T_{\infty,1} - T_{\infty,2})}{2k^2 h_1 + 2kh_1 h_2 L + 2k^2 h_2}$$

$$C_2 = T_{\infty,1} + \frac{2k\dot{g}L + h_2 \dot{g}L^2 - 2kh_2(T_{\infty,1} - T_{\infty,2})}{2kh_1 + 2h_1 h_2 L + 2kh_2}.$$

(8.51)

Temperatures at the left and right sides of the wall are

$$T(0) = C_2$$

$$T(L) = -\frac{\dot{g}}{2k}L^2 + C_1 L + C_2.$$

Two special cases are a symmetric wall where $h_1 = h_2 = h$ and $T_{\infty,1} = T_{\infty,2} = T_\infty$ and a wall in which one side is adiabatic. **For a symmetric wall**:

$$T(x) = -\frac{\dot{g}}{2k}x^2 + \frac{\dot{g}L}{k}\left[\frac{1 + hL/2k}{1 + hL/k}\right]x + \frac{\dot{g}L}{h}\left[\frac{1 + hL/2k}{1 + hL/k}\right] + T_\infty. \quad (8.52)$$

The temperature distribution **for a wall in which the left side is adiabatic** is obtained by setting $h_1 = 0$ in Equation 8.51, and replacing C_1 and C_2 in Equation 8.46,

$$T = -\frac{\dot{g}}{2k}(x^2 - L^2) + \frac{\dot{g}}{h_2}L + T_{\infty,2}. \quad (8.53)$$

Heat Conduction Equation

8.4.2 Two-Dimensional, Steady-State Heat Conduction

The solution of a two-dimensional, steady-state heat conduction equation, in a general situation, is usually obtained by an appropriate thermal simulation tool. However, analytical solutions are available for many simple cases. A description of solution techniques for two- and three-dimensional heat conduction equations is beyond the scope and objective of this book and the interested reader is referred to text books on this subject [1,2]. Instead, a few simple cases will be introduced and their corresponding analytical solutions will be given without going through the solution process.

Consider a thin rectangular plate or the cross section of a long rectangular rod as shown in Figure 8.13a. Three sides of this rectangular object are kept at zero temperature and the temperature at $y = W$ is a general function of x; $T_1(x,W) = f_1(x)$. Temperature distribution inside this object is given by:

$$T_1(x,y) = \frac{2}{L}\sum_{n=1}^{\infty} \frac{\sin\frac{n\pi x}{L}\sinh\frac{n\pi y}{L}}{\sinh\frac{n\pi W}{L}} \int_0^L f_1(x)\sin\frac{n\pi x}{L}dx. \quad (8.54)$$

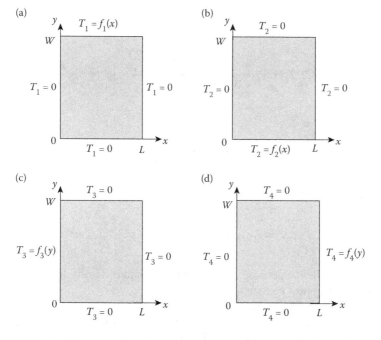

FIGURE 8.13 A thin rectangular plate or a long rectangular rod with four different sets of boundary conditions.

If $f_1(x) = T_0$ is constant, the temperature distribution will be given by

$$T_1(x,y) = \frac{2T_0}{\pi} \sum_{n=1}^{\infty} \frac{1-(-1)^n}{n} \frac{\sin\frac{n\pi x}{L}\sinh\frac{n\pi y}{L}}{\sinh\frac{n\pi W}{L}}. \qquad (8.55)$$

The solutions for other cases shown in Figure 8.13 are given by the following expressions.

Figure 8.13b: $$T_2(x,y) = \frac{2}{L}\sum_{n=1}^{\infty} \frac{\sin\frac{n\pi x}{L}\sinh\frac{n\pi(b-y)}{L}}{\sinh\frac{n\pi W}{L}} \int_0^L f_2(x)\sin\frac{n\pi x}{L}dx, \qquad (8.56)$$

Figure 8.13c: $$T_3(x,y) = \frac{2}{W}\sum_{n=1}^{\infty} \frac{\sinh\frac{n\pi(L-x)}{W}\sin\frac{n\pi y}{W}}{\sinh\frac{n\pi L}{W}} \int_0^W f_3(y)\sin\frac{n\pi y}{W}dy, \qquad (8.57)$$

and

Figure 8.13d: $$T_4(x,y) = \frac{2}{W}\sum_{n=1}^{\infty} \frac{\sinh\frac{n\pi x}{W}\sin\frac{n\pi y}{W}}{\sinh\frac{n\pi L}{W}} \int_0^W f_4(y)\sin\frac{n\pi y}{W}dy. \qquad (8.58)$$

The importance of simple cases shown in Figure 8.13 and their corresponding temperature distributions is that the solution of heat conduction equation for an object with general temperature boundary conditions, shown in Figure 8.14, is given by the sum of these four solutions:

$$T(x,y) = T_1(x,y) + T_2(x,y) + T_3(x,y) + T_4(x,y). \qquad (8.59)$$

This is the so-called superposition principle that applies to any linear differential equation including heat conduction equation.

Another case for which the analytical solution of two-dimensional heat conduction equation can be obtained is shown in Figure 8.15. The sides at $x = 0$ and $y = 0$ are adiabatic, the side at $x = L$ is kept at constant temperature, and the side at $y = W$

Heat Conduction Equation

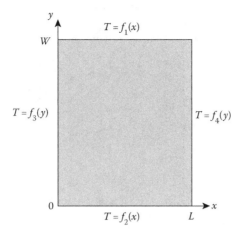

FIGURE 8.14 A thin rectangular plate or long rectangular rod with general temperature distribution on its four sides.

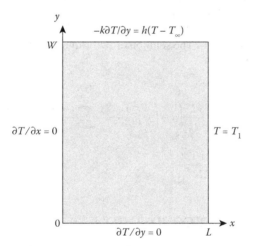

FIGURE 8.15 A thin rectangular plate or long rectangular rod with convection, adiabatic, and constant temperature boundary condition.

is exposed to convection heat transfer with an ambient at T_∞. The solution of heat conduction equation for this domain and boundary conditions is [1,2]

$$T(x,y) = T_\infty + 2(T_1 - T_\infty) \sum_{n=1}^{\infty} \frac{\sin \lambda_n W}{(\lambda_n W + \sin \lambda_n W \cos \lambda_n W) \cosh \lambda_n L} \cos \lambda_n y \cosh \lambda_n x,$$

(8.60)

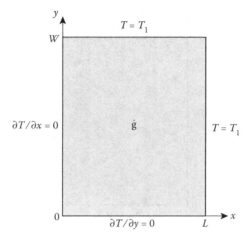

FIGURE 8.16 A two-dimensional plate with internal heat generation.

where λ_n is the solution of the equation $\lambda_n W \tan \lambda_n W = hW/k$ for $n = 1, 2, 3, \ldots$

Figure 8.16 shows a thin rectangular plate or a long rectangular rod with uniform internal heat generation. The sides at $x = 0$ and $y = 0$ are adiabatic and the sides at $x = L$ and $y = W$ are at $T = T_1$. The solution of two-dimensional heat conduction equation for this domain is [1,2]

$$T(x,y) = T_1 + \frac{\dot{g}L^2}{2k}\left(1 - \left(\frac{x}{L}\right)^2\right) - \frac{2\dot{g}}{Lk}\sum_{n=0}^{\infty}\frac{(-1)^n}{\lambda_n^3}\frac{\cos\lambda_n x \cosh\lambda_n y}{\cosh\lambda_n W} \tag{8.61}$$

where $\lambda_n = \dfrac{(2n+1)\pi}{2L}$.

8.5 TRANSIENT HEAT CONDUCTION

Consider a section of a thin plane wall shown in Figure 8.17. Let's assume that the initial temperature distribution in the wall is given by $T(x,0) = f(x)$ and the left and right sides of the wall are suddenly exposed to temperatures $T(0,t) = T_1$ and $T(L,t) = T_2$ for $t \geq 0$. Temperature distribution inside this wall for any time $t \geq 0$ is given by [1,2]

$$T(x,t) = T_1 + (T_2 - T_1)\frac{x}{L} + \sum_{n=1}^{\infty} C_n e^{-(n\pi/L)^2 \alpha t} \sin\frac{n\pi}{L}x, \tag{8.62}$$

where $C_n = \dfrac{2}{n\pi}\left[(-1)^n T_2 - T_1\right] + \dfrac{2}{L}\displaystyle\int_0^L f(x)\sin\frac{n\pi}{L}x\,dx.$

Heat Conduction Equation

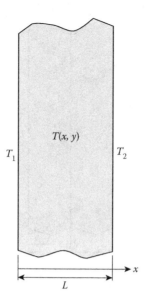

FIGURE 8.17 A thin plane wall with constant temperature of sides.

In a special case where $T(x,0) = T_i$ and $T_1 = T_2 = T_s$,

$$\int_0^L f(x)\sin\frac{n\pi}{L}x\,dx = \int_0^L T_i \sin\frac{n\pi}{L}x\,dx = -T_i \frac{L}{n\pi}\cos\frac{n\pi}{L}x\bigg|_0^L$$

$$= -\frac{LT_i}{n\pi}(\cos n\pi - 1) = \frac{LT_i}{n\pi}\left[1-(-1)^n\right]$$

and

$$C_n = \frac{2}{n\pi}\left[(-1)^n T_2 - T_1\right] + \frac{2}{L}\int_0^L f(x)\sin\frac{n\pi}{L}x\,dx = \frac{2T_s\left[(-1)^n - 1\right]}{n\pi} + \frac{2T_i\left[1-(-1)^n\right]}{n\pi}$$

$$= \frac{2(T_s - T_i)\left[(-1)^n - 1\right]}{n\pi}.$$

Therefore,

$$T(x,t) = T_s + \frac{2(T_s - T_i)}{\pi}\sum_{n=1}^{\infty}\frac{\left[(-1)^n - 1\right]}{n}e^{-\left(\frac{n\pi}{L}\right)^2 \alpha t}\sin\frac{n\pi}{L}x. \quad (8.63)$$

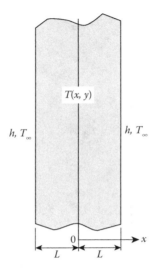

FIGURE 8.18 A thin plane wall with convection on both sides.

Another possible boundary condition is convection boundary condition. Consider a long wall of thickness $2L$, constant thermal conductivity k, uniform initial temperature T_i, whose sides are suddenly exposed to convection heat transfer with a fluid at temperature $T_\infty \neq T_i$ as shown in Figure 8.18. The solution of the heat conduction equation for this wall is [1,2]

$$T(x,t) = T_\infty + 2(T_i - T_\infty) \sum_{n=1}^{\infty} \frac{\sin \lambda_n L}{\lambda_n L + \sin \lambda_n L \cos \lambda_n L} \cos \lambda_n x \cdot e^{-\lambda_n^2 \alpha t} \tag{8.64}$$

where $\lambda_n L \tan \lambda_n L = \dfrac{hL}{k}$.

8.6 LUMPED SYSTEMS

A special case of transient systems is one in which temperature, at any time, is uniform over the entire system (i.e., temperature is a function of time only). These are called **lumped systems**. Examples of lumped systems are a small metal ball, a copper wire, a silicon die, and a small electronic package.

8.6.1 Simple Lumped System Analysis

Consider a lumped system with mass m, volume $\mathcal{V}$, external surface area A, density ρ, and specific heat C_p as shown in Figure 8.19. Let's assume that this system is originally at temperature T_i when it is exposed to an ambient with constant temperature

Heat Conduction Equation

T_∞, which is different from T_i, and constant heat transfer coefficient h. The heat balance equation for this system can be written as

$$\dot{Q}_{in} - \dot{Q}_{out} = mC_p \frac{dT}{dt} \quad \Rightarrow \quad -\dot{Q}_{comb} = mC_p \frac{dT}{dt}. \tag{8.65}$$

Replacing Newton's law of cooling for combined heat transfer rate and substituting $m = \rho \mathcal{V}$ gives

$$-hA(T - T_\infty) = \rho \mathcal{V} C_p \frac{dT}{dt}, \tag{8.66}$$

where T is the system's uniform temperature at time t. If we define:

$$\theta = T - T_\infty \quad \text{and} \quad b = \frac{hA}{\rho \mathcal{V} C_p}, \tag{8.67}$$

Equation 8.66 can be written as

$$\frac{d\theta}{\theta} = -b\,dt. \tag{8.68}$$

If b is constant, this equation can easily be solved to give

$$\ln \frac{\theta}{\theta_i} = -bt \quad \text{or} \quad \theta = \theta_i e^{-bt} \tag{8.69}$$

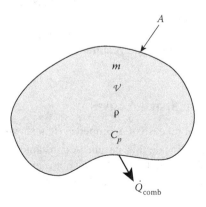

FIGURE 8.19 A typical lumped system.

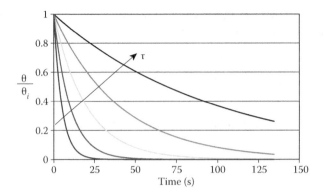

FIGURE 8.20 Temperature variation of simple lumped systems with various time constants.

where $\theta_i = T_i - T_\infty$. The final solution in terms of temperature variable is

$$T - T_\infty = (T_i - T_\infty)e^{-bt}. \tag{8.70}$$

As time passes temperature of the system approaches the ambient temperature. The rate at which the system temperature approaches the ambient temperature depends on the value of constant b or its reciprocal $\tau = 1/b$. Figure 8.20 shows that a system takes a longer time to reach the ambient temperature if τ is larger and vise versa. That is why τ is called the time constant of the system. Note that:

$$\tau = \left(\frac{1}{hA}\right)(\rho V C_p) = R_{sys} C_{sys}, \tag{8.71}$$

where R_{sys} and C_{sys} are thermal resistance from the external surface of the system and thermal capacitance of the system, respectively. A large thermal resistance or a large thermal capacitance reduces heat transfer rate from a system to its ambient and therefore increases the time required for the system to reach the ambient temperature.

8.6.2 General Lumped System Analysis

Let's consider a more general case of a lumped system shown in Figure 8.21. Heat is generated with the rate $\dot{G}$ inside this system. A constant heat flux $\dot{q}$ is applied over a portion of the system's external area, A_h, and convection and radiation heat fluxes $\dot{q}_{conv}$ and $\dot{q}_{rad}$ happen over areas A_c and A_r, respectively. Note that these surface

Heat Conduction Equation

areas are not necessarily exclusive portions of the system's external area. The heat balance equation for this system can be written as

$$\dot{Q}_{in} - \dot{Q}_{out} = mC_p \frac{dT}{dt} \quad \Rightarrow \quad \dot{G} + \dot{q}A_h - \dot{q}_{conv}A_c - \dot{q}_{rad}A_r = mC_p \frac{dT}{dt}, \qquad (8.72)$$

$$\dot{G} + \dot{q}A_h - h_{conv}A_c(T - T_\infty) - \varepsilon\sigma A_r(T^4 - T^4_{surr}) = mC_p \frac{dT}{dt}. \qquad (8.73)$$

A special case of this system in which there was no volumetric heat generation and no incoming heat flux, and radiation heat transfer was negligible or was combined with convection heat transfer was studied in Section 8.6.1. Another interesting case is one in which radiation heat transfer can be neglected or combined with convection heat transfer and heat generation, external heat flux and convection heat transfer coefficient are independent of time [3]. The heat balance equation in this case is

$$\dot{G} + \dot{q}A_h - h_{conv}A_c(T - T_\infty) = mC_p \frac{dT}{dt}. \qquad (8.74)$$

Introducing the new variable $\theta = T - T_\infty$ and defining parameters:

$$b = \frac{h_{conv}A_c}{mC_p} \quad \text{and} \quad a = \frac{\dot{G} + \dot{q}A_h}{mC_p} \qquad (8.75)$$

reduces this equation to the following simple form:

$$\frac{d\theta}{dt} + b\theta - a = 0. \qquad (8.76)$$

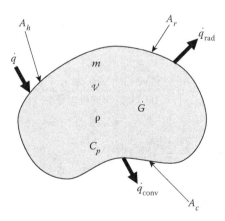

FIGURE 8.21 A general lumped system with various boundary conditions.

The solution of this equation is

$$\theta = \theta_i e^{-bt} + \frac{a}{b}(1-e^{-bt}), \quad (8.77)$$

or in terms of temperature and the original parameters:

$$T - T_\infty = (T_i - T_\infty)e^{-bt} + \frac{\dot{G} + \dot{q}A_h}{h_{conv} A_c}(1-e^{-bt}). \quad (8.78)$$

Note that after a long enough time this equation reduces to the steady-state energy balance equation:

$$\dot{G} + \dot{q}A_h = h_{conv} A_c (T - T_\infty). \quad (8.79)$$

This equation simply states that the rate of energy transfer into the system by heat generation and incoming heat flux is equal to convection heat transfer rate from the system.

Example 8.5

Consider an electronic device with a convection surface area of 20 mm × 20 mm. The mass of this device is 20 g, its average specific heat is 900 J/kg°C and it is exposed to an ambient at 25°C with convection heat transfer coefficient of 25 W/m²°C. If this device dissipates 5 W, determine how long after it is turned on it will reach 85°C.

Solution

The equation that gives the variation of the device temperature versus time is obtained by substituting the given values into Equation 8.78 gives

$$b = \frac{h_{conv} A_c}{mC_p} = \frac{25\,W/m^2°C \times 0.02^2\,m^2}{0.02\,kg \times 900\,J/kg°C} = \frac{1}{1800}\,s^{-1}$$

$$T - 25 = \frac{5}{25 \times 0.02^2}(1 - e^{-t/1800})$$

$$T = 25 + 500(1 - e^{-t/1800}).$$

This equation shows that the steady-state device temperature is 525°C. Figure 8.22 shows the variation of the device temperature with time. It is seen that the steady-

Heat Conduction Equation

state condition is almost reached at around $t = 10{,}000$ seconds. Temperature equation can be rearranged to the form:

$$t = -1800 \ln\left[1 - \frac{T-25}{500}\right],$$

which can be used to calculate the time required for the device to reach a certain temperature. Plugging $T = 85°C$ into this equation gives $t = 230s$ as the time it takes for the device to reach 85°C after it is turned on.

Another special case of the general lumped system heat balance equation, Equation 8.73, is a case in which radiation heat transfer is the only mode of heat transfer from the external surfaces of the system [3]:

$$-\varepsilon \sigma A_r (T^4 - T_{\text{surr}}^4) = mC_p \frac{dT}{dt}. \quad (8.80)$$

This equation can be rearranged and put into the following integral form:

$$\frac{\varepsilon \sigma A_r}{mC_p} \int_0^t dt = \int_{T_i}^{T} \frac{dT}{(T_{\text{surr}}^4 - T^4)}. \quad (8.81)$$

Taking the integrals on both sides gives

$$t = \frac{mC_p}{4\varepsilon \sigma A_r T_{\text{surr}}^3} \left\{ \ln\left|\frac{T_{\text{surr}} + T}{T_{\text{surr}} - T}\right| - \ln\left|\frac{T_{\text{surr}} + T_i}{T_{\text{surr}} - T_i}\right| + 2\left[\tan^{-1}\left(\frac{T}{T_{\text{surr}}}\right) - \tan^{-1}\left(\frac{T_i}{T_{\text{surr}}}\right)\right]\right\}. \quad (8.82)$$

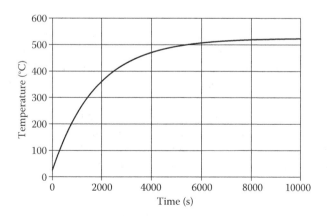

FIGURE 8.22 Variation of device temperature with time.

This is an implicit equation for temperature versus time. The best way to use this equation is to plot time versus temperature and use that plot to estimate temperature at any given time.

Examples 8.6

Consider an electronic device with an exposed surface area of 20 mm × 20 mm. The mass of the device is 20 g, its average specific heat is 900 J/kg°C, and its surface emissivity is 0.7. This device is originally at 125°C when it is exposed to a surrounding temperature of 25°C. Calculate the time it will take for the device to reach 40°C.

Solution

Substituting the given values into Equation 8.82 gives

$$t = \frac{0.02 \times 900}{4 \times 0.7 \times 5.67 \times 10^{-8} \times 0.02^2 \times 298^3} \left\{ \ln\left|\frac{298+T}{298-T}\right| - \ln\left|\frac{298+398}{298-398}\right| + 2\left[\tan^{-1}\left(\frac{T}{298}\right) - \tan^{-1}\left(\frac{398}{298}\right)\right]\right\}$$

$$t = 10710.82\left\{\ln\left|\frac{298+T}{298-T}\right| - 1.94 + 2\left[\tan^{-1}\left(\frac{T}{298}\right) - 0.9281\right]\right\}.$$

Plugging $T = (40 + 273) = 313 K$ in this equation, gives $t = 16,395$ seconds for the device to reach this temperature. A plot of temperature variation with time is shown in Figure 8.23. It is seen that radiation heat transfer, with the parameters given in this example, is not a very effective method for cooling this device.

8.6.3 Validity of Lumped System Analysis

Lumped system analysis is a simple and useful method for analyzing transient heat transfer problems. However, it may give erroneous results if there is appreciable special temperature variation inside the medium. Therefore, it is necessary to verify the validity of lumped system analysis before using it in a specific situation.

Let's consider a thin plane wall with thermal conductivity k and thickness L that is insulated on the left side and, on the right side, is exposed to convection heat transfer to an ambient with temperature T_∞ and heat transfer coefficient h. If temperature at the left and right sides of this wall are T_1 and T_2 and $T_1 > T_2 > T_\infty$, it can be shown that, in steady-state condition,

$$k\frac{T_1 - T_2}{L} = h(T_2 - T_\infty). \qquad (8.83)$$

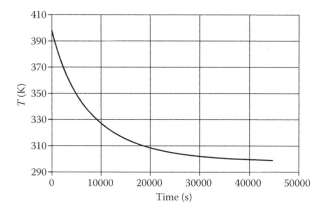

FIGURE 8.23 Temperature variation with time for the device considered in Example 8.6.

This equation equates conduction heat flux inside the wall with convection heat flux from the right side of the wall. It can be rearranged as

$$\frac{T_1 - T_2}{T_2 - T_\infty} = \frac{hL}{k} = \text{Bi}. \tag{8.84}$$

The quantity on the right side of this equation is dimensionless and is called the Biot number. This equation shows that if the Biot number is very small, the difference between temperatures at the left and right sides of the wall is much smaller than the difference between the temperature at the right side of the wall and outside ambient. This means that the wall is close to a lumped system [3,4].

It can also be shown that the Biot number is the ratio of conduction thermal resistance inside the wall to convection thermal resistance from the wall to its ambient:

$$\text{Bi} = \frac{hL}{k} = \frac{L/kA}{1/hA} = \frac{\text{conduction resistance within the system}}{\text{convection resistance from system boundary}}. \tag{8.85}$$

A small Biot number means that conduction thermal resistance inside a system and therefore temperature variation across the system is small.

The Biot number is used as a criterion to validate the lumped system assumption. For a system with a general shape, the Biot number is defined as

$$\text{Bi} = \frac{hL_c}{k}, \tag{8.86}$$

where h is the combined convection and radiation heat transfer coefficient and L_c is the system characteristic length defined as the ratio of the system's volume to its external heat transfer area; $L_c = V/A$. It has been observed that if $\text{Bi} < 0.1$, the error associated with the lumped system assumption is very small [3,4].

PROBLEMS

8.1 When is heat transfer called transient and when is it called steady state?

8.2 What is the temperature boundary condition at a symmetry plane?

8.3 Write down the expression for the following boundary conditions for the heat conduction equation:
   a. Specified temperature at $x = 0$.
   b. Specified heat flux at $x = L$.
   c. Insulated boundary at $x = 0$.
   d. Thermal symmetry at $x = L/2$.
   e. Convection boundary at $x = L$.
   f. Radiation boundary at $x = 0$.

8.4 What are the appropriate boundary conditions at the interface between two layers with different thermal conductivities?

8.5 What is called a lumped system?

8.6 Which object is more likely to be considered a lumped system: a square block of copper with a side length of 10 cm or a square block of copper with a side length of 50 cm?

8.7 Which object is more likely to be considered a lumped system: a block of wood or the same size and shape block made of aluminum?

8.8 Two copper balls are heated to 200°C in an oven. Then one of them is left in front of a fan and the second one is left to be cooled naturally. For which copper ball is the lumped system analysis more likely to be applicable? Why?

8.9 Two same size aluminum blocks are heated Inside separate ovens. One of them is left in a still air over while the other one is located in an oven where air circulates with a very high velocity. Which aluminium block is more likely to be considered a lumped system? Why?

8.10 Two similar copper balls are heated inside an oven. Then one of them is left in still air while the other one is inside still water. Which copper ball is more likely to be considered a lumped system? Why?

8.11 Consider a 10 mm × 10 mm in size and 2 mm thick silicon die attached to a ceramic substrate. The heat dissipated uniformly in the die is 2 W. The side of the die attached to the substrate is kept at 110°C, and the other side is exposed to convection heat transfer to air at 25°C. The convection heat transfer coefficient is $h = 12$ W/m²°C. Obtain the temperature distribution in the die and the heat transfer rate to the substrate and to the air. The thermal conductivity of silicon is 148 W/m°C.

8.12 Consider a 12.5 mm × 12.5 mm and 3 mm thick silicon chip. There is 5 W of heat generated by the circuitry on the bottom of the chip. This heat is transferred to the top of this chip and finally to the air. Thermal resistance between the top of the chip and air is 10°C/W, and air temperature is 25°C. Obtain the temperature distribution in the chip and its top surface temperature. Thermal conductivity of silicon is 148 W/m°C.

8.13 Consider a layer with thickness L, area A, and thermal conductivity k. The heat is generated uniformly through this layer with the rate $\dot{g}$. The left side of this layer is exposed to air with temperature T_∞ and heat transfer coefficient h, and the right side of it is attached to another layer with thickness L_1 and thermal conductivity k_1. If the temperature at the right side of the second layer is T_s, derive the expression governing temperature variation through the first layer. What is the rate of heat

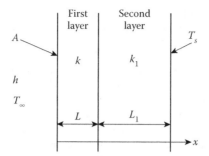

transfer through the second layer? What are the temperatures at the left and right sides of the first layer?

8.14 Consider a layer with thickness L, area A, and thermal conductivity k. The heat is generated uniformly through this layer with the rate $\dot{g}$. The left side of this layer is completely insulated. The right side of this layer is exposed to solar load $\dot{q}_s$ and convection heat transfer to air at temperature T_∞ and convection heat transfer coefficient h. Derive the expression governing temperature variation through this layer. What is the heat transfer through this layer? What are the temperatures at the left and right sides of this layer? Assume the solar absorptivity of the wall is one and therefore absorbs all the solar load.

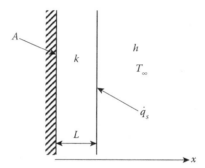

8.15 Consider a layer with thickness L, area A, and thermal conductivity k. The heat is generated uniformly through this layer with the rate $\dot{g}$. The left side of this layer is completely insulated and the right side of it is attached to another layer with thickness L_1 and thermal conductivity

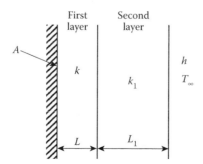

k_1. If the air temperature and heat transfer coefficient on the right side of the second layer are T_∞ and h, respectively, derive the expression governing temperature variation through the first layer. What is the rate of heat transfer through the second layer? What are the temperatures at the left and right sides of the second layer?

8.16 Consider two layers in perfect contact with each other. The first layer has a thickness L_1, thermal conductivity k_1, and uniform heat generation $\dot{g}_1$ and the second layer has a thickness L_2, thermal conductivity k_2, and uniform heat generation $\dot{g}_2$. The left side of the first layer is insulated and the right side of the second layer is exposed to a fluid with temperature T_∞ and convection heat transfer coefficient h.

a. Write down the heat conduction equation and the boundary conditions for the second layer and solve it to obtain temperature distribution in this layer.
b. What is the heat flux on the right side of the second layer?
c. What are the temperatures at the left and right sides of the second layer?

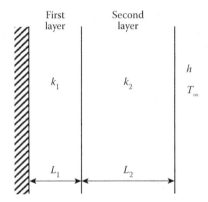

8.17 Consider layer 1 with thickness L_1 and thermal conductivity k_1 sandwiched between layer 2 with thickness L_2 and thermal conductivity k_2 and layer 3 with thickness L_3 and thermal conductivity k_3. The surface areas of all these layers are the same. Heat is generated uniformly inside layer 1 with the rate $\dot{g}$. This composite layer is exposed to the air with temperature $T_{\infty,1}$ and h_1 on the left side and $T_{\infty,2}$ and h_2 on the right side.

a. Write down the differential equation and boundary conditions governing the variation of temperature inside layer 1.
b. What are the temperatures and heat transfer rates at the left and right sides of layer 1?

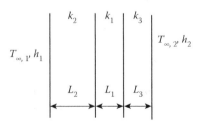

Heat Conduction Equation

8.18 Consider a plane wall with thickness $L = 100$ mm, uniform heat generation $\dot{g}$, and thermal conductivity k. The surface emissivity at the left side of this wall is $\varepsilon = 0.8$, its temperature is $T_1 = 50°C$ and is exposed to a surrounding with temperature $T_{surr} = 30°C$. The right side of this wall is at temperature $T_2 = 70°C$ and is separated from air with convection heat transfer coefficient $h = 20$ W/m²°C and temperature $T_\infty = 25°C$ by another plane wall with thickness $L_1 = 200$ mm and thermal conductivity $k_1 = 0.5$ W/m°C.
  a. Calculate heat flux from the left and right sides of this wall.
  b. Calculate the volumetric heat generation rate $\dot{g}$.
  c. Write down the heat conduction equation and appropriate boundary conditions for this wall and solve it to obtain the temperature distribution inside the wall and the thermal conductivity of the wall.
  d. What is the maximum wall temperature?

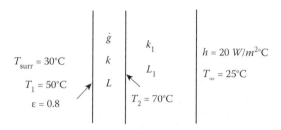

8.19 Consider a one-dimensional heat conduction in the plane composite wall shown below: $k_A = 10$ W/m°C, $k_C = 5$ W/m°C, $L_A = 20$ cm, $L_B = 30$ cm and $L_C = 25$ cm. The outer surfaces of this wall are exposed to a fluid at 25°C and convection heat transfer coefficient of 60 W/m²°C. Heat is uniformly generated with the rate of $\dot{g}$ W/m³ in the middle wall. The temperatures at the left and right sides of the composite wall are measured as $T_1 = 120°C$ and $T_2 = 100°C$.
  a. Calculate heat flux through the left and right sides of the middle wall.
  b. Calculate temperatures at the left and right sides of the middle wall.
  c. Obtain temperature distribution inside the middle wall in terms of $\dot{g}$ and k_B and then determine these two parameters.

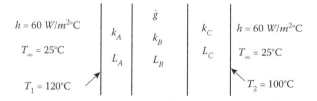

8.20 Consider a plane wall of thickness L and thermal conductivity k. One surface of the wall is kept at a constant temperature T_0 while the other surface is insulated. Heat is generated in the wall at a rate given by $\dot{g} = \dot{g}_0 \exp(-x/L)$ where x is in the thickness direction of the wall and is measured from the insulated surface. Find expressions for the temperature of the insulated surface and the heat flux from the constant temperature surface.

8.21 Consider a plane wall of thickness L and thermal conductivity k. One surface of the wall is kept at constant temperature T_0 while the other surface is insulated. Heat is generated in the wall at a rate given by $\dot{g} = \dot{g}_0 \sin(\pi x/2L)$ where x is in the thickness direction and is measured from the constant temperature surface. Find expressions for the temperature of the insulated surface and the heat flux from the constant temperature surface.

8.22 The junction to air thermal resistance of an electronic device is 25°C/W. The junction is made of 5 mm × 5 mm × 2 mm silicon and dissipates 5 W in an ambient temperature of 25°C. Let's assume that the temperature distribution in the junction of this device is uniform at any given time.
   a. What is the steady-state junction temperature of this device?
   b. How long will it take for the junction temperature of this device to reach 95% of its steady-state value after the device is turned on?

8.23 Consider a 13 mm × 13 mm and 0.7 mm thick silicon die that dissipates 10 W. Let's assume that this heat is generated uniformly inside the die volume. The silicon die is packaged such that thermal resistance between the top of the die and the top of the package is 2°C/W and thermal resistance between the bottom of the die and the bottom of the package is 10°C/W. Temperature at the top and bottom of the package are 80°C and 60°C, respectively. Obtain temperature distribution inside the die and heat transfer rate to the top and bottom of the package, respectively. Assume heat transfer inside the die is one-dimensional and thermal conductivity of silicon is 148 W/m°C.

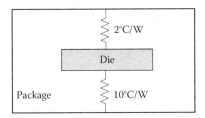

REFERENCES

1. Arpaci, V. S. 1966. *Conduction heat transfer.* Reading, MA: Addison-Wesley.
2. Ozisik, M. N. 1993. *Heat conduction.* New York, NY: John Wiley & Sons, Inc.
3. Incropera, F. P., D. P. DeWitt, T. L. Bergman, and A. S. Lavine. 2006. *Fundamentals of heat and mass transfer.* Sixth Edition. Hoboken, NJ: John Wiley & Sons, Inc.
4. Holman, J. P. 2010. *Heat transfer.* New York, NY: McGraw-Hill.

9 Fundamentals of Convection Heat Transfer

So far we have used Newton's law of cooling, with some knowledge of convection heat transfer coefficient, to calculate convection heat transfer from a surface. However, convection heat transfer coefficient varies from one flow to another and depends on fluid properties, flow geometry, surface dimensions and roughness, and flow type. The relationship between convection heat transfer coefficient and flow variables are illustrated in this chapter, and specific correlations for calculating convection heat transfer coefficients in common flows will be given in Chapters 10 through 12.

9.1 TYPE OF FLOWS

Although every flow has its unique features, many flows can be grouped together based on their common characteristics. This allows engineers to extend the knowledge they gain on a specific flow to other flows with similar features. Common classifications of flows are based on whether the flow is bounded by surfaces around it, whether the flow is generated by an external source or is naturally generated, whether the flow is smooth and orderly or turbulent and random, and whether the flow is time dependent or not. These classifications are described in the rest of this section.

9.1.1 EXTERNAL AND INTERNAL FLOWS

Consider the airflow over a car or around the body of a bird and the flow of smoke out of a chimney. These and similar flows in which the fluid has partial or no contact with a solid surface on part of the flow and extends to infinity with no interference by a solid surface on the rest of the flow are called **external flows**. Other examples are flow over a flat plate such as wind flow over the roof and walls of a building and airflow over the surface of an electronic package, and flow across a circular cylinder such as airflow across an electric wire and airflow around the fins of a pin-fin heat sink.

There are also flows that are bounded by solid surfaces on all sides. Two familiar examples are airflow inside air conditioning ducts and water flow inside pipes. These flows are called **internal flows**. Common electronic-related examples are airflow inside desktop and laptop computers, airflow inside servers, and airflow between electronic cards that are inserted in telecommunication and networking equipment. Figure 9.1 shows two examples of internal flow in electronics equipment.

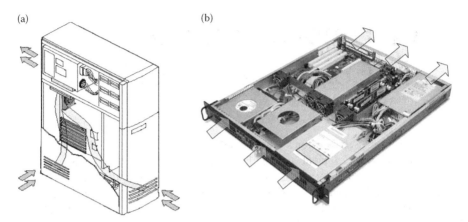

FIGURE 9.1 Internal flow inside (a) a desktop computer and (b) a server.

9.1.2 Forced and Natural Convection Flows

If the fluid motion is generated by a device such as a fan or a pump, the flow is called **forced convection**. Classical examples of forced convection flow are fan-generated flows over a flat plate, inside a pipe or duct, over a circular cylinder or an array of circular cylinders such as a pin-fin heat sink, over a general-shaped object such as an electronic device or an arbitrarily shaped heat sink. Examples of electronics that are cooled by forced convection are desktop and notebook CPUs and GPUs, high output power supplies, and boards in the networking and telecommunication equipment such as high-end servers, routers, and switches.

If the fluid motion is generated by a density difference due to the temperature difference, the flow is called **natural** or **free convection**. Common examples are natural flow of air around a hot plate or an array of plates such as a heat sink, and flow of air around a horizontal cylinder or an array of cylinders such as pin-fin heat sink. Examples of electronics that are cooled by natural convection are cell phones; most handheld computers; televisions; radios; video, CD, and DVD players; and home-use wireless routers.

9.1.3 Laminar and Turbulent Flows

There are two forces that act against each other in every flow; a force that tends to move the flow forward and a frictional or viscous force that tends to slow down the flow. The driving force is the inertia force in forced convection and buoyancy force in natural convection. If the viscous force is large enough, the flow will be slow and orderly with smooth streamlines as shown in Figure 9.2. These flows are called **laminar flows**.

If the force that generates the flow (i.e., inertia or buoyancy force) is much larger than viscous force, the viscous force can not hold the flow orderly and smooth. The flow becomes disordered and fluctuating with bulks of fluids moving from one

Fundamentals of Convection Heat Transfer 211

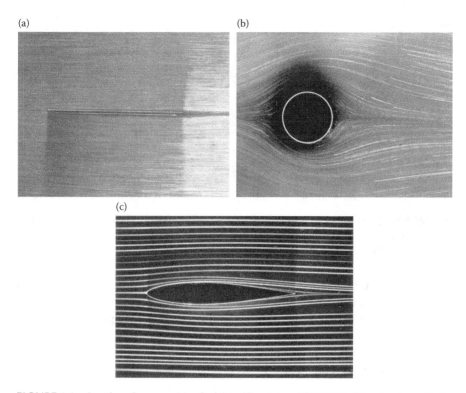

FIGURE 9.2 Laminar flow over (a) a flat plate (Courtesy of ONERA), (b) a circular cylinder (Courtesy of Sadatoshi Taneda), and (c) an airfoil (Courtesy of ONERA). (From Van Dyke, M., *An Album of Fluid Motion*. Stanford, CA: The Parabolic Press, 1982.)

region to another and mixing with each other in a random manner. These flows are called **turbulent flows**. Figure 9.3 shows an instance of a turbulent flow over a flat plate and a turbulent water jet entering from the top of the picture. The nice streamlines that are seen in laminar flows of Figure 9.2 have disappeared and random movements of bulk fluid are seen in both these pictures. Fluid velocity in a turbulent flow is time-dependent and three-dimensional with time-dependent fluctuating components in all directions. This means that the velocity of fluid at a given point in space changes with time in both magnitude and direction. Therefore, turbulent flows are transient flows and a snapshot of a turbulent flow at one time will be different from another snapshot at a different time. However, some general features of the flow and mean values of velocity and temperature over a long time may show steady-state behavior.

Since of the random motion of bulks of fluid and the strong mixing that are present in turbulent flows, these flows are more efficient than the similar laminar flows in transferring heat. However, the same phenomenon will increase the friction in these flows and therefore turbulent flows require more energy to generate and sustain.

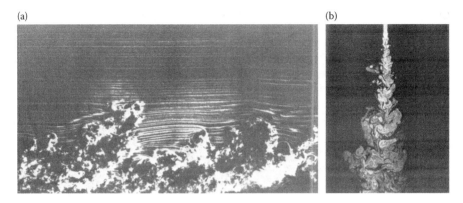

FIGURE 9.3 Turbulent flow over (a) a flat plate (Courtesy of Corke et al.) and (b) a turbulent water jet (Courtesy of Dimotakis et al.). (From Van Dyke, M., *An Album of Fluid Motion*. Stanford, CA: The Parabolic Press, 1982.)

It must be mentioned that a flow may be laminar in one region and turn to turbulent in another region. Transition from laminar to turbulent flow depends on surface geometry, surface roughness, flow velocity, surface temperature, and type of fluid.

9.1.4 Steady-State and Transient Flows

If flow variables do not change with time, it is called **steady-state flow** and if they change with time, it is called **transient flow**. Turbulent flows are transient flows although time-averaged values of some variables may show steady-state behavior. On the other hand, many laminar flows are steady-state flows. Transient flows are generated due to transient driving force, transient velocity, and thermal boundary conditions, or a change in internal heat generation.

9.2 VISCOUS FORCE, VELOCITY BOUNDARY LAYER, AND FRICTION COEFFICIENT

As it is mentioned earlier, there is a frictional or viscous force, between the moving fluid and the surface it is flowing over, that acts against the flow. This force is also present between a layer of a moving fluid and its adjacent faster or slower layer. The amount of this force, at the surface, per unit area of the surface is called **viscous sheer stress**. It is denoted by τ_s and in the SI system of units is measured by N/m^2 or Pa. If U is the component of fluid velocity parallel to the surface and the y axis of the coordinate system starts at the surface and is normal to it, as shown in Figure 9.4, the viscous sheer stress is give by

$$\tau_s = \mu \frac{\partial U}{\partial y}\bigg|_{y=0}. \qquad (9.1)$$

Fundamentals of Convection Heat Transfer

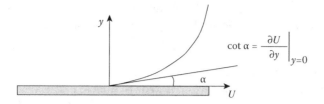

FIGURE 9.4 Viscous sheer stress is proportional to velocity gradient at the surface.

FIGURE 9.5 (**See color insert following page 240.**) Velocity boundary layer over a flat plate.

Equation 9.1 shows that viscous sheer stress is proportional to velocity gradient at the surface. The μ is called dynamic viscosity of fluid and is a measure of its resistance to flow. It is harder to make engine oil to flow than to make water flow and a larger force is needed to move water than to move air. This is because viscosity of oil is larger than that of water and viscosity of water is larger than that of air. The unit for dynamic viscosity in the SI system of units is kg/m·s or N.s/m². Dynamic viscosity of a fluid has similar behavior as its thermal conductivity. Dynamic viscosity of a gas increases with increase in temperature and decrease in molar mass. On the other hand, dynamic viscosity of liquids decreases with an increase in their temperatures.

Consider a flow with uniform velocity U_∞ approaching a surface. The fluid velocity right at the surface is zero. This is the so-called no-slip boundary condition, which implies that fluids do not slide over a surface. The zero-velocity particles slow down the fluid particles in the adjacent layer and those fluid particles in turn slow down the particles next to them. This generates a velocity profile, as shown in Figure 9.5, where fluid velocity continuously changes from zero to U_∞. The fluid layer near the surface in which this velocity change happens is called **velocity boundary layer**. Since viscous force is proportional to velocity gradient, velocity boundary layer is the only region in the flow where viscous shear is not zero. It may be argued that the fluid velocity reaches U_∞ at an infinite distance from the plate and therefore shear stress is present throughout the entire flow over the surface. However, velocity gradient and shear stress at some finite distance from a plate become very small and hard to recognize. Thickness of a velocity boundary layer, δ_v, is defined as the distance from the surface to the point where fluid velocity is $U = 0.99 U_\infty$. Note that as flow moves downstream, the continuous action of viscous force reduces the fluid velocity in layers near the wall and these slower moving layers slow down fluid layers above them. That is why, as shown in Figure 9.5, the thickness of velocity boundary layer increases as flow moves downstream the plate.

Equation 9.1 can be used to calculate viscous sheer stress if the variation of velocity U with distance from the surface, y, is known. A more practical approach that is used in most engineering calculations is to define a **local friction coefficient**, C_f, that relates viscous sheer stress to easily measurable quantities:

$$C_f \equiv \frac{\tau_s}{\frac{1}{2}\rho U_\infty^2}. \tag{9.2}$$

ρ is fluid density and U_∞ is the free stream velocity. Note that ρU_∞^2 has the same unit as τ_s and therefore the friction coefficient is a dimensionless parameter. It can be shown that if a fluid with velocity U_∞ is brought to rest in a frictionless process, its pressure will increase by $\rho U_\infty^2 / 2$. That is why $\rho U_\infty^2 / 2$ is called dynamic pressure of the flow.

Substituting for τ_s from Equation 9.1 in Equation 9.2, the following expression for local friction coefficient is obtained:

$$C_f \equiv \frac{\mu \frac{\partial U}{\partial y}\bigg|_{y=0}}{\frac{1}{2}\rho U_\infty^2}. \tag{9.3}$$

Either this equation, with some analytical or numerical data for velocity profile, or Equation 9.2, with some test data for viscous sheer stress, is used to obtain friction coefficient for different flows.

9.3 TEMPERATURE BOUNDARY LAYER AND CONVECTION HEAT TRANSFER COEFFICIENT

Consider the flow over a flat plate as shown in Figure 9.5 and assume that the incoming fluid has a uniform temperature that is less than the plate's surface temperature. The fluid particles that are in contact with the surface come to thermal equilibrium with that surface and their temperatures will be the same as the surface temperature. The heat transfer from the fluid particles next to the surface to fluid particles in the adjacent layer increases the internal energy and temperature of the particles in the adjacent layer. These particles transfer part of their energy downstream the plate and part of it to the particles in the layer above them. This will generate a smooth temperature gradient in the fluid next to the surface. The layer of a fluid next to a solid surface in which there is a significant temperature gradient is called **thermal boundary layer**. If T_∞ is the incoming fluid temperature and T_s is the surface temperature, the thickness of thermal boundary layer, δ_t, is defined as the distance from

Fundamentals of Convection Heat Transfer

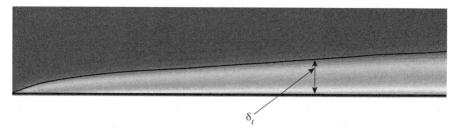

FIGURE 9.6 (See color insert following page 240.) Temperature boundary layer over a surface.

the surface to the point where $T - T_s = 0.99(T_\infty - T_s)$. Figure 9.6 shows temperature boundary layer over a flat plate.

The fluid layer at the surface has a zero velocity. Therefore, the only heat transfer mechanism in this layer is conduction, and the heat flux is given by

$$\dot{q}_{\text{cond}} = -k_f \frac{\partial T}{\partial y}\bigg|_{y=0}, \qquad (9.4)$$

where k_f is the thermal conductivity of fluid. Notice the similarity between this equation and Equation 9.1 for viscous sheer stress. The heat that is conducted to the layer next to the surface is transferred by the fluid motion, otherwise called convection,

$$\dot{q}_{\text{conv}} = h(T_s - T_\infty). \qquad (9.5)$$

An expression for **local convection heat transfer coefficient** is obtained by equating Equations 9.4 and 9.5:

$$h = \frac{-k_f \dfrac{\partial T}{\partial y}\bigg|_{y=0}}{T_s - T_\infty}. \qquad (9.6)$$

This expression, with some analytical or numerical data for fluid temperature near a surface or Equation 9.5 with measured heat flux data, is used to obtain convection heat transfer coefficient.

9.4 CONSERVATION EQUATIONS

Mass, momentum and energy conservation principles can be used to derive differential equations for the fluid's velocity components, temperature, and pressure. The process is very similar to the one used in Chapter 8 to develop the heat conduction equation. These derivations can be seen in many fluid mechanics and heat transfer text books [2,3] and will not be repeated here. Instead, a form of these equations that are valid for incompressible fluids with constant material properties and are the relevant equations for most electronics thermal management applications, will be presented in this section.

Let's assume U, V, and W are components of fluid velocity in x, y, and z directions, P and T are fluid pressure and temperature, and ρ, μ, and C_p are fluid density, dynamic viscosity, and specific heat at constant pressure. The **mass conservation equation**, also called **continuity equation**, for an **incompressible fluid** is

$$\frac{\partial U}{\partial x} + \frac{\partial V}{\partial y} + \frac{\partial W}{\partial z} = 0. \tag{9.7}$$

The three **momentum equations** for an **incompressible fluid** with **constant dynamic viscosity** are

$$\rho\left(\frac{\partial U}{\partial t} + U\frac{\partial U}{\partial x} + V\frac{\partial U}{\partial y} + W\frac{\partial U}{\partial z}\right) = -\frac{\partial P}{\partial x} + \mu\left(\frac{\partial^2 U}{\partial x^2} + \frac{\partial^2 U}{\partial y^2} + \frac{\partial^2 U}{\partial z^2}\right) + X, \tag{9.8}$$

$$\rho\left(\frac{\partial V}{\partial t} + U\frac{\partial V}{\partial x} + V\frac{\partial V}{\partial y} + W\frac{\partial V}{\partial z}\right) = -\frac{\partial P}{\partial y} + \mu\left(\frac{\partial^2 V}{\partial x^2} + \frac{\partial^2 V}{\partial y^2} + \frac{\partial^2 V}{\partial z^2}\right) + Y, \tag{9.9}$$

$$\rho\left(\frac{\partial W}{\partial t} + U\frac{\partial W}{\partial x} + V\frac{\partial W}{\partial y} + W\frac{\partial W}{\partial z}\right) = -\frac{\partial P}{\partial z} + \mu\left(\frac{\partial^2 W}{\partial x^2} + \frac{\partial^2 W}{\partial y^2} + \frac{\partial^2 W}{\partial z^2}\right) + Z. \tag{9.10}$$

The X, Y, and Z are components of body force per unit volume of the fluid in x, y, and z directions, respectively. An example of body force is gravitational force or the weight of the fluid that is important in natural convection flows. Note that the three momentum equations plus the continuity equation constitute a set of four simultaneous equations for four flow variables; three velocity components U, V, and W and pressure P. This is a complex set of nonlinear equations and does not have a general solution. Analytical solutions had been derived for a few simple flows but the solutions for most flows are obtained numerically.

The **energy equation** for an **incompressible fluid** with **constant thermal conductivity** is

$$\rho C_p\left(\frac{\partial T}{\partial t} + U\frac{\partial T}{\partial x} + V\frac{\partial T}{\partial y} + W\frac{\partial T}{\partial z}\right) = k\left(\frac{\partial^2 T}{\partial x^2} + \frac{\partial^2 T}{\partial y^2} + \frac{\partial^2 T}{\partial z^2}\right) + \mu\Phi + \dot{g}, \tag{9.11}$$

Fundamentals of Convection Heat Transfer

where $\dot{g}$ is heat generation per unit volume of the fluid and $\mu\Phi$ is viscous dissipation that accounts for the rate at which kinetic energy of fluid is converted to internal energy due to viscous effects in the fluid. Φ is given by

$$\Phi = 2\left[\left(\frac{\partial U}{\partial x}\right)^2 + \left(\frac{\partial V}{\partial y}\right)^2 + \left(\frac{\partial W}{\partial z}\right)^2\right] + \left(\frac{\partial U}{\partial x} + \frac{\partial U}{\partial y}\right)^2 + \left(\frac{\partial W}{\partial y} + \frac{\partial V}{\partial z}\right)^2 + \left(\frac{\partial U}{\partial z} + \frac{\partial W}{\partial x}\right)^2. \tag{9.12}$$

The viscous dissipation term is important only in very high speed flows (i.e., speeds close to and higher than the speed of sound). Note that if the fluid is motionless, the energy equation reduces to the three-dimensional heat conduction equation.

9.5 BOUNDARY LAYER EQUATIONS

Drag force and convection heat transfer rate are the two most important characteristics of a flow in most engineering applications. They are obtained if the friction coefficient and convection heat transfer coefficient are known. Friction coefficient is proportional to velocity gradient at the surface and convection heat transfer coefficient is proportional to temperature gradient at the surface. That is why fluid mechanics and heat transfer researchers have been interested in a simpler subset of the conservation equations that is valid inside the velocity and thermal boundary layers and are called **boundary layer equations**.

Let's consider a two-dimensional boundary layer such as those shown in Figures 9.5 and 9.6. The following observations can be made about these boundary layers.

- Thicknesses of velocity and temperature boundary layers are much smaller than the flow length; if L is the flow length, δ_v and $\delta_t \ll L$.
- The component of fluid velocity parallel to the surface is much larger than its component normal to the surface; $U \gg V$.
- Gradient of any flow variable in the direction normal to the surface is much larger than the gradient of that variable in the direction along the surface; $\partial U/\partial y \gg \partial U/\partial x$, $\partial T/\partial y \gg \partial T/\partial x$, $\partial P/\partial y \gg \partial P/\partial x$, and so forth.

A comparison between the magnitude of different terms of the continuity, momentum, and energy equations in a boundary layer shows that some of the terms are much smaller than others and can be eliminated without introducing significant errors [2]. The results are the **two-dimensional boundary layer equations**:

$$\frac{\partial U}{\partial x} + \frac{\partial V}{\partial y} = 0, \tag{9.13}$$

$$\frac{\partial U}{\partial t} + U\frac{\partial U}{\partial x} + V\frac{\partial U}{\partial y} = -\frac{1}{\rho}\frac{\partial P}{\partial x} + \nu\frac{\partial^2 U}{\partial y^2} + \frac{X}{\rho}, \tag{9.14}$$

$$\frac{\partial P}{\partial y} = Y, \tag{9.15}$$

$$\frac{\partial T}{\partial t} + U\frac{\partial T}{\partial x} + V\frac{\partial T}{\partial y} = \alpha\frac{\partial^2 T}{\partial y^2} + \frac{\nu}{C_p}\Phi + \frac{\dot{g}}{\rho C_p} \quad \text{where} \quad \Phi = \left(\frac{\partial U}{\partial y}\right)^2. \tag{9.16}$$

The $\nu = \mu/\rho$ is kinematic viscosity or momentum diffusivity and $\alpha = k/\rho C_p$ is thermal diffusivity of the fluid. Both are measured with the unit m²/s in the SI system of units. Note that the only terms remaining from the y-momentum equation are pressure and body force terms and the only term remaining from the viscous dissipation term is $(\partial U/\partial y)^2$.

REFERENCES

1. Van Dyke, M. 1982. *An album of fluid motion*. Stanford, CA: The Parabolic Press.
2. Schlichting, H. 1979. *Boundary layer theory*. New York, NY: McGraw-Hill.
3. Incropera, F. P., D. P. DeWitt, T. L. Bergman, and A. S. Lavine. 2006. *Fundamentals of heat and mass transfer*. Sixth Edition. Hoboken, NJ: John Wiley & Sons, Inc.

10 Forced Convection Heat Transfer: External Flows

10.1 NORMALIZED BOUNDARY LAYER EQUATIONS

Conservation equations have been derived for physical variables such as velocity components, pressure, and temperature. However, a lot can be learned about the solution of these equations if those variables are normalized. Consider the two-dimensional, steady-state boundary layer equations [1]:

$$\frac{\partial U}{\partial x} + \frac{\partial V}{\partial y} = 0. \tag{10.1}$$

$$U\frac{\partial U}{\partial x} + V\frac{\partial U}{\partial y} = -\frac{1}{\rho}\frac{\partial P}{\partial x} + \nu\frac{\partial^2 U}{\partial y^2} + \frac{X}{\rho}, \tag{10.2}$$

$$\frac{\partial P}{\partial y} = Y, \tag{10.3}$$

$$U\frac{\partial T}{\partial x} + V\frac{\partial T}{\partial y} = \alpha\frac{\partial^2 T}{\partial y^2} + \frac{\nu}{C_p}\left(\frac{\partial U}{\partial y}\right)^2 + \frac{\dot{g}}{\rho C_p}. \tag{10.4}$$

If L is a characteristic length of the flow, the normalized coordinates will be defined as

$$x^* = \frac{x}{L} \quad \text{and} \quad y^* = \frac{y}{L}. \tag{10.5}$$

Also, if U_∞ and T_∞ are free stream velocity and temperature, respectively, and T_s is a constant surface temperature, the following normalized variables can be defined:

$$U^* = \frac{U}{U_\infty}, \quad V^* = \frac{V}{U_\infty}, \quad T^* = \frac{T - T_s}{T_\infty - T_s}, \quad P^* = \frac{P}{\rho U_\infty^2}, \tag{10.6}$$

219

$$X^* = \frac{XL}{\rho U_\infty^2}, \quad Y^* = \frac{YL}{\rho U_\infty^2}, \quad \dot{g}^* = \frac{\dot{g}L}{\rho U_\infty C_p (T_\infty - T_s)}. \tag{10.7}$$

Substituting these normalized variables into the boundary layer equations results in the following set of **normalized two-dimensional, steady-state boundary layer equations**:

$$\frac{\partial U^*}{\partial x^*} + \frac{\partial V^*}{\partial y^*} = 0, \tag{10.8}$$

$$U^* \frac{\partial U^*}{\partial x^*} + V^* \frac{\partial U^*}{\partial y^*} = -\frac{\partial P^*}{\partial x^*} + \frac{\nu}{U_\infty L} \frac{\partial^2 U^*}{\partial y^{*2}} + X^*, \tag{10.9}$$

$$\frac{\partial P^*}{\partial y^*} = Y^*, \tag{10.10}$$

$$U^* \frac{\partial T^*}{\partial x^*} + V^* \frac{\partial T^*}{\partial y^*} = \frac{\alpha}{U_\infty L} \frac{\partial^2 T^*}{\partial y^{*2}} + \frac{\nu U_\infty}{C_p L (T_s - T_\infty)} \left(\frac{\partial U^*}{\partial y^*} \right)^2 + \dot{g}^*. \tag{10.11}$$

It is also helpful to rewrite the expressions for the local friction coefficient and the local convection heat transfer coefficient,

$$C_f = \frac{\mu \left. \frac{\partial U}{\partial y} \right|_{y=0}}{\frac{1}{2} \rho U_\infty^2} \quad \text{and} \quad h = \frac{-k_f \left. \frac{\partial T}{\partial y} \right|_{y=0}}{T_s - T_\infty}, \tag{10.12}$$

in terms of the normalized variables:

$$C_f = \frac{2\nu}{U_\infty L} \left. \frac{\partial U^*}{\partial y^*} \right|_{y^*=0}, \tag{10.13}$$

$$\frac{hL}{k_f} = \left. \frac{\partial T^*}{\partial y^*} \right|_{y^*=0}. \tag{10.14}$$

Four groups of parameters appear in the normalized boundary layer equations, Equations 10.8 through 10.11, and the expressions for friction coefficient, Equation 10.13, and convection heat transfer coefficient, Equation 10.14. These are $\nu/U_\infty L$, $\alpha/U_\infty L$, $\nu U_\infty/C_p L(T_s - T_\infty)$ and hL/k_f. It can be easily verified that these groups are all dimensionless. For example, the units for ν, U_∞, and L are m²/s, m/s, and m, respectively, and therefore $\nu/U_\infty L$ is just a number with no unit. Three of these

Forced Convection Heat Transfer: External Flows   221

dimensionless groups appear as coefficients of different terms of the boundary layer equations and therefore affect the solution of these equations. The last one, hL/k_f, can be considered as a dimensionless convection heat transfer coefficient. Each of these dimensionless groups has an important physical meaning that will be described in the next section.

10.2 REYNOLDS NUMBER, PRANDTL NUMBER, ECKERT NUMBER, AND NUSSELT NUMBER

Let's consider the first dimensionless group, $\nu/U_\infty L$, that appears in the x-momentum equation. The inverse of this dimensionless group is called the **Reynolds number**,

$$\text{Re}_L \equiv \frac{U_\infty L}{\nu} = \frac{\rho U_\infty L}{\mu}. \tag{10.15}$$

The index L for Re indicates that characteristic length L has been used in the calculation of the Reynolds number. Different flows have different characteristic lengths. The characteristic length for flow over a flat plate is the distance from the leading edge of the plate, and for flow across a circular cylinder is the diameter of the cylinder. It can be shown that Reynolds number is a measure of the ratio of inertia force that moves the flow to viscous force that acts against the flow. Inertia force is equal to mass times acceleration and inertia force per unit volume is equal to density times acceleration. If velocity is scaled by U_∞ and length is scaled by L, it is seen that

$$\text{Inertia force/volume} = \rho \frac{dU}{dt} = \rho \frac{dU}{dx/U} \propto \rho \frac{\Delta U}{\Delta x/U} \propto \rho \frac{U_\infty}{L/U_\infty} \propto \frac{\rho U_\infty^2}{L}.$$

Note that we used $U = dx/dt$ and assumed $\Delta U \propto U_\infty$ and $\Delta x \propto L$. On the other hand, the viscous force applied to a fluid is equal to viscous shear stress, τ_s, times surface area A and viscous force per unit volume of fluid is

$$\text{Viscous force/volume} = \frac{\tau_s A}{\text{volume}} = \frac{A}{\text{volume}}\left(\mu \frac{\partial U}{\partial y}\right) \propto \frac{A}{\text{volume}}\left(\mu \frac{\Delta U}{\Delta y}\right)$$

$$\propto \frac{L^2}{L^3}\mu \frac{U_\infty}{L} \propto \frac{\mu U_\infty}{L^2}.$$

Therefore,

$$\frac{\text{inertia force}}{\text{viscous force}} \propto \frac{\rho U_\infty^2/L}{\mu U_\infty/L^2} = \frac{\rho U_\infty L}{\mu} \equiv \text{Re}_L.$$

A large value for the Reynolds number indicates that inertia force is much larger than viscous force. A flow in which inertia force is much larger than viscous force is more likely to be turbulent. Therefore, the value of Reynolds number for a flow gives an indication of whether that flow is laminar or turbulent. In fact, for every flow, there is a specific value of the Reynolds number beyond which the flow is inherently turbulent. This value, which is different for different flows, is called the **critical Reynolds number**.

The first dimensionless group that appears on the right side of the normalized energy equation is $\alpha/U_\infty L$. It can be written as the product of two dimensionless groups:

$$\frac{\alpha}{U_\infty L} = \frac{\nu}{U_\infty L} \cdot \frac{\alpha}{\nu}.$$

The first group is the inverse of the Reynolds number and the second one is the inverse of a new dimensionless group called the **Prandtl number**,

$$\Pr \equiv \frac{\nu}{\alpha}. \tag{10.16}$$

The Prandtl number is the ratio of momentum diffusivity to thermal diffusivity of a fluid. It gives an indication of how well momentum, compared to energy, is transported by molecules of a fluid. Therefore, it is a measure of relative thickness of velocity and thermal boundary layers. Gases such as air have a Prandtl number close to one. Therefore, momentum and energy diffusions in a gas are comparable and velocity and thermal boundary layers have almost the same thicknesses. Liquid metals have very low Prandtl numbers meaning that thermal diffusion goes much deeper than momentum diffusion and thermal boundary layers are much thicker than velocity boundary layers. On the other hand, oils have very high Prandtl numbers indicating that momentum diffusion is more effective than thermal diffusion and velocity boundary layers are much thicker than thermal boundary layers.

The second dimensionless group in the normalized energy equation, $\nu U_\infty / C_p L (T_s - T_\infty)$, can be broken into two dimensionless groups,

$$\frac{\nu U_\infty}{C_p L (T_s - T_\infty)} = \frac{\nu}{U_\infty L} \times \frac{U_\infty^2}{C_p (T_s - T_\infty)}.$$

The first group is the inverse of the Reynolds number. The second group is twice the ratio of the kinetic energy per unit mass of the free stream flow to the difference in enthalpy per unit mass of the fluid across the thermal boundary layer. This dimensionless group is called the **Eckert number**,

$$\mathrm{Ec} \equiv \frac{U_\infty^2}{C_p (T_s - T_\infty)}. \tag{10.17}$$

Note that the coefficient of the viscous dissipation term in the energy equation is the ratio of the Eckert to the Reynolds number, Ec/Re. This ratio gives an indication of the importance of this term.

Forced Convection Heat Transfer: External Flows

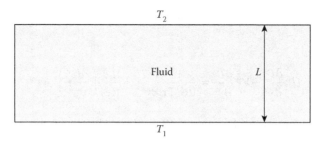

FIGURE 10.1 A layer of fluid with thickness L and different temperatures on its two sides.

The dimensionless group that appears in Equation 10.14 is a dimensionless convection heat transfer coefficient and is called the **Nusselt number**,

$$\mathrm{Nu}_L \equiv \frac{hL}{k_f}. \tag{10.18}$$

The index L indicates that the characteristic length L was used in the Nusselt number calculation. Equation 10.14 shows that the Nusselt number is equal to the normalized temperature gradient at the surface. It can also be shown that the Nusselt number is the ratio of the convection to the conduction heat flux in a fluid layer. Consider a fluid layer with thickness L over a surface that is at temperature T_1 as shown in Figure 10.1. Let's assume that the temperature on the other side of the fluid layer is T_2. If the fluid is not moving, the heat transfer through the fluid layer is through conduction and heat flux is given by $\dot{q}_{\mathrm{cond}} = k_f (T_1 - T_2)/L$. However, if the fluid moves, convection heat transfer becomes important and heat flux is given by $\dot{q}_{\mathrm{conv}} = h(T_1 - T_2)$. The ratio of convection heat flux to conduction heat flux through this layer is

$$\frac{\dot{q}_{\mathrm{conv}}}{\dot{q}_{\mathrm{cond}}} = \frac{h(T_1 - T_2)}{k_f(T_1 - T_2)/L} = \frac{hL}{k_f} \equiv \mathrm{Nu}_L.$$

Therefore, the Nusselt number is a measure of the increase in heat transfer due to fluid motion.

10.3 FUNCTIONAL FORMS OF FRICTION COEFFICIENT AND CONVECTION HEAT TRANSFER COEFFICIENT

Let's consider the normalized boundary layer equations,

$$\frac{\partial U^*}{\partial x^*} + \frac{\partial V^*}{\partial y^*} = 0, \tag{10.19}$$

$$U^* \frac{\partial U^*}{\partial x^*} + V^* \frac{\partial U^*}{\partial y^*} = -\frac{\partial P^*}{\partial x^*} + \frac{1}{\mathrm{Re}_L} \frac{\partial^2 U^*}{\partial y^{*2}} + X^*, \tag{10.20}$$

$$\frac{\partial P^*}{\partial y^*} = Y^*, \qquad (10.21)$$

$$U^* \frac{\partial T^*}{\partial x^*} + V^* \frac{\partial T^*}{\partial y^*} = \frac{1}{\mathrm{Re}_L \mathrm{Pr}} \frac{\partial^2 T^*}{\partial y^{*2}} + \frac{\mathrm{Ec}}{\mathrm{Re}_L} \left(\frac{\partial U^*}{\partial y^*} \right)^2 + \dot{g}^*, \qquad (10.22)$$

and the expressions for local friction coefficient and Nusselt number,

$$C_f = \frac{2}{\mathrm{Re}_L} \left. \frac{\partial U^*}{\partial y^*} \right|_{y^*=0} \qquad (10.23)$$

and

$$\mathrm{Nu}_L = \left. \frac{\partial T^*}{\partial y^*} \right|_{y^*=0}. \qquad (10.24)$$

These equations will be used to understand what flow variables affect friction coefficient and convection heat transfer coefficient.

The *x*-momentum equation shows that the normalized *x*-component of fluid velocity, U^*, depends on x^*, y^*, Re_L, $\partial P^*/\partial x$, and X^*. This can be expressed by a general functional form,

$$U^* = f_1\left(x^*, y^*, \mathrm{Re}_L, \frac{\partial P^*}{\partial x^*}, X^*\right). \qquad (10.25)$$

Equation 10.25 shows that

$$\left. \frac{\partial U^*}{\partial y^*} \right|_{y^*=0} = \left. \frac{\partial f_1}{\partial y^*} \right|_{y^*=0} = f_2\left(x^*, \mathrm{Re}_L, \frac{\partial P^*}{\partial x^*}, X^*\right). \qquad (10.26)$$

Using Equation 10.26 in Equation 10.23 indicates that the local friction coefficient has the functional form,

$$C_f = f\left(x^*, \mathrm{Re}_L, \frac{\partial P^*}{\partial x^*}, X^*\right). \qquad (10.27)$$

Equation 10.27 shows that the value of the local friction coefficient depends on the location in the flow direction, the Reynolds number, the pressure gradient in the flow direction, and the body force. This is a very important result and has helped fluid mechanics researchers in obtaining experimental correlations for the friction coefficient. Note that **for a flow with no pressure gradient and no body force in the flow direction** the friction coefficient is a function of the Reynolds number and the location in the flow direction;

$$C_f = f(x^*, \mathrm{Re}_L). \qquad (10.28)$$

Forced Convection Heat Transfer: External Flows

Engineers are mainly interested in the total friction or drag force applied to a surface. The total drag force applied to a surface is calculated by integrating the local viscous shear force over the entire area of the surface. If C_f is the local friction coefficient over the area dA, the total drag force is

$$F_D = \int_A C_f \frac{1}{2}\rho U_\infty^2 dA = \frac{1}{2}\rho U_\infty^2 A \left[\frac{1}{A}\int_A C_f dA\right].$$

The quantity inside the bracket is called the **average friction coefficient** over the entire surface area,

$$\bar{C}_f = \frac{1}{A}\int_A C_f dA, \qquad (10.29)$$

and the equation for total drag force becomes

$$F_D = \bar{C}_f \frac{1}{2}\rho U_\infty^2 A. . \qquad (10.30)$$

Substituting from Equation 10.27 into Equation 10.29 and carrying out the integration shows that

$$\bar{C}_f = \bar{f}\left(\text{Re}_L, \frac{\partial P^*}{\partial x^*}, X^*\right), \qquad (10.31)$$

where $\bar{f}$ is the average of function f over the entire surface area. It is seen that, **for flows with no pressure gradient and no body force in the flow direction**, the average friction coefficient is a function of the Reynolds number only,

$$\bar{C}_f = \bar{f}(\text{Re}_L). \qquad (10.32)$$

Now, let's consider the normalized boundary layer energy equation. The solution to this equation will have a general functional form,

$$T^* = g_1\left(x^*, y^*, \text{Re}_L, \text{Pr}, \text{Ec}, \frac{\partial P^*}{\partial x^*}, \dot{g}^*, X^*\right). \qquad (10.33)$$

The dependency on the pressure gradient and the body force is due to the fact that fluid temperature depends on its velocity that, in general, is a function of those variables. Temperature gradient at the wall, and therefore the local Nusselt number, has the following functional form,

$$\text{Nu}_L = \frac{hL}{k_f} = \frac{\partial T^*}{\partial y^*}\bigg|_{y^*=0} = \frac{\partial g_1}{\partial y^*}\bigg|_{y^*=0} = g\left(x^*, \text{Re}_L, \text{Pr}, \text{Ec}, \frac{\partial P^*}{\partial x^*}, \dot{g}^*, X^*\right). \qquad (10.34)$$

Equation 10.34 shows that the local Nusselt number is a function of the location in the flow direction, the Reynolds number, the Prandtl number, the Eckert number, the pressure gradient in the flow direction, the heat generation, and the body force in the flow direction. Note that for a low speed flow with no pressure gradient, no heat generation, and no body force,

$$\text{Nu}_L = g(x^*, \text{Re}_L, \text{Pr}). \tag{10.35}$$

The local Nusselt number is used to calculate the local convection heat transfer coefficient h and local convection heat flux,

$$\dot{q}_x = h(T_s - T_\infty).$$

The total convection heat transfer rate is obtained by integrating the local convection heat flux over the entire surface area,

$$\dot{Q}_{conv} = \int_A \dot{q}_x dA = \int_A h(T_s - T_\infty) dA = A(T_s - T_\infty)\left[\frac{1}{A}\int_A h \, dA\right].$$

The terms inside the bracket define the average convection heat transfer coefficient,

$$\bar{h} = \left[\frac{1}{A}\int_A h \, dA\right], \tag{10.36}$$

and the equation for the total convection heat transfer rate becomes the well-known Newton's law of cooling,

$$\dot{Q}_{conv} = \bar{h}A(T_s - T_\infty). \tag{10.37}$$

The average convection heat transfer coefficient is used for defining the average Nusselt number,

$$\overline{\text{Nu}}_L = \frac{\bar{h}L}{k_f}. \tag{10.38}$$

It can be easily shown that

$$\overline{\text{Nu}}_L = \frac{\bar{h}L}{k_f} = \frac{L}{k_f}\left[\frac{1}{A}\int_A h \, dA\right] = \frac{L}{k_f}\left[\frac{1}{A}\int_A \text{Nu}_L \frac{k_f}{L} dA\right] = \frac{1}{A}\int_A \text{Nu}_L dA, \tag{10.39}$$

which will have the following functional form,

$$\overline{\text{Nu}}_L = \bar{g}\left(\text{Re}_L, \text{Pr}, \text{Ec}, \frac{\partial P^*}{\partial x^*}, \dot{g}^*, X^*\right). \tag{10.40}$$

Forced Convection Heat Transfer: External Flows

It is worthwhile to mention that the Nusselt number **in a low speed flow with no pressure gradient along the flow direction, no heat generation, and no body force along the flow direction** is a function of the Reynolds and the Prandtl numbers only.

$$\overline{Nu_L} = \overline{g}(Re_L, Pr). \tag{10.41}$$

10.4 FLOW OVER FLAT PLATES

Flow over a flat plate is probably the most widely studied fluid flow. One reason is certainly the simplicity of this flow. However, another important reason is that it represents a good approximation to many real flow situations. Examples are flow over the top and sides of a refrigeration truck, flow over a printed circuit board, flow over the top surface of an electronic package and flow around widely spaced fins of a plate-fin heat sink.

Consider a flow parallel to a flat plate as shown in Figure 10.2. If the incoming flow stream is not turbulent, the flow at the leading edge of the plate will be laminar. However, at some point downstream the plate, velocity fluctuations and flow instabilities develop in the flow. Bursts of turbulent flow may appear and disappear in this region, which is called the transition region. Further downstream, the flow becomes fully turbulent. The start of the turbulent region corresponds to the location where the local Reynolds number, $Re_x = \rho U_\infty x/\mu$, is equal to 5×10^5 where x is the distance from the leading edge of the plate. This is the critical Reynolds number for parallel flow over a flat plate. Note that the characteristic length L in the definition of the Reynolds number for this flow is the distance x from the leading edge of the plate.

Two-dimensional boundary layer equations apply to fluid flow near a flat plate. This flow will be studied in detail and appropriate correlations for the friction coefficient and the convection heat transfer coefficient will be given in Sections 10.4.1 to 10.4.3.

10.4.1 LAMINAR FLOW OVER A FLAT PLATE WITH CONSTANT TEMPERATURE

Consider a low-speed laminar flow over a flat plate with no pressure gradient, no heat generation, and no body force. The boundary layer equations for this flow are

$$\frac{\partial U}{\partial x} + \frac{\partial V}{\partial y} = 0, \tag{10.42}$$

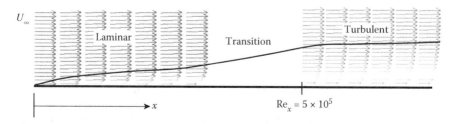

FIGURE 10.2 Laminar and turbulent flow regions over a flat plate.

$$U\frac{\partial U}{\partial x} + V\frac{\partial U}{\partial y} = \nu\frac{\partial^2 U}{\partial y^2}, \tag{10.43}$$

$$U\frac{\partial T}{\partial x} + V\frac{\partial T}{\partial y} = \alpha\frac{\partial^2 T}{\partial y^2}, \tag{10.44}$$

and the boundary conditions are

$$U(x,0) = V(x,0) = 0, \; U(x,\infty) = U_\infty, \; U(0,y) = U_\infty,$$
$$T(x,0) = T_s, \; T(x,\infty) = T_\infty, \; T(0,y) = T_\infty. \tag{10.45}$$

A solution for these equations was obtained by Blasius [1]. First, a continuous and twice differentiable function $\psi(x,y)$ is defined such that the velocity components U and V are

$$U = \frac{\partial \psi}{\partial y} \quad \text{and} \quad V = -\frac{\partial \psi}{\partial x}. \tag{10.46}$$

Substituting these into Equation 10.42 shows that the continuity equation is automatically satisfied. Then, new independent and dependent variables and normalized temperature were defined as

$$\eta = y\sqrt{\frac{U_\infty}{\nu x}} \quad f(\eta) = \frac{\psi}{U_\infty\sqrt{\nu x / U_\infty}}, \quad T^* = \frac{T - T_s}{T_\infty - T_s}. \tag{10.47}$$

These will reduce partial differential Equations 10.43 and 10.44 to the following ordinary differential equations,

$$2\frac{d^3 f}{d\eta^3} + f\frac{d^2 f}{d\eta^2} = 0, \quad \text{Boundary conditions:} \quad \left.\frac{df}{d\eta}\right|_{\eta=0} = 0, \; f(0) = 0, \; \left.\frac{df}{d\eta}\right|_{\eta=\infty} = 1 \tag{10.48}$$

$$\frac{d^2 T^*}{d\eta^2} + \frac{\Pr}{2} f \frac{dT^*}{d\eta} = 0, \quad \text{Boundary conditions:} \; T^*(0) = 0 \quad T^*(\infty) = 1. \tag{10.49}$$

Note that $df/d\eta = U/U_\infty$ and the above two boundary conditions on $df/d\eta$ at $\eta = 0$ and $\eta = \infty$ come from $U = 0$ at the wall and $U = U_\infty$ at a distance far from the surface. Figure 10.3 shows U/U_∞ and T^* obtained by the numerical solution of Equations 10.48 and 10.49. It is seen that the profiles for normalized velocity, $U/U_\infty = df/d\eta$,

Forced Convection Heat Transfer: External Flows

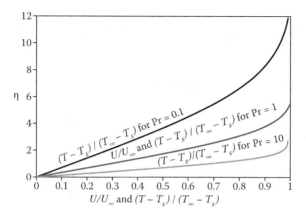

FIGURE 10.3 Normalized velocity and temperature profiles for laminar flow over flat plate.

and normalized temperature, T^*, are the same for $Pr = 1$. This should be expected since differential equations for $df/d\eta$ and T^*, are exactly the same if $Pr = 1$. It is also seen that the temperature boundary layer thickness is larger than velocity boundary layer thickness if $Pr < 1$ and is smaller than velocity boundary layer thickness if $Pr > 1$.

The numerical solutions of Equations 10.48 and 10.49 were used to obtain the following expressions for the thicknesses of the velocity and thermal boundary layers, the local friction coefficient and the local Nusselt number for low-speed laminar flow over a flat plate with no pressure gradient, no body force, and no heat generation [1]:

$$\delta_v = 5x\,\mathrm{Re}_x^{-1/2} \quad \text{and} \quad \delta_t = \delta_v\,\mathrm{Pr}^{-1/3}, \tag{10.50}$$

$$C_f = 0.664\,\mathrm{Re}_x^{-1/2}, \tag{10.51}$$

$$\mathrm{Nu}_x = \frac{hx}{k_f} = 0.332\,\mathrm{Re}_x^{1/2}\,\mathrm{Pr}^{1/3} \quad \mathrm{Pr} \geq 0.6. \tag{10.52}$$

Note that the index x in Re_x and Nu_x indicates that the distance x from the leading edge of the plate was used as the characteristic length of the flow when calculating these dimensionless groups.

It is seen that the thickness of the velocity and temperature boundary layers, and the Nusselt number, are proportional to $x^{1/2}$. On the other hand, the friction coefficient and the convection heat transfer coefficient are proportional to $x^{-1/2}$. Figure 10.4 shows variations of these variables along a flat plate over which air flows with free stream velocity of $U_\infty = 1$ m/s. Similar behaviors can be seen for other fluids and flow velocities.

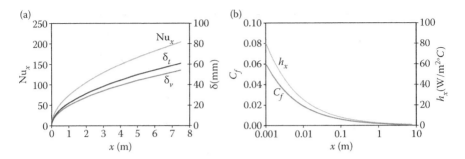

FIGURE 10.4 Variations of the Nusselt number, the velocity and temperature boundary layers thicknesses, the friction coefficient, and the convection heat transfer coefficient in airflow over a flat plate ($U_\infty = 1$ m/s).

The average friction coefficient and Nusselt number over the length L from the leading edge of a plate are calculated through the following integrals:

$$\bar{C}_f = \frac{1}{A}\int_A C_f dA = \frac{1}{x}\int_0^L C_f dx, \tag{10.53}$$

$$\overline{Nu}_L = \frac{\bar{h}L}{k_f} = \frac{L}{k_f}\left[\frac{1}{A}\int_A h dA\right] = \frac{L}{k_f}\left[\frac{1}{L}\int_0^L h dx\right] = \frac{1}{k_f}\int_0^L \frac{Nu_x k_f}{x}dx, \tag{10.54}$$

$$\overline{Nu}_L = \int_0^L \frac{Nu_x}{x}dx. \tag{10.55}$$

Substituting Equations 10.51 and 10.52 into Equations 10.53 and 10.55, respectively, and performing the integration results in the following expressions for the average friction coefficient and the Nusselt number:

$$\bar{C}_f = 1.328\,Re_L^{-1/2} \tag{10.56}$$

$$\overline{Nu}_L = \frac{\bar{h}L}{k_f} = 0.664\,Re_L^{1/2}\,Pr^{1/3} \quad Pr \geq 0.6. \tag{10.57}$$

Note that the average friction coefficient and the average Nusselt number over the length L of a plate are twice the local values of these quantities at distance L from the leading edge of the plate.

Fluid properties are required to calculate the Reynolds and the Prandtl numbers and therefore the friction coefficient and the convection heat transfer coefficient. These properties are usually functions of fluid temperature and, in some case, its pressure. However, fluid temperature changes from T_s at the surface to T_∞ outside the

Forced Convection Heat Transfer: External Flows

boundary layer. Unless otherwise specified, fluid properties have to be calculated at the average temperature between the surface and the free stream. This temperature is called the **film temperature**, $T_f = 0.5(T_s + T_\infty)$.

Example 10.1

Consider the plate-fin heat sink shown in Figure 10.5 and assume that each of its fins can be treated as a flat plate at a uniform temperature of 80°C. Calculate total drag force and convection heat transfer rate per fin if air with 2 m/s velocity, atmospheric pressure, and 20°C temperature flows between these fins. Note that each fin has two sides that contribute to both friction force against the fluid flow and convection heat transfer rate to the fluid.

Solution

Since total friction force and total convection heat transfer rate from a fin have to be calculated, average values of the friction coefficient and the convection heat transfer coefficient over the entire length of the fin, $x = L = 0.06$ m, are needed.

The film temperature in this case is $T_f = (80 + 20)/2 = 50°C$ and air properties at this temperature and atmospheric pressure are

$$\rho = 1.0926 \, kg/m^3, \quad \mu = 1.961 \times 10^{-5} \, kg/m \cdot s, \quad k = 0.0277 \, W/m \cdot °C \quad \text{and}$$

$Pr = 0.7124$.

Reynolds number for this flow is

$$Re_L = \frac{\rho U_\infty L}{\mu} = \frac{1.0926 \, kg/m^3 \times 2 \, m/s \times 0.06 \, m}{1.961 \times 10^{-5} \, kg/m \cdot s} = 6686,$$

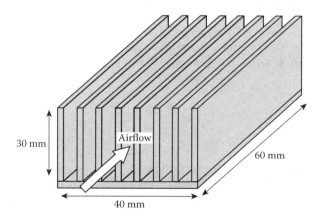

FIGURE 10.5 Plate-fin heat sink of Example 10.1.

which is less than the critical Reynolds number and therefore flow is laminar. The average friction coefficient and Nusselt number over the entire length of the fin are

$$\bar{C}_f = 1.328 \text{Re}_L^{-1/2} = 1.328 \times 6686^{-1/2} = 0.016$$

and

$$\overline{\text{Nu}}_L = \frac{\bar{h}L}{k_f} = 0.664 \text{Re}_L^{1/2} \text{Pr}^{1/3} = 0.664 \times 6686^{1/2} \times 0.7124^{1/3} = 48.5.$$

The average convection heat transfer coefficient is

$$\bar{h} = \overline{\text{Nu}}_L \frac{k_f}{L} = 48.5 \times \frac{0.0277 \text{ W/m°C}}{0.06 \text{ m}} = 22.4 \text{ W/m}^2\text{°C}.$$

Total friction force and total convection heat transfer rate per fin are

$$F_D = \bar{C}_f \frac{1}{2}\rho U_\infty^2 A = 0.016 \times \frac{1}{2} \times 1.0926 \text{ kg/m}^3 \times (2 \text{ m/s})^2$$
$$\times (2 \times 0.06 \text{ m} \times 0.03 \text{ m}) = 0.00013 \text{ N},$$

$$\dot{Q}_{conv} = \bar{h}A(T_s - T_\infty) = 22.4 \text{ W/m}^2\text{°C} \times (2 \times 0.06 \text{ m} \times 0.03 \text{ m})$$
$$\times (80 - 20)\text{°C} = 4.84 \text{ W}.$$

Example 10.2

Let's assume that the power dissipating components of a printed circuit board are attached to a common heat spreader as shown in Figure 10.6. The printed circuit board and the heat spreader are 600 mm long and 400 mm wide. Air with 3 m/s velocity, 40°C temperature, and 70 kPa pressure blows over the top of the heat spreader. If maximum allowable temperature of the heat spreader is 100°C, determine maximum allowable heat flux from the electronic

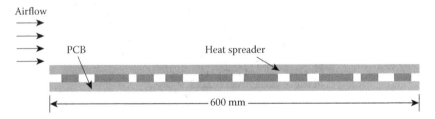

FIGURE 10.6 Printed circuit board and heat spreader of Example 10.2.

Forced Convection Heat Transfer: External Flows

components as a function of distance from the leading edge of the printed circuit board.

Solution

The film temperature is $T_f = (100 + 40)/2 = 70°C$ and air properties at this temperature and atmospheric pressure, 101.3 kPa, are

$$\rho(\text{at } 101.3\,\text{kPa}) = 1.0289\,\text{kg/m}^3, \quad \mu = 2.051\times10^{-5}\,\text{kg/m·s}, \quad k = 0.0292\,\text{W/m°C}$$
$$\text{and} \quad Pr = 0.7101.$$

Air is an ideal gas and its density is proportional to pressure as well. Therefore,

$$\rho(\text{at } 70\,\text{kPa}) = 1.0289\,\text{kg/m}^3 \times \frac{70\,\text{kPa}}{101.3\,\text{kPa}} = 0.711\,\text{kg/m}^3.$$

The local Reynolds number at distance x from leading edge of this heat spreader is

$$Re_x = \frac{\rho U_\infty x}{\mu} = \frac{0.711\,\text{kg/m}^3 \times 3\,\text{m/s} \times x}{2.051\times10^{-5}\,\text{kg/m·s}} = 1.04\times10^5\,x.$$

This will be less than the critical Reynolds number throughout the length of the heat spreader because $x \leq 0.6$ m and therefore flow will be laminar. The local Nusselt number at distance x from the leading edge of the heat spreader is

$$Nu_x = \frac{hx}{k_f} = 0.332 Re_x^{1/2} Pr^{1/3} = 0.332 \times (1.04\times10^5\,x)^{1/2} \times 0.7101^{1/3} = 95.5 x^{1/2},$$

and the local convection heat transfer coefficient at that location is

$$h = Nu_x \frac{k_f}{x} = 95.5 x^{1/2} \times \frac{0.0292\,\text{W/m°C}}{x} = 2.79 x^{-1/2}\,\text{W/m}^2\text{°C}.$$

The local allowable heat flux is

$$\dot{q}_{conv,x} = h(T_s - T_\infty) = 2.79 x^{-1/2}\,\text{W/m}^2\text{°C} \times (100 - 40)°C = 167.4 x^{-1/2}\,\text{W/m}^2.$$

The maximum allowable heat dissipation by the electronic components of this printed circuit board is obtained by integrating the local allowable heat flux over the length of the heat spreader

$$\dot{Q}_{conv,max} = \int_A \dot{q}_{conv,x} dA = \int_A 167.4 x^{-1/2} dA = \int_0^{0.6} 167.4 x^{-1/2} \times 0.4 dx$$
$$= 133.92 x^{1/2}\Big|_0^{0.6} = 103.7\,\text{W}.$$

The same results could have been obtained by using the average convection heat transfer coefficient over the entire length of the heat spreader, $\bar{h}$, and using the Newton's law of cooling for the entire heat spreader area,

$$\dot{Q}_{conv,max} = \bar{h}A(T_s - T_\infty).$$

10.4.2 Turbulent Flow over a Flat Plate with Uniform Temperature

If the local Reynolds number is larger than 5×10^5, flow will be turbulent. There are no closed form solutions for the boundary layer equations of turbulent flows over a flat plate. However, turbulent flows have been extensively studied, experimentally and numerically. Researchers have developed many models to predict friction force and convection heat transfer rate in turbulent flows. These models are called turbulence models and will be described in more details in Chapter 14. They can be used to simplify boundary layer equations for turbulent flows. However, even the simplified forms of the turbulent boundary layer equations do not result in closed form solutions and most available correlations for turbulent flows are semiempirical (i.e., they have been developed based on a combination of experimental observations and flow modeling).

It has been observed that the following semiempirical correlations predict fairly accurately the boundary layer thickness, the friction coefficient, and the Nusselt number in turbulent flow over a flat plate with constant temperature [1]:

$$\delta_v \approx \delta_t = 0.37x \operatorname{Re}_x^{-1/5} \quad 5 \times 10^5 \leq \operatorname{Re}_x \leq 10^7, \tag{10.58}$$

$$C_f = 0.0592 \operatorname{Re}_x^{-1/5} \quad 5 \times 10^5 \leq \operatorname{Re}_x \leq 10^7, \tag{10.59}$$

$$\operatorname{Nu}_x = \frac{hx}{k_f} = 0.0296 \operatorname{Re}_x^{4/5} \operatorname{Pr}^{1/3} \quad 5 \times 10^5 \leq \operatorname{Re}_x \leq 10^7 \quad 0.6 \leq \operatorname{Pr} \leq 60. \tag{10.60}$$

Note that the thicknesses of velocity and temperature boundary layers in turbulent flows over a flat plate are almost equal. These equations show that the boundary layer thickness and the Nusselt number are proportional to $x^{4/5}$, and the friction coefficient and the convection heat transfer coefficient are proportional to $x^{-1/5}$. Compared to laminar flow, the boundary later thickness and the Nusselt number increase faster, and the friction coefficient and the convection heat transfer coefficient drop slower with the distance from the leading edge of the plate. Figures 10.7 and 10.8 show variations of these parameters in laminar and turbulent boundary layers of air with a free stream velocity of 1 m/s. Sudden increase in the values of these parameters, as flow turns turbulent, and different rates of change of those in the turbulent region compared to the laminar region are seen in these figures.

Let's consider flow over a flat plate with length L and assume that the flow is laminar at the leading edge of the plate and turns turbulent at location x_c where the

Forced Convection Heat Transfer: External Flows

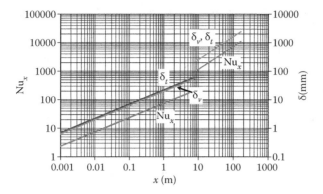

FIGURE 10.7 Variation of the Nusselt number and the boundary layer thickness in airflow over a flat plate ($U_\infty = 1$ m/s).

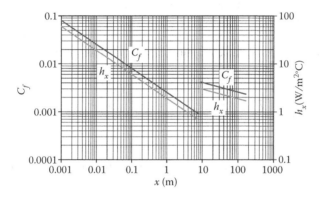

FIGURE 10.8 Variation of the friction coefficient and the convection heat transfer coefficient in airflow over a flat plate ($U_\infty = 1$ m/s).

Reynolds number is equal to the critical value of 5×10^5. The average friction coefficient over the length L of the plate is obtained as follows:

$$\bar{C}_f = \frac{1}{L}\int_0^L C_f dx = \frac{1}{L}\left(\int_0^{x_c} C_{f,\text{laminar}} dx + \int_{x_c}^L C_{f,\text{turbulent}} dx\right), \tag{10.61}$$

$$\bar{C}_f = \frac{1}{L}\left(\int_0^{x_c} 0.664 \text{Re}_x^{-1/2} dx + \int_{x_c}^L 0.0592 \text{Re}_x^{-1/5} dx\right),$$

$$\bar{C}_f = 0.074 \text{Re}_L^{-1/5} - 1742 \text{Re}_L^{-1} \quad 5 \times 10^5 \leq \text{Re}_L \leq 10^7. \tag{10.62}$$

Similarly, the average Nusselt number over the length L of the plate is

$$\overline{Nu}_L = \frac{\overline{h}L}{k_f} = \int_0^L \frac{Nu_x}{x} dx = \int_0^{x_c} \frac{Nu_{x,\text{laminar}}}{x} dx + \int_{x_c}^L \frac{Nu_{x,\text{turbulent}}}{x} dx, \quad (10.63)$$

$$\overline{Nu}_L = \frac{\overline{h}L}{k_f} = \int_0^{x_c} \frac{0.332 Re_x^{1/2} Pr^{1/3}}{x} dx + \int_{x_c}^L \frac{0.0296 Re_x^{4/5} Pr^{1/3}}{x} dx,$$

$$\overline{Nu}_L = \frac{\overline{h}L}{k_f} = (0.037 Re_L^{4/5} - 871) Pr^{1/3} \quad 0.6 \leq Pr \leq 60 \quad \text{and} \quad 5 \times 10^5 \leq Re_L \leq 10^7.$$

(10.64)

Equations 10.62 and 10.64 for the average friction coefficient and the average Nusselt number over a flat plate will reduce to the following simpler forms if Re_L is much larger than Re_c or if the flow over the entire length of the plate is turbulent. The latter situation will exist if the incoming flow is already turbulent or if it is perturbed and turned turbulent by surface roughness, small objects on the surface, and so on, at the leading edge of the plate.

$$\overline{C}_f = 0.074 Re_L^{-1/5} \quad 5 \times 10^5 \leq Re \leq 10^7, \quad (10.65)$$

$$\overline{Nu}_L = \frac{\overline{h}L}{k_f} = 0.037 Re_L^{4/5} Pr^{1/3} \quad 0.6 \leq Pr \leq 60 \quad \text{and} \quad 5 \times 10^5 \leq Re_L \leq 10^7.$$

(10.66)

Example 10.3

Consider a 600 mm by 400 mm printed circuit board. All the electronic components are mounted on the top of this board and dissipate 200 W. This heat transfers through the board to the other side where water flows with a velocity of 5 m/s, along the 600 mm side, and temperature of 50°C. What will be the board temperature if it can be approximated as a uniform surface temperature plate? Let's assume that the back of the board is rough enough to cause the flow to become turbulent at the leading edge of the plate.

Solution

Since the board temperature, T_s, is not known, the film temperature can not be calculated. Let's use 50°C instead of the film temperature and check the validity of this assumption after T_s is obtained. Properties of saturated water at 50°C are

$\rho = 988 \, \text{kg/m}^3$, $\mu = 5.47 \times 10^{-4} \, \text{kg/m} \times \text{s}$, $k = 0.644 \, \text{W/m} \times °\text{C}$ and $Pr = 3.55$.

Forced Convection Heat Transfer: External Flows

Reynolds number for this flow is

$$Re_L = \frac{\rho U_\infty L}{\mu} = \frac{988\,\text{kg/m}^3 \times 5\,\text{m/s} \times 0.6\,\text{m}}{5.47 \times 10^{-4}\,\text{kg/m} \times \text{s}} = 5.42 \times 10^6,$$

which is larger than the critical Reynolds number for flow over a flat plate. If we assume that the flow is turbulent over the entire surface of the board,

$$\overline{Nu}_L = \frac{\overline{h}L}{k_f} = 0.037 Re_L^{4/5} Pr^{1/3} = 0.037 \times (5.42 \times 10^6)^{4/5} \times 3.55^{1/3} = 13766.$$

The average convection heat transfer coefficient is

$$\overline{h} = \overline{Nu}_L \frac{k_f}{L} = 13766 \times \frac{0.644\,\text{W/m} \times °C}{0.6\,\text{m}} = 14775.5\,\text{W/m}^2°C.$$

The surface temperature is obtained using the Newton's law of cooling,

$$\dot{Q}_{conv} = \overline{h}A(T_s - T_\infty)$$

$$T_s = T_\infty + \frac{\dot{Q}_{conv}}{\overline{h}A} = 50°C + \frac{200\,\text{W}}{14775.5\,\text{W/m}^2°C \times (0.6\,\text{m} \times 0.4\,\text{m})} = 50.06°C.$$

It is seen that the board temperature is only slightly higher than the water temperature and that is due to the very large convection heat transfer coefficient. In general, convection heat transfer rates in liquid flows are much larger than in similar gas flows. It is also seen that the film temperature $T_f = 0.5(T_s + T_\infty)$ is almost the same as the fluid temperature of 50°C. Therefore, the use of water properties at 50°C to solve this problem is justified.

10.4.3 FLOW OVER A FLAT PLATE WITH UNIFORM SURFACE HEAT FLUX

The correlations given in Sections 10.4.1 and 10.4.2 for the Nusselt number have been derived for a plate with uniform surface temperature. However, there may be situations in which surface heat flux rather than its temperature is uniform. The local Nusselt number for laminar flow over a flat plate with a uniform surface heat flux is given by

$$Nu_x = \frac{hx}{k_f} = 0.453 Re_x^{1/2} Pr^{1/3} \quad Pr \geq 0.6, \tag{10.67}$$

and for turbulent flow over such a plate, where $Re_x \geq 5 \times 10^5$, is given by

$$Nu_x = \frac{hx}{k_f} = 0.0308 Re_x^{4/5} Pr^{1/3} \quad 0.6 \leq Pr \leq 60. \tag{10.68}$$

The local convection heat transfer coefficient is used to correlate the convection heat flux with the local surface temperature $T_{s,x}$:

$$\dot{q}_{conv} = h(T_{s,x} - T_\infty). \tag{10.69}$$

Example 10.4

Let's consider the printed circuit board and the heat spreader of Example 10.2 and assume that the electronic components on the board are distributed such that the heat spreader can be approximated as a uniform heat flux surface. If the maximum allowable spreader temperature is 100°C, what will the total allowable heat dissipation from this board be?

Solution

Equations 10.67 and 10.68 show that h decreases with x and Equation 10.69 shows that $T_{s,x}$ increases with x. Therefore, the maximum spreader temperature occurs at the end of the board, $x = 0.6$ m. The maximum allowable heat flux for this board is the heat flux that results in a heat spreader temperature of 100°C at the end of the board. From Example 10.2, the air properties at film temperature of $T_f = (100 + 40)/2 = 70°C$ and 70 kPa air pressure are

$$\rho = 0.711 \text{ kg/m}^3, \quad \mu = 2.051 \times 10^{-5} \text{ kg/m·s}, \quad k = 0.0292 \text{ W/m°C} \quad \text{and}$$

$$\text{Pr} = 0.7101.$$

The local Reynolds number at distance $x = 0.6$ m from the leading edge of this heat spreader is

$$\text{Re}_x = \frac{\rho U_\infty x}{\mu} = \frac{0.711 \text{ kg/m}^3 \times 3 \text{ m/s} \times 0.6}{2.051 \times 10^{-5} \text{ kg/m·s}} = 6.24 \times 10^4.$$

This is less than the critical Reynolds number and therefore flow is laminar. The local Nusselt number at the end of the heat spreader is

$$\text{Nu}_x = \frac{hx}{k_f} = 0.453 \text{Re}_x^{1/2} \text{Pr}^{1/3} = 0.453 \times (6.24 \times 10^4)^{1/2} \times 0.7101^{1/3} = 101,$$

and the local convection heat transfer coefficient at that location is

$$h = \text{Nu}_x \frac{k_f}{x} = 101 \times \frac{0.0292 \text{ W/m°C}}{0.6 \text{ m}} = 4.92 \text{ W/m}^2\text{°C}.$$

The maximum allowable heat flux is

$$\dot{q}_{\text{conv}} = h(T_{s,x} - T_\infty) = 4.92 \text{ W/m}^2\text{°C} \times (100 - 40)\text{°C} = 295.2 \text{ W/m}^2,$$

and the maximum allowable heat dissipation from the board is

$$\dot{Q}_{\text{conv}} = \dot{q}_{\text{conv}} A = 295.2 \text{ W/m}^2 \times (0.6 \text{ m} \times 0.4 \text{ m}) = 70.8 \text{ W}.$$

Note that the uniform heat flux assumption, with the same maximum allowable heat spreader temperature, resulted in less total allowable power dissipation in the board.

10.5 FLOW ACROSS CYLINDERS

Another external flow that has many practical applications is the flow across a long cylinder. Examples include heat transfer from hot wire anemometers used for velocity measurement, circular or square pin fins, and tubes in a cross flow heat exchanger.

Let's consider flow across a circular cylinder with diameter D, as shown in Figure 10.9, and assume that the free stream velocity is U_∞. As fluid particles approach the cylinder, their velocities diminish. In fact, the fluid particles that move normal to the axis of the cylinder will come to complete rest in front of the cylinder. This point is called the forward stagnation point. A close look at the region near the surface of the cylinder shows that boundary layers develop on this surface. However, there are two major differences between this boundary layer and the one over a flat plate.

1. There is a nonzero pressure gradient applied to this boundary layer. If x is the coordinate along the surface of the cylinder, $\partial P/\partial x < 0$ on the front side and $\partial P/\partial x > 0$ on the rear side of the cylinder.
2. The local free stream velocity, $U_{\infty,x}$, is not constant. It can be shown that $U_{\infty,x}$ increases wherever $\partial P/\partial x < 0$ and decreases wherever $\partial P/\partial x > 0$.

The Reynolds number for flow over a circular cylinder is defined based on its diameter and upstream free stream velocity,

$$\mathrm{Re}_D \equiv \frac{\rho U_\infty D}{\mu} = \frac{U_\infty D}{\nu}. \qquad (10.70)$$

Experimental observations have shown that the critical Reynolds number for this flow is $\mathrm{Re}_c = 2 \times 10^5$. The boundary layer over the surface of the cylinder will remain laminar if $\mathrm{Re}_D < \mathrm{Re}_c$ and will transition to turbulent if $\mathrm{Re}_D > \mathrm{Re}_c$.

Figure 10.10 shows four different laminar flows around a circular cylinder [2]. Figure 10.10a shows a very viscous flow with $\mathrm{Re}_D = 1.54$. Flow streamlines were

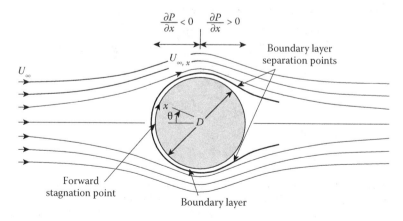

FIGURE 10.9 Typical characteristics of flow across a circular cylinder.

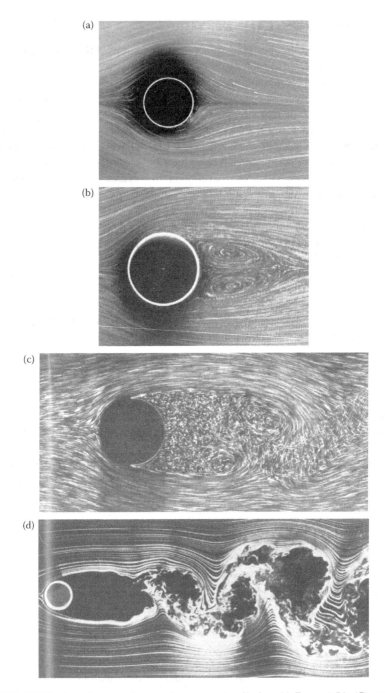

FIGURE 10.10 Four laminar flow regimes over a cylinder: (a) $Re_D = 1.54$, (Courtesy of Sadatoshi Taneda); (b) $Re_D = 26$, (Courtesy of Sadatoshi Taneda); (c) $Re_D = 2,000$, (Courtesy of ONERA); and (d) $Re_D = 10,000$, (Courtesy of Corke and Nagib). (From Van Dyke, M., *An Album of Fluid Motion*. Stanford, CA: The Parabolic Press, 1982.)

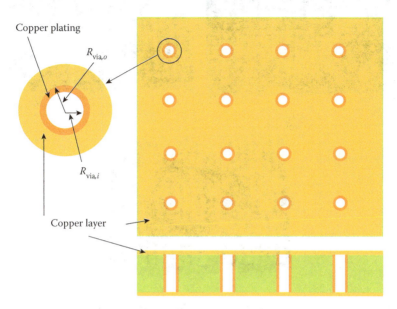

FIGURE 5.29 A printed circuit board with thermal vias.

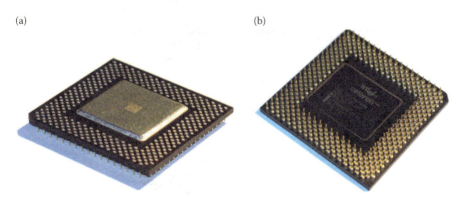

FIGURE 6.8 Top (a) and bottom (b) of a Pin Grid Array package. (http://en.wikipedia.org/wiki/Pin_grid_array.)

FIGURE 6.10 Bottom sides of a 23 mm × 23 mm 169 ball BGA (a), a 27 mm × 27 mm 256 ball BGA (b), and a 27 mm × 27 mm 292 ball BGA (c).

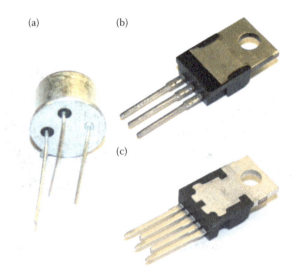

FIGURE 6.13 Three different TO packages: (a) TO-92, (b) three-lead TO-220, and (c) five-lead TO-220.

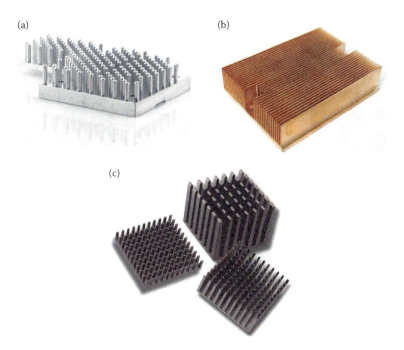

FIGURE 7.14 Examples of (a) cylindrical pin fin heat sink. (Courtesy of Cool Innovations.) (b) rectangular plate fin heat sink. (Courtesy of Auras Technology Co. Ltd.) (c) rectangular pin fin heat sink. (Courtesy of Aavid Thermalloy.)

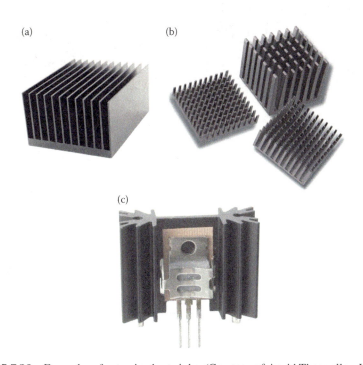

FIGURE 7.20 Examples of extrusion heat sinks. (Courtesy of Aavid Thermalloy, LLC.)

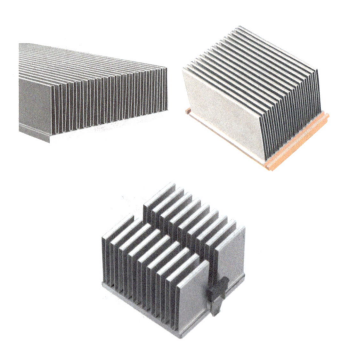

FIGURE 7.23 Examples of folded-fin heat sinks. (Courtesy of Aavid Thermalloy, LLC.)

FIGURE 7.25 Examples of skived fin heat sinks. (Courtesy of Auras Technology Co. Ltd.)

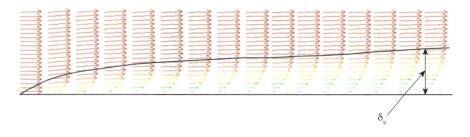

FIGURE 9.5 Velocity boundary layer over a flat plate.

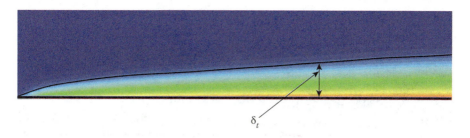

FIGURE 9.6 Temperature boundary layer over a surface.

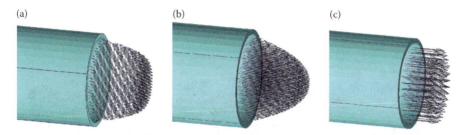

FIGURE 11.1 Pipes with actual velocity profiles (a and b) and their equivalent with a uniform velocity (c).

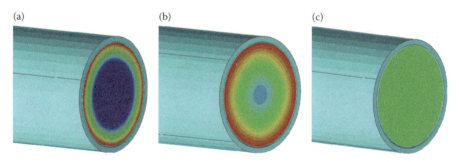

FIGURE 11.2 Pipes with actual temperature profiles (a and b) and their equivalent pipe with uniform temperature (c).

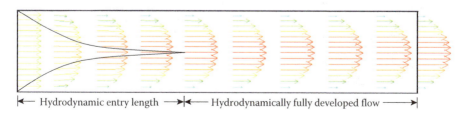

|← Hydrodynamic entry length →|← Hydrodynamically fully developed flow →|

FIGURE 11.3 Hydrodynamic entry length and fully developed flow in a pipe.

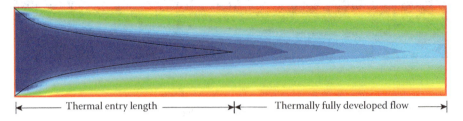

FIGURE 11.4 Thermal entry length and fully developed flow inside a pipe.

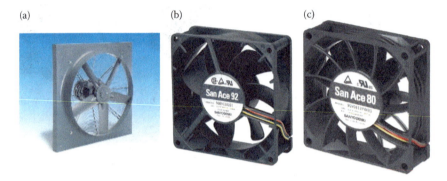

FIGURE 11.8 Examples of (a) propeller fan (Courtesy of Delta Electronics, Inc.), (b) tube-axial fan (Courtesy of Sanyo Denki), and (c) vane-axial fan (Courtesy of Sanyo Denki).

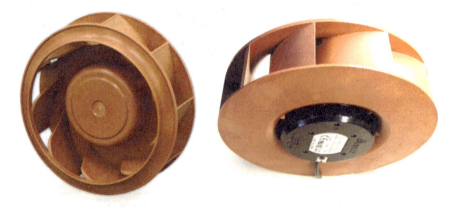

FIGURE 11.9 Two examples of radial fans or impellers. (Courtesy of Delta Electronics, Inc.)

FIGURE 11.10 Examples of blowers. (Courtesy of Delta Electronics, Inc.)

FIGURE 11.11 A mixed flow fan. (Courtesy of Delta Electronics, Inc.)

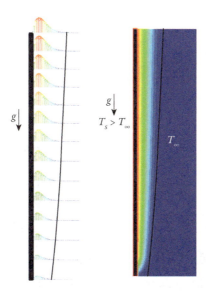

FIGURE 12.3 Natural convection velocity and temperature boundary layers near a vertical plate.

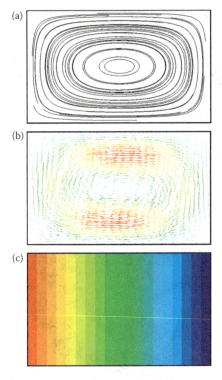

FIGURE 12.11 (a) Particle path, (b) velocity vectors, and (c) temperature contours in an enclosure at the conduction limit.

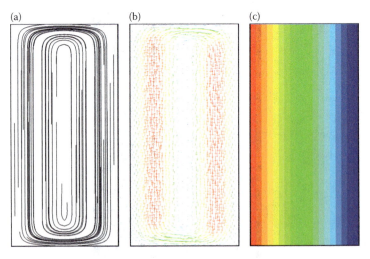

FIGURE 12.12 (a) Particle path, (b) velocity vectors, and (c) temperature contours in an enclosure at the tall limit.

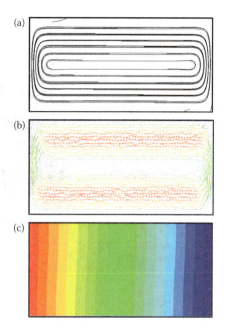

FIGURE 12.13 (a) Particle path, (b) velocity vectors, and (c) temperature contours in an enclosure at the shallow limit.

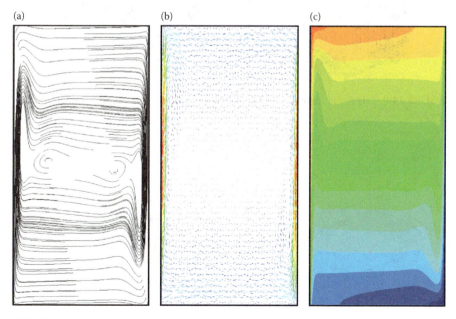

FIGURE 12.15 (a) Particle path, (b) velocity vectors, and (c) temperature contours in an enclosure for the boundary layer regime.

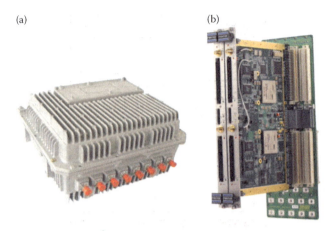

FIGURE 12.17 (a) An outdoor electronic enclosure with vertical fins and (b) a pair of printed circuit boards mounted on a backplane.

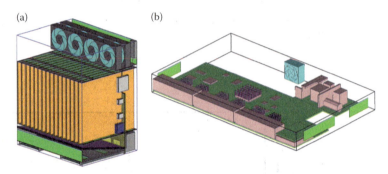

FIGURE 14.8 Thermal models of two electronic systems created inside a thermal simulation tool.

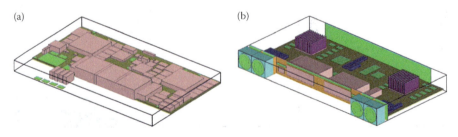

FIGURE 14.9 (a) An imported CAD file and (b) the thermal model built based on that.

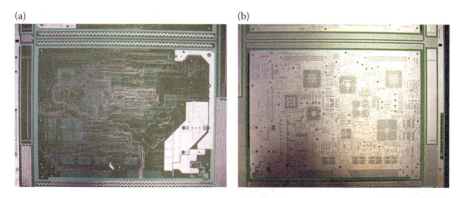

FIGURE 14.10 (a) A trace or signal layer and (b) a power layer.

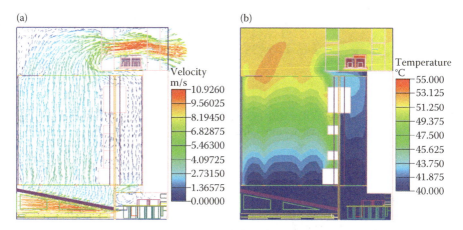

FIGURE 14.19 Predicted air velocity and air temperature in a telecommunication system.

FIGURE 15.3 A commercial flow bench. (Courtesy of Airflow Measurement Systems.)

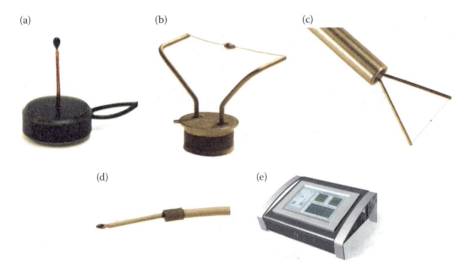

FIGURE 15.12 Examples of hot-wire probes (a, b, c, and d) and data acquisition unit (e) (Courtesy of Advanced Thermal Solutions, Inc.).

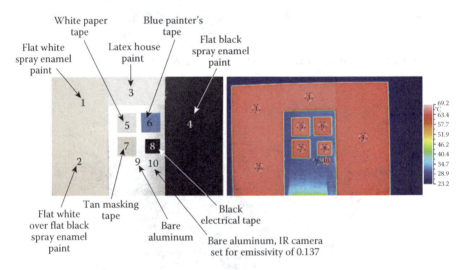

FIGURE 15.15 An aluminum plate covered with different paints and tapes and the corresponding infrared temperature measurement. (Courtesy of Dave Dlouhy and Doug Twedt, Rockwell Collins.)

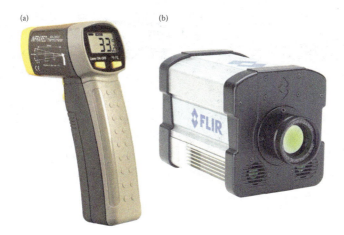

FIGURE 15.16 (a) A handheld spot radiation thermometer (Courtesy of Omega Engineering, Inc.) and (b) a thermal imaging camera. (Courtesy of FLIR systems.)

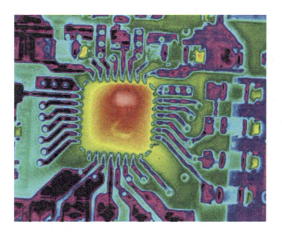

FIGURE 15.17 A two-dimensional temperature map obtained by a thermal imaging camera. (Courtesy of FLIR Systems.)

FIGURE 15.18 The interior of an anechoic chamber. (Courtesy of Eckel Industries, Inc.)

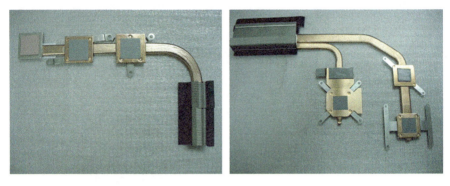

FIGURE 16.6 Examples of notebook cooling solutions with heat pipes. (Courtesy of Auras Technology Co. Ltd.)

FIGURE 16.7 Examples of desktop CPU coolers with heat pipes. (Courtesy of Auras Technology Co. Ltd.)

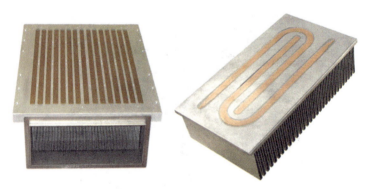

FIGURE 16.8 Examples of heat sink base with embedded heat pipes. (Courtesy of Aavid Thermalloy.)

FIGURE 16.20 Liquid-cooled Cray-2 supercomputer. (http://en.wikipedia.org/wiki/cray-2)

FIGURE 16.21 A typical module of Cray-2. (http://en.wikipedia.org/wiki/cray-2)

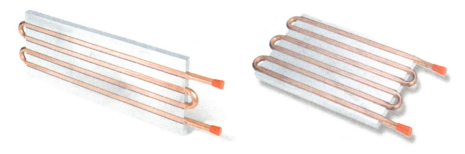

FIGURE 16.23 Examples of tubed cold plates. (Courtesy of Lytron Inc.)

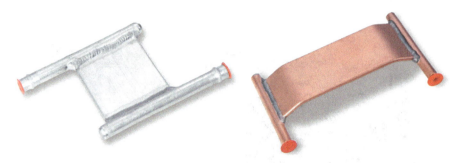

FIGURE 16.24 Two flat tube cold plates. (Courtesy of Lytron Inc.)

FIGURE 16.27 A fin-and-tube heat exchanger used in indirect liquid cooling systems. (Courtesy of Lytron Inc.)

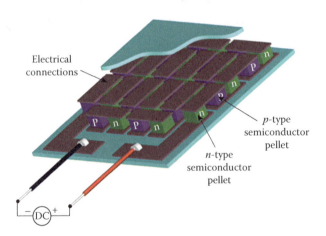

FIGURE 16.29 A thermoelectric module. (Courtesy of Marc Hodes, Tuft University.)

Forced Convection Heat Transfer: External Flows

made visible by aluminum powders in water. It is seen that the flow stays attached to the entire surface of the cylinder. However, the flow is not completely front-to-back symmetric. Figure 10.10b shows a flow with $Re_D = 26$. The flow has separated from the surface of the cylinder at about $\theta = 120°$ and a pair of recirculating eddies has formed downstream of the cylinder. At this Reynolds number the flow is still steady. The length of the wake region increases and the separation point moves forward as the Reynolds number increases. Figure 10.10c shows a flow at $Re_D = 2,000$. The flow has separated from the surface at about $\theta = 80°$. Although the flow around the cylinder is laminar, the downstream wake is turbulent. At a much higher $Re_D = 10,000$, shown in Figure 10.10d the separation still happens at the same location and the flow pattern is very similar to the one at $Re_D = 2,000$. Note that the wake region is a low pressure region and larger wake region corresponds to larger pressure difference between the front and back of the cylinder. The resultant force due to this pressure difference adds to the viscous drag force and increases the total drag force applied to the cylinder.

Figure 10.11 shows flow around a cylinder at $Re_D = 30,000$ [2]. The incoming flow is turbulent in this case. Therefore, the boundary layer that develops over the surface of the cylinder is turbulent although the Reynolds number is less than the critical Reynolds number. It is seen that the separation point has moved downstream of the cylinder and the wake region is smaller. This is due to the fact that turbulent velocity profiles have steeper gradients near the surface. It takes more work, viscous, and pressure forces applied over a longer distance, to change the flow direction and to separate a turbulent flow from the surface. A smaller wake region is equivalent to less surface area of the cylinder being exposed to low pressure wake and smaller pressure force applied to the cylinder. Although viscous force is larger in turbulent flows, it is usually smaller than the pressure force.

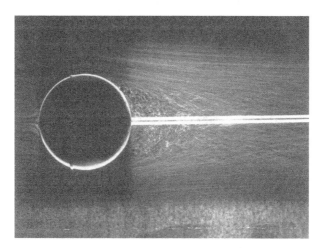

FIGURE 10.11 Turbulent flow over a cylinder with $Re_D = 30,000$ (Courtesy of ONEAR). Turbulent boundary layer is generated by a trip wire. (From Van Dyke, M., *An Album of Fluid Motion*. Stanford, CA: The Parabolic Press, 1982.)

Therefore, reduction of pressure force due to the late separation in turbulent flows, and therefore a smaller wake region, compensates for the larger viscous force and the resultant drag force will be less. This is the main reason golf and tennis balls are made with rough surfaces. The rough surface augments transition of flow form laminar to turbulent, delays separation point, and reduces the total drag force against the motion of the ball.

Two parameters of interest for flow across a cylinder are the force applied to the cylinder, called the drag force, and the total convection heat transfer rate from the cylinder. The drag force applied to a cylinder in cross flow is calculated by

$$F_D = C_D A_F \frac{1}{2} \rho U_\infty^2. \tag{10.71}$$

C_D is called the drag coefficient and $A_F = LD$, where L and D are the length and diameter of the cylinder, respectively, is called the cylinder frontal area. It is the projected area of the cylinder in a plane normal to the flow direction. C_D is a function of the Reynolds number and is given by the curve shown in Figure 10.12 [1].

The total convection heat transfer rate from a circular cylinder is calculated using the Newton's law of cooling,

$$\dot{Q}_{conv} = \bar{h} A (T_s - T_\infty). \tag{10.72}$$

$A = \pi DL$ is the surface area exposed to fluid flow and $\bar{h}$ is the average convection heat transfer coefficient given by the following semiempirical correlation [3–6]

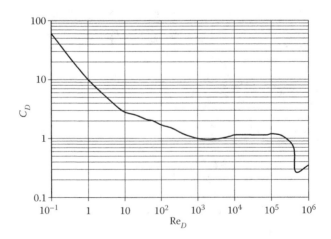

FIGURE 10.12 Variation of drag coefficient with Reynolds number for flow across a circular cylinder. (Adapted from Schlichting, H., *Boundary Layer Theory*. New York, NY: McGraw-Hill, 1979.)

Forced Convection Heat Transfer: External Flows

$$\overline{\mathrm{Nu}}_D = \frac{\overline{h}D}{k_f} = C\,\mathrm{Re}_D^m\,\mathrm{Pr}^{1/3}. \qquad (10.73)$$

The constants C and m depend on the Reynolds number and are given in Table 10.1 [3].

Cylinders with noncircular cross sections have many applications in electronics cooling including heat sinks and heat exchangers. Equation 10.73 for the average Nusselt number, along with values for C and m given in Table 10.2,

TABLE 10.1
Constants for Nusselt Number Equation for Circular Cylinder [3]

Re_D	C	m
0.4–4	0.989	0.330
4–40	0.911	0.385
40–4,000	0.683	0.466
4,000–40,000	0.193	0.618
40,000–400,000	0.027	0.805

TABLE 10.2
Constants for Nusselt Number Equation for Noncircular Cylinders [3]

Flow	Re_D	C	m
$U_\infty \rightarrow$ ◇ (diamond, D)	$5 \times 10^3 - 10^5$	0.246	0.588
$U_\infty \rightarrow$ ▭ (square, D)	$5 \times 10^3 - 10^5$	0.102	0.675
$U_\infty \rightarrow$ ⬡ (hexagon, D)	$5 \times 10^3 - 1.95 \times 10^4$	0.160	0.638
	$1.95 \times 10^4 - 10^5$	0.0385	0.782
$U_\infty \rightarrow$ ⬣ (hexagon, D)	$5 \times 10^3 - 10^5$	0.153	0.638
$U_\infty \rightarrow$ ▌ (vertical plate, D)	$4 \times 10^3 - 1.5 \times 10^4$	0.228	0.731

can be used to obtain convection heat transfer coefficient for **gas flow across noncircular cylinders** [3].

10.6 CYLINDRICAL PIN-FIN HEAT SINK

Many engineering applications include fluid flow across a series of cylinders. Examples are airflow through the coil of a car radiator, water flow across the tubes of a shell-and-tube heat exchanger, and air or liquid flow across fins of a pin-fin heat sink or cold plate. There are many empirical correlations for convection heat transfer coefficient and drag coefficient in these flows [3 and 5]. These correlations will not be presented here. Instead, a special case of these flows (i.e., the flow over cylindrical pin-fin heat sink that has many applications in electronics thermal management) will be considered.

Figure 10.13 shows two pin-fin heat sinks. The heat sink width, length, and base thickness are W, H, and t_b, respectively, and the fin height and diameter are L and D. The pin fins may be arranged either in an in-line or a staggered arrangement with longitudinal and transverse pitches S_L and S_T, and N_L and N_T number of fins in the length and width directions.

Khan et al. [7] proposed separate correlations for the convection heat transfer coefficient for the fins, $\bar{h}_f$, and the convection heat transfer coefficient for the heat sink base, $\bar{h}_{hsb}$. Let's assume that the fluid velocity approaching this heat sink is U. It is also assumed that the heat sink is inside a duct whose flow cross section is W by $L + t_b$. This assures that the flow that approaches the heat sink does not bypass through the sides or top of the heat sink. The Reynolds number for this flow is defined as

$$\mathrm{Re}_D \equiv \frac{\rho U_{max} D}{\mu}, \qquad (10.74)$$

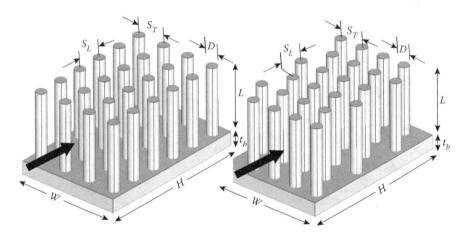

FIGURE 10.13 Schematic of an in-line (left) and a staggered (right) pin-fin heat sink.

Forced Convection Heat Transfer: External Flows

where ρ and μ are fluid density and dynamic viscosity and U_{max} is the maximum fluid velocity between the fins given by

$$U_{max} = \max\left(\frac{S_T}{S_T - D}U, \frac{S_T}{2(S_D - D)}U\right), \quad (10.75)$$

where $S_D = \sqrt{S_L^2 + (S_T/2)^2}$ is the diagonal pitch in the staggered arrangement.

The average Nusselt numbers for the fins and the heat sink base are

$$\overline{Nu}_{D,f} = \frac{\overline{h}_f D}{k_f} = C_1 Re_D^{1/2} Pr^{1/3}, \quad (10.76)$$

and

$$\overline{Nu}_{D,hsb} = \frac{\overline{h}_{hsb} D}{k_f} = \frac{3}{4}\sqrt{\frac{S_T D - D^2}{N_L S_L S_T}} Re_D^{1/2} Pr^{1/3}. \quad (10.77)$$

Here

$$C_1 = \left[0.2 + \exp\left(-0.55\frac{S_L}{D}\right)\right]\left(\frac{S_T}{D}\right)^{0.285}\left(\frac{S_L}{D}\right)^{0.212} \quad (10.78)$$

for in-line arrangement and

$$C_1 = \frac{0.61\left(\frac{S_T}{D}\right)^{0.091}\left(\frac{S_L}{D}\right)^{0.053}}{\left[1 - 2\exp\left(-1.09\frac{S_L}{D}\right)\right]} \quad (10.79)$$

for staggered arrangement.

The heat sink convection thermal resistance is given by

$$\frac{1}{R_{hs,c}} = \frac{N_f}{R_{f,c}} + \frac{1}{R_{unfinned,c}}, \quad (10.80)$$

where

$$N_f = N_L \times N_T, \quad R_{f,c} = \frac{1}{\eta_f A_f \overline{h}_f}, \quad R_{unfinned,c} = \frac{1}{\overline{h}_{hsb} A_{unfinned}} \quad (10.81)$$

and

$$A_f = \pi D L_c \quad \text{and} \quad A_{unfinned} = \left(HW - N_f \frac{\pi D^2}{4}\right). \quad (10.82)$$

Note that $L_c = L + D/4$ is the corrected length of the fin and η_f is the fin efficiency.

The force that the pin-fin heat sink applies to the air results in a pressure drop in the air that has to be overcome by the fan that generates the airflow. The total pressure drop through this heat sink is

$$\Delta P = \frac{1}{2}(k_c + k_e + fN_L)\rho U_{max}^2, \qquad (10.83)$$

where k_c and k_e are the pressure drop coefficients due to contraction and expansion of air at inlet and exhaust of the pin-fin heat sink, respectively,

$$k_c = -0.0311\sigma^2 - 0.3722\sigma + 1.0676,$$
$$k_e = 0.9301\sigma^2 - 2.5746\sigma + 0.973, \qquad (10.84)$$

$$\sigma = \frac{S_T - D}{S_T}. \qquad (10.85)$$

The f is the friction factor through the pin-fin array. For in-line arrangement

$$f = 1.009\left(\frac{S_T - D}{S_L - D}\right)^{1.09/Re_D^{0.0553}}\left[0.233 + \frac{45.78}{(S_T/D-1)^{1.1}Re_D}\right] \qquad (10.86)$$

and for staggered arrangement,

$$f = \left[1.175(S_L/S_T Re_D^{0.3124}) + 0.5 Re_D^{0.0807}\right]\left[\frac{378.6}{(S_T/D)^{13.1D/S_T} Re_D^{0.68/(S_T/D)^{1.29}}}\right]. \qquad (10.87)$$

10.7 PROCEDURE FOR SOLVING EXTERNAL FORCED CONVECTION PROBLEMS

Let's consider an external forced convection flow such as flow over a flat plate or flow across a cylinder. The objective is to find the total drag force applied to the object and either the heat transfer rate from the object or its surface temperature. This requires finding the friction coefficient and the convection heat transfer coefficient or the Nusselt number. The following procedure and the use of a spreadsheet program are strongly recommended. This will reduce the time to repeat these calculations for similar problems and will allow quick sensitivity analyses when one is interested in understanding how the results vary if the inputs change.

1. **Determine fluid properties**
   The first step is to find fluid properties. Unless otherwise specified, all the correlations are based on fluid properties at the film temperature,

Forced Convection Heat Transfer: External Flows

which is the average temperature between the surface and the free stream fluid,

$$T_f = (T_s + T_\infty)/2. \tag{10.88}$$

If either the surface or the free stream temperature is not known, and the objective of the problem is to find it, the solution can start with a guess for that unknown temperature. For example, if the surface temperature of a cylinder is not known but we know that it is hotter than the air crossing over it, the surface temperature can be assumed to be a few degrees higher than the free stream temperature. In many cases the solution can start by assuming $T_f = T_s$ or $T_f = T_\infty$ depending on which one is known.

2. **Calculate the Reynolds number based on the appropriate length scale**
   If the objective is to find the local heat flux or the local surface temperature of a plate, use the local Reynolds number based on the distance x from the leading edge of the plate. If the total convection heat transfer rate or the average surface temperature is unknown, use the Reynolds number based on the total length of the plate, L. For flow across a circular cylinder use its diameter D and for noncircular cylinders use D as suggested in Table 10.2.

3. **Calculate the friction coefficient and the Nusselt number**
   Depending on the value of the Reynolds number, which shows whether the flow is laminar or turbulent, find the appropriate correlations to calculate the friction or drag coefficient and the Nusselt number.

4. **Calculate the drag force if required**
   For flow over a flat plate,

$$F_D = \bar{C}_f A \frac{1}{2} \rho U_\infty^2.$$

   For flow across a cylinder,

$$F_D = C_D A_F \frac{1}{2} \rho U_\infty^2.$$

5. **Calculate the heat transfer rate, the heat flux, or the unknown temperature**
   If the local values are of interest, use

$$\dot{q}_{conv,x} = h(T_{s,x} - T_\infty)$$

to calculate the heat flux at distance x from the leading edge of the plate, the local surface temperature, or the fluid-free stream temperature. If the average quantities are needed, use

$$\dot{Q}_{conv} = \bar{h}A(T_s - T_\infty)$$

to obtain the total heat transfer rate, the average surface temperature, or the free stream temperature.

If either the surface temperature or the fluid free-stream temperature has been guessed in step 1 and is calculated in step 5, the calculated value has to be compared with the guessed value. If the difference between the guessed and the calculated values is not negligible, the calculations have to be repeated with a new guess. The average of the originally guessed value and the recently calculated value provides a good guess for the next step. Steps 1 though 5 shall be repeated until the difference between the calculated and the guessed values are less than the expected error of the calculations.

PROBLEMS

10.1 What is velocity boundary layer?
10.2 Briefly explain how does the thickness of velocity boundary layer change with fluid viscosity?
10.3 What is a laminar flow? What is a turbulent flow?
10.4 The density and viscosity of air are $\rho = 1.177 \text{ kg/m}^3$ and $\mu = 1.85 \times 10^{-5}$ kg/m.s and of water are $\rho = 997 \text{ kg/m}^3$ and $\mu = 8.91 \times 10^{-4}$ kg/m.s. At the same velocity over the same length plate, which flow is more likely to be turbulent? Why?
10.5 Consider the flow of air over a flat plate. If the air temperature increases, will it be more likely for the flow to become laminar or turbulent? Why?
10.6 What is the physical significance of the Nusselt number? How is it defined for a flat plate with length L?
10.7 What is the value of the Nusselt number for pure conduction?
10.8 What is the physical significance of the Reynolds number? How is it defined for a flat plate of length L?
10.9 Consider laminar flow over a flat plate. How does the heat transfer coefficient change with the distance from the beginning of the plate?
10.10 How does the average forced convection heat transfer coefficient, $\bar{h}$, and total convection heat transfer rate, $\dot{Q}_{conv}$, change with the plate length?
10.11 What is the value of critical Reynolds number for flow over a flat plate?
10.12 How does convection thermal resistance change with air velocity? Why?
10.13 Derive Equations 10.62 and 10.64 for the average friction coefficient and the average Nusselt number in combined laminar and turbulent flows over flat plate.
10.14 Plot variations of average friction coefficient and average Nusselt number versus Reynolds number (for Prandtl numbers 0.7, 5, and 50) for a plate with turbulent flow over the entire length of the plate, Equation 10.65 and 10.66, and a plate with combined laminar and turbulent flow over the plate, Equations 10.62 and 10.64. For what range of Reynolds number the laminar flow region can be neglected?
10.15 Consider a house that is maintained at 22°C at all times. The thickness and thermal conductivity of the walls are 0.15 m and 0.04 W/m°C.

During a cold winter night, the outside air temperature is 4°C and wind at 50 km/h is blowing parallel to a 3 m high and 8 m long wall of the house. If the heat transfer coefficient on the interior surface of the wall is 8 W/m²°C, determine the rate of heat loss from that wall of the house. Disregard any radiation heat transfer.

10.16 Consider a 1.5 m × 1.5 m double-pane window consisting of two 5-mm thick layers of glass separated by a layer of stagnant air. The inside convection heat transfer coefficient and air temperature are 7 W/m²°C and 25°C, respectively. The outside air at 5°C temperature and 10 m/s is blowing over the glass window. The external surface temperature of the outer glass layer is 6.5°C, and the conductivity of glass is 0.78 W/m°C.
  a. Determine the rate of heat loss through the window.
  b. What is the internal surface temperature of the inner glass layer and temperature difference across each glass layer?
  c. Determine the temperature at either side of the air layer.
  d. Using the thermal conductivity of the air at its average temperature, determine the thickness of the air layer.

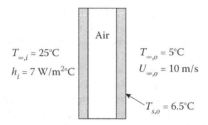

10.17 Electronic components on the top of a 20 cm × 20 cm printed circuit board dissipate 20 W heat. This heat is transferred to the air at the bottom of the board. The air velocity is 2 m/s and its temperature is 25°C. The board is made of two 0.035 mm thick copper layers ($k = 390$ W/m°C) sandwiched between three 0.5 mm thick glass epoxy layers ($k = 0.35$ W/m°C). Determine the temperature of the top and bottom of the board. Use air properties at 25°C in all your calculations.

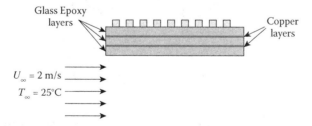

10.18 Consider an electronic device that is mounted on a 50 mm × 50 mm printed circuit board and cooled by air. The junction-to-case, junction-to-board, and case-to-air thermal resistances of this device are 5°C/W, 10°C/W and 20°C/W, respectively. The printed circuit board is 2 mm thick and is made of four 0.035-mm thick copper layers sandwiched between glass epoxy layers. Thermal conductivity of copper is 390 W/m°C

and thermal conductivity of glass epoxy is 0.35 W/m°C. Air velocity on the back side of the board is 2 m/s.
 a. Sketch the thermal resistance network that shows the heat transfer path from the junction of this device to air.
 b. Determine the normal thermal resistance of the printed circuit board.
 c. Determine the convection thermal resistance between the board and air (use air properties at 25°C).
 d. Use the results in (b) and (c) above and determine the board-to-air thermal resistance.
 e. Determine the junction-to-air thermal resistance of this device.

10.19 Consider a 40 mm × 40 mm × 2 mm electronic package that dissipates 5 W. The manufacturer's data sheet shows that the junction-to-case thermal resistance is $R_{jc} = 5°C/W$ but no information is given for the case-to-air thermal resistance, R_{ca}. Air at a velocity of 2 m/s and temperature of 25°C is blowing over the package.
 a. Calculate the case-to-air thermal resistance of the package.
 b. What is the junction temperature of the package if all the power is dissipated through the case of the package?
 c. If the junction temperature is not to exceed 105°C, determine the maximum thermal resistance of a heat sink that has to be placed on this package.

Use the air properties at 25°C in all your calculations.

10.20 A straight fin aluminum heat sink ($k = 230$ W/m°C) has a base area of 50 mm × 50 mm. There are eight 20-mm high and 2-mm thick straight fins on the base. Air with a velocity of 2 m/s and temperature of 25°C is blowing into this heat sink between the fins.
 a. What is the convection heat transfer coefficient around each fin?
 b. What is the fin efficiency?
 c. What is the thermal resistance of this heat sink?
Note: Use air properties at 25°C in all your calculations.

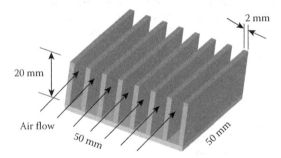

10.21 Consider the aluminum pin-fin heat sink shown below. The base area is 40 mm × 40 mm. There are 25 fins on this heat sink. The fins are 20 mm long with 3 mm diameter. If the air velocity over the fins is 2 m/s and the air temperature is 25°C,
 a. Calculate the heat transfer coefficient for one fin assuming each fin can be treated as an isolated cylinder.

Forced Convection Heat Transfer: External Flows

b. What is the fin efficiency?
c. What is the efficiency of the heat sink? Assume that the heat transfer coefficient for the base is the same as that for the fins.
d. What is the thermal resistance of the heat sink?
e. Calculate thermal resistance of the heat sink using correlations given in Section 10.6.

Assume that the thermal conductivity of aluminum is 230 W/m°C and use the air properties at 25°C in all the calculations.

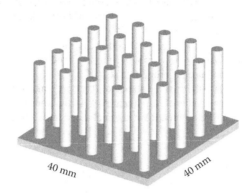

10.22 An aluminum die cast heat sink with 40 mm × 40 mm base consists of 25 fins with square cross sections. The fins are 20 mm long with 4 mm × 4 mm cross section. Air with a velocity of 3 m/s is flowing into this heat sink parallel to the heat sink base.
   a. What is the convection heat transfer coefficient around each fin if each fin is treated as an isolated cylinder?
   b. What is the fin efficiency?
   c. What is the thermal resistance of this heat sink?

Note: Assume that the thermal conductivity of aluminum die cast is 170 W/m°C and use air properties at 25°C in all the calculations.

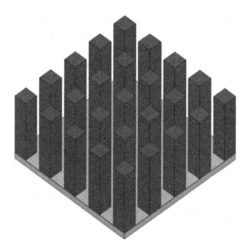

10.23 A straight fin heat sink has to be designed with thermal resistance of $R_{hs} = 3°C/W$ at 1.5 m/s air velocity. The heat sink is made of aluminum with thermal conductivity of 230 W/m°C, its base is 50 mm × 50 mm, and fins are 20 mm long and 2 mm thick. How many fins are required for this heat sink? Assume total heat sink efficiency is equal to the fin efficiency and check this assumption at the end.
Use air properties at 25°C in all the calculations.

10.24 Consider an aluminum plate-fin heat sink with 80 mm × 80 mm × 5 mm base and 20 fins that are 40 mm tall and 1 mm thick. Air with a velocity of 2 m/s is passing over the fins of this heat sink. Thermal conductivity of aluminum is 170 W/m°C.
   a. What is the average convection heat transfer coefficient over each fin? Consider flow over the fin as forced convection over a flat plate and use air properties at 25°C.
   b. Calculate thermal resistance of this heat sink. Note that the thermal resistance of this heat sink is the sum of the convection thermal resistance, base thermal resistance, and caloric thermal resistance.

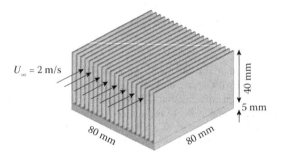

10.25 Consider a package with junction-to-case, junction-to-board, and board-to-ambient thermal resistances of 2°C/W, 12°C/W, and 27°C/W, respectively. Maximum power dissipation and allowable junction temperature of this package are 20 W and 125°C.
   a. What is the maximum allowable thermal resistance of a heat sink needed to cool the junction of this package below 125°C at an ambient temperature of 50°C?
   b. Let's assume that an aluminum pin-fin heat sink with 50 mm × 50 mm × 5 mm base and 30 mm long and 2 mm diameter fins have to be designed to cool this package in air with a velocity of 3 m/s and temperature of 50°C. Determine minimum number of fins needed for this heat sink. Neglect heat transfer rate from the base and assume the total heat sink efficiency is the same as the fin efficiency. Also, neglect spreading and caloric resistance of the heat sink. Thermal conductivity of aluminum is 170 W/m°C. Use air properties at 50°C in your calculations.
   Hint: Use correlations for cross flow over a single cylinder to obtain convection heat transfer coefficient.

10.26 Consider a 50 mm × 50 mm × 3 mm electronic package that dissipates 20 W. Junction-to-case, junction-to-board, and board-to-air thermal

resistances of this package are 2°C/W, 10°C/W, and 15°C/W, respectively. Air with a velocity of 2 m/s and temperature of 45°C is flowing over this package.
a. Calculate case-to-air thermal resistance of this package?
b. What is the junction temperature of this package?
c. A straight-fin heat sink will be designed to reduce the junction temperature of this package. Let's assume this heat sink is made of pure copper, its base is 50 mm × 50 mm × 2 mm and fins are 30 mm tall and 2 mm thick. Calculate the minimum number of fins required to keep the junction temperature of this package less than 125°C. Assume total heat sink efficiency is 100% and neglect the no-fin areas of the heat sink base compared to the total area of fins ($A_t = n_f A_f$).

Use air properties at 45°C in all the calculations.

REFERENCES

1. Schlichting, H. 1979. *Boundary layer theory.* New York, NY: McGraw-Hill.
2. Van Dyke, M. 1982. *An album of fluid motion.* Stanford, CA: The Parabolic Press.
3. Incropera, F. P., D. P. DeWitt, T. L. Bergman, and A. S. Lavine. 2006. *Fundamentals of heat and mass transfer.* Sixth Edition. Hoboken, NJ: John Wiley & Sons, Inc.
4. Kays, W., M. Crawford, and B. Weigand. 2005. *Convective heat and mass transfer.* New York, NY: McGraw-Hill.
5. Holman, J. P. 2010. *Heat Transfer.* New York, NY: McGraw-Hill.
6. Bejan, A. 2004. *Convection heat transfer.* Hoboken, NJ: John Wiley & Sons, Inc.
7. Khan, W. A., J. R. Culham, and M. M. Yovanovich. 2005, March 15–17. Modeling of cylindrical pin-fin heat sinks for electronic packaging. Proceedings of 21st IEEE Semiconductor Thermal Measurement and Management (SEMI-THERM) Symposium, San Jose, California.

# 11 Forced Convection Heat Transfer: Internal Flows

Forced convection internal flows have many engineering applications including those in electronic thermal management. Examples are airflow between two adjacent cards of a rack-mount telecommunication or networking equipment, airflow between two closely spaced memory modules, airflow inside a notebook, liquid flow between tightly spaced fins of a cold plate, and water flow inside pipes of a heat exchanger. Convection heat transfer in these flows will be discussed in detail in this chapter.

11.1 MEAN VELOCITY AND MEAN TEMPERATURE

External flows are usually characterized by a free-stream velocity and temperature. These are fluid velocity and temperature outside the corresponding velocity and thermal boundary layers where the fluid is not affected by a surface. However, every fluid particle in an internal flow is somehow affected by the bounding surfaces. Therefore, concepts such as free-stream velocity and temperature are not applicable in internal flows. Instead, the mean velocity and the mean temperature are used to characterize internal flows. These will be defined in this section.

Consider fluid flow inside a pipe as shown in Figure 11.1a and b. Fluid velocity changes from zero at the pipe surface to a maximum value at the center of the pipe. The infinitesimal mass flow rate through an infinitesimal cross section area dA_c where fluid velocity and density are U and ρ is given by $\rho U dA_c$. The total mass flow rate through any cross section of the pipe is obtained by integrating this over the entire cross section of the pipe, A_c

$$\dot{m} = \int_{A_c} \rho U dA_c. \quad (11.1)$$

Now, consider a pipe with a uniform velocity U_m, as shown in Figure 11.1c, and the same total mass flow rate $\dot{m}$. If fluid is incompressible and therefore its density is constant across the cross section of the pipe,

$$\dot{m} = \rho U_m A_c. \quad (11.2)$$

Equating Equations 11.1 and 11.2 for $\dot{m}$ gives

$$U_m = \frac{1}{A_c} \int_{A_c} U dA_c. \quad (11.3)$$

U_m is called the **mean fluid velocity** at cross section A_c.

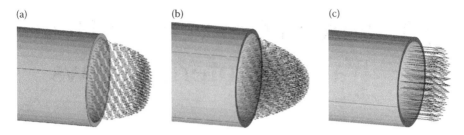

FIGURE 11.1 (See color insert following page 240.) Pipes with actual velocity profiles (a and b) and their equivalent with a uniform velocity (c).

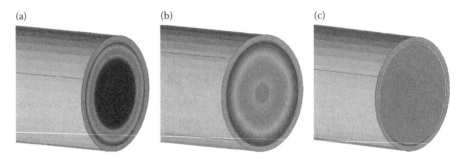

FIGURE 11.2 (See color insert following page 240.) Pipes with actual temperature profiles (a and b) and their equivalent pipe with uniform temperature (c).

Now, let's assume that fluid enters a pipe with a temperature less than the pipe surface temperature. The fluid temperature increases as it flows through the pipe as a result of heat transfer from the pipe surface to the fluid. Fluid temperature at any cross section of the pipe changes from a maximum value at the pipe surface to a minimum at the pipe center. This is shown in Figure 11.2a and b, which correspond to velocity profiles of Figure 11.1a and b. If ρ and C_p are density and specific heat of the fluid in an infinitesimal cross section area dA_c where fluid velocity and temperature are U and T, respectively, mass flow rate through this area is $d\dot{m} = \rho U dA_c$ and energy transfer rate by this mass flow rate is $d\dot{E} = d\dot{m} C_p T = \rho U C_p T dA_c$. The total energy transfer rate by the mass flow rate through this cross section of the pipe is obtained by integrating this over the entire cross section of the pipe:

$$\dot{E} = \int_{A_c} \rho C_p U T dA_c. \qquad (11.4)$$

Now, consider a flow with a uniform velocity U_m, as shown in Figure 11.1c, a uniform temperature T_m, as shown in Figure 11.2c, and the same total energy transfer rate $\dot{E}$. If fluid density and specific heat are constant across the cross section of the pipe,

$$\dot{E} = \dot{m} C_p T_m = \rho U_m A_c C_p T_m. \qquad (11.5)$$

Forced Convection Heat Transfer: Internal Flows

Equations 11.4 and 11.5 for $\dot{E}$ gives

$$T_m = \frac{1}{U_m A_c} \int_{A_c} UT dA_c. \tag{11.6}$$

T_m is called the **mean fluid temperature** at cross section A_c.

11.2 LAMINAR AND TURBULENT PIPE FLOWS

The Reynolds number for flow inside a circular pipe is defined based on the mean velocity and the pipe diameter,

$$\text{Re}_D \equiv \frac{\rho U_m D}{\mu} = \frac{U_m D}{\nu}. \tag{11.7}$$

The characteristic length for noncircular pipes, ducts, and channels is called the **hydraulic diameter** and is defined as

$$D_h = \frac{4 A_c}{P}, \tag{11.8}$$

where A_c and P are the area and perimeter of the pipe cross section, respectively. The Reynolds number for a pipe with a general shape cross section is defined as

$$\text{Re}_{D_h} \equiv \frac{\rho U_m D_h}{\mu} = \frac{U_m D_h}{\nu}. \tag{11.9}$$

The hydraulic diameter for a circular pipe is the same as its diameter. Therefore, Equations 11.7 and 11.9 for the Reynolds number are consistent. Since $\dot{m} = \rho U_m A_c$, the Reynolds number can also be expressed in terms of the mass flow rate,

$$\text{Re}_{D_h} \equiv \frac{\dot{m} D_h}{\mu A_c}. \tag{11.10}$$

Flow inside a pipe is laminar if $\text{Re}_{D_h} < 2300$ and is fully turbulent if $\text{Re}_{D_h} > 10{,}000$. It will be in transition between laminar and turbulent flows if $2300 < \text{Re}_{D_h} < 10{,}000$.

11.3 ENTRY LENGTH AND FULLY DEVELOPED FLOW

Consider a flow with a uniform velocity entering a pipe as shown in Figure 11.3. Boundary layers develop over the inner surfaces of the pipe. Fluid velocity right at the pipe surface is zero and increases to a maximum somewhere between the surface and the center of the pipe. These boundary layers form a fluid shell with

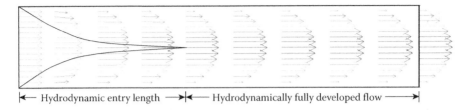

|← Hydrodynamic entry length →|← Hydrodynamically fully developed flow →|

FIGURE 11.3 (See color insert following page 240.) Hydrodynamic entry length and fully developed flow in a pipe.

continuously increasing thickness along the flow direction. The thickness of this fluid shell, increases until the boundary layers merge at the center of the pipe. From this point on, the velocity profile does not change along the flow direction. If U is fluid velocity along the pipe and x and y are coordinate axes along and normal to the axis of the pipe, $\partial U/\partial x = 0$ after boundary layers merge. The length of a pipe from its entrance to the point at which boundary layers merge is called the **hydrodynamic entry length**. The region beyond the entry length where velocity profile does not change is called the **hydrodynamically fully developed region**.

The hydrodynamic entry length, $x_{fd,h}$ for a **laminar flow** is given by [1]:

$$x_{fd,h} = 0.05 D_h \, \mathrm{Re}_{D_h}. \tag{11.11}$$

The hydrodynamic entry length in turbulent flow is independent of the Reynolds number and is approximately given by [1]:

$$10 D_h \leq x_{fd,h} \leq 60 D_h. \tag{11.12}$$

Let's assume that the fluid temperature entering a pipe, T, is less than the pipe surface temperature, T_s. Temperature boundary layers develop over the inner surface of the pipe as shown in Figure 11.4. The thickness of these thermal boundary layers increases until they merge at the center of the pipe. Although fluid temperature continues to increase along the flow, until it reaches an equilibrium with the surface temperature, the normalized temperature difference, $(T_s - T)/(T_s - T_m)$, does not change after boundary layers merge. The length of the pipe from its inlet to the point where thermal boundary layers merge is called the **thermal entry length**. The region beyond the entry length where the normalized temperature does not change along the flow is called the **thermally fully developed region**.

The thermal entry length, $x_{fd,t}$ in a **laminar flow** is given by [1]:

$$x_{fd,t} = 0.05 D_h \, \mathrm{Re}_{D_h} \, \mathrm{Pr}. \tag{11.13}$$

The thermal entry length in turbulent flow is independent of the Reynolds the and Prandtl numbers and is approximately equal to the hydrodynamic entry length [1]:

$$10 D_h \leq x_{fd,t} \leq 60 D_h. \tag{11.14}$$

Forced Convection Heat Transfer: Internal Flows

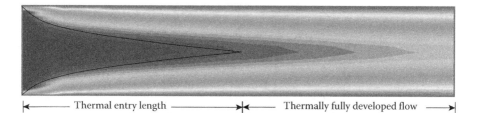

|←——— Thermal entry length ———→|←——— Thermally fully developed flow ———→|

FIGURE 11.4 (See color insert following page 240.) Thermal entry length and fully developed flow inside a pipe.

The thermally fully developed condition can be expressed as follows:

$$\frac{\partial}{\partial x}\left(\frac{T_s - T}{T_s - T_m}\right) = 0. \tag{11.15}$$

Expanding Equation 11.15 gives

$$\frac{\partial T_s}{\partial x} - \frac{\partial T}{\partial x} - \frac{(T_s - T)}{(T_s - T_m)}\frac{\partial T_s}{\partial x} + \frac{(T_s - T)}{(T_s - T_m)}\frac{\partial T_m}{\partial x} = 0. \tag{11.16}$$

11.4 PUMPING POWER AND CONVECTION HEAT TRANSFER IN INTERNAL FLOWS

Two important engineering parameters in any forced convection internal flow are the pumping power, which is required by a fan or pump to create the internal flow, and the convection heat transfer rate from or to the fluid. The total pumping power is equal to the product of the pressure drop from inlet to exhaust of the flow, ΔP, and the volumetric flow rate, $\dot{V}$,

$$\dot{W} = \dot{V}\Delta P = U_m A_c \Delta P. \tag{11.17}$$

U_m and A_c are the fluid mean velocity and the flow cross section area, respectively. The **pressure drop in a fully developed flow** is given by

$$\Delta P = f\frac{L}{D_h}\frac{1}{2}\rho U_m^2. \tag{11.18}$$

f is an analytically or experimentally determined dimensionless parameter called the **friction factor** that is defined by

$$f \equiv \frac{-(dP/dx)D_h}{\rho U_m^2/2}, \tag{11.19}$$

where dP/dx is the pressure gradient in the flow direction. Both dP/dx and f are constant in a fully developed flow.

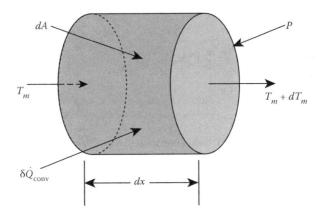

FIGURE 11.5 A small section of an internal flow.

Consider a small section of an internal flow with length dx, perimeter P, and surface area $dA = Pdx$ as shown in Figure 11.5. Fluid with the mean temperature T_m enters this volume and leaves it with the mean temperature $T_m + dT_m$. An energy balance gives the convection heat transfer rate to the fluid passing through this small control volume,

$$\delta \dot{Q}_{\text{conv}} = \dot{m} C_p \, dT_m. \tag{11.20}$$

The total convection heat transfer rate from the inlet to the exhaust can be calculated by integrating this over the entire length of the flow. If T_i and T_e are the fluid mean temperatures at inlet and exhaust, and $C_{p,\text{ave}}$ is the fluid's average specific heat at constant pressure, the total convection heat transfer rate is

$$\dot{Q}_{\text{conv}} = \dot{m} C_{p,\text{ave}} (T_e - T_i). \tag{11.21}$$

The local convection heat flux in an internal flow is given by

$$\dot{q}_{\text{conv}} = h(T_s - T_m), \tag{11.22}$$

where $\dot{q}_{\text{conv}}$, h, T_s, and T_m are the local convection heat flux, the convection heat transfer coefficient, the surface temperature, and the fluid mean temperature at distance x from the inlet and may change with x. Note that there is an important difference between the reference temperature T_∞ used in Newton's law of cooling for flow over a flat plate and T_m used in the above equation. While T_∞ is constant, T_m changes continuously with x as long as there is heat transfer to or from the fluid. If heat transfer is from the surface to the fluid, T_m will increase with x and if heat transfer is from the fluid to the surface, T_m will decrease with x until $T_m = T_s$.

The convection heat transfer rate to the fluid passing through the control volume shown in Figure 11.5 can be calculated using the local convection heat flux,

$$\delta \dot{Q}_{\text{conv}} = \dot{q}_{\text{conv}} \, dA = \dot{q}_{\text{conv}} Pdx. \tag{11.23}$$

Forced Convection Heat Transfer: Internal Flows

Equating Equations 11.20 and 11.23 for $\delta \dot{Q}_{conv}$ gives the following important expression:

$$\frac{dT_m}{dx} = \frac{\dot{q}_{conv} P}{\dot{m} C_p}. \tag{11.24}$$

This expression shows that if the surface heat flux, the flow perimeter, and the fluid specific heat at constant pressure are constant, the fluid mean temperature changes linearly with x,

$$T_m(x) = T_i + \frac{\dot{q}_{conv} P}{\dot{m} C_p} x, \tag{11.25}$$

where $\dot{q}_{conv}$, C_p, and P are constant along the flow.

The local convection heat flux can also be calculated using Fourier's law of conduction in the region near the wall. If y_s is the y coordinate of the surface,

$$\dot{q}_{conv} = -k_f \left.\frac{\partial T}{\partial y}\right|_{y=y_s}. \tag{11.26}$$

Equating Equations 11.22 and 11.26 for the local convection heat flux gives the definition of the convection heat transfer coefficient for internal flows:

$$h \equiv \frac{-k_f \left.\frac{\partial T}{\partial y}\right|_{y=y_s}}{T_s - T_m}. \tag{11.27}$$

Let's consider a thermally fully developed flow in which the dimensionless temperature $(T_s - T)/(T_s - T_m)$, and therefore its derivative with respect to y, are independent of x:

$$\frac{\partial}{\partial y}\left(\frac{T_s - T}{T_s - T_m}\right)_{y=y_s} \neq f(x) \quad \Rightarrow \quad \frac{(\partial T/\partial y)_{y=y_s}}{T_s - T_m} \neq f(x). \tag{11.28}$$

Using Equation 11.28 in Equation 11.27 shows that the ratio of the local convection heat transfer coefficient to the fluid thermal conductivity, in a thermally fully developed flow, is independent of x:

$$\frac{h}{k_f} \neq f(x). \tag{11.29}$$

If fluid thermal conductivity is constant, the local convection heat transfer coefficient in a thermally fully developed flow will not change along the flow.

The appropriate form of the Newton's law of cooling for an internal flow is obtained by considering the thermal boundary condition at the surface. For the **uniform surface heat flux** boundary condition, when $\dot{q}_{conv}$ does not depend on x, the total convection heat transfer rate is calculated by the simple integration of Equation 11.23 using Equation 11.22 for the local convection heat flux:

$$\dot{Q}_{conv} = \int_A \dot{q}_{conv}\, dA = \dot{q}_{conv} A = hA(T_s - T_m).$$

If P is constant along the flow length L, $A = PL$ and

$$\dot{Q}_{conv} = hPL(T_s - T_m). \tag{11.30}$$

Since both $\dot{q}_{conv}$ and h are constant, the derivative of Equation 11.30 with respect to x gives

$$\frac{\partial T_s}{\partial x} = \frac{\partial T_m}{\partial x}. \tag{11.31}$$

Inserting this into Equation 11.16 gives the following relationship between the fluid local temperature, the surface temperature, and the fluid mean temperature in a **fully developed flow with uniform surface heat flux**:

$$\frac{\partial T}{\partial x} = \frac{\partial T_s}{\partial x} = \frac{\partial T_m}{\partial x}. \tag{11.32}$$

Now, let's consider an internal flow with a **uniform surface temperature** T_s that does not change with x. The convection heat transfer rate to the fluid passing through the control volume shown in Figure 11.5 is

$$\delta \dot{Q}_{conv} = \dot{q}_{conv} P dx = h(T_s - T_m) P dx. \tag{11.33}$$

Equating Equations 11.33 and 11.20 for $\delta \dot{Q}_{conv}$ gives

$$\dot{m} C_p dT_m = hP(T_s - T_m) dx. \tag{11.34}$$

Since T_s is constant $dT_m = -d(T_s - T_m)$, and Equation 11.34 can be written as

$$\frac{d(T_s - T_m)}{(T_s - T_m)} = -\frac{hP}{\dot{m} C_p} dx. \tag{11.35}$$

Assuming P is constant, taking the average specific heat at constant pressure as a constant value over the length x of the flow, and integrating both sides of Equation 11.35 gives

$$\ln \frac{(T_s - T_m)}{(T_s - T_i)} = -\frac{Px\overline{h}}{\dot{m} C_{p,ave}}, \tag{11.36}$$

Forced Convection Heat Transfer: Internal Flows 263

where

$$\bar{h} = \frac{1}{x}\int_0^x h\,dx \tag{11.37}$$

is the average convection heat transfer coefficient over the length x of the flow. Taking the exponential of Equation 11.36 gives the following relationship between the mean temperatures at the inlet and at a distance x from the inlet:

$$T_m(x) = T_s + (T_i - T_s)\exp\left(-\frac{Px\bar{h}}{\dot{m}C_{p,\text{ave}}}\right). \tag{11.38}$$

Equation 11.38 shows that the difference between the surface and the fluid mean temperatures reduces exponentially with distance x from the inlet. The exhaust temperature T_e is obtained by substituting $x = L$ in Equation 11.38:

$$T_e = T_s + (T_i - T_s)\exp\left(-\frac{PL\bar{h}}{\dot{m}C_{p,\text{ave}}}\right). \tag{11.39}$$

Here $\bar{h}$ is the average convection heat transfer coefficient over the entire length of the flow.

Solving Equation 11.39 for $\dot{m}C_{p,\text{ave}}$ and substituting it in Equation 11.21 gives the appropriate form of the Newton's law of cooling for an internal flow with **uniform surface temperature**,

$$\dot{Q}_{\text{conv}} = \bar{h}A\Delta T_{\ln}. \tag{11.40}$$

Here, A is the total surface area and

$$\Delta T_{\ln} = \frac{T_i - T_e}{\ln\left[(T_s - T_e)/(T_s - T_i)\right]} \tag{11.41}$$

is called the logarithmic mean temperature difference. Note that $A = PL$ if perimeter P is constant.

For a **fully developed flow with uniform surface temperature** Equation 11.16 reduces to

$$\frac{\partial T}{\partial x} = \frac{(T_s - T)}{(T_s - T_m)}\frac{\partial T_m}{\partial x}. \tag{11.42}$$

11.5 VELOCITY PROFILES AND FRICTION FACTOR CORRELATIONS

Consider a two-dimensional, steady-state, and fully developed laminar flow between two large parallel plates that are separated by distance $2b$ from each other as shown in Figure 11.6. Let's select the coordinate system such that its center is located halfway

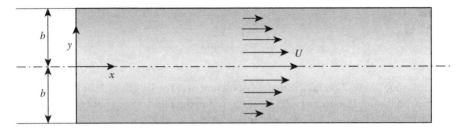

FIGURE 11.6 Cross section of an internal flow between two large parallel plates.

between the two plates, its x direction is along the fluid flow, and its y direction is normal to the plates. The choice of the x axis along the flow implies that the fluid velocity has no component in the y direction (i.e., $V = 0$). If we neglect any body force (i.e., $X = Y = 0$), the x and y components of the boundary layer momentum equation, Equations 9.14 and 9.15 in Chapter 9, will reduce to the following simple forms:

$$\frac{\partial P}{\partial x} = \mu \frac{\partial^2 U}{\partial y^2}, \tag{11.43}$$

$$\frac{\partial P}{\partial y} = 0. \tag{11.44}$$

Equation 11.44 implies that the fluid pressure is a function of x only. Therefore, its partial derivative with respect to x in Equation 11.43 can be replaced by an ordinary derivative. On the other hand, since the flow is fully developed, the fluid velocity is independent of x and the second-order partial derivative of U with respect to y can be converted to a second-order ordinary derivative. Therefore, the x-momentum equation for this flow becomes

$$\frac{dP}{dx} = \mu \frac{d^2 U}{dy^2}. \tag{11.45}$$

The left side of Equation 11.45 is a function of x only while its right side is a function of y only. This is possible only if both sides are equal to a constant,

$$\mu \frac{\partial^2 U}{\partial y^2} = \frac{dP}{dx} = C. \tag{11.46}$$

Equation 11.46 is a second-order ordinary differential equation for the fluid velocity $U = U(y)$. Two boundary conditions are obtained from the no-slip conditions on the top and bottom plates; $U(b) = 0$ and $U(-b) = 0$. Another boundary condition that can be used instead of one of these two boundary conditions is the symmetry condition, or zero velocity gradient, halfway between the two plates; $(dU/dy)_{y=0} = 0$. The solution of the differential Equation 11.46 subject to these boundary conditions is

$$U(y) = \frac{1}{2\mu} \frac{dP}{dx}(y^2 - b^2). \tag{11.47}$$

Forced Convection Heat Transfer: Internal Flows

This is a parabolic velocity profile with the maximum velocity occurring halfway between the two plates,

$$U_{max} = -\frac{b^2}{2\mu}\frac{dP}{dx}. \tag{11.48}$$

The mean velocity for this flow is

$$U_m = \frac{1}{A_c}\int_{A_c} U\,dA_c = \frac{1}{2b}\int_{-b}^{b}\frac{1}{2\mu}\frac{dP}{dx}(y^2 - b^2)\,dy = -\frac{b^2}{3\mu}\frac{dP}{dx}. \tag{11.49}$$

Note that the maximum fluid velocity is 1.5 times its mean velocity. Solving Equation 11.49 for pressure gradient gives

$$\frac{dP}{dx} = -\frac{3\mu U_m}{b^2}. \tag{11.50}$$

Substituting Equation 11.50 for pressure gradient into Equation 11.47 gives the velocity profile in terms of the mean fluid velocity,

$$U = \frac{3}{2}U_m\left(1 - \frac{y^2}{b^2}\right). \tag{11.51}$$

Since the flow is fully developed, pressure gradient is independent of x and the total pressure drop in the pipe with length L is

$$\Delta P = -\frac{3\mu U_m L}{b^2}. \tag{11.52}$$

Therefore, the pumping power required to generate this flow is

$$\dot{W} = \dot{V}\Delta P = U_m A_c \Delta P = -\frac{3\mu U_m^2 A_c L}{b^2}. \tag{11.53}$$

For the flow between two large plates $D_h = 4b$ and the friction factor is obtained by substituting the expression for dP/dx from Equation 11.50 in Equation 11.19. This results in

$$f = \frac{96}{Re_{D_h}}. \tag{11.54}$$

Note that Equation 11.54 is valid for the laminar fully developed flow between two large parallel plates.

Similarly, it can be shown that the laminar fully developed velocity profile in a circular pipe with radius r_0 is

$$U(r) = 2U_m\left[1 - \frac{r^2}{r_0^2}\right], \tag{11.55}$$

where the mean velocity U_m is given by

$$U_m = -\frac{r_0^2}{8\mu}\frac{dP}{dx}. \tag{11.56}$$

By substituting dP/dx from this equation into Equation 11.19, it can be shown that

$$f = \frac{64}{Re_D} \tag{11.57}$$

for the laminar fully developed flow in a circular pipe. Correlations for the friction factors for fully developed laminar flows inside smooth pipes with various cross sections are given in Table 11.1 [1].

TABLE 11.1
Friction Factor and Nusselt Number for Fully Developed Laminar Internal Flows

Cross Section	b/a	Constant Heat Flux	Constant Temperature	fRe_{Dh}
circle	–	4.36	3.66	64
a, b square	1	3.61	2.98	57
a, b rectangle	1.43	3.73	3.08	59
a, b rectangle	2.0	4.12	3.39	62
a, b rectangle	3.0	4.79	3.96	69
a, b rectangle	4.0	5.33	4.44	73
a, b rectangle	8.0	6.49	5.60	82
a, b parallel plates	∞	8.23	7.54	96
a Heated, b Insulated	∞	5.39	4.86	96

where $Nu_D = hD_h/k$

Source: From Kays, Q., M. Crawford, and B. Weigand., *Convection Heat and Mass Transfer.* New York, NY: McGraw-Hill, 2005.

Forced Convection Heat Transfer: Internal Flows

The friction factor for the **fully developed turbulent flow** inside a smooth pipe with an arbitrary cross section area can be calculated using the following semiempirical correlations [2]:

$$f = 0.316 \text{Re}_{D_h}^{-1/4} \quad \text{Re}_{D_h} \leq 2 \times 10^4$$
$$f = 0.184 \text{Re}_{D_h}^{-1/5} \quad \text{Re}_{D_h} \geq 2 \times 10^4. \tag{11.58}$$

Alternatively, the following single correlation that covers a larger Reynolds number range and has been developed by Petukhov [3] can be used

$$f = (0.79 \ln \text{Re}_{D_h} - 1.64)^{-2} \quad 3000 \leq \text{Re}_{D_h} \leq 5 \times 10^6. \tag{11.59}$$

11.6 TEMPERATURE PROFILES AND CONVECTION HEAT TRANSFER CORRELATIONS

Consider the two-dimensional, steady-state and fully developed laminar flow between two large parallel plates shown in Figure 11.6. The boundary layer form of the energy equation applies to this flow. If there is no heat generation and the flow speed is low enough such that viscous dissipation can be neglected, the boundary layer energy equation, Equation 9.16, will reduce to

$$U \frac{\partial T}{\partial x} = \alpha \frac{\partial^2 T}{\partial y^2}. \tag{11.60}$$

Let's consider the case with a **uniform surface heat flux** first. Replacing for U from Equation 11.51 gives

$$\frac{3}{2} U_m \left(1 - \frac{y^2}{b^2}\right) \frac{\partial T}{\partial x} = \alpha \frac{\partial^2 T}{\partial y^2}. \tag{11.61}$$

Equation 11.32 shows that $\partial T / \partial x$ is not a function of y. Therefore, integrating Equation 11.61 twice gives

$$T(x,y) = \frac{3}{2} \frac{U_m}{\alpha} \frac{\partial T}{\partial x} \left(\frac{1}{2} y^2 - \frac{1}{12} \frac{y^4}{b^2}\right) + C_1 y + C_2. \tag{11.62}$$

The symmetry boundary condition at $y = 0$ gives $C_1 = 0$ and the surface temperature condition, $T(x, \pm b) = T_s$ gives

$$C_2 = T_s - \frac{15}{24} \frac{U_m b^2}{\alpha} \frac{\partial T}{\partial x}. \tag{11.63}$$

Therefore, temperature profile is

$$T(x,y) = T_s + \frac{3}{2}\frac{U_m}{\alpha}\frac{\partial T}{\partial x}\left(\frac{1}{2}y^2 - \frac{1}{12}\frac{y^4}{b^2} - \frac{5}{12}b^2\right). \qquad (11.64)$$

Fluid mean temperature is obtained as follows:

$$T_m = \frac{1}{U_m A_c}\int_{A_c} UT dA_c$$

$$= \frac{1}{2U_m b}\int_{-b}^{b}\left[\frac{3}{2}U_m\left(1-\frac{y^2}{b^2}\right)\right]\left[T_s + \frac{3}{2}\frac{U_m}{\alpha}\frac{\partial T}{\partial x}\left(\frac{1}{2}y^2 - \frac{1}{12}\frac{y^4}{b^2} - \frac{5}{12}b^2\right)\right]dy$$

$$T_m(x) = T_s - \frac{17}{35}\frac{U_m b^2}{\alpha}\frac{\partial T}{\partial x}. \qquad (11.65)$$

But, Equations 11.24 and 11.32 show that for a uniform surface heat flux internal flow:

$$\frac{\partial T}{\partial x} = \frac{\partial T_m}{\partial x} = \frac{\dot{q}_{conv} P}{\dot{m} C_p}. \qquad (11.66)$$

Substituting this into Equation 11.65 for T_m and noting that $P = 2W$, where W is the plate width, and $\dot{m} = \rho U_m A_c = 2\rho U_m bW$ gives

$$T_s - T_m = \frac{17}{35}\frac{U_m b^2}{\alpha}\frac{\dot{q}_{conv} P}{\dot{m} C_p}, \qquad (11.67)$$

$$T_s - T_m = \frac{17}{35}\frac{b}{k_f}\dot{q}_{conv}. \qquad (11.68)$$

The convection heat transfer coefficient is defined as

$$h = \frac{\dot{q}_{conv}}{T_s - T_m} = \frac{35}{17}\frac{k_f}{b}. \qquad (11.69)$$

Hydraulic diameter for this geometry is $D_h = 4b$. Therefore, the **Nusselt number for laminar fully developed flow between two large parallel plates with uniform surface heat flux** is

$$\mathrm{Nu}_{D_h} = \frac{hD_h}{k_f} = 8.23, \quad D_h = 4b. \qquad (11.70)$$

Forced Convection Heat Transfer: Internal Flows

A similar, but more complex procedure, shows that the **Nusselt number for fully developed laminar flow between two large parallel plates with uniform surface temperature** is

$$\mathrm{Nu}_{D_h} = \frac{hD_h}{k_f} = 7.54, \quad D_h = 4b. \tag{11.71}$$

The Nusselt numbers for fully developed laminar internal flows with other cross sectional geometries are given in Table 11.1 [1].

The Nusselt number in fully developed turbulent internal flows with arbitrary cross sectional area can be calculated using the Dittus-Boelter equation [2]:

$$\mathrm{Nu}_{D_h} = 0.023 \mathrm{Re}_{D_h}^{0.8} \mathrm{Pr}^n, \tag{11.72}$$

where $n = 0.4$ if $T_s > T_m$ and $n = 0.3$ if $T_s < T_m$. This equation gives results with less than 25% error if $0.7 \leq \mathrm{Pr} \leq 160$, $\mathrm{Re}_{D_h} \geq 10{,}000$ and $L/D_h \geq 10$.

More accurate results can be obtained by using the following correlation recommended by Gnielinski [4]:

$$\mathrm{Nu}_{D_h} = \frac{(f/8)(\mathrm{Re}_{D_h} - 1{,}000)\mathrm{Pr}}{1 + 12.7(f/8)^{1/2}(\mathrm{Pr}^{2/3} - 1)}, \tag{11.73}$$

where f is the friction factor given by Equation 11.59. This equation gives results with less than 10% error if $0.5 \leq \mathrm{Pr} \leq 2{,}000$ and $3{,}000 \leq \mathrm{Re}_{D_h} \leq 5 \times 10^6$. Equations 11.72 and 11.73 can be used for both uniform surface temperature and uniform surface heat flux boundary conditions.

Fluid properties are needed when using correlations for the friction factor and the Nusselt number. These correlations were derived assuming that the change in fluid properties due to the change in fluid temperature is small. Since fluid temperature changes from the surface to the center and from the inlet to the exhaust of a pipe, the average of the fluid mean temperatures at the inlet and exhaust is used to obtain the required fluid properties. This temperature is called the bulk fluid mean temperature; $T_b = (T_i + T_e)/2$.

It must be noted that the correlations given for the friction factor and the Nusselt number are valid for fully developed flows only. The convection heat transfer coefficient is larger in the entry region and reduces to its fully developed value at a sufficiently large distance from the inlet. Therefore, if the flow length is not long enough such that the flow is either not fully developed or the entry length makes a major portion of the flow, the use of the fully developed correlations will underestimate the heat transfer rate. The following correlation gives the average Nusselt number for the combined entry and the fully developed laminar flow between two uniform surface temperature parallel plates of length L [5]:

$$\overline{\mathrm{Nu}}_{D_h} = 7.54 + \frac{0.03(D_h/L)\mathrm{Re}_{D_h}\mathrm{Pr}}{1 + 0.016[(D_h/L)\mathrm{Re}_{D_h}\mathrm{Pr}]^{2/3}}. \tag{11.74}$$

Note that as $L \to \infty$, the average Nusselt number approaches to its fully developed value of 7.54.

The average Nusselt number for the combined entry and fully developed laminar flow inside a circular pipe at uniform surface temperature can be determined from [5]:

$$\overline{\mathrm{Nu}_D} = 3.66 + \frac{0.065(D/L)\mathrm{Re}_D \mathrm{Pr}}{1 + 0.04[(D/L)\mathrm{Re}_D \mathrm{Pr}]^{2/3}}. \tag{11.75}$$

Again, it is seen that as $L \to \infty$, the average Nusselt number approaches to its fully developed value of 3.66.

Example 11.1

Consider an electronic enclosure shaped as a duct with a cross section of 350 mm × 50 mm and length of 400 mm as shown in Figure 11.7. Let's assume that all the heat dissipating components are located on two 400 mm × 350 mm similar boards that are attached to the top and bottom surfaces of the duct. Let's approximate the enclosure as a duct with constant surface temperature and assume the flow is fully developed. Air with a mass flow rate of 0.05 kg/s and inlet temperature of $T_i = 40°C$ enters the duct and leaves it with a temperature of $T_e = 50°C$.

   a. What is the total heat transfer rate to the air inside the duct?
   b. Calculate the temperature of the boards.

Solution

The total heat transfer rate is obtained by using energy balance between inlet and exhaust:

$$\dot{Q} = \dot{m} C_{p,\mathrm{ave}} (T_e - T_i)$$

$$T_b = 0.5(T_e + T_i) = 0.5(50 + 40) = 45°C \Rightarrow C_{p,\mathrm{ave}} = 1007.7 \ \mathrm{J/kg°C}$$

$$\dot{Q} = 0.05 \ \mathrm{kg/s} \times 1007.7 \ \mathrm{J/kg°C} \times (50°C - 40°C) = 503.8 \ \mathrm{W}.$$

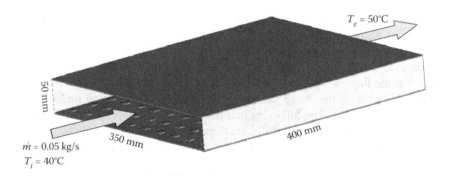

FIGURE 11.7 The electronic enclosure of Example 11.1.

Forced Convection Heat Transfer: Internal Flows

Since this is a constant surface temperature duct, $\dot{Q} = \bar{h}A\Delta T_{\ln}$. The air properties at T_b are $\mu = 1.938 \times 10^{-5}$ kg/m.s, $\rho = 1.11$ kg/m³, $k = 0.0274$ W/m°C and Pr = 0.713:

$$D_h = \frac{4A}{P} = \frac{4 \times 0.35\,\text{m} \times 0.05\,\text{m}}{2(0.35\,\text{m} + 0.05\,\text{m})} = 0.0875\,\text{m}$$

$$\text{Re}_{D_h} = \frac{\dot{m}D_h}{\mu A} = \frac{0.05\,\text{kg/s} \times 0.0875\,\text{m}}{1.938 \times 10^{-5}\,\text{kg/m.s} \times (0.35\,\text{m} \times 0.05\,\text{m})} = 12{,}900.$$

Flow is turbulent. Therefore, the Nusselt number and convection heat transfer coefficient are

$$\text{Nu}_{D_h} = 0.023\,\text{Re}_{D_h}^{0.8}\,\text{Pr}^{0.4} = 0.023 \times 12{,}900^{0.8} \times 0.713^{0.4} = 39$$

$$\bar{h} = \text{Nu}_{D_h}\frac{k}{D_h} = 39 \times \frac{0.0274\,\text{W/m°C}}{0.0875\,\text{m}} = 12.2\,\text{W/m}^2\text{°C}.$$

Since $\dot{Q}$, $\bar{h}$, and A are known $\Delta T_{\ln}$ and board surface temperature are obtained as follows:

$$\dot{Q} = \bar{h}A\Delta T_{\ln} \Rightarrow \Delta T_{\ln} = \frac{\dot{Q}}{\bar{h}A} = \frac{503.8\,\text{W}}{12.2\,\text{W/m}^2\text{°C} \times 2(0.35\,\text{m} \times 0.4\,\text{m})} = 147.5\,\text{°C}$$

$$\Delta T_{\ln} = \frac{T_i - T_e}{\ln[(T_s - T_e)/(T_s - T_i)]} \Rightarrow 147.5 = \frac{40 - 50}{\ln[(T_s - 50)/(T_s - 40)]}$$

$$T_s = 192.6\,\text{°C}.$$

11.7 FANS AND PUMPS

Fans and pumps are fluid moving devices that are used to generate forced convection flows. Many air-cooled electronic equipment such as desktop and laptop computers, external hard drives, power supplies, computing and data servers, and networking routers and switches include one or more fans that generate the required airflow. On the other hand, the liquid-cooled computers and servers include at least one pump to generate the liquid flow. Selecting the appropriate fan and pump is an important task for the system designers. In addition to the required air or liquid flow rate, parameters such as the available volume, the acceptable acoustic noise, the power consumption, and the cost have to be considered.

11.7.1 Types of Fans

Based on the shape of their blades and housing and the direction of inlet and exhaust airflows, fans are classified into three major categories; axial, radial, and mixed-flow fans.

Airflow enters and leaves an axial fan parallel to its axis of rotation. Three different kinds of axial fans are propeller fans, tube-axial fans, and vane-axial fans as

FIGURE 11.8 (See color insert following page 240.) Examples of (a) propeller fan (Courtesy of Delta Electronics, Inc.), (b) tube-axial fan (Courtesy of Sanyo Denki), and (c) vane-axial fan (Courtesy of Sanyo Denki).

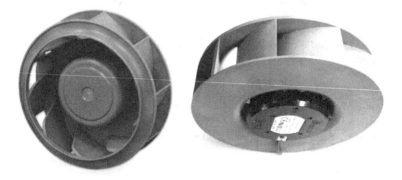

FIGURE 11.9 (See color insert following page 240.) Two examples of radial fans or impellers. (Courtesy of Delta Electronics, Inc.)

shown in Figure 11.8. Propeller fans consist of the blade and motor only. They are able to generate a large amount of airflow at a small pressure rise across the blade and therefore are used in low flow resistance applications. One drawback of the propeller fan is the presence of tip vortices due to the air movement from the high pressure to the low pressure side of the blade. Examples of propeller fans are ceiling and ventilation fans. Tube-axial fans consist of blade, motor, and housing with a small gap between the blade tip and the housing to reduce the tip vortices. Tube-axial fans are the most common type of axial fans used in electronic systems. Vane-axial fans are similar to tube-axial fans with additional fixed vanes to straighten the swirling flow out of the fan blade.

Two typical blade assemblies of radial fans are shown in Figure 11.9. They are called wheels or impellers. The motor is located at the center of the impeller. Air enters the impeller parallel to its axis of rotation and leaves it in the radial direction. If the impeller blades are angled in the same direction as its rotation, it will be called a forward curved impeller. On the other hand, if the blades are angled in the direction opposite its rotation, it will be called a backward curved impeller. Backward

Forced Convection Heat Transfer: Internal Flows

curved impellers, usually called impellers, come without any housing and the end user designs appropriate housings for those. On the other hand, forward curved impellers are usually caged in appropriate housings, as shown in Figure 11.10, and are called centrifugal blowers or simply blowers.

Although mixed flow fans look very similar to axial fans their airflow characteristics can be considered somewhere between axial and radial fans. An example of a mixed flow fan is shown in Figure 11.11. Similar to axial and radial fans, air enters the blades in an axial direction. However, the exhaust airflow is neither axial nor radial but rather diagonal. Therefore, the exhaust flow has a conical shape.

Fans are also classified by the type of their motors as either an AC or a DC fan, and by the bearing type as either a sleeve or ball bearing fan.

FIGURE 11.10 (See color insert following page 240.) Examples of blowers. (Courtesy of Delta Electronics, Inc.)

FIGURE 11.11 (See color insert following page 240.) A mixed flow fan. (Courtesy of Delta Electronics, Inc.)

11.7.2 FAN CURVE AND SYSTEM IMPEDANCE CURVE

The amount of air a fan moves in a system depends on the characteristics of both the fan and the system. Any system acts as a flow resistance to the fan that is used to move air in that system. The air pressure rise across the fan balances the air pressure drop through the system. The performance of a fan is given by **fan performance curve** or simply called **fan curve**. It gives the pressure rise across the fan as a function of the flow rate through the fan. Figure 11.12 shows typical fan curves for different fan types of the same size running at the same speed. It is seen that axial fans generate large amounts of airflow at relatively low pressure rises. On the other hand, centrifugal blowers can overcome large pressure rises at relatively small airflow rates. Backward curved impellers and mixed flow fans fit somewhere between these two limits and are capable of generating medium airflow rates at medium pressure rises.

A fan's performance curve depends on parameters such as the fan diameter and speed, and the air density. Fan laws give relationships between airflow rate, pressure rise, fan noise, and its power output and these parameters. If D (m), N (RPM), and ρ (kg/m³) are the fan diameter, its rotational speed, and the air density, respectively, and $\dot{V}$ (m³/s), ΔP (Pa), SPL (dB) and P (W) are the flow rate, the pressure rise, the sound pressure level, and the fan output power, respectively, the fan laws will be expressed as follows:

$$\frac{\dot{V}_2}{\dot{V}_1} = \left(\frac{D_2}{D_1}\right)^3 \left(\frac{N_2}{N_1}\right), \tag{11.76}$$

$$\frac{\Delta P_2}{\Delta P_1} = \left(\frac{D_2}{D_1}\right)^2 \left(\frac{N_2}{N_1}\right)^2 \left(\frac{\rho_2}{\rho_1}\right), \tag{11.77}$$

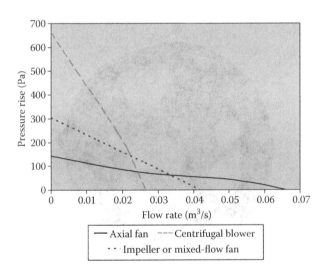

FIGURE 11.12 Typical characteristic curves of same size axial fans, centrifugal blowers, impellers, and mixed-flow fans.

Forced Convection Heat Transfer: Internal Flows

$$\frac{P_2}{P_1} = \left(\frac{D_2}{D_1}\right)^5 \left(\frac{N_2}{N_1}\right)^3 \left(\frac{\rho_2}{\rho_1}\right), \tag{11.78}$$

$$SPL_2 = SPL_1 + 10\log\left(\frac{N_2}{N_1}\right)^5 \left(\frac{\rho_2}{\rho_1}\right)^2. \tag{11.79}$$

The following observations can be made:

- The fan flow rate is independent of density while its pressure rise is proportional to it. Air density decreases with altitude and therefore the performance curve of a fan will be different at various altitudes. Figure 11.13 compares performance curves of a typical fan at different altitudes. It must be noted that this is true for constant speed fans such as the AC fans. For the constant voltage DC fans, a reduction in air density may cause an increase in the fan speed, which partially compensates for the loss in the pressure rise and increases the flow rate.
- The fan flow rate increases faster than fan pressure rise if its dimensions increase.
- The acoustic noise emitted by a fan is a strong function of its rotational speed. For example, if the fan speed is halved, noise level will be reduced by about 15 dB. It is also seen that the sound pressure level goes down at higher altitudes (i.e., lower air densities).

As mentioned in Sections 11.4 and 11.5, the flow through a system goes through a pressure drop as it moves from the inlet to the exhaust of that system. Any system can be characterized by a curve that gives pressure drop through the system as a function of the flow rate. This curve is called the **system impedance curve**. The correlations

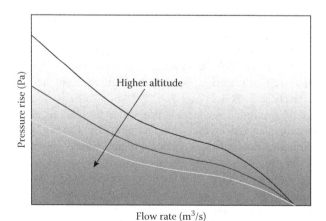

FIGURE 11.13 Effect of altitude (air density) on fan curve. Note that higher altitude corresponds to lower air density.

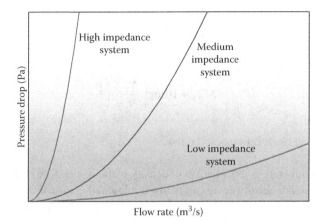

FIGURE 11.14 Examples of system impedance curves.

for friction factor and pressure drop in internal flows show that the pressure drop is proportional to the flow rate in laminar flows and is almost proportional to the square of the flow rate in the fully turbulent flows. In general, a system impedance curve will have the form:

$$\Delta P = C_1 \dot{V} + C_2 \dot{V}^n, \tag{11.80}$$

where n takes a value close to 2. Figure 11.14 shows examples of system impedance curves for typical low impedance, medium impedance, and high impedance systems. Note that low, medium, and high, when used to refer to the system impedance, are relative terms and should be only used to compare different systems with each other.

11.7.3 Fan Selection

If the impedance curve of a system and the characteristic curve of a fan are plotted on the same graph, the intersection of the two curves gives the flow rate that the fan generates in that system. Figure 11.15 shows intersections of a fan curve with impedance curves of three different systems. It is seen that this fan generates about 0.011, 0.029, and 0.056 m³/s airflow through the high, medium, and low impedance systems, respectively.

The pressure drop through a system is proportional to the air density and therefore, the system impedance decreases at higher altitudes. The fan laws state that the pressure rise through a fan is also proportional to air density. Therefore, as shown in Figure 11.16, a fan running at a fixed speed generates the same volumetric flow rate in a system at all altitudes. On the other hand, the correlations for the Nusselt number show that the convection heat transfer coefficient decreases at a lower air density (higher altitude). Therefore, higher velocity (more volumetric flow rate) is required to compensate for the reduction in the convection heat transfer coefficient due to the reduction in air density. This means that the airflow rate required to cool an electronic system at a higher altitude is more than what is needed at the lower altitude.

Forced Convection Heat Transfer: Internal Flows

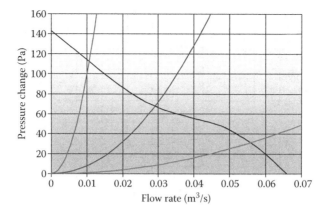

FIGURE 11.15 A typical fan curve plotted with three system impedance curves.

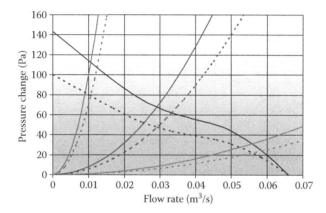

FIGURE 11.16 Fan and system impedance curves at two altitudes. The dashed line curves are at a higher altitude.

Let's assume that the airflow generated in a system, by a specific fan, is not enough to cool it appropriately. The following options are available to increase the airflow rate through this system.

1. Use the same fan at a higher speed: The fan laws state that the fan airflow rate is proportional to its speed and the fan pressure rise is proportional to its speed squared. Therefore, as shown in Figure 11.17, a fan at a higher speed generates more flow rate in the same system. However, this comes with an increase in the noise generated by the fan.
2. Use a larger fan: The fan laws show that the airflow rate and the pressure rise through a fan are proportional to the cube and the square of its diameter, respectively. Therefore, as shown in Figure 11.18, a larger fan generates higher airflow rate in a system compared to a similar smaller fan. Larger fans are sometimes used at lower speeds to generate the same airflow rate at a lower acoustic noise level.

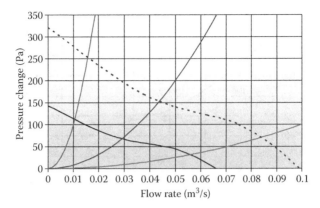

FIGURE 11.17 Typical system impedance curves and fan curves at speed N_1 (solid curve) and $N_2 = 1.5\ N_1$ (dashed-line curve).

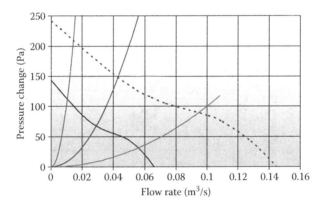

FIGURE 11.18 Typical system impedance curves and fan curves at diameters D_1 (solid curve) and $D_2 = 1.3\ D_1$ (dashed-line curve).

3. Use multiple fans: If increasing the fan speed or using a larger fan is not possible, multiple fans may be used to generate the required airflow. Multiple fans may be used either in a side-by-side, also called parallel, or in a tandem, also called series, arrangement. Parallel arrangement of two fans doubles the airflow rate, at any given pressure rise, compared to a single fan. On the other hand, a series arrangement of two fans has double the pressure rise capability of a single fan at any given airflow rate. Figure 11.19 shows system impedance curves for three different systems along with the fan curve of a single fan and the equivalent fan curves for two parallel and two series fans. It is seen that the parallel fan configuration is most effective for the low impedance systems while the series fans are effective for the high impedance systems. Not that the doubling of the airflow rate for two parallel fans, and the doubling of the pressure rise for two fans in series, happens only if the fans do not interfere with each other's performance. This is a fairly valid assumption for parallel fans but is not acceptable for fans in series if they are too close to each other.

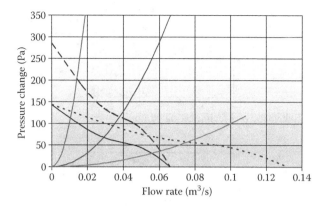

FIGURE 11.19 Typical system impedance curves and fan curves for a single fan (solid curve), two parallel fans (dashed-line curve), and two series fans (long dashed-line curve).

11.7.4 Types of Pumps

Pumps can be classified based on their pumping mechanism as positive displacement pumps or dynamic pumps. All positive displacement pumps work based on the same principle. They draw fluid in through an intake, during suction cycles, and push it through an exhaust, during discharge cycles. Positive displacement pumps are used in high pressure applications. Dynamic pumps operate by first increasing fluid velocity and then converting part of its kinetic energy to pressure in a diffuser shape flow passage. Dynamic pumps usually have lower efficiencies than positive displacement pumps. However, they are able to operate at fairly high speeds and to generate high flow rates.

Examples of positive displacement pumps are reciprocating pumps, diaphragm pumps, and rotary pumps. A reciprocating pump is basically a cylinder-piston combination with intake and exhaust valves as shown in Figure 11.20. A volume of liquid is drawn into the cylinder through the intake valve on the suction stroke and is discharged under positive pressure through the exhaust valve on the discharge stroke. Diaphragm pumps are similar to reciprocating pumps in operation principle but use a flexible diaphragm to move the fluid. Figure 11.21 shows the schematic of a diaphragm pump. A diaphragm can be moved back and forth by a mechanical mean, like the one shown in Figure 11.21, or electrically such as in piezoelectric pumps. A rotary pump traps the fluid at the suction inlet and moves it to the discharge outlet. It traps the liquid, pushes it around inside a closed casing, and discharges it in a continuous flow. They can handle almost any liquid that does not contain hard and abrasive solids, including viscous liquids. Types of rotary pumps include cam-and-piston, gear, lobe, screw, and vane pumps. Figure 11.22 shows schematics of an external and an internal gear pump as well as a lobe pump. Rotary pumps find a wide use for viscous liquids. In these cases, rotary pumps must operate at low speeds because at high speeds the liquid cannot flow into the casing fast enough to fill it. Unlike centrifugal pumps, a rotary pump delivers a capacity that is not affected by pressure variations

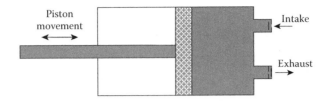

FIGURE 11.20 Schematic of a reciprocating pump.

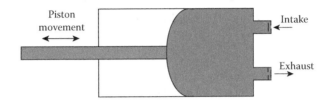

FIGURE 11.21 Schematic of a diaphragm pump in suction cycle.

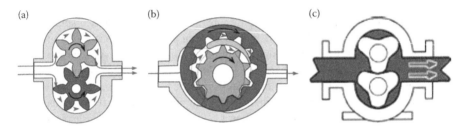

FIGURE 11.22 Schematic of (a) external, (b) internal gear pumps, and a (c) lobe pump.

on either the suction or the discharge end. Therefore, they are good candidates for situations where large changes in pressure are anticipated.

Examples of dynamic pumps are centrifugal pumps and axial flow pumps. A centrifugal pump is very similar to a centrifugal blower and consists of an impeller with an intake at its center. As the impeller rotates, liquid is drawn through the intake and is discharged by the centrifugal force into the casing surrounding the impeller. The casing is designed such that the fluid velocity decreases gradually and its kinetic energy is converted to the pressure that is needed to discharge the fluid. Some of the advantages of centrifugal pumps are smooth flow through the pump, uniform pressure in the discharge pipe, and low cost. Axial flow pumps, also called propeller pumps, develop most of their pressure by the propelling or lifting action of the vanes on the liquid. These pumps are often used in low pressure and large flow rate applications.

Pumps that are used in electronic cooling applications have to be small, cheap, highly reliable, and run with DC power. Many centrifugal, rotary, and piezoelectric pumps have been introduced for this application and the development and testing of these pumps have been an active research and engineering topic in the last few years [6]. In addition to the classical pump types, electroosmotic pumps and electromagnetic

Forced Convection Heat Transfer: Internal Flows

pumps have recently received attention as well. Electroosmotic pumps work based on the bulk motion of an electrolyte caused by the electric forces acting on regions of net charge in a microchannel or a porous media. Electromagnetic pumps can be used only to pump magnetic fluids that are also good electrical conductors. The pipe carrying the fluid is placed in a magnetic field and a current is passed crosswise through the fluid, so that the fluid is subjected to an electromagnetic force in the direction of the flow.

11.8 PLATE-FIN HEAT SINKS

Consider the plate-fin heat sink shown in Figure 11.23. The flow between the fins of this heat sink can be considered as an internal flow. The heat sink width and length are W and H, fin height is L, and fin gap and thickness are s and t. Let's assume that the fluid approaches this heat sink with volumetric flow rate $\dot{V}$ and fluid density, dynamic viscosity, specific heat at constant pressure, thermal conductivity, and Prandtl number are ρ_f, μ_f, $C_{p,f}$, k_f, and Pr, respectively, where the subscript f stands for fluid. Teertstra et al. [7] developed the following expression for the convection heat transfer coefficient in a plate-fin heat sink, by combining the two limiting cases of the fully developed and the developing flows between two isothermal parallel plates. Approximating the fins as isothermal is acceptable when they are made of conductive materials such as aluminum or copper and the fin efficiency is relatively high.

$$\text{Nu}_s = \frac{h_{\text{conv}} s}{k_f} = \left[\left(\frac{\text{Re}_s^* \text{Pr}}{2} \right)^{-3} + \left(0.664 \sqrt{\text{Re}_s^*}\, \text{Pr}^{1/3} \sqrt{1 + \frac{3.65}{\sqrt{\text{Re}_s^*}}} \right)^{-3} \right]^{-1/3}. \quad (11.81)$$

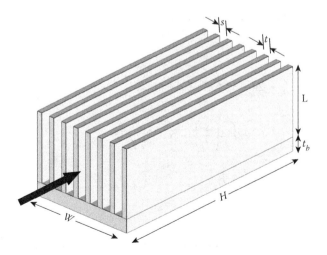

FIGURE 11.23 Schematic of a plate-fin heat sink with flow parallel to its length H.

Here

$$\text{Re}_s^* = \text{Re}_s \frac{s}{H} = \frac{\rho_f U_f s^2}{\mu_f H} \qquad (11.82)$$

is called the channel Reynolds number and U_f is the fin velocity given by

$$U_f = \frac{\dot{V}}{(n_f - 1)sL}. \qquad (11.83)$$

The heat sink convection thermal resistance is given by

$$R_{hs,conv} = \frac{1}{\eta_t h_{conv} A_t}. \qquad (11.84)$$

The pressure drop through the plate fin heat sink shown in Figure 11.23 is given by [1]:

$$\Delta P = \left(f_{app} \frac{H}{D_h} + K_c + K_e \right) \frac{1}{2} \rho_f U_f^2. \qquad (11.85)$$

The D_h is the hydraulic diameter of the channel created between the two neighboring fins. The first term in Equation 11.85 is the frictional pressure loss and f_{app} is called the apparent friction factor. Muzychka and Yovanovich [8] proposed the following correlation for the apparent friction factor:

$$f_{app} \text{Re}_{D_h} = 4 \left[\left(\frac{3.44}{\sqrt{H/(D_h \text{Re}_{D_h})}} \right)^2 + \left(f \text{Re}_{D_h} \right)^2 \right]^{0.5}, \qquad (11.86)$$

$$f \text{Re}_{D_h} = 24 - 32.527 \left(\frac{s}{L} \right) + 46.721 \left(\frac{s}{L} \right)^2 - 40.829 \left(\frac{s}{L} \right)^3$$

$$+ 22.954 \left(\frac{s}{L} \right)^4 - 6.089 \left(\frac{s}{L} \right)^5, \qquad (11.87)$$

$$\text{Re}_{D_h} = \frac{\rho_f U_f D_h}{\mu_f}. \qquad (11.88)$$

The second and third terms in Equation 11.85 are the contraction and expansion pressure drops at the inlet and exhaust, respectively, with K_c and K_e given by [9]:

Forced Convection Heat Transfer: Internal Flows

$$K_c = 0.42(1 - \sigma^2),$$
$$K_e = (1 - \sigma^2)^2, \qquad (11.89)$$
$$\sigma = 1 - \frac{N_f t}{W}.$$

PROBLEMS

11.1 How is the Reynolds number defined for a duct with a rectangular cross section with lengths a and b? What is the Reynolds number above which the flow in a pipe is turbulent?

11.2 Consider water flow inside two similar pipes. The Reynolds number in one pipe is 550 and in the other pipe is 15,000. A small amount of a colored fluid is injected into each of these pipes. Explain what happens to the colored fluid in each pipe.

11.3 A 0.6 m × 0.45 m printed circuit board is mounted in a 0.6 m × 0.45 m × 0.15 m rectangular enclosure as shown in the figure below. Air at temperature $T_i = 25°C$ enters this enclosure through fan(s) mounted at the inlet. The printed circuit board dissipates 100 W uniformly across its surface. The maximum allowable board temperature, which happens at the exit, is 80°C.
   a. Determine the minimum airflow rate and the air temperature at the exit of the enclosure.
   b. If the maximum allowable air velocity at the intake of the fans is 4 m/s, how many 92 mm fans are needed? Assume that the air velocity is uniform at the intake of the fans.
   Use air properties at 25°C in all the calculations.

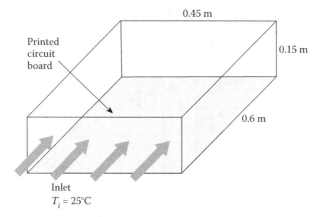

11.4 The components of an electronic system are mounted on the two 0.6 m × 0.45 m sides of the enclosure shown below. Air with a uniform velocity of $U_m = 1.5$ m/s and temperature of $T_i = 25°C$ enters the enclosure. The electronic components dissipate 200 W uniformly across these two surfaces. Determine the air temperature at the exit of the enclosure, and the surface temperature of the enclosure at the exit.

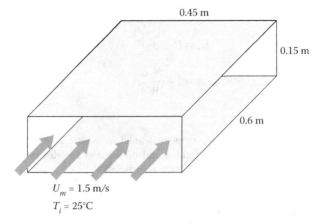

$U_m = 1.5$ m/s
$T_i = 25°C$

11.5 A 0.6 m × 0.45 m printed circuit board is mounted in a 0.6 m × 0.45 m × 0.15 m rectangular enclosure as shown in the figure below. Air with a uniform velocity of $U_m = 1$ m/s and temperature of $T_i = 25°C$ enters the enclosure. The printed circuit board dissipates 120 W uniformly across its surface. Determine the air temperature at the exit of the enclosure, and the surface temperature of the printed circuit board at the exit. Neglect natural convection and radiation.

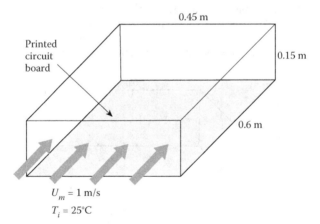

$U_m = 1$ m/s
$T_i = 25°C$

11.6 Components of an electronic system are mounted on two 300 mm × 400 mm printed circuit boards. Each board is mounted on a 300 mm × 400 mm wall of a 300 mm × 20 mm water channel as shown below. Let's assume each board dissipates 350 W and all the heat generated by the electronic components is transferred to the water in the channel. Water enters the channel with the mean velocity of 1 m/s and the mean temperature of 40°C. Determine the mean temperature of the water at the exit of the channel and the maximum surface temperature of the channel assuming:

Forced Convection Heat Transfer: Internal Flows

a. Constant heat flux condition at the walls of the channel.
b. Walls of the channel are at constant temperature.
   Use water properties at 40°C in all the calculations.

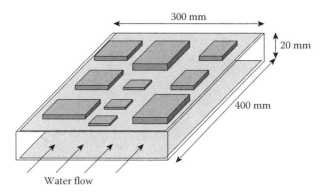

REFERENCES

1. Kays, Q., M. Ceawford and B. Weigand, 2005. *Convection heat and mass transfer.* New York, NY: McGraw-Hill.
2. Incropera, F. P., D. P. DeWitt, T. L. Bergman, and A. S. Lavine. 2006. *Fundamentals of heat and mass transfer.* Sixth Edition. Hoboken, NJ: John Wiley & Sons.
3. Petukhov, B. S. 1970. *Advances in heat transfer*, ed. T. F. Irvine and J. P. Hartnett. Vol. 6. New York: Academic Press.
4. Gnielinski, V. 1976. New equations for heat and mass transfer in turbulent pipe and channel flow. *International Chemical Engineering* 16:359–68.
5. Edwards, D. K., V. E. Denny, and A. F. Mills. 1979. *Transfer processes.* Washington, DC: Hemisphere Publishing Corporation.
6. Iverson, B. D., and S. V. Garimella. 2008. Recent advances in microscale pumping technologies; A review and evaluation. Microfluidics and Nanofluidics 5: 145–74.
7. Teertstra, P., M. M. Yovanovich, and J. R. Culham. 2000. Analytical forced convection modeling of plate fin heat sinks. *Journal of Electronics Manufacturing* 10 (4): 253–61.
8. Muzychka, Y. S., and M. M. Yovanovich. 1998, June 15–18. Modeling friction factors in non-circular ducts for developing laminar flow. Proceedings of 2nd AIAA Theoretical Fluid Mechanics Meeting, Albuquerque, New Mexico.
9. White, F. M. 1987. *Fluid mechanics.* New York, NY: McGraw-Hill.

12 Natural Convection Heat Transfer

Many electronic systems dissipate a relatively small amount of heat and may not require forced air or liquid flow for cooling. On the other hand, forced air or liquid flows are generated by fans or pumps that are noisy and have shorter life compared to most electronic devices. Therefore, it is better to avoid forced convection cooling in systems that are designed for low maintenance, long life, or low acoustic noise. Examples of these systems are televisions, video cassette recorders (VCRs), compact disk (CD) and digital video disk (DVD) players, personal modems and routers and firewalls, cell phones, handheld computers, portable game consoles, and wireless outdoor base stations and radio units. Heat transfer inside these systems and from their external surfaces to their ambient is partly through natural (free) convection and partly through radiation. These two modes of heat transfer will be studied in this and the next chapter, respectively.

12.1 BUOYANCY FORCE AND NATURAL CONVECTION FLOWS

We have all heard that "hot air rises and cold air sinks." We have also observed that a piece of wood floats on the surface of a pool while a coin or a rock sinks to the bottom. The physical mechanism behind these was first discovered by the Greek mathematician and physicist Archimedes (287–212 BC). Consider an object that is left over the surface of a fluid as shown in Figure 12.1. The weight of this object forces it to go downward in the fluid. But, as it moves down, the fluid exerts a force to it in the opposite direction. This force is called **buoyancy force**. Archimedes principle states that the buoyancy force on a submerged object is equal to the weight of the fluid that is displaced by that object. If ρ_f is fluid density, g is the gravitational acceleration and $\mathcal{V}_{submerged}$ is the submerged portion of the object, the buoyancy force is given by

$$F_{buoyancy} = \rho_f g \mathcal{V}_{submerged}. \qquad (12.1)$$

The net vertical force applied to this object, F_{net}, is the difference between its weight, W, and the buoyancy force. If ρ is the density of the object and $\mathcal{V}$ is its volume, $W = \rho g \mathcal{V}$, and the net vertical force applied to this object is

$$\begin{aligned}F_{net} &= W - F_{buoyancy} \\ &= \rho g \mathcal{V} - \rho_f g \mathcal{V}_{submerged}.\end{aligned} \qquad (12.2)$$

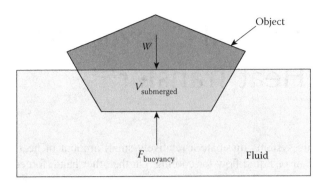

FIGURE 12.1 Forces applied on an object that is floating over a fluid.

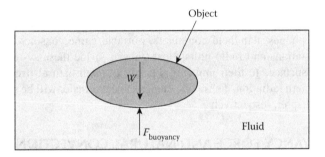

FIGURE 12.2 An object that is completely submerged in a fluid.

Note that F_{net} is downward if it is positive and is upward if it is negative.

Now consider an object that is completely submerged in a fluid as shown in Figure 12.2. In this case $\mathcal{V}_{submerged} = \mathcal{V}$ and the net force applied to this object is

$$F_{net} = W - F_{buoyancy}$$
$$= \rho g \mathcal{V} - \rho_f g \mathcal{V} \qquad (12.3)$$
$$= (\rho - \rho_f) g \mathcal{V}.$$

Let's assume that this submerged object is the same fluid but at a different temperature. If it is hotter than the rest of the fluid and its density is $\rho_{f,h}$, the net force applied to it is

$$F_{net} = (\rho_{f,h} - \rho_f) g \mathcal{V}. \qquad (12.4)$$

Density of most fluids, such as air and water, decreases as their temperatures increase. Therefore, $\rho_{f,h} - \rho_f < 0$ and the net force is negative. This means that the hot fluid is pushed upward. Similarly, it can be seen that, the net force on a volume of cold fluid

Natural Convection Heat Transfer

surrounded by warmer fluid is downward and pushes the cold fluid down. Equation 12.4 shows that natural convection is due to the density difference between different regions of a fluid. It also shows that the net force is directly proportional to the gravitational acceleration. Since the gravitational acceleration on the earth's surface is larger than that on the moon, natural convection flow on the earth is stronger than that on the moon.

As explained above, natural convection flows are generated as a result of a density difference in a fluid and in the presence of a gravitational field. It will also be helpful to understand the relationship between natural convection and the temperature variation in a fluid. This is achieved by considering a fluid property called **volumetric thermal expansion coefficient**, β. It is defined as the rate of change of a unit volume of the fluid per unit temperature change at constant pressure. If v is the fluid specific volume,

$$\beta = \frac{1}{v}\left(\frac{\partial v}{\partial T}\right)_p. \tag{12.5}$$

Since fluid density $\rho = 1/v$, volumetric thermal expansion coefficient can also be defined as

$$\beta = -\frac{1}{\rho}\left(\frac{\partial \rho}{\partial T}\right)_p. \tag{12.6}$$

The unit for β is the inverse of the absolute unit for temperature. If temperature is measured in Kelvin (K) or Rankin (R), β will be measured in 1/K or 1/R, respectively.

For an ideal gas

$$Pv = RT \implies v = \frac{RT}{P} \implies \left(\frac{\partial v}{\partial T}\right)_p = \frac{R}{P}.$$

Therefore, the volumetric thermal expansion coefficient for an ideal gas is

$$\beta = \frac{1}{v}\left(\frac{\partial v}{\partial T}\right)_p = \frac{1}{v}\frac{R}{P} = \frac{R}{Pv} \implies \beta = \frac{1}{T}. \tag{12.7}$$

The net vertical force applied to the volume $\mathcal{V}$ of a fluid with density ρ, which is different from the bulk fluid density, ρ_f, is

$$F_{net} = \Delta\rho g V, \tag{12.8}$$

where $\Delta\rho = \rho - \rho_f$. If we assume that the fluid is incompressible but its density changes with temperature, and the variation of the density with temperature is linear, the volumetric thermal expansion coefficient can be used to correlate the density difference to the temperature difference

$$\beta = -\frac{1}{\rho}\left(\frac{\partial \rho}{\partial T}\right)_p = -\frac{1}{\rho}\frac{\Delta\rho}{\Delta T}$$

$$\Delta\rho = -\rho\beta\Delta T. \tag{12.9}$$

This is known as the Boussinesq approximation. Substituting Equation 12.9 into 12.8 gives

$$F_{net} = -\rho\beta g \mathcal{V}\Delta T. \tag{12.10}$$

Equation 12.10 shows that the net vertical force applied to a hot fluid with a positive volumetric thermal expansion coefficient is negative and therefore upward. It also shows that the hotter the fluid is, the larger the vertical force, and the stronger the natural convection flow will be.

12.2 NATURAL CONVECTION VELOCITY AND TEMPERATURE BOUNDARY LAYERS

Consider a vertical plate exposed to a fluid with temperature T_∞, and assume that the plate temperature is $T_s > T_\infty$. The fluid layer next to the surface of the plate reaches a temperature equilibrium with the plate and its temperature is T_s. The fluid temperature decreases from T_s at the surface to T_∞ at some distance from the surface. The layer near the plate where this temperature change happens is called the **temperature boundary layer**. Now, let's consider a small volume of the fluid inside the temperature boundary layer. Since its temperature is higher than the fluid outside the boundary layer, it is subjected to an upward force that pushes the fluid up. The closer the volume of the fluid to the plate, the larger the temperature difference between that volume of the fluid and the fluid outside the boundary layer, and the stronger the upward force. However, for volumes of the fluid very close to the plate, the viscous shear force against the upward motion of the fluid is large as well. This force will dominate in the region very close to the plate and reduces the upward velocity of the fluid volume. Therefore, as seen in Figure 12.3, the maximum fluid velocity happens at some distance from the plate surface although the maximum temperature difference is at the surface. The layer near the plate where the fluid velocity increases to a maximum value and the decreases to zero is called the **velocity boundary layer**.

Natural Convection Heat Transfer 291

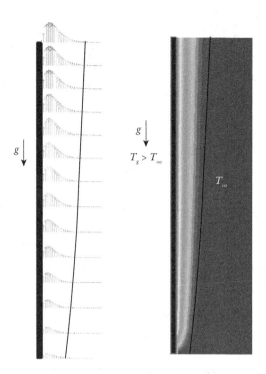

FIGURE 12.3 (See color insert following page 240.) Natural convection velocity and temperature boundary layers near a vertical plate.

12.3 NORMALIZED NATURAL CONVECTION BOUNDARY LAYER EQUATIONS

Many insights into natural convection flows and the important factors that influence convection heat transfer in these flows are obtained by studying a normalized form of the boundary layer equations. Consider the two-dimensional, steady-state natural convection flow along a vertical plate. Let's assume that there is no heat generation in the fluid and, since the velocity in natural convection flows is very low, the viscous dissipation term in the energy equation can be neglected. If we choose the x axis to be along the length of the plate and the gravity in the negative x direction, as shown in Figure 12.4, the body force in the x direction will be equal to the weight of the fluid and the body force per unit volume will be $X = -\rho g$. The boundary layer approximations show that

$$\frac{\partial P}{\partial y} = Y. \tag{12.11}$$

Since there is no force in the y direction, the pressure gradient in that direction is zero and

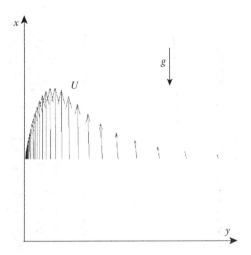

FIGURE 12.4 Natural convection velocity profile along a vertical plate.

$$\frac{\partial P}{\partial x} = \frac{dP}{dx} = -\rho_\infty g. \qquad (12.12)$$

This equation states that the pressure gradient along the length of a vertical plate is the same inside and outside the boundary layer, and is equal to the pressure gradient in the still fluid outside the boundary layer.

Using Equation 12.12 and setting $Y = \dot{g} = \Phi = 0$ in the steady-state forms of Equations 9.13 through 9.16 in Chapter 9 results in the boundary layer equations for steady-state natural convection flows with no heat generation:

$$\frac{\partial U}{\partial x} + \frac{\partial V}{\partial y} = 0, \qquad (12.13)$$

$$U\frac{\partial U}{\partial x} + V\frac{\partial U}{\partial y} = \frac{(\rho_\infty - \rho)g}{\rho} + \nu\frac{\partial^2 U}{\partial y^2}, \qquad (12.14)$$

$$U\frac{\partial T}{\partial x} + V\frac{\partial T}{\partial y} = \alpha\frac{\partial^2 T}{\partial y^2}. \qquad (12.15)$$

The first term on the right side of the momentum equation, Equation 12.14, can be expressed in terms of the temperature difference by applying the Boussinesq approximation, Equation 12.9

$$\Delta\rho = -\rho\beta\Delta T \Rightarrow \rho - \rho_\infty = -\rho\beta(T - T_\infty) \Rightarrow \frac{\rho_\infty - \rho}{\rho} = \beta(T - T_\infty). \qquad (12.16)$$

Natural Convection Heat Transfer

Substituting into the momentum equation gives

$$U\frac{\partial U}{\partial x}+V\frac{\partial U}{\partial y}=g\beta(T-T_\infty)+\nu\frac{\partial^2 U}{\partial y^2}. \quad (12.17)$$

Let's assume L is a characteristic length of the flow, T_∞ is the free-stream fluid temperature, and U_0 is a reference velocity. The normalized coordinates x^* and y^*, normalized velocity components U^* and V^*, and normalized temperature T^* are defined as

$$x^*=\frac{x}{L}, \quad y^*=\frac{y}{L}, \quad U^*=\frac{U}{U_0}, \quad V^*=\frac{V}{U_0}, \quad T^*=\frac{T-T_\infty}{T_s-T_\infty}. \quad (12.18)$$

Substituting these normalized variables into Equations 12.13, 12.15, and 12.17 results in the following set of normalized boundary layer equations for two-dimensional, and steady-state natural convection flows.

$$\frac{\partial U^*}{\partial x^*}+\frac{\partial V^*}{\partial y^*}=0, \quad (12.19)$$

$$U^*\frac{\partial U^*}{\partial x^*}+V^*\frac{\partial U^*}{\partial y^*}=\frac{g\beta(T_s-T_\infty)L}{U_0^2}T^*+\frac{\nu}{U_0 L}\frac{\partial^2 U^*}{\partial y^{*2}}, \quad (12.20)$$

$$U^*\frac{\partial T^*}{\partial x^*}+V^*\frac{\partial T^*}{\partial yv}=\frac{\alpha}{U_0 L}\frac{\partial^2 T^*}{\partial y^{*2}}. \quad (12.21)$$

12.3.1 GRASHOF AND RAYLEIGH NUMBERS

Three dimensionless groups appear in the normalized boundary layer equations for natural convection. The ones appearing in the viscous term of the momentum equation and in the conduction term of the energy equation are the inverses of the Reynolds number and the Peclet number, respectively;

$$\mathrm{Re}_L=\frac{U_0 L}{\nu}, \quad (12.22)$$

and

$$\mathrm{Pe}_L=\mathrm{Re}_L\,\mathrm{Pr}=\frac{U_0 L}{\nu}\frac{\nu}{\alpha}=\frac{U_0 L}{\alpha}. \quad (12.23)$$

The coefficient of T^* in Equation 12.20 can be written as

$$\frac{g\beta(T_s-T_\infty)L}{U_0^2}=\frac{g\beta(T_s-T_\infty)L^3}{\nu^2}\frac{\nu^2}{U_0^2 L^2}=\frac{g\beta(T_s-T_\infty)L^3}{\nu^2}\frac{1}{\mathrm{Re}_L^2}. \quad (12.24)$$

The new dimensionless group in Equation 12.24 is called the **Grashof** number,

$$\text{Gr}_L \equiv \frac{g\beta(T_s - T_\infty)L^3}{\nu^2}. \tag{12.25}$$

The Grashof number plays the same role in natural convection flows as the Reynolds number plays in forced convection flows. It can be shown that the Grashof number is a measure of the ratio of the net buoyancy force to the viscous force acting on a volume of fluid. It was shown in Chapter 10 that viscous force per unit volume is proportional to $\mu U_0/L^2$;

$$\frac{\text{viscous force}}{\text{volume}} = \frac{\tau A}{\text{volume}} = \frac{A}{\text{volume}}\left(\mu \frac{\partial U}{\partial y}\right) \propto \frac{A}{\text{volume}}\left(\mu \frac{\Delta U}{\Delta y}\right)$$

$$\propto \frac{L^2}{L^3}\mu \frac{U_0}{L} \propto \frac{\mu U_0}{L^2}.$$

Pure natural convection flows are usually slower than forced convection flows. Therefore, it is not unrealistic to assume that the Reynolds number for these flows is of the order of one. This will allow us to obtain an estimate for the reference velocity in terms of the fluid viscosity and the reference length;

$$\text{Re}_L = \frac{U_0 L}{\nu} \propto 1 \Rightarrow U_0 \propto \frac{\nu}{L} \Rightarrow \frac{\text{viscous force}}{\text{volume}} = \frac{\mu\nu}{L^3}.$$

On the other hand, Equation 12.10 shows that the net buoyancy force per unit volume of a fluid is equal to $\rho\beta g\Delta T$. Therefore, the ratio of the net buoyancy force to the viscous force is

$$\frac{\text{net buoyancy force}}{\text{viscous force}} \propto \frac{\rho g\beta\Delta T}{\mu\nu/L^3} \propto \frac{g\beta\Delta T L^3}{\nu^2} \equiv \text{Gr}_L. \tag{12.26}$$

Note that the index L in Gr_L indicates that the characteristic length L is used in the definition of the Grashof number.

Another dimensionless parameter that appears in many natural convection correlations is the **Rayleigh** number, which is the product of the Grashof and Prandtl numbers,

$$\text{Ra}_L \equiv \text{Gr}_L \, \text{Pr} = \frac{g\beta\Delta T L^3}{\nu\alpha}. \tag{12.27}$$

The Rayleigh number plays an important role in determining whether natural convection flow is laminar or turbulent.

Natural Convection Heat Transfer

12.3.2 Functional Form of the Convection Heat Transfer Coefficient

Using definitions of Gr_L, Re_L, and Pr in Equations 12.19, 12.20, and 12.21 gives

$$\frac{\partial U^*}{\partial x^*} + \frac{\partial V^*}{\partial y^*} = 0, \tag{12.28}$$

$$U^*\frac{\partial U^*}{\partial x^*} + V^*\frac{\partial U^*}{\partial y^*} = \frac{Gr_L}{Re_L^2}T^* + \frac{1}{Re_L}\frac{\partial^2 U^*}{\partial y^{*2}}, \tag{12.29}$$

$$U^*\frac{\partial T^*}{\partial x^*} + V^*\frac{\partial T^*}{\partial y^*} = \frac{1}{Re_L Pr}\frac{\partial^2 T^*}{\partial y^{*2}}. \tag{12.30}$$

Notice that the ratio Gr_L/Re_L^2 appears as the coefficient of T^* in the momentum equation. This ratio is a measure of the relative magnitude of the buoyancy force to the inertia force. In a general flow, where both forced and natural convection flows are present, this ratio gives an indication of whether any of those two is dominant or both are of the same order. Generally, if $Gr_L/Re_L^2 \gg 1$, natural convection is dominant. On the other hand, if $Gr_L/Re_L^2 \ll 1$, natural convection can be neglected compared to forced convection. However, if $Gr_L / Re_L^2 \approx 1$, both natural and forced convections are as important.

Because of the appearance of T^* in Equation 12.29, the momentum, continuity, and energy equations are coupled. This means that these equations have to be solved simultaneously to obtain the velocity components and temperature. The normalized boundary layer equations show that the normalized x component of velocity, U^*, and the normalized temperature, T^*, have the following functional forms:

$$U^* = f_1(x^*, y^*, Gr_L, Re_L, Pr), \tag{12.31}$$

$$T^* = g_1(x^*, y^*, Gr_L, Re_L, Pr). \tag{12.32}$$

The local Nusselt number is equal to the normalized temperature gradient at the surface

$$Nu_L = \frac{hL}{k_f} = \frac{\dot{q}_x L}{k_f(T_s - T_\infty)} = -k_f \frac{\partial T}{\partial y}\bigg|_{y=0} \frac{L}{k_f(T_s - T_\infty)} = -\frac{\partial T^*}{\partial y^*}\bigg|_{y^*=0}.$$

Therefore, the functional form of the local Nusselt number is

$$Nu_L = g_2(x^*, Gr_L, Re_L, Pr), \tag{12.33}$$

and the functional form of the average Nusselt number over the entire length of the plate is

$$\overline{Nu}_L = \frac{\overline{h}L}{k_f} = \overline{g}_2(Gr_L, Re_L, Pr). \tag{12.34}$$

The above expressions for the functional forms of the normalized velocity and temperature and the Nusselt number are for a combined forced and natural convection flow. **If the flow is predominantly a natural convection flow**, Re_L becomes irrelevant and

$$Nu_L = \frac{hL}{k_f} = g_2(x^*, Gr_L, Pr) \quad \text{and} \quad \overline{Nu_L} = \frac{\overline{h}L}{k_f} = \overline{g}_2(Gr_L, Pr). \tag{12.35}$$

12.4 LAMINAR AND TURBULENT NATURAL CONVECTION OVER A VERTICAL FLAT PLATE

Consider the natural convection boundary layer along a vertical flat plate with the uniform surface temperature $T_s > T_\infty$. Experimental observations and theoretical analyses have shown that this flow is laminar if $Re_L < 10^9$ and turns to turbulent for the higher Rayleigh numbers.

The boundary layer equations for the steady-state laminar natural convection flow are are given by Equations 12.13, 12.15, and 12.17. The boundary conditions are

$$U(y = 0) = V(y = 0) = 0 \quad \text{and} \quad T(y = 0) = T_s,$$

$$U(y \to \infty) = 0 \quad \text{and} \quad T(y \to \infty) = T_\infty \tag{12.36}$$

A similarity solution to these equations was obtained [1] by defining a similarity variable:

$$\eta = \frac{y}{x}\left(\frac{Gr_x}{4}\right)^{1/4}, \tag{12.37}$$

and a stream function of the form:

$$\psi(x, y) = f(\eta)\left[4\nu\left(\frac{Gr_x}{4}\right)^{1/4}\right], \tag{12.38}$$

such that

$$U = \frac{\partial \psi}{\partial y} \quad \text{and} \quad V = -\frac{\partial \psi}{\partial x}$$

automatically satisfy the continuity Equation 12.13. Using the similarity variable η and the stream function $\psi(x,y)$, partial differential Equations 12.15 and 12.17 are converted to the following set of ordinary differential equations,

$$f - + 3ff'' - 2f'^2 + T^* = 0,$$
$$T^{*''} + 3 Pr\, fT^{*'} = 0, \tag{12.39}$$

Natural Convection Heat Transfer

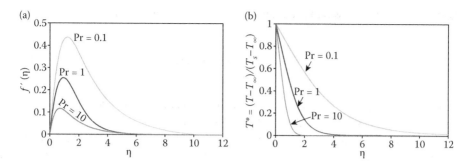

FIGURE 12.5 Normalized velocity and temperature profiles in the laminar natural convection flow along a vertical flat plate at a uniform surface temperature.

with the boundary conditions

$$f(\eta = 0) = f'(\eta = 0) = 0 \quad \text{and} \quad T^*(\eta = 0) = 1,$$

$$f'(\eta \to \infty) = 0 \quad \text{and} \quad T^*(\eta \to \infty) = 0. \tag{12.40}$$

The numerical solutions of these equations for Pr = 0.1, 1, and 10 are shown in Figure 12.5. It is seen that the thermal and momentum diffusions are deeper, and the boundary layers are thicker, for the smaller Prandtl number fluids.

The local Nusselt number at a distance x from the leading edge of the plate is

$$\mathrm{Nu}_x = \frac{hx}{k_f} = \frac{hL}{k_f} \times \frac{x}{L} = -\left.\frac{\partial T^*}{\partial y^*}\right|_{y^*=0} \times \frac{x}{L} = -\left.\frac{\partial y}{\partial y^*}\frac{\partial \eta}{\partial y}\frac{\partial T^*}{\partial \eta}\right|_{\eta=0} \times \frac{x}{L}.$$

Note that $(\partial T^*/\partial \eta)_{\eta=0}$ is a function of the Prandtl number only. Therefore, if $g(\mathrm{Pr}) = -(\partial T^*/\partial \eta)_{\eta=0}$,

$$\mathrm{Nu}_x = \left(\frac{\mathrm{Gr}_x}{4}\right)^{1/4} g(\mathrm{Pr}). \tag{12.41}$$

The following interpolation formulae was recommended for $g(\mathrm{Pr})$ [2]:

$$g(\mathrm{Pr}) = \frac{0.75\,\mathrm{Pr}^{1/2}}{(0.609 + 1.221\,\mathrm{Pr}^{1/2} + 1.238\,\mathrm{Pr})^{1/4}}. \tag{12.42}$$

The average Nusselt number over the entire length of the plate is obtained as follows:

$$\overline{\mathrm{Nu}_L} = \frac{\overline{h}L}{k_f} = \frac{L}{k_f}\left(\frac{1}{L}\int_0^L h\,dx\right) = \frac{1}{k_f}\int_0^L \frac{\mathrm{Nu}_x k_f}{x}\,dx = \int_0^L \frac{\mathrm{Nu}_x}{x}\,dx$$

$$= \int_0^L \frac{1}{x}\left(\frac{\mathrm{Gr}_x}{4}\right)^{1/4} g(\mathrm{Pr})\,dx.$$

Substituting the expression for Gr_x and integrating with respect to x gives

$$\overline{Nu}_L = \frac{\overline{h}L}{k_f} = \frac{4}{3}\left(\frac{Gr_L}{4}\right)^{1/4} g(Pr). \qquad (12.43)$$

Equation 12.43 shows that the average Nusselt number over the entire length L of a vertical plate is 4/3 of the local Nusselt number at $x = L$.

A simpler empirical correlation for the average Nusselt number in the laminar natural convection flow along the entire length L of a vertical plate at a uniform temperature is [3]:

$$\overline{Nu}_L = \frac{\overline{h}L}{k_f} = 0.59 Ra_L^{1/4}. \qquad (12.44)$$

The above correlations are valid for the laminar natural convection along a vertical flat plate at a uniform temperature. If $Ra_L > 10^9$, the flow becomes turbulent and these correlations will underestimate the convection heat transfer rate. Instead, the following empirical correlation was recommended for the **turbulent natural convection heat transfer over a vertical flat plate** [3]:

$$\overline{Nu}_L = \frac{\overline{h}_L L}{k_f} = 0.1 Ra_L^{1/3}. \qquad (12.45)$$

It must be noted that the fluid properties used in all these correlations have to be evaluated at the film temperature $T_f = (T_s + T_\infty)/2$. It is also important to mention that the above correlations were derived for a uniform surface temperature plate. However, these correlations can also be used for a uniform heat flux plate if the surface temperature at the midpoint of the plate, T_s at $x = L/2$, is used in calculating the Grashof and Rayleigh numbers. The convection heat flux in this case is

$$\dot{q} = \overline{h}(T_{s,L/2} - T_\infty). \qquad (12.46)$$

Example 12.1

An electronic device with 50 mm × 50 mm surface area is mounted vertically on a printed circuit board. This device dissipates 3 W to the air at 25°C. Assume that all the power is dissipated through the case of this device and calculate its case to ambient thermal resistance.

Solution

Since we do not know the case temperature, T_s, we can not calculate the film temperature. Let's start with a guess for the case temperature; $T_s = 155°C$. Then

$$T_f = 0.5(T_s + T_\infty) = 0.5(155 + 25) = 90° \text{ C},$$

Natural Convection Heat Transfer

and air properties at this temperature are

$$\beta = \frac{1}{T_f} = \frac{1}{273.15+90} = 0.00275 \, \text{K}^{-1}, \rho = 0.9722 \, \text{kg/m}^3,$$

$$\mu = 2.138 \times 10^{-5} \, \text{N.s/m}^2, \quad k = 0.0305 \, \text{W/m}° \, \text{C}, \quad \text{Pr} = 0.7081.$$

The Rayleigh number is

$$Ra_L = \frac{\rho^2 g \beta (T_s - T_\infty) L^3}{\mu^2} \text{Pr}$$

$$Ra_L = \frac{(0.9722 \, \text{kg/m}^3)^2 \times 9.8 \, \text{m/s}^2 \times 0.00275 \, \text{K}^{-1}(155-25)° \, \text{C} \times 0.05^3 \, \text{m}^3}{(2.138 \times 10^{-5} \, \text{N} \times \text{s/m}^2)^2} \times 0.7081$$

$$= 641{,}213,$$

which shows the flow is laminar. Therefore, the average Nusselt number and convection heat transfer coefficient are

$$\overline{Nu}_L = 0.59 Ra_L^{1/4} = 0.59 \times 641{,}213^{1/4} = 16.7,$$

$$\bar{h} = \overline{Nu}_L \frac{k_f}{L} = 16.7 \times \frac{0.0305 \, \text{W/m°C}}{0.05 \, \text{m}} = 10.19 \, \text{W/m}^2°\text{C}.$$

Case-to-ambient thermal resistance of this device and its case temperature are

$$R_{ca} = \frac{1}{\bar{h}A} = \frac{1}{10.19 \, \text{W/m}^2°\text{C} \times 0.05 \, \text{m} \times 0.05 \, \text{m}} = 39.2°\text{C/W},$$

$$T_s = T_a + \dot{Q}R_{ca} = 25°\text{C} + 3\, \text{W} \times 39.2°\text{C/W} = 142.6°\text{C}.$$

Note that the calculated case temperature is different from the original guess. The above calculations shall be repeated using $T_s = 142.6°\text{C}$ as the new guess. The new film temperature is

$$T_f = 0.5(T_s + T_\infty) = 0.5(142.6 + 25) = 83.8°\text{C},$$

and the air properties at this temperature are

$$\beta = \frac{1}{T_f} = \frac{1}{273.15 + 83.8} = 0.0028 \, \text{K}^{-1},$$

$$\rho = 0.9893 \, \text{kg/m}^3, \quad \mu = 2.111 \times 10^{-5} \, \text{N.s/m}^2, \quad k = 0.0301 \, \text{W/m°C},$$

$$\text{Pr} = 0.087.$$

Therefore,

$$Ra_L = \frac{\rho^2 g \beta (T_s - T_\infty) L^3}{\mu^2} \text{Pr} = 627{,}832$$

$$\overline{Nu}_L = 0.59\, Ra_L^{1/4} = 16.61,$$

$$\overline{h} = \overline{Nu}_L \frac{k_f}{L} = 10\ \text{W/m}^2{}^\circ\text{C},$$

$$R_{ca} = \frac{1}{hA} = 40^\circ\text{C/W},$$

$$T_s = T_a + \dot{Q} R_{ca} = 145^\circ\text{C}.$$

This value for the case temperature is close enough to the assumed value of 142.6°C and therefore another iteration is not necessary.

12.5 NATURAL CONVECTION AROUND INCLINED AND HORIZONTAL PLATES

There are situations in which a plate is not in the vertical orientation and instead, as shown in Figure 12.6, makes an angle θ with the direction of gravity. In this situation the buoyancy force has a component parallel to the plate surface (i.e., $g \cos \theta$) that creates the natural convection flow. It has been suggested that, for $0 \le \theta \le 60^\circ$, convection heat transfer rate from the bottom of a heated inclined plate, $T_s > T_\infty$, or the top of a cooled inclined plate, $T_s < T_\infty$, can be calculated by replacing g by $g \cos \theta$, in the equations given in Section 12.4 for the vertical plates [3].

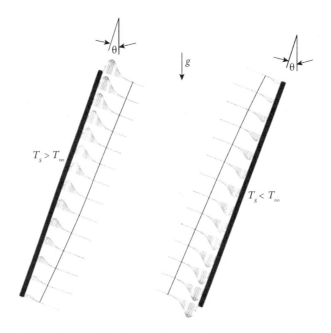

FIGURE 12.6 Natural convection flow on the bottom of a hot (left) and the top of a cold (right) inclined plate.

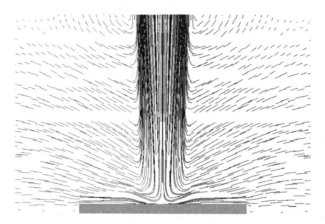

FIGURE 12.7 Flow field around a horizontal plate with heat transfer from its top surface.

If the plate is horizontal, the buoyancy force is normal to the plate surface. Consider a horizontal plate whose top is hotter than the ambient fluid around it while its bottom is either insulated or not exposed to a fluid. The fluid near the top surface of this plate is heated and a natural convection flow, similar to that shown in Figure 12.7, is created. The average Nusselt number over the entire area of this plate is given by the following empirical correlations [3]:

$$\overline{\mathrm{Nu}}_{L_c} = 0.54 \mathrm{Ra}_{L_c}^{1/4} \quad 10^4 < \mathrm{Ra}_{L_c} < 10^7$$
$$\overline{\mathrm{Nu}}_{L_c} = 0.15 \mathrm{Ra}_{L_c}^{1/3} \quad 10^7 < \mathrm{Ra}_{L_c} < 10^{11}.$$
(12.47)

Here, the Nusselt and Rayleigh numbers are calculated based on the characteristic length $L_c = A/P$ where A and P are the surface area and the perimeter of the plate, respectively.

$$\overline{\mathrm{Nu}}_{L_c} = \frac{\bar{h} L_c}{k_f} \quad \text{and} \quad \mathrm{Ra}_{L_c} = \frac{g\beta(T_s - T_\infty)L_c^3}{\nu\alpha}.$$

It must be noted that these correlations can also be used for calculating the heat transfer rate to the bottom of a cold plate from the hotter fluid around it.

If the hot surface is face down, the fluid right under the plate gets hotter and lighter than the bulk of the fluid around the plate and a natural convection flow similar to the one shown in Figure 12.8 is created. The average Nusselt number over the entire area of this plate is given by the following empirical correlation:

$$\overline{\mathrm{Nu}}_{L_c} = 0.27 \mathrm{Ra}_{L_c}^{1/4} \quad 10^5 < \mathrm{Ra}_{L_c} < 10^{11}.$$
(12.48)

FIGURE 12.8 Flow field around a horizontal plate with heat transfer from its bottom surface.

Again, this equation can be used to calculate the convection heat transfer coefficient over the top surface of a cold plate surrounded by a hotter fluid. Comparing Equation 12.48 with 12.47 shows that the convection heat transfer is more efficient in the latter case.

Example 12.2

Consider the electronic enclosure of Example 11.1 shown again in Figure 12.9 and assume that the surface temperature of the duct is the same as the temperature of the boards. Calculate the natural convection heat transfer rate from the top and bottom surfaces of the duct to the air at 25°C.

Solution

Using the calculated surface temperature of $T_s = 192.6°C$ and the air temperature of $T_\infty = 25°C$, the film temperature is $T_f = 0.5(T_s + T_\infty) = 0.5(25 + 192.6) = 108.8°C$ and air properties at this temperature are $\rho = 0.9244$ kg/m³, $k = 0.0318$ W/m°C, $\nu = 2.4 \times 10^{-5}$ m²/s, $\beta = 1/T_f = 0.0026$ and $Pr = 0.7065$.

The characteristic length and Rayleigh number are

$$L_c = \frac{A}{P} = \frac{0.35\,\text{m} \times 0.4\,\text{m}}{2(0.35\,\text{m} + 0.4\,\text{m})} = 0.093\,\text{m},$$

$$Ra_{L_c} = \frac{g\beta(T_s - T_\infty)L_c^3}{\nu^2}Pr = \frac{9.8\,\text{m/s}^2 \times 0.0026\,\text{K}^{-1}(192.6-25)\text{K} \times 0.093^3\,\text{m}^3}{(2.4 \times 10^{-5}\,\text{m}^2/\text{s})^2} \times 0.7065$$

$$= 4,213,199.$$

For the top surface

$$\overline{Nu}_{L_c} = 0.54 Ra_{L_c}^{1/4} = 0.54(4,213,199)^{1/4} = 24.5$$

$$\overline{h}_{top} = \overline{Nu}_{L_c}\frac{k}{\delta} = 24.5 \times \frac{0.0318\,\text{W/m°C}}{0.093\,\text{m}} = 8.38\,\text{W/m}^2\,°\text{C}$$

$$\dot{Q}_{top} = \overline{h}A(T_s - T_\infty) = 8.38\,\text{W/m}^2\,°\text{C} \times (0.35\,\text{m} \times 0.4\,\text{m}) \times (192.6°\text{C} - 25°\text{C})$$

$$= 196.6\,\text{W}.$$

Natural Convection Heat Transfer

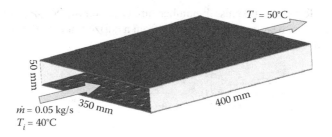

FIGURE 12.9 The electronic enclosure of Example 12.2.

Comparing Equation 12.48 with the first of 12.47, it is seen that the average Nusselt number for the bottom surface is half of that for the top surface. Therefore, the heat transfer rate from the bottom surface is

$$\dot{Q}_{bot} = 0.5\dot{Q}_{top} = 98.3 \text{ W}.$$

12.6 NATURAL CONVECTION AROUND VERTICAL AND HORIZONTAL CYLINDERS

Both vertical and horizontal cylinders are important geometries in electronic cooling applications. Some transistors and many resistors and capacitors are packaged in cylindrical shapes, pin-fin heat sinks have cylindrical fins, and power cables that carry current into electronic equipment are long cylinders.

Consider a vertical cylinder with the height L and diameter D that is immersed in a fluid, and let's assume that the temperature difference between the surface of this cylinder and the fluid away from it is ΔT. A natural convection boundary layer will develop along the surface of this cylinder. If the boundary layer thickness is much smaller than the diameter of the cylinder, the surface curvature will not have much effect on the fluid inside the boundary layer. This means that, the fluid around the cylinder does not distinguish between the slightly curved surface of this cylinder and a perfectly flat plate. Therefore, the correlations for natural convection along a vertical plate can be used to calculate the convection heat transfer coefficient from the surface of this cylinder. This condition is met if

$$\frac{D}{L} \geq \frac{35}{\text{Gr}_L^{1/4}}. \qquad (12.49)$$

If this condition is not met, the cylinder is considered a thin one and the boundary layer thickness is comparable to its diameter. LeFevre and Ede [4] recommended the following correlation for this case:

$$\overline{\text{Nu}}_L = \frac{\bar{h}L}{k_f} = \frac{4}{3}\left[\frac{7\text{Ra}_L\text{Pr}}{5(20+21\text{Pr})}\right]^{1/4} + \frac{4(272+315\text{Pr})L}{35(64+63\text{Pr})D}. \qquad (12.50)$$

Note that the Rayleigh and Nusselt numbers are based on the height of the cylinder.

The convection heat transfer coefficient from a horizontal cylinder with diameter D is obtained from Churchill and Chu [5]:

$$\overline{\mathrm{Nu}}_D = \frac{\overline{h}D}{k_f} = \left\{ 0.6 + \frac{0.387 \mathrm{Ra}_D^{1/6}}{\left[1 + (0.559/\mathrm{Pr})^{9/16}\right]^{8/27}} \right\}^2, \qquad (12.51)$$

where both Rayleigh and Nusselt numbers are based on the diameter of the cylinder.

12.7 NATURAL CONVECTION IN ENCLOSURES

Natural convection heat transfer inside closed enclosures, or internal natural convection, is a major heat transfer mechanism inside the closed electronic systems with no forced airflow. The heat generated by the electronics inside these systems is transferred to the surfaces of the enclosure by internal natural convection, conduction, and radiation. Then, it is transferred to the outside ambient by external natural convection and radiation. Examples are cell phones, handheld computers, phone chargers, outdoor radio units, and microcell base stations.

Consider an enclosure with height H, length L, and depth W as shown in Figure 12.10. Let's assume that the depth of the enclosure is long enough such that its effects can be neglected and the enclosure can be treated as a two-dimensional enclosure.

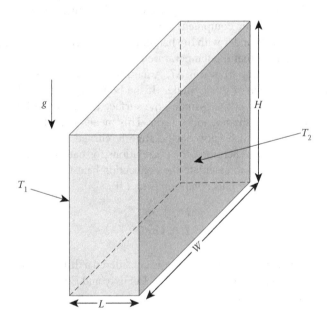

FIGURE 12.10 A typical enclosure with different vertical wall temperatures and insulated horizontal walls.

Natural Convection Heat Transfer

Let's also assume that the two vertical side walls are kept at temperatures T_1 and T_2 where $T_1 > T_2$ and the horizontal walls are insulated. The air inside the enclosure is at temperature $T_1 \leq T \leq T_2$. The convection heat transfer rates between the wall at temperature T_1 and the air inside the enclosure, and between the air inside the enclosure and the wall at temperature T_2, result in a net convection heat transfer rate between the two side walls. The Newton's law of cooling for this enclosure is written as

$$\dot{q}_{conv} = \overline{h}(T_1 - T_2), \qquad (12.52)$$

where $\dot{q}_{conv}$ is the net convection heat flux between the side walls of the enclosure and $\overline{h}$ is the average convection heat transfer coefficient. The Nusselt number is equal to the ratio of the convection heat flux to the conduction heat flux through this enclosure;

$$\overline{Nu}_L = \frac{\dot{q}_{conv}}{\dot{q}_{cond}} = \frac{\overline{h}(T_1 - T_2)}{k(T_1 - T_2)/L} = \frac{\overline{h}L}{k}. \qquad (12.53)$$

Bejan [6] identified four different flow regimes inside an enclosure. These flow regimes are described below.

- Conduction Limit: This happens if $Ra_H \leq 1$ where

$$Ra_H = \frac{g\beta(T_1 - T_2)H^3}{\alpha\nu}. \qquad (12.54)$$

Temperature varies linearly across the enclosure. Therefore, the heat transfer rate between the two side walls is equal to $kHW(T_1 - T_2)/L$. The horizontal temperature gradient creates a slow clockwise circulation as shown in Figure 12.11. However, the heat transfer contribution of this flow is negligible.
- Tall Enclosure Limit: This limit is defined by the criterion

$$\frac{H}{L} > Ra_H^{1/4}. \qquad (12.55)$$

For most of the enclosure height, as shown in Figure 12.12, temperature varies linearly between the two side walls and conduction heat transfer is dominant. The heat transfer rate is equal to $kHW(T_1 - T_2)/L$.
- Shallow Enclosure Limit: This limit is specified by the criterion

$$\frac{H}{L} < Ra_H^{-1/4}. \qquad (12.56)$$

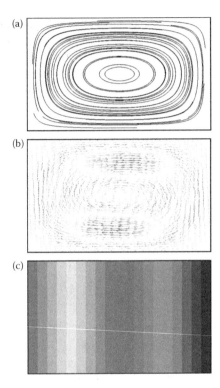

FIGURE 12.11 (**See color insert following page 240.**) (a) Particle path, (b) velocity vectors, and (c) temperature contours in an enclosure at the conduction limit.

The heat transfer mechanism is dominated by the presence of vertical thermal layers shown in Figure 12.13. The heat transfer rate scales with $kHW(T_1 - T_2)/\delta_t$, where δ_t is the thermal boundary layer thickness.

Bejan [6] gave the following expression for the average Nusselt number in the shallow enclosure limit:

$$\overline{\mathrm{Nu}}_L = \frac{\bar{h}L}{k} = K_1 + \frac{K_1^3}{362{,}880}\left(\frac{H}{L}\mathrm{Ra}_H\right)^2. \qquad (12.57)$$

Here, K_1 is a function of H/L and Ra_H, and approaches 1 as $(H/L)^2 \mathrm{Ra}_H \to 0$. Figure 12.14 shows the variation of the average Nusselt number versus $\mathrm{Ra}_H(H/L)$ for the shallow enclosure limit.

- Boundary Layer Regime or High-Ra_H Limit: This is the regime between the tall and shallow enclosure limits,

$$\mathrm{Ra}_H^{-1/4} < \frac{H}{L} < \mathrm{Ra}_H^{1/4}. \qquad (12.58)$$

Natural Convection Heat Transfer

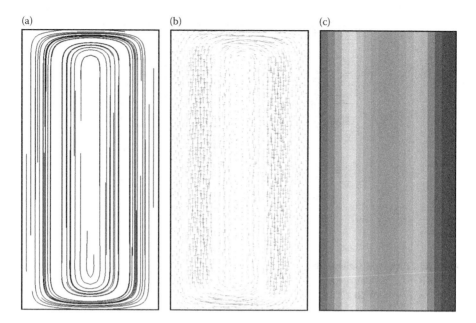

FIGURE 12.12 (**See color insert following page 240.**) (a) Particle path, (b) velocity vectors, and (c) temperature contours in an enclosure at the tall limit.

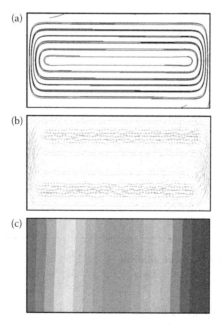

FIGURE 12.13 (**See color insert following page 240.**) (a) Particle path, (b) velocity vectors, and (c) temperature contours in an enclosure at the shallow limit.

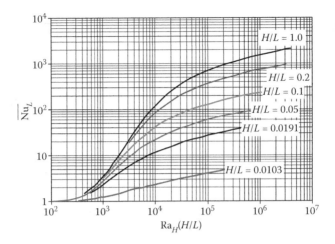

FIGURE 12.14 Nusselt number in a shallow enclosure with constant temperature sidewalls. (From Bejan, A., *Convection Heat Transfer,* Third Edition. New York, NY: John Wiley & Sons, Inc., 2004. With permission.)

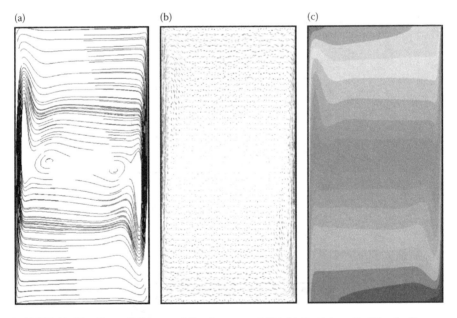

FIGURE 12.15 (**See color insert following page 240.**) (a) Particle path, (b) velocity vectors, and (c) temperature contours in an enclosure for the boundary layer regime.

Vertical boundary layers form distinctly along both vertical walls and most of the fluid in the core of the enclosure is relatively stagnant and thermally stratified as shown in Figure 12.15. The heat transfer rate is proportional to $kHW(T_1 - T_2)/\delta_t$.

Natural Convection Heat Transfer

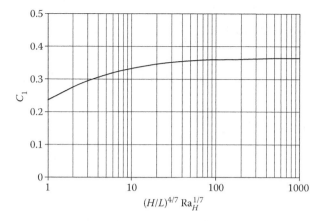

FIGURE 12.16 Coefficient C_1 for the Nusselt number correlation for boundary layer regime in vertical enclosures. (From Bejan, A., *Convection Heat Transfer*, Third Edition. New York, NY: John Wiley & Sons, Inc., 2004. With permission.)

Bejan [6] has derived the following expression for the average Nusselt number in the boundary layer regime:

$$\overline{\mathrm{Nu}}_L = \frac{\bar{h}L}{k} = C_1 \frac{L}{H} \mathrm{Ra}_H^{1/4}. \qquad (12.59)$$

The constant C_1 can be found from the curve in Figure 12.16. It approaches 0.364 at very large values of Ra_H.

Other correlations for the laminar natural convection heat transfer in the enclosures at boundary layer regime are those given by Berkovsky and Polevikov [7]:

$$\begin{cases} \overline{\mathrm{Nu}}_L = 0.22 \left(\dfrac{\mathrm{Pr}}{0.2 + \mathrm{Pr}} \mathrm{Ra}_H \right)^{0.28} \left(\dfrac{L}{H} \right)^{1.09} \\[2pt] \text{if } 2 < \dfrac{H}{L} < 10 \quad \mathrm{Pr} < 10^5 \quad 10^3 < \mathrm{Ra}_H \left(\dfrac{L}{H} \right)^3 < 10^{10} \\[2pt] \overline{\mathrm{Nu}}_L = 0.18 \left(\dfrac{\mathrm{Pr}}{0.2 + \mathrm{Pr}} \mathrm{Ra}_H \right)^{0.29} \left(\dfrac{L}{H} \right)^{0.87} \\[2pt] \text{if } 1 < \dfrac{H}{L} < 2 \quad 10^{-3} < \mathrm{Pr} < 10^5 \quad 10^3 < \dfrac{\mathrm{Pr}}{0.2 + \mathrm{Pr}} \mathrm{Ra}_H \left(\dfrac{L}{H} \right)^3 \end{cases} \qquad (12.60)$$

and those proposed by MacGregor and Emery [8]

$$\begin{cases} \overline{\mathrm{Nu}}_L = 0.42\,\mathrm{Ra}_H^{1/4}\,\mathrm{Pr}^{0.012}\left(\dfrac{H}{L}\right)^{-1.05} \\[2mm] \text{if } 10 \le \dfrac{H}{L} \le 40 \quad 1 \le \mathrm{Pr} < 2\times 10^4 \quad 10^4 \le \mathrm{Ra}_H\left(\dfrac{L}{H}\right)^3 \le 10^7 \end{cases}$$

$$\begin{cases} \overline{\mathrm{Nu}}_L = 0.046\,\mathrm{Ra}_H^{1/3}\left(\dfrac{H}{L}\right)^{-1} \\[2mm] \text{if } 1 \le \dfrac{H}{L} \le 40 \quad 1 \le \mathrm{Pr} \le 20 \quad 10^6 \le \mathrm{Ra}_H\left(\dfrac{L}{H}\right)^3 \le 10^9 \end{cases} \tag{12.61}$$

For an enclosure with side walls heated and cooled with a uniform heat flux $\dot{q}$, the Nusselt number is given by

$$\overline{\mathrm{Nu}}_L = 0.34\,\mathrm{Ra}_H^{*2/9}\left(\dfrac{H}{L}\right)^{-8/9}, \tag{12.62}$$

where $\mathrm{Ra}_H^* = g\beta\dot{q}H^4/\alpha\nu k$ is the modified Rayleigh number based on heat flux [6].

12.8 NATURAL CONVECTION FROM ARRAY OF VERTICAL PLATES

Natural convection from an array of vertical plates has many practical applications in electronic cooling. For examples, the fins on the external surface of many outdoor mounted electronic enclosures such as the one shown in Figure 12.17a can be approximated as an array of vertical plates with uniform surface temperature. On the other hand, similar printed circuit boards mounted vertically on a backplane, as shown in Figure 12.17b, can be approximated as an array of vertical plates with uniform heat flux.

Consider two uniform surface temperature vertical plates that are spaced from each other by distance S. If S is large enough, the natural convection boundary layers that develop on theses plates do not interact with each other as shown in Figure 12.18a. This is similar to the entry length region of a forced convection internal flow and is called the developing natural convection flow. As the distance between these plates decreases, the boundary layers merge and the flow becomes fully developed as shown in Figure 12.18b. This flow is less efficient than the developing flow as far as convection heat transfer from these plates is concerned. Therefore, the average convection heat transfer coefficient in the fully developed flow is less than that in a developing flow. The convection heat flux from these plates is calculated as

$$\dot{q}_{\mathrm{conv}} = \overline{h}(T_s - T_\infty), \tag{12.63}$$

Natural Convection Heat Transfer

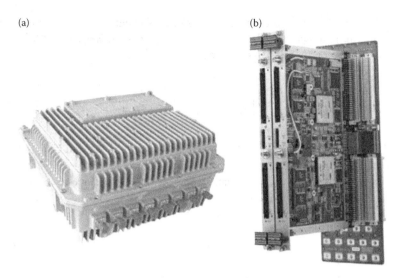

FIGURE 12.17 (**See color insert following page 240.**) (a) An outdoor electronic enclosure with vertical fins and (b) a pair of printed circuit boards mounted on a backplane.

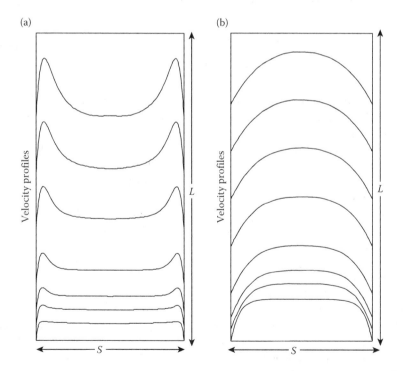

FIGURE 12.18 Natural convection velocity profiles between two vertical plates; (a) developing flow and (b) developed flow.

where $\bar{h}$, T_s, and T_∞ are the average heat transfer coefficient, the surface temperature and the inlet fluid temperature, respectively. Bar-Cohen and Rohsenow [9] gave the following expression for the Nusselt number in this flow:

$$\overline{\text{Nu}_S} = \frac{\bar{h}S}{k_f} = \left[\frac{C_1}{(\text{Ra}_S\, S/L)^2} + \frac{2.87}{(\text{Ra}_S\, S/L)^{1/2}}\right]^{-1/2}. \qquad (12.64)$$

Here $C_1 = 576$ if both plates are at the same uniform temperature and $C_1 = 144$ if one plate is at a uniform temperature and the other plate is adiabatic.

Now, let's consider two vertical plates with the uniform surface heat flux $\dot{q}$. The convection heat transfer coefficient and the temperature at the end of each plate, h and $T_{s,L}$, are correlated by the Newton's law of cooling:

$$\dot{q} = h(T_{s,L} - T_\infty). \qquad (12.65)$$

Note that h in Equation 12.65 is the local heat transfer coefficient at the end of the plate where $x = L$, while $\bar{h}$ in Equation 12.63 is the average heat transfer coefficient over the length of the plate. Bar-Cohen and Rohsenow [9] gave the following expression for the Nusslet number at the end of the plates with uniform heat flux:

$$\text{Nu}_S = \frac{hS}{k_f} = \left[\frac{C_2}{(\text{Ra}_S^*\, S/L)} + \frac{2.51}{(\text{Ra}_S^*\, S/L)^{2/5}}\right]^{-1/2}. \qquad (12.66)$$

Here $C_2 = 48$ if both plates are at the same uniform heat flux $\dot{q}$ and $C_2 = 24$ if one plate is at the uniform heat flux $\dot{q}$ and the other plate is adiabatic. The heat flux based Rayleigh number is defined as $\text{Ra}_S^* = g\beta\, \dot{q} S^4/k\alpha\nu$.

Let's consider an array of constant temperature vertical plates with thickness t and dimensions L and H that are spaced from each other by distance S, and are mounted on a common base with width W as shown in Figure 12.19. Maximum heat transfer rate from a single plate is achieved if the spacing between the plates is large enough such that the boundary layers do not merge. However, this may require a large spacing between the plates and result in only a few plates mounted on the base. Therefore, the total surface area, and consequently the total heat transfer rate from all the plates, may not be very large. On the other hand, for smaller spacing between the plates, the flow becomes fully developed and the average heat transfer coefficient is lower. But, since more plates will be mounted on the base and the heat transfer rate is proportional to the product of the convection heat transfer coefficient and the surface area, the total heat transfer rate from all the plates may increase. This implies that there is an optimum plate spacing that corresponds to the maximum heat transfer rate from an array of constant temperature vertical plates. It was shown that this optimum plate spacing is [9]

$$S_{\text{opt}} = 2.71(\text{Ra}_L/L^4)^{-1/4}, \qquad (12.67)$$

Natural Convection Heat Transfer

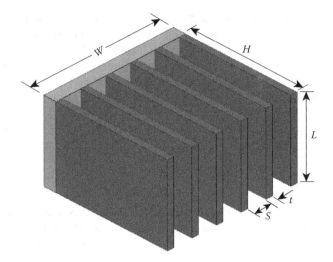

FIGURE 12.19 An array of vertical plates mounted on a common base.

where Ra_L is the Rayleigh number based on the length L. It was also shown that the average Nusselt number corresponding to this plate spacing is

$$\overline{Nu}_{S_{opt}} = \frac{\overline{h} S_{opt}}{k} = 1.307. \tag{12.68}$$

Now, consider an array of uniform heat flux vertical plates as shown in Figure 12.19. The highest heat transfer coefficient and therefore the lowest plate temperature at $x = L$ is achieved if the plate spacing is large enough such that the flow there is a developing flow. As plate spacing is reduced, the number of plates and therefore the total heat transfer area and the heat transfer rate increase proportionally. However, this results in a higher plate temperature at $x = L$, $T_{s,L}$. The optimum plate spacing in this case is defined as the spacing that maximizes total heat transfer rate, $\dot{Q}_{conv}$, per unit temperature difference at the end of each plate, $T_{s,L} - T_\infty$. This plate spacing is given by Bar-Cohen and Rohsenow [9] as

$$S_{opt} = 2.12(Ra_L^* / L^5)^{-1/5}, \tag{12.69}$$

where $Ra_L^* = g\beta \, \dot{q} L^4 / k\alpha\nu$.

12.9 MIXED CONVECTION

It was shown in Section 12.3.2 that the ratio Gr_L / Re_L^2 appears as the coefficient of the buoyancy term in the normalized form of the momentum equation. This ratio is a measure of the relative magnitude of the buoyancy force to the inertia force and gives an indication of whether forced convection or natural convection is dominant or both are of the same order. Generally, if $Gr_L / Re_L^2 \gg 1$, natural convection flow is dominant. On the other hand, if $Gr_L / Re_L^2 \ll 1$, natural convection flow can be neglected compared to forced convection flow. However, if $Gr_L/Re_L^2 \approx 1$, both natural and forced convection flows must be considered. Natural and forced convection flows interact in three different ways: (1) assisting flows in which the buoyancy driven natural convection flow is in the same direction as the forced convection flow, (2) opposing flows where the natural convection and forced convection flows are in opposite directions, and (3) transverse flows where the natural convection and forced convection flows are normal to each other.

The mixed convection Nusselt number, Nu, is usually related to the forced convection Nusselt number, Nu_f, and the natural convection Nusselt number, Nu_n, by the expression

$$Nu^m = Nu_f^m \pm Nu_n^m, \qquad (12.70)$$

where the plus sign is used for assisting and transverse flows and the negative sign is used for opposing flows. It was shown that the best correlation with data would be obtained with $m = 3$ for vertical plates and $m = 3.5$ for horizontal plates.

PROBLEMS

12.1 Physically, what does the Grashof number represent? What does the Nusselt number represent?

12.2 Explain how natural convection flow is created around a hot object?

12.3 Some objects that sink in fresh water may float in salty sea water. Why?

12.4 If the temperature difference between a surface and its ambient increases, will it be more likely for the natural convection flow to become laminar or turbulent? Why?

12.5 The electrical components on a 50 cm × 30 cm plate dissipate 50 W power. The plate is mounted horizontally. The bottom of the plate is exposed to air with velocity $U_\infty = 2$ m/s and temperature $T_\infty = 25°C$. The top of the plate is exposed to air at temperature $T_\infty = 25°C$. Assuming the top and bottom surface temperatures are the same, and neglecting any radiation heat transfer, determine the average surface temperature of the plate. What fraction of the total power is removed from the top of the plate?

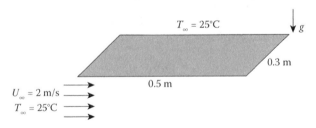

12.6 Consider an electronic package mounted on a horizontal printed circuit board. The package dissipates 3 W and its dimensions are 60 mm × 60 mm. The junction-to-case thermal resistance of this package is $R_{jc} = 5°C/W$.
   a. Calculate the case-to-air thermal resistance, R_{ca}, of this package in natural convection.
   b. What will be the junction temperature of this package if the air temperature is 25°C?

Natural Convection Heat Transfer

12.7 The components of an electronic system dissipating 150 W are installed on four sides of a 0.5 m long horizontal duct whose cross section is 0.2 m × 0.2 m. Air enters the duct at a velocity of 3 m/s and temperature of 30°C. Outside the duct air temperature is 20°C. The temperature is the same and constant on all the duct surfaces. The conduction thermal resistance of the duct is negligible and therefore the internal and external temperatures of the duct surfaces are the same. Calculate the surface temperature of the duct and the heat dissipated to the air outside the duct and the air inside it?
Use air properties at 25°C in all your calculations.

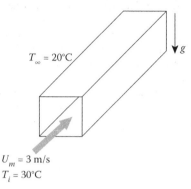

12.8 Consider an electronic enclosure with dimensions 50 mm × 400 mm × 400 mm as shown below. The electronics inside this enclosure generate 500 W heat. This heat is partially removed by air moving inside the enclosure and partially by natural convection from the two 400 mm × 400 mm vertical walls of the enclosure (let's neglect any heat transfer from other surfaces of the enclosure). Let's assume the surface temperature of this enclosure is 65°C and the outside air temperature is 25°C.
  a. Calculate the total heat transfer rate from two vertical sides of this enclosure to the outside air.
  b. If the maximum allowable temperature rise for the air moving inside the enclosure is 15°C, what is the minimum required airflow rate?

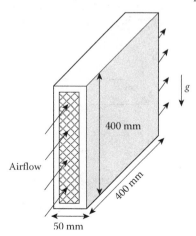

12.9 All the heat dissipating components of an electronic board are connected to a 600 mm × 400 mm heat spreader that is cooled by forced convection airflow over its surface. Total heat dissipation through the heat spreader is 200 W and the air velocity and temperature are 2 m/s and 25°C, respectively.
   a. There are two possible choices for airflow direction; one along the 600 mm side and the other along the 400 mm side. Explain the airflow direction results in lower heat spreader temperature.
   b. Calculate the heat spreader temperature for the airflow direction you chose in part a.
   c. Now, let's consider natural convection heat transfer from this heat spreader in addition to the forced convection heat transfer. Using the spreader temperature calculated in part b, calculate the natural convection heat transfer coefficient assuming the heat spreader is in the horizontal orientation and the heat transfers from the top.
   d. Assume that the forced convection and natural convection heat transfer rates from this heat spreader simply add together and find the resultant heat spreader temperature.

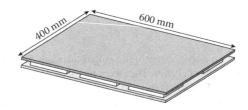

12.10 Consider an electronic enclosure with dimensions 400 mm × 600 mm × 200 mm as shown below. The electronics inside this enclosure generate 200 W heat. This heat will be dissipated through six sides of the enclosure to the outside air, which is at temperature 40°C. Assume all the surfaces of this enclosure are at the same temperature and calculate this temperature as well as the heat transfer from each side of the enclosure. Neglect any radiation heat transfer from the enclosure surfaces.

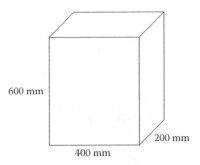

12.11 The roof of a house is 8 m long by 5 m wide and is made of 20 cm thick brick with thermal conductivity of 0.7 W/m°C. This house is air conditioned such that the inside of the house is kept at 15°C at all times. The outside air temperature is 45°C and wind at 36 km/h is blowing along the 8 m long side of the roof.
  a. Determine the rate of heat gain from the roof of this house. Neglect any radiation heat transfer rate.
  b. What are the temperatures at the interior and exterior sides of this roof?
  c. If the emissivity of the roof is 0.7, calculate the interior and exterior radiation heat transfer coefficients and comment on whether neglecting radiation heat transfer was a reasonable assumption or not. Assume the surrounding interior and exterior temperatures are 15°C and 45°C, respectively.
  Use properties of air at 25°C in your calculations.

12.12 The electronic devices in the system shown below dissipate their heat partially to the air moving inside the system and partially through natural convection and radiation from the two vertical surfaces of this system. Air with an average temperature of 25°C and mass flow rate of 0.05 kg/s enters this system and leaves it with an average temperature of 30°C.
  a. What is the heat transfer to the air passing through the system?
  b. If we approximate the system as a duct with uniform surface temperature, determine what that temperature will be. Assume the flow is fully developed inside the system.
  c. If the emissivity of the external surface of the system is 0.8 and outside air and surrounding temperatures is 25°C, determine natural convection and radiation heat transfer rate from the two vertical sidewalls of this system.
  d. What is the total heat generated by the electronics inside this system?
  Note: Use air properties at 25°C in all your calculations.

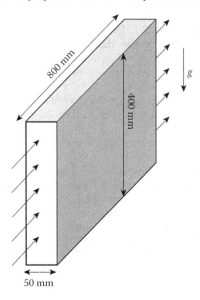

12.13 Consider the electronic equipment shown below that is mounted on a pole. The heat generated by the internal electronics of this equipment is dissipated through one of its 500 mm × 300 mm side walls. The total heat dissipated by these components is 90 W and the air temperature is 40°C.
  a. Calculate the average temperature of this side neglecting radiation heat transfer.
  b. Let's assume that wind with a velocity of 2 m/s blows along the 500 mm side of this wall. Calculate the temperature of this side neglecting radiation and natural convection.

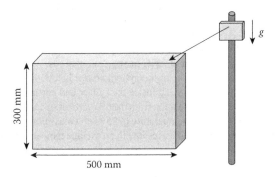

REFERENCES

1. Ostrach, S. 1953. An analysis of laminar free convection flow and heat transfer about a flat plate parallel to the direction of the generating body force. National Advisory Committee for Aeronautics, Report 1111. Report can be viewed at http://aerade.cranfield.ac.uk/ara/1953/naca-report-1111.pdf (accessed June 2009).
2. LeFevre, E. J. 1956. Laminar free convection from a vertical plane surface. Proceedings of the 9th International Congress on Applied Mechanics, Brussels, Vol. 4.
3. Incropera, F. P., D. P. DeWitt, T. L. Bergman, and A. S. Lavine. 2006. *Fundamentals of heat and mass transfer*. Sixth Edition. Hoboken, NJ: John Wiley & Sons, Inc.
4. LeFevre, E. J., and A. J. Ede. 1956. Laminar free convection from the outer surface of a vertical circular cylinder. Proceedings of the 9th International Congress on Applied Mechanics, Brussels, Vol. 4, 175–83.
5. Churchill, S. W., and H. H. S. Chu. 1975. Correlating equations for laminar and turbulent free convection from a horizontal cylinder. *International Journal of Heat Mass Transfer* 18:1049–1053.
6. Bejan, A. 2004. *Convection heat transfer*. Third Edition. New York, NY: John Wiley & Sons, Inc.
7. Berkovsky, B. M., and V. K. Polevikov. 1977. Numerical study of problems on high-intensive free convection. In *Heat transfer and turbulent buoyant convections,* ed. D. B. Spalding and N. Afgan, 443–445. Hemisphere.
8. MacGregor, R. K., and A. P. Emery. 1969. Free convection through vertical plane layers: Moderate and high Prantl number fluids. *Journal of Heat Transfer* 91:391–403.
9. Bar-Cohen, A., and W. M. Rohsenow. 1984. Thermally optimum spacing of vertical, natural convection cooled, parallel plates. *Journal of Heat Transfer* 106:116–23.

13 Radiation Heat Transfer

Radiation is the only mode of heat transfer between two objects that are at different temperatures if there is no material between them. However, it can also be present along, and in parallel, with other modes of heat transfer such as conduction and convection. For example, our bodies dissipate heat through convection and radiation and we feel warm because of the radiation heat received from a fire or from the sun. Every object radiates heat to other objects in its surroundings and in turn receives radiation energy from them. Net radiation heat transfer between any two objects depends on their temperatures, dimensions, radiation properties, and the relative orientation of those with respect to each other.

Radiation heat transfer occurs through the emission of electromagnetic waves or photons from an object. An oscillating electric field generates an oscillating magnetic field. The magnetic field in turn generates an oscillating electric field and so on. These oscillating fields form an electromagnetic wave. Electromagnetic waves are characterized by their frequency, ν, or wavelength, λ, that are related to each other through,

$$c = \lambda \nu, \tag{13.1}$$

where c is the speed of light. Energy of an electromagnetic wave with frequency ν is given by

$$E_\nu = h\nu, \tag{13.2}$$

where $h = 6.626 \times 10^{-34}$ J.s is the Planck constant.

Figure 13.1 shows the electromagnetic wave spectrum. It is seen that the wavelength of electromagnetic waves can be as low as 10^{-9} μm and as high as 10^{10} μm, and their properties and applications are significantly different. Large wavelength radio waves are generated by an electric current that alternates at a radio frequency in an antenna and are used for wireless transmission of information. Microwaves are electromagnetic waves with wavelengths between 0.1 mm and 10 cm and are used in household microwave ovens, short range radio and television broadcasting, cable TV and internet access on coax, radars, wireless protocols such as Bluetooth, and metropolitan area networks such as WiMax. Gamma rays have the shortest wavelength in the electromagnetic spectrum. They are generated by particle interactions in subatomic scales and the energy released through radioactive decays. They have the highest frequency, and therefore the highest energy, among the electromagnetic waves and can penetrate a long distance into materials before being absorbed significantly. X-rays have a longer wavelength and therefore

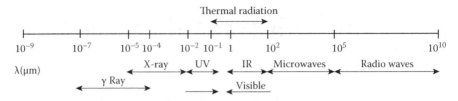

FIGURE 13.1 Electromagnetic wave spectrum.

lower frequency and less energy than gamma rays. They are generated by the collision of accelerating electrons with a metal target. If the electrons have a high enough energy, they can knock out electrons from the inner shells of the metal atoms and create voids. Then, electrons from the higher energy levels will move to these lower energy shells and emit the difference between their original and final energies as X-rays. These waves can penetrate through soft materials such as tissues or blood but are absorbed by harder materials such as bone. This property is used to create X-rays images.

The electromagnetic waves in the middle portion of the spectrum include Infrared (IR), visible, and Ultraviolet (UV) waves. These electromagnetic waves are generated as a result of the atomic and molecular interactions inside a material due to its own temperature and are collectively called the **thermal radiation**. Although the radiation emitted from a material at a certain temperature covers all wavelengths of the thermal radiation, the wavelength at which the maximum radiation intensity happens is a function of the material temperature. For example, the maximum radiation intensity happens at the lowest wavelength UV region if the material temperature is significantly high and at the highest wavelength portion of the IR section if the material temperature is very low.

13.1 RADIATION INTENSITY AND EMISSIVE POWER

Consider a general-shaped object as shown in Figure 13.2a. Every small element on the surface of this object can be considered as a flat surface that radiates in all directions to a hemisphere surrounding it as shown in Figure 13.2b. The strength of radiation from these surface elements is not the same in all directions. The **radiation intensity** defines the magnitude of radiation from a surface in a specific direction and the **emissive power** describes the radiation heat flux (i.e., the total radiation energy per unit time per unit surface). The relationship between these two is described in this section.

Since the radiation rays from any point on a surface travel along radii of a hemisphere, a spherical coordinate system is the most suitable for the mathematical treatment of radiation. Figure 13.3a shows that a point P in a spherical coordinate system is specified by its radius r (i.e., the distance from the origin of the coordinate system to that point), the angle θ from the z axis to the radius r, and the angle ϕ between the x axis and the projection of the radius r on the xy plane. Note that for every r, a full sphere is swept if $0 \leq \theta \leq \pi$ and $0 \leq \phi \leq 2\pi$. Similarly a hemisphere is swept if $0 \leq \theta \leq \pi/2$ and $0 \leq \phi \leq 2\pi$.

Radiation Heat Transfer

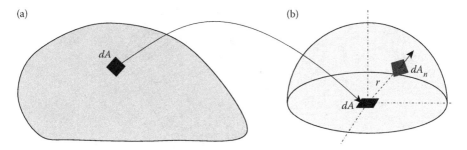

FIGURE 13.2 Every small element of a surface radiates in all directions to a hemisphere around it.

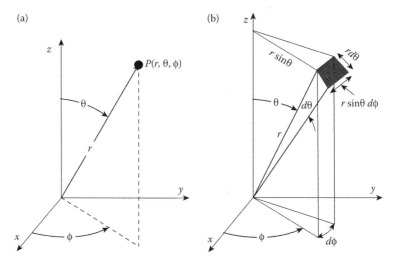

FIGURE 13.3 (a) Coordinates of a point and (b) dimensions of a surface element in a spherical coordinate system.

The surface element dA_n of Figure 13.2 is shown in Figure 13.3b. This element is obtained by sweeping the point $P(r,\theta,\phi)$ by angles $d\theta$ and $d\phi$ and its area is

$$dA_n = r^2 \sin\theta \, d\theta \, d\phi. \tag{13.3}$$

Note that the total surface area of a sphere is obtained by integrating Equation 13.3 over the whole range of θ and ϕ.

$$A_{\text{sphere}} = \int_{\theta=0}^{\pi} \int_{\phi=0}^{2\pi} r^2 \sin\theta \, d\theta \, d\phi = r^2 \int_{\theta=0}^{\pi} \sin\theta \, d\theta \int_{\phi=0}^{2\pi} d\phi = 4\pi r^2.$$

Consider the radiation rays initiating at a point P and passing through an area dA_n, at a distance r from P, as shown in Figure 13.4. The conical region that is covered by

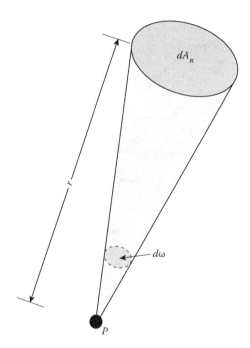

FIGURE 13.4 Differential solid angle $d\omega$ covered by area dA_n from point P.

these radiation rays defines a differential **solid angle** that is measured by the ratio of the surface area dA_n divided by the radius squared,

$$d\omega = \frac{dA_n}{r^2}. \tag{13.4}$$

Substituting dA_n from Equation 13.3 gives

$$d\omega = \sin\theta \, d\theta \, d\phi. \tag{13.5}$$

Note that the definition of solid angle by Equation 13.4 is an extension of the definition of planar angle $d\alpha = ds/r$ in two-dimensional space, where ds is the length of an arc obtained by sweeping the radius r of a circle through the angle $d\alpha$, to three-dimensional space. The unit for the solid angle is steradian (sr).

Let's assume that the rate at which the radiation energy at wavelength λ is emitted from surface dA in Figure 13.2 is $d\dot{Q}_{e,\lambda}$. The rate of radiation energy at wavelength λ per unit area of the emitting surface is $d\dot{Q}_{e,\lambda}/dA$, and the rate of radiation energy at wavelength λ per unit area of the emitting surface normal to the direction of radiation θ is $d\dot{Q}_{e,\lambda}/dA\cos\theta$. The **spectral intensity** is defined as the rate of radiation energy at wavelength λ, per unit area of the emitting surface normal to the direction of radiation θ, and per unit solid angle with this direction

Radiation Heat Transfer

$$I_{e,\lambda} = \frac{d\dot{Q}_{e,\lambda}}{dA\cos\theta\, d\omega}. \tag{13.6}$$

Substituting for $d\omega$ from Equation 13.5 gives

$$I_{e,\lambda} = \frac{d\dot{Q}_{e,\lambda}}{dA\cos\theta\, \sin\theta\, d\theta\, d\phi}. \tag{13.7}$$

Rearranging this expression results in

$$d\dot{Q}_{e,\lambda} = I_{e,\lambda} dA \cos\theta\, \sin\theta\, d\theta\, d\phi. \tag{13.8}$$

An equation for the **spectral radiation flux** of dA is obtained by dividing this equation by dA

$$d\dot{q}_{e,\lambda} = \frac{d\dot{Q}_{e,\lambda}}{dA} = I_{e,\lambda} \cos\theta\, \sin\theta\, d\theta\, d\phi. \tag{13.9}$$

The **spectral emissive power** is defined as the rate at which the radiation of wavelength λ is emitted in all directions per unit area of the radiating surface. This is the integral of the spectral radiation flux, $d\dot{q}_{e,\lambda}$, over all directions in a hemisphere surrounding dA,

$$E_\lambda = \dot{q}_{e,\lambda} = \int_{\phi=0}^{2\pi} \int_{\theta=0}^{\pi/2} I_{e,\lambda} \cos\theta\, \sin\theta\, d\theta\, d\phi. \tag{13.10}$$

If $I_{e,\lambda}$ is the same in all directions, the surface is called a **diffuse emitter**. In this case,

$$E_\lambda = \pi I_{e,\lambda}. \tag{13.11}$$

The **total emissive power** is defined as the rate at which the radiation at all wavelengths and in all directions is emitted per unit area of the radiating surface,

$$E = \int_0^\infty E_\lambda d\lambda. \tag{13.12}$$

If the surface is a diffuse emitter,

$$E = \pi I_e, \tag{13.13}$$

where I_e is the total intensity of the emitted radiation at all wavelengths,

$$I_e = \int_0^\infty I_{e,\lambda} d\lambda. \tag{13.14}$$

13.2 BLACKBODY RADIATION

A blackbody is a conceptual surface defined as the perfect emitter and absorber of radiation energy. It is defined by the following three characteristics:

1. A blackbody emits maximum radiation energy at a given temperature and wavelength. No surface can emit more radiation than a blackbody at the same temperature.
2. A blackbody absorbs all the radiation energy that reaches its surface.
3. A blackbody emits radiation energy uniformly in all directions. This means that a blackbody is a diffuse emitter. Figure 13.5 compares the radiation rays from a blackbody with those from a real surface. The arrow size is representative of the strength of radiation in that direction. It is seen that the strength of the radiation rays from a blackbody is the same in all directions while it may be different for a real surface.

The **blackbody spectral intensity** was first determined by Planck [1] and is given by

$$I_{\lambda,b}(\lambda,T) = \frac{2hc_0^2}{\lambda^5[\exp(hc_0/\lambda kT - 1)]}, \tag{13.15}$$

where $h = 6.626 \times 10^{-34}$ J.s is the universal Planck constant, $k = 1.381 \times 10^{-23}$ J/K is the Boltzmann constant, $c_0 = 2.998 \times 10^8$ m/s is the speed of light in vacuum and T is the absolute temperature of the blackbody in Kelvin. Since the blackbody is a diffuse emitter, its **spectral emissive power** is

$$E_{\lambda,b}(\lambda,T) = \pi I_{\lambda,b}(\lambda,T) = \frac{3.742 \times 10^8}{\lambda^5[\exp(1.439 \times 10^4 / \lambda T) - 1]}. \tag{13.16}$$

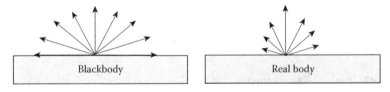

FIGURE 13.5 Directional uniformity of radiation from a blackbody compared to the directional dependence of radiation from a real surface.

Radiation Heat Transfer

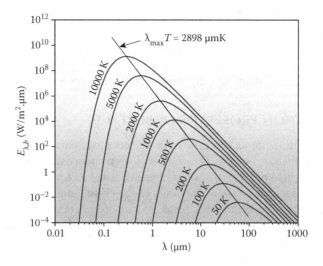

FIGURE 13.6 Blackbody spectral emissive power.

Note that the spectral emissive power of a blackbody is a function of its temperature and the radiation wavelength. Equation 13.16 is also known as the **Planck distribution** and is plotted in Figure 13.6 for several blackbody temperatures.

The following observations can be made from the Planck distribution.

a. Most of the thermal radiation energy emitted from an object is at wavelengths between 0.1 and 100 μm.
b. The amount of radiation energy emitted from the blackbody at a certain temperature varies with wavelength.
c. As the blackbody temperature increases, most of the radiation energy emitted from it will be at shorter wavelengths.
d. The amount of radiation energy at a given wavelength increases as the temperature of the blackbody increases.

Figure 13.6 shows that the Planck distribution at any temperature has a local maximum. The locus of these local maxima can be obtained by finding the roots of the derivative of the Planck distribution, Equation 13.16,

$$\frac{dE_{\lambda,b}}{d\lambda} = 0 \Rightarrow \lambda_{max}T = 2898 \ \mu\text{m.K}. \quad (13.17)$$

Equation 13.17 is known as the **Wein's displacement law**.

The **total emissive power of a blackbody** is obtained by integrating its spectral emissive power over the whole wavelength spectrum,

$$E_b = \int_0^\infty E_{\lambda,b}(\lambda,T) = \int_0^\infty \frac{3.742 \times 10^8}{\lambda^5[\exp(1.439 \times 10^4 / \lambda T) - 1]}. \quad (13.18)$$

This results in **the Stefan-Boltzmann law**,

$$E_b = \sigma T^4, \quad (13.19)$$

where $\sigma = 5.67 \times 10^{-8}$ W/m².K⁴ is the **Stefan-Boltzmann constant**.

13.3 RADIATION PROPERTIES OF SURFACES

Every object emits radiation energy to other objects surrounding it and receives radiation energy from those objects. The rate of radiation energy, in all directions and all wavelengths, emitted per unit area of an object is called its **total emissive power**, $\dot{q}_e$ or E (W/m²). It depends on the temperature of the object and its surface properties. Since a blackbody emits the maximum rate of radiation energy at any temperature, it offers a base with which the emissive properties of real surfaces are compared. The surface property that is used to compare different surfaces with a blackbody, and each other, is called **emissivity**.

The rate of radiation energy, from all directions and at all wavelengths, incident per unit area of an object is called the **total irradiation**, $\dot{q}_i$ or G (W/m²). Depending on what an object does to the incident radiation rays, it is classified as a transparent, semitransparent, or opaque object. A **transparent** object transmits all the incident radiation through it as shown in Figure 13.7a. A **semitransparent** object transmits part of the incident radiation and absorbs or reflects the rest as shown in Figure 13.7b. An **opaque object** absorbs or reflects all the incident radiation and does not transmit any as shown in Figure 13.7c. Note that the blackbody is an opaque object that absorbs all and does not reflect any of the radiation that reaches its surface. The surface properties that determine how an object treats the incident radiation are called **absorptivity**, **reflectivity**, and **transmissivity**.

13.3.1 SURFACE EMISSIVITY

A blackbody emits uniformly in all directions and emits the maximum amount of radiation at any wavelength and temperature. The **surface emissivity** is defined as the ratio of the radiation energy emitted from a real surface to the radiation energy emitted from a blackbody under the same conditions. Since the rate of the emitted radiation from a surface depends on the direction and wavelength of the radiation as well as the surface temperature, its emissivity depends on these parameters as well.

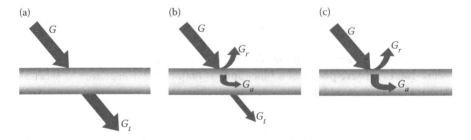

FIGURE 13.7 A (a) transparent, (b) semitransparent, and (c) opaque object.

… Radiation Heat Transfer

The **spectral-directional emissivity** of a surface is defined as the ratio of the spectral intensity in the direction of θ and ϕ to the spectral intensity of a blackbody at the same wavelength and temperature [2],

$$\varepsilon_{\lambda,\theta}(\lambda,\theta,\phi,T) = \frac{I_{\lambda,e}(\lambda,\theta,\phi,T)}{I_{\lambda,b}(\lambda,T)}. \tag{13.20}$$

The subscripts λ and θ for ε indicate that the emissivity is for a certain wavelength and at a certain direction. The variables in the parentheses indicate that the emissivity depends on λ, θ, ϕ, and T. If $\varepsilon_{\lambda,\theta}$ is independent of direction, the surface is called a **diffuse** surface.

The **spectral-hemispherical emissivity**, or simply called the **spectral emissivity**, of a surface at temperature T and wavelength λ is defined as the ratio of its spectral emissive power at wavelength λ to the spectral emissive power of a blackbody at the same wavelength and temperature,

$$\varepsilon_{\lambda}(\lambda,T) = \frac{E_{\lambda}(\lambda,T)}{E_{\lambda,b}(\lambda,T)}. \tag{13.21}$$

Substituting the expressions for the spectral emissive powers, in terms of the spectral intensity, and considering that the spectral intensity of a blackbody is independent from direction results in the following relationship between the spectral, directional emissivity, and the spectral emissivity of a surface [2],

$$\begin{aligned}\varepsilon_{\lambda}(\lambda,T) &= \frac{\int_{\phi=0}^{2\pi}\int_{\theta=0}^{\pi/2} I_{\lambda,e}(\lambda,T)\cos\theta\sin\theta\, d\theta\, d\phi}{\pi I_{\lambda,b}(\lambda,T)} \\ &= \frac{1}{\pi}\int_{\phi=0}^{2\pi}\int_{\theta=0}^{\pi/2} \varepsilon_{\lambda,\theta}(\lambda,\theta,\phi,T)\cos\theta\sin\theta\, d\theta\, d\phi.\end{aligned} \tag{13.22}$$

If ε_{λ} is independent of wavelength, the surface is called a **gray** surface.

The **total-hemispherical emissivity**, or simply called the **total emissivity**, of a surface at temperature T is defined as the ratio of its total emissive power to the total emissive power of a blackbody at the same temperature,

$$\varepsilon(T) = \frac{E(T)}{E_b(T)}. \tag{13.23}$$

Since

$$E(T) = \int_0^\infty E_{\lambda}(\lambda,T) = \int_0^\infty \varepsilon_{\lambda}(\lambda,T)E_{\lambda,b}(\lambda,T)\,d\lambda, \tag{13.24}$$

and $E_b(T) = \sigma T^4$, the expression for the total emissivity can be written as

$$\varepsilon(T) = \frac{1}{\sigma T^4} \int_0^\infty \varepsilon_\lambda(\lambda,T) E_{\lambda,b}(\lambda,T) d\lambda. \quad (13.25)$$

The total emissivity of polished metals is about 0.05–0.2, oxidized metal is about 0.25–0.8, anodized aluminum is about 0.8, glass and PCB material is about 0.8–0.9, building materials and paint is about 0.8–0.9, water and body skin is about 0.9, and white paper is about 0.9. Table 13.2 gives the total emissivity of some common surfaces.

If the spectral-directional emissivity, spectral emissivity, or total emissivity of a surface is known, its spectral intensity, spectral emissive power, or total emissive power can be calculated using one of the equations,

$$I_{\lambda,e}(\lambda,\theta,\phi,T) = \varepsilon_{\lambda,\theta}(\lambda,\theta,\phi,T) I_{\lambda,b}(\lambda,T), \quad (13.26)$$

$$E_\lambda(\lambda,T) = \varepsilon_\lambda(\lambda,T) E_{\lambda,b}(\lambda,T), \quad (13.27)$$

$$E(T) = \varepsilon(T) \sigma T^4. \quad (13.28)$$

Equation 13.28 uses an average total surface emissivity that is sufficient for most engineering applications.

13.3.2 Surface Absorptivity

Consider the semitransparent object shown in Figure 13.7b. A portion of the total irradiation G is absorbed by this object. The **total-hemispherical absorptivity**, or simply called the **total absorptivity**, of this object is defined as the ratio of the total absorbed irradiation to the total irradiation,

$$\alpha = \frac{G_a}{G}. \quad (13.29)$$

Note that $\alpha = 0$ for a transparent or a perfectly reflecting opaque object and $\alpha = 1$ for a blackbody.

Surface absorptivity can also be defined for the irradiation at a certain wavelength and in a certain direction. The **spectral-hemispherical absorptivity**, or simply called the **spectral absorptivity**, is defined as the fraction of the spectral irradiation at wavelength λ that is absorbed by the surface,

$$\alpha_\lambda(\lambda) = \frac{G_{\lambda,a}(\lambda)}{G_\lambda(\lambda)}. \quad (13.30)$$

Radiation Heat Transfer

The **spectral-directional absorptivity** is defined as the fraction of the spectral intensity at wavelength λ incident in the direction of θ and ϕ, $I_{\lambda,i}(\lambda,\theta,\phi)$, that is absorbed by the surface,

$$\alpha_{\lambda,\theta}(\lambda,\theta,\phi) = \frac{I_{\lambda,a}(\lambda,\theta,\phi)}{I_{\lambda,i}(\lambda,\theta,\phi)}. \tag{13.31}$$

The relationships between the spectral-directional absorptivity, spectral absorptivity, and total absorptivity are

$$\alpha_\lambda(\lambda) = \frac{\int_{\phi=0}^{2\pi}\int_{\theta=0}^{\pi/2} \alpha_{\lambda,\theta}(\lambda,\theta,\phi) I_{\lambda,i}(\lambda,\theta,\phi)\cos\theta \sin\theta\, d\theta\, d\phi}{\int_{\phi=0}^{2\pi}\int_{\theta=0}^{\pi/2} I_{\lambda,i}(\lambda,\theta,\phi)\cos\theta \sin\theta\, d\theta\, d\phi}, \tag{13.32}$$

$$\alpha = \frac{\int_0^\infty \alpha_\lambda(\lambda) G_\lambda(\lambda) d\lambda}{G}. \tag{13.33}$$

Equations 13.29 through 13.31 shows that the surface absorptivity of an object is independent from its own temperature. However, because of its dependence on the irradiation, it is strongly dependent on the temperature of the source of the incident radiation. For example, the absorptivity of a surface for the radiation originating from the sun that is at an apparent temperature of about 5800 K is different from its absorptivity for the radiation originating from an object at room temperature.

13.3.3 Surface Reflectivity

Consider the semitransparent object shown in Figure 13.7b. A portion of the total irradiation G is reflected by this object. The **total-hemispherical reflectivity**, or simply called the **total reflectivity**, of an object is defined as the ratio of the total reflected irradiation to the total irradiation,

$$\rho = \frac{G_r}{G}. \tag{13.34}$$

Note that $\rho = 0$ for a transparent object or a blackbody and $\rho = 1$ for a perfectly reflecting opaque object.

The **spectral-hemispherical reflectivity**, or simply called the **spectral reflectivity** is defined as the fraction of the spectral irradiation at wavelength λ that is reflected by the surface,

$$\rho_\lambda(\lambda) = \frac{G_{\lambda,r}(\lambda)}{G_\lambda(\lambda)}. \tag{13.35}$$

The **spectral-directional reflectivity** is defined as the fraction of the spectral intensity at wavelength λ incident in the direction of θ and ϕ, $I_{\lambda,i}(\lambda,\theta,\phi)$, that is reflected by the surface,

$$\rho_{\lambda,\theta}(\lambda,\theta,\phi) = \frac{I_{\lambda,r}(\lambda,\theta,\phi)}{I_{\lambda,i}(\lambda,\theta,\phi)}. \tag{13.36}$$

The relationships between the spectral-directional reflectivity, spectral reflectivity, and total reflectivity are

$$\rho_\lambda(\lambda) = \frac{\int_{\phi=0}^{2\pi}\int_{\theta=0}^{\pi/2} \rho_{\lambda,\theta}(\lambda,\theta,\phi) I_{\lambda,i}(\lambda,\theta,\phi) \cos\theta \sin\theta \, d\theta \, d\phi}{\int_{\phi=0}^{2\pi}\int_{\theta=0}^{\pi/2} I_{\lambda,i}(\lambda,\theta,\phi) \cos\theta \sin\theta \, d\theta \, d\phi}, \tag{13.37}$$

$$\rho = \frac{\int_0^\infty \rho_\lambda(\lambda) G_\lambda(\lambda) d\lambda}{G}. \tag{13.38}$$

13.3.4 Surface Transmissivity

Consider again the semitransparent object shown in Figure 13.7b. A portion of the total irradiation G is transmitted through the object. The **total-hemispherical transmissivity**, or simply called the **total transmissivity**, of an object is defined as the ratio of the total transmitted irradiation to the total irradiation,

$$\tau = \frac{G_t}{G}. \tag{13.39}$$

Note that $\tau = 0$ for an opaque object or a blackbody and $\tau = 1$ for a transparent object.

The **spectral-hemispherical transmissivity**, or simply called the **spectral transmissivity**, is defined as the fraction of the spectral irradiation at wavelength λ that is transmitted through the object,

$$\tau_\lambda(\lambda) = \frac{G_{\lambda,t}(\lambda)}{G_\lambda(\lambda)}. \tag{13.40}$$

The **spectral-directional transmissivity** is defined as the fraction of the spectral intensity at wavelength λ incident in the direction of θ and ϕ, $I_{\lambda,i}(\lambda,\theta,\phi)$, that is transmitted through the object,

$$\tau_{\lambda,\theta}(\lambda,\theta,\phi) = \frac{I_{\lambda,t}(\lambda,\theta,\phi)}{I_{\lambda,i}(\lambda,\theta,\phi)}. \tag{13.41}$$

Radiation Heat Transfer

The relationships between the spectral-directional transmissivity, spectral transmissivity, and total transmissivity are

$$\tau_\lambda(\lambda) = \frac{\int_{\phi=0}^{2\pi}\int_{\theta=0}^{\pi/2} \tau_{\lambda,\theta}(\lambda,\theta,\phi) I_{\lambda,i}(\lambda,\theta,\phi) \cos\theta \sin\theta \, d\theta \, d\phi}{\int_{\phi=0}^{2\pi}\int_{\theta=0}^{\pi/2} I_{\lambda,i}(\lambda,\theta,\phi) \cos\theta \sin\theta \, d\theta \, d\phi}, \quad (13.42)$$

$$\tau = \frac{\int_0^\infty \tau_\lambda(\lambda) G_\lambda(\lambda) d\lambda}{G}. \quad (13.43)$$

A relationship between the total absorptivity, total reflectivity, and total transmissivity of a surface is obtained by noting that

$$G = G_a + G_r + G_t, \quad (13.44)$$

and dividing this by the total irradiation G. This results in

$$\alpha + \rho + \tau = 1. \quad (13.45)$$

Equation 13.45 shows that not all of these three surface properties are independent from each other. For an opaque surface $\tau = 0$ and therefore,

$$\alpha + \rho = 1. \quad (13.46)$$

13.3.5 Kirchhoff's Law

Consider a small object with surface area A, total emissivity ε, total absorptivity α, and temperature T inside a large enclosure at the same temperature as shown in Figure 13.8. The rate of total radiation energy emitted by this object is

$$\dot{Q}_e = AE(T) = \varepsilon(T) A E_b(T) = \varepsilon(T) A \sigma T^4.$$

A large isothermal enclosure behaves like a blackbody. Therefore, the rate of radiation that is incident on the surface of the object inside the enclosure shown in Figure 13.8 is $AG = AE_b(T)$. The amount of incident radiation energy absorbed by this object is

$$\dot{Q}_a = \alpha(T) A E_b(T) = \alpha(T) A \sigma T^4,$$

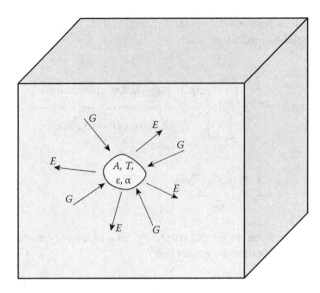

FIGURE 13.8 An object with temperature T inside an enclosure at the same temperature.

where the dependence on temperature T in $\alpha(T)$ refers to the dependence of the surface absorptivity to the temperature of the source of the incident radiation. Since both the object and the enclosure are at the same temperature, the net radiation exchange between them is zero. Therefore, $\dot{Q}_e = \dot{Q}_a$ and

$$\varepsilon(T) = \alpha(T). \qquad (13.47)$$

Equation 13.47 is known as **Kirchhoff's law** and states that the total emissivity of a surface at temperature T is the same as its total absorptivity if the irradiation originates from a blackbody at the same temperature. However, Kirchhoff's law can be applied with a reasonable accuracy if the difference between the surface temperature and the temperature of the source of irradiation is less than a few hundred degrees. This condition is satisfied, and therefore Kirchhoff's law applies, in many engineering applications including electronic systems.

It can also be shown that Kirchhoff's law is valid if the surface is gray (ε_λ and α_λ are independent of wavelength λ) and either the irradiation is diffuse ($I_{\lambda,i}$ is independent of direction) or the surface is diffuse ($\varepsilon_{\lambda,\theta}$ and $\alpha_{\lambda,\theta}$ are independent of direction) [3].

13.4 SOLAR AND ATMOSPHERIC RADIATIONS

The sun is an important source of radiation energy that reaches the surface of the earth. Spectral or Planck distribution of solar radiation is very close to that of a blackbody at 5800 K. This temperature is known as the apparent temperature of the sun. The total radiation energy leaving the surface of the sun is $4\pi r_{sun}^2 \sigma T_{sun}^4$ where r_{sun} is the radius of the sun and $T_{sun} = 5800$ K is the apparent temperature of the sun. If

Radiation Heat Transfer

the distance between the center of the sun and the edge of the earth's atmosphere is $r_{\text{sun-earth}}$, the solar irradiation on the edge of the earth's atmosphere is

$$S_c = \frac{4\pi r_{\text{sun}}^2 \sigma T_{\text{sun}}^4}{4\pi r_{\text{sun-earth}}^2} = \sigma T_{\text{sun}}^4 \left(\frac{r_{\text{sun}}}{r_{\text{sun-earth}}}\right)^2. \tag{13.48}$$

S_c is called the **solar constant** and is equal to 1373 W/m². Note that the solar constant is the solar radiation flux on a plane normal to the radiation rays as shown in Figure 13.9. If a plane makes an angle θ with the direction of solar rays, the solar irradiation on the surface of this plane is

$$G_s = S_c \cos\theta. \tag{13.49}$$

As the solar radiation passes through the earth's atmosphere, it is partially absorbed and partially reflected by the gas molecules in the atmosphere. Therefore, both its magnitude and its direction change. The notation G_{solar} will be used for the solar irradiation that reaches the earth's surface. The magnitude of the solar irradiation reaching the earth's surface on a cloudy day is less than that on a clear day. Also, at sunrise and sunset the solar rays have to travel a longer distance to reach the earth's surface and therefore are attenuated more than in midday when the sun is straight above. Table 13.1 gives the solar irradiation on the earth's surface at different solar altitude angles and under average atmospheric conditions [4].

The amount of solar irradiation that is absorbed by a surface depends on its absorptivity. As mentioned in Section 13.3.2, the absorptivity of a surface depends on the temperature of the source of irradiation. Since the sun's apparent temperature of 5800 K is significantly higher than any typical temperature encountered on the earth, the notation α_s will be used to refer to the absorptivity of a surface for the solar irradiation or simply called the **solar absorptivity**. Using the definition of total absorptivity given in Section 13.3.2 and substituting $G \approx E_b(5800 \text{ K})$ gives

$$\alpha_s \approx \frac{\int_0^\infty \alpha_\lambda(\lambda) E_{\lambda,b}(\lambda, 5800 \text{ K}) d\lambda}{E_b(5800 \text{ K})}. \tag{13.50}$$

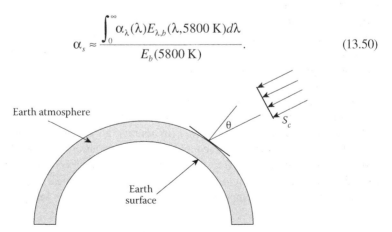

FIGURE 13.9 Solar radiation flux on the edge of earth's atmosphere.

TABLE 13.1
Solar Irradiation under Average Atmospheric Conditions

Solar Altitude Angle (degrees)	Solar Radiation Flux, G_{solar} (W/m²)
5	41.9
10	112.8
15	200.0
20	290.7
25	381.4
30	472.1
35	554.6
40	636.0
45	710.4
50	781.4
60	901.1
70	991.8
80	1043
90	1063

Source: From Holman, J. P., *Heat Transfer*. New York, NY: McGraw-Hill, 2010. With permission.

The solar irradiation absorbed by a surface with the solar absorptivity α_s is

$$G_{solar,a} = \alpha_s G_{solar}. \quad (13.51)$$

Table 13.2 gives the solar absorptivity and the surface emissivity of some common surfaces [5]. It is seen that, for most surfaces, $\alpha_s \neq \varepsilon$ and that is expected since the sun's apparent temperature is significantly higher than the room temperature and Kirchhoff's law does not apply. A quick observation of Table 13.2 shows that white paint has very low solar absorptivity and very high surface emissivity. Therefore, white and other light colors are the best candidates for the external surfaces of the electronic equipment and enclosures that will be exposed to the sun during the day. This minimizes the amount of solar irradiation absorbed by these equipments, keeps radiation heat emitted from their surfaces close to that of a blackbody, and therefore results in lower temperatures for the electronics inside them.

Other sources of irradiation in outdoor environment are the gas molecules, mainly CO_2 and H_2O, in the atmosphere. These molecules emit and absorb radiation like any other material. Although radiation from these particles, collectively called the **atmospheric radiation**, G_{sky}, is not similar to that of a blackbody, it is convenient to treat that as radiation from a blackbody at temperature T_{sky}, called the **effective sky temperature**,

$$G_{sky} = \sigma T_{sky}^4. \quad (13.52)$$

TABLE 13.2
Solar Absorptivity and Surface Emissivity at Room Temperature

Surface	α_s	ε
Aluminum		
Polished	0.09	0.03
Anodized	0.14	0.84
Foil	0.15	0.05
Copper		
Polished	0.18	0.03
Tarnished	0.65	0.75
Stainless Steel		
Polished	0.37	0.60
Dull	0.50	0.21
Plated Metals		
Black nickel oxide	0.92	0.08
Black chrome	0.87	0.09
Concrete	0.60	0.88
White marble	0.46	0.95
Red brick	0.63	0.93
Asphalt	0.90	0.90
Black paint	0.97	0.97
White paint	0.14	0.93
Snow	0.28	0.97

Source: From Cengel, Y. A., *Heat and Mass Transfer: A Practical Approach.* New York, NY: McGraw-Hill, 2007. With permission.

The value of T_{sky} depends on the atmospheric conditions. It was found to be as low as 230 K on a cold clear night and as high as 285 K on a hot cloudy day. Since these temperatures are very close to the room temperature, the absorptivity of a surface for the sky radiation is the same as its emissivity; $\alpha = \varepsilon$.

Let's consider an object with temperature T_s located in an outdoor environment and is exposed to the solar as well as the atmospheric (sky) irradiation as shown in Figure 13.10. It emits radiation flux E and receives irradiation G_{solar} and G_{sky}. Portions of the solar and atmospheric irradiations that are absorbed by this surface are $\alpha_s G_{solar}$ and αG_{sky}, respectively. The net radiation heat flux from this surface is

$$\dot{q}_{rad} = E - \alpha G_{sky} - \alpha_s G_{solar}. \tag{13.53}$$

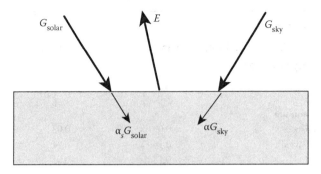

FIGURE 13.10 An object in an outdoor environment.

Substituting $\alpha = \varepsilon$, $G_{sky} = \sigma T_{sky}^4$ and $E = \varepsilon \sigma T_s^4$ gives

$$\dot{q}_{rad} = \varepsilon \sigma (T_s^4 - T_{sky}^4) - \alpha_s G_{solar}. \tag{13.54}$$

13.5 RADIOSITY

An object may reflect a portion of the irradiation that reaches its surface. Therefore, the radiation energy that leaves a surface is the sum of the radiation emitted from that surface because of its own temperature and the irradiation that is reflected from the surface. The rate of radiation energy that leaves the unit area of a surface is called the **radiosity**. If the total reflectivity of a surface is ρ, the total reflected irradiation is ρG and the total radiosity of the surface is

$$\begin{aligned} J &= E + \rho G, \\ &= \varepsilon \sigma T^4 + \rho G. \end{aligned} \tag{13.55}$$

Note that, similar to the emissive power and irradiation, the spectral radiosity can be defined as

$$J_\lambda(\lambda) = \int_{\phi=0}^{2\pi} \int_{\theta=0}^{\pi/2} I_{\lambda,e+r}(\lambda,\theta,\phi) \cos\theta \sin\theta \, d\theta \, d\phi, \tag{13.56}$$

where $I_{\lambda,e+r}(\lambda,\theta,\phi)$ is the intensity of the emitted and reflected radiation. The total radiosity J is obtained by integrating the spectral radiosity over all wavelengths,

$$J = \int_0^\infty J_\lambda(\lambda) \, d\lambda. \tag{13.57}$$

13.6 VIEW FACTORS

Consider surfaces i with area A_i and j with area A_j as shown in Figure 13.11. The rate of radiation energy that leaves the surface i is $A_i J_i$ where J_i is the total radiosity of surface i. A portion of this radiation energy, $\dot{Q}_{ij}$, reaches surface j. The **view factor** between surfaces i and j is defined as the ratio of radiation energy that leaves surface i and reaches surface j,

$$F_{ij} = \frac{\dot{Q}_{ij}}{A_i J_i}. \tag{13.58}$$

Note that $0 \leq F_{ij} \leq 1$. It can be shown that view factor is a purely geometrical parameter and can be found using the expressions similar to those in Tables 13.3 and 13.4 [4,5].

The following important correlations exist among view factors between different objects exchanging radiation energy.

Summation Rule: Consider an object i that is completely surrounded by N other objects that constitute a closed enclosure. The sum of the radiation energies that leave surface i and reach each of its surrounding surfaces is equal to the total radiation leaving surface i. This results in the following relationship among the view factors between surface i and all the other surfaces surrounding it

$$F_{i1} + F_{i2} + F_{i3} + \cdots + F_{iN} = \sum_{j=1}^{N} F_{ij} = 1. \tag{13.59}$$

Reciprocity Rule: If surface i has the area A_i and surface j has the area A_j, the view factors between surfaces i and j, F_{ij}, and between surfaces j and i, F_{ji}, are correlated by the reciprocity rule,

$$A_i F_{ij} = A_j F_{ji}. \tag{13.60}$$

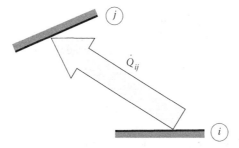

FIGURE 13.11 A portion of the radiation energy that leaves surface i reaches surface j.

TABLE 13.3
View Factor for Some Two-Dimensional Geometries

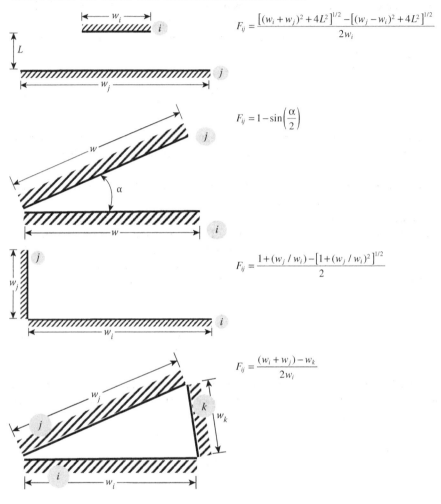

Superposition Rule: The radiation energy that leaves surface i and reaches surfaces j and k is equal to the sum of the radiation energy that leaves surface i and reaches surface j and the radiation energy that leaves surface i and reaches surface k. This results in the superposition rule,

$$F_{i(jk)} = F_{ij} + F_{ik}. \tag{13.61}$$

Symmetry Rule: Consider symmetric geometries such as the isosceles triangle or the equilateral polygon shown in Figure 13.12. Surfaces 2 and 3 of each of these

Radiation Heat Transfer

TABLE 13.4
View Factor for Some Three-Dimensional Geometries

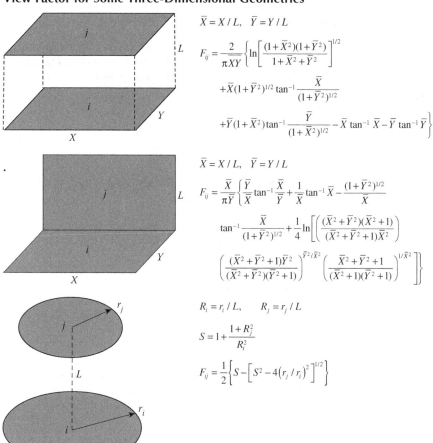

$\bar{X} = X/L, \quad \bar{Y} = Y/L$

$F_{ij} = \dfrac{2}{\pi \bar{X}\bar{Y}} \left\{ \ln \left[\dfrac{(1+\bar{X}^2)(1+\bar{Y}^2)}{1+\bar{X}^2+\bar{Y}^2} \right]^{1/2} \right.$

$+ \bar{X}(1+\bar{Y}^2)^{1/2} \tan^{-1} \dfrac{\bar{X}}{(1+\bar{Y}^2)^{1/2}}$

$\left. + \bar{Y}(1+\bar{X}^2)^{1/2} \tan^{-1} \dfrac{\bar{Y}}{(1+\bar{X}^2)^{1/2}} - \bar{X}\tan^{-1}\bar{X} - \bar{Y}\tan^{-1}\bar{Y} \right\}$

$\bar{X} = X/L, \quad \bar{Y} = Y/L$

$F_{ij} = \dfrac{\bar{X}}{\pi\bar{Y}} \left\{ \dfrac{\bar{Y}}{\bar{X}} \tan^{-1}\dfrac{\bar{X}}{\bar{Y}} + \dfrac{1}{\bar{X}}\tan^{-1}\bar{X} - \dfrac{(1+\bar{Y}^2)^{1/2}}{\bar{X}} \right.$

$\tan^{-1}\dfrac{\bar{X}}{(1+\bar{Y}^2)^{1/2}} + \dfrac{1}{4}\ln\left[\left(\dfrac{(\bar{X}^2+\bar{Y}^2)(\bar{X}^2+1)}{(\bar{X}^2+\bar{Y}^2+1)\bar{X}^2}\right)\right.$

$\left.\left.\left(\dfrac{(\bar{X}^2+\bar{Y}^2+1)\bar{Y}^2}{(\bar{X}^2+\bar{Y}^2)(\bar{Y}^2+1)}\right)^{\bar{Y}^2/\bar{X}^2}\left(\dfrac{\bar{X}^2+\bar{Y}^2+1}{(\bar{X}^2+1)(\bar{Y}^2+1)}\right)^{1/\bar{X}^2}\right]\right\}$

$R_i = r_i/L, \quad R_j = r_j/L$

$S = 1 + \dfrac{1+R_j^2}{R_i^2}$

$F_{ij} = \dfrac{1}{2}\left\{S - \left[S^2 - 4(r_j/r_i)^2\right]^{1/2}\right\}$

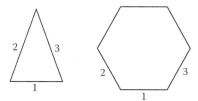

FIGURE 13.12 Symmetry rule for view factors applies between surface 1 and surfaces 2 and 3 in these symmetric geometries.

geometries are viewed the same from surface 1 in each geometry. The symmetry rules states that $F_{12} = F_{13}$ in these situations.

13.7 RADIATION HEAT TRANSFER BETWEEN BLACK BODIES

Consider the blackbodies i and j with surface areas A_i and A_j and temperatures T_i and T_j as shown in Figure 13.13. The net radiation heat transfer from surface i to surface j is equal to the difference between the radiation that leaves surface i and reaches surface j and the radiation that leaves surface j and reaches surface i,

$$\dot{Q}_{rad,ij} = \dot{Q}_{ij} - \dot{Q}_{ji}. \tag{13.62}$$

Since surfaces i and j are blackbodies, the rate of radiation energy emitted from their surfaces are $A_i E_{bi}$ and $A_j E_{bj}$, respectively. Therefore,

$$\dot{Q}_{ij} = A_i E_{bi} F_{ij} \quad \text{and} \quad \dot{Q}_{ji} = A_j E_{bj} F_{ji}.$$

Substituting these into Equation 13.62 gives

$$\dot{Q}_{rad,ij} = A_i E_{bi} F_{ij} - A_j E_{bj} F_{ji}.$$

Using the reciprocity rule, $A_i F_{ij} = A_j F_{ji}$, and the Stefan–Boltzmann law for blackbody emissive power, $E_b = \sigma T^4$, gives the following correlation for the net radiation exchange between blackbodies i and j,

$$\dot{Q}_{rad,ij} = A_i F_{ij} \sigma (T_i^4 - T_j^4). \tag{13.63}$$

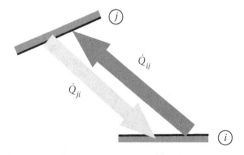

FIGURE 13.13 Radiation exchange between two blackbodies.

Radiation Heat Transfer

13.8 RADIATION HEAT TRANSFER BETWEEN NONBLACK BODIES

Consider an opaque surface and let's assume that Kirshhoff's law applies and its surface emissivity and absorptivity are the same. This assumption is valid if the surface is gray and either the surface or the irradiation is diffuse. The rate of radiation energy leaving a surface per unit area of the surface (i.e., its radiosity) is

$$J = \varepsilon E_b + \rho G$$
$$= \varepsilon E_b + (1-\alpha)G \qquad (13.64)$$
$$= \varepsilon E_b + (1-\varepsilon)G.$$

Solving Equation 13.64 for irradiation G gives

$$G = \frac{J - \varepsilon E_b}{1-\varepsilon}. \qquad (13.65)$$

The net rate of radiation heat transfer from a surface i, $\dot{Q}_i$, is equal to the difference between its radiosity and irradiation times its area,

$$\dot{Q}_i = A_i(J_i - G_i). \qquad (13.66)$$

Substituting for G from Equation 13.65 gives

$$\dot{Q}_i = \frac{A_i \varepsilon_i}{1-\varepsilon_i}(E_{bi} - J_i). \qquad (13.67)$$

Equation 13.67 can be written as

$$\dot{Q}_i = \frac{E_{bi} - J_i}{R_i}, \qquad (13.68)$$

where

$$R_i = \frac{1-\varepsilon_i}{A_i \varepsilon_i} \qquad (13.69)$$

is called the **surface resistance to radiation**. Equations 13.68 and 13.69 can be represented by the resistance analogy shown in Figure 13.14.

Let's consider the real objects i and j with surface areas A_i and A_j and temperatures T_i and T_j, respectively. The net radiation heat transfer from surface i to surface j is equal to the difference between the radiation that leaves surface i and reaches surface j and the radiation that leaves surface j and reaches surface i,

$$\dot{Q}_{rad,ij} = \dot{Q}_{ij} - \dot{Q}_{ji}. \qquad (13.62)$$

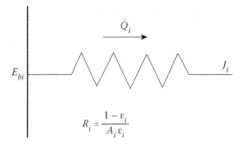

FIGURE 13.14 Resistance analogy for the surface resistance to radiation.

The rate of radiation energy leaving surfaces i and j are $A_i J_i$ and $A_j J_j$, respectively. Therefore,

$$\dot{Q}_{ij} = A_i J_i F_{ij} \quad \text{and} \quad \dot{Q}_{ji} = A_j J_j F_{ji}. \tag{13.70}$$

Substituting these into Equation 13.62 and using the reciprocity rule, $A_i F_{ij} = A_j F_{ji}$, gives

$$\dot{Q}_{\text{rad},ij} = A_i F_{ij} (J_i - J_j). \tag{13.71}$$

Equation 13.71 can be written as

$$\dot{Q}_{\text{rad},ij} = \frac{J_i - J_j}{R_{ij}}, \tag{13.72}$$

where

$$R_{ij} = \frac{1}{A_i F_{ij}} \tag{13.73}$$

is called the **space resistance to radiation**. Figure 13.15 shows the resistance network for the radiation heat transfer between two objects. Note that the net rate of radiation heat transfer between surfaces i and j is

$$\dot{Q}_{\text{rad},ij} = \frac{\sigma(T_i^4 - T_j^4)}{\dfrac{1-\varepsilon_i}{A_i \varepsilon_i} + \dfrac{1}{A_i F_{ij}} + \dfrac{1-\varepsilon_j}{A_j \varepsilon_j}}. \tag{13.74}$$

Two special cases of Equation 13.74 are of particular interest in many engineering applications including electronic cooling; radiation between two infinite parallel plates, and radiation between a small object and its large surrounding.

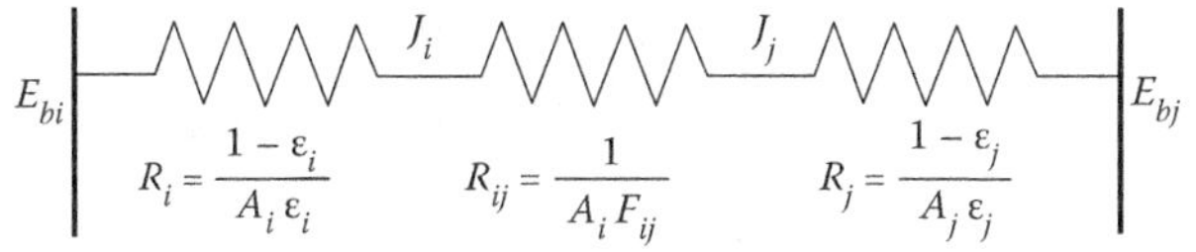

Radiation Heat Transfer

$$E_{bi} \;—\!\!\!\bigvee\!\!\!— \;J_i\; —\!\!\!\bigvee\!\!\!— \;J_j\; —\!\!\!\bigvee\!\!\!—\; E_{bj}$$

$$R_i = \frac{1-\varepsilon_i}{A_i \varepsilon_i} \qquad R_{ij} = \frac{1}{A_i F_{ij}} \qquad R_j = \frac{1-\varepsilon_j}{A_j \varepsilon_j}$$

FIGURE 13.15 Resistance network for radiation heat transfer between two objects.

Consider two infinite parallel plates i and j that are exchanging radiation heat transfer with each other. Since $A_i = A_j = A$ and $F_{ij} = 1$, Equation 13.74 gives

$$\dot{q}_{\mathrm{rad},ij} = \frac{\sigma(T_i^4 - T_j^4)}{\dfrac{1}{\varepsilon_i} + \dfrac{1}{\varepsilon_j} - 1} \tag{13.75}$$

for the nest radiation heat flux, $\dot{Q}_{\mathrm{rad},ij}/A$, between these two plates. Equation 13.75 can also be used to approximate the net radiation heat flux between two large parallel plates that are very close to each other. Examples are the radiation heat transfer between closely spaced boards of a telecommunication equipment and between a board and the enclosure it is mounted on.

Let's consider the radiation heat transfer between a small object and its large surrounding. In this case A_i is much smaller than A_j such that $A_i/A_j \approx 0$ and $F_{ij} \approx 1$ and Equation 13.74 will reduce to

$$\dot{Q}_{\mathrm{rad},ij} = A_i \varepsilon_i \sigma(T_i^4 - T_j^4), \tag{13.76}$$

which is the equation first introduced in Chapter 4.

13.9 RADIATION HEAT TRANSFER FROM A PLATE-FIN HEAT SINK

Radiation heat transfer can be a significant portion of the total heat transfer from a heat sink particularly in natural convection flows. The enclosures for most natural convection-cooled indoor and outdoor electronic equipments, such as pole or wall-mount base stations and radio units, include external fins to improve the natural convection and radiation heat transfer rates.

Consider the plate-fin heat sink shown in Figure 13.16. The total radiation heat transfer rate from this heat sink is equal to the sum of the radiation heat transfer rates from the U-shaped channels between pairs of neighboring fins, the fin tips, the fin front and rear sections, the outside surfaces of the outer fins, and the peripheral surface of the heat sink base. If we assume that the heat sink surface is diffuse and gray, the surrounding medium is very large and/or is a blackbody at temperature T_{surr}, and the heat sink is at uniform temperature T_s, the total radiation heat transfer rate from this heat sink is

$$\dot{Q}_{\mathrm{total},r} = (N_f - 1)\dot{Q}_{ch,r} + \left[N_f t(L + 2H) + 2HL + 2t_b(L+W)\right]\sigma\varepsilon(T_s^4 - T_{\mathrm{surr}}^4), \tag{13.77}$$

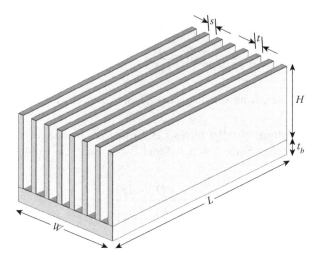

FIGURE 13.16 A typical plate-fin heat sink.

where $\dot{Q}_{ch,r}$ is the radiation heat transfer rate from a U-shaped channel, N_f is the number of fins, σ is the Stefan–Boltzmann constant, ε is the surface emissivity, and t, t_b, H, L, W, and S are heat sink dimensions as shown in Figure 13.16. Rea and West [6] and Bilitzky [7] defined "apparent" or "effective" channel emittance, $\hat{\varepsilon}$, such that the channel radiation heat transfer rate can be calculated as

$$\dot{Q}_{ch,r} = \hat{\varepsilon}\sigma SL(T_s^4 - T_{\text{surr}}^4). \tag{13.78}$$

The effective emittance accounts for emission as well as reflection from the plate fins. Since channel base area rather than the whole channel area is used in this equation, the effective emittance is always larger than the surface emissivity and may even be larger than one. A slightly different approach was used by Ellison [8] who defined the gray body shape factor, $\hat{F}$, such that

$$\dot{Q}_{ch,r} = \hat{F}\sigma(S + 2H)L(T_s^4 - T_{\text{surr}}^4). \tag{13.79}$$

However, no explicit analytical formulation is available for either the effective emittance or the gray body shape factor, and their calculations require the simultaneous solution of a set of three equations. In these cases, numerical results were obtained and presented in graphical forms as functions of the channel surface emissivity and dimensions. A different approach will be presented here that results in a simple and explicit analytical correlation for $\dot{Q}_{ch,r}$ [9].

A typical channel created by any two adjacent fins of a plate-fin heat sink is shown in Figure 13.17. If the channel surface is diffuse and gray and the surrounding medium is very large and/or is a blackbody, the net radiation heat transfer rate from this channel to its surrounding is

Radiation Heat Transfer

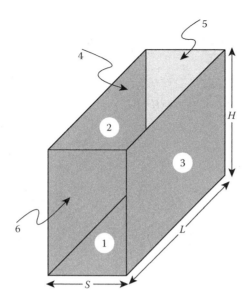

FIGURE 13.17 A typical channel created by two adjacent fins of a plate-fin heat sink.

$$\dot{Q}_{ch} = \frac{\sigma(S+2H)L(T_s^4 - T_{surr}^4)}{\dfrac{1-\varepsilon}{\varepsilon} + \dfrac{1}{F_{s\text{-surr}}}} \tag{13.80}$$

where $F_{s\text{-surr}}$, henceforth called the channel view factor, is the total view factor between the walls and the base of the channel and its surrounding. It can be shown that the channel view factor is related to the simpler view factors from the walls 1, 2, and 3 that constitute this channel to the virtual walls 4, 5, and 6 that represent the surrounding for this channel [9],

$$F_{s\text{-surr}} = \frac{F_{14} + 2F_{15} + 2\bar{H}(F_{24} + 2F_{25})}{1 + 2\bar{H}}, \tag{13.81}$$

where the expressions for F_{14}, F_{15}, F_{24}, and F_{25} in terms of the normalized length $\bar{L} = L/S$ and normalized height $\bar{H} = H/S$ can be found using the correlations given in Table 13.4:

$$F_{14} = \frac{2\bar{H}^2}{\pi\bar{L}}\left\{\ln\left[\frac{(\bar{H}^2+1)(\bar{H}^2+\bar{L}^2)}{(\bar{H}^2+\bar{L}^2+1)\bar{H}^2}\right]^{1/2} + \frac{(\bar{H}^2+\bar{L}^2)^{1/2}}{\bar{H}^2}\tan^{-1}\frac{1}{(\bar{H}^2+\bar{L}^2)^{1/2}}\right.$$

$$\left. + \frac{\bar{L}(\bar{H}^2+1)^{1/2}}{\bar{H}^2}\tan^{-1}\frac{\bar{L}}{(\bar{H}^2+1)^{1/2}} - \frac{1}{\bar{H}}\tan^{-1}\frac{1}{\bar{H}} - \frac{\bar{L}}{\bar{H}}\tan^{-1}\frac{\bar{L}}{\bar{H}}\right\}, \tag{13.82}$$

$$F_{24} = \frac{\bar{L}}{\pi \bar{H}} \left\{ \frac{\bar{H}}{\bar{L}} \tan^{-1} \frac{\bar{L}}{\bar{H}} + \frac{1}{\bar{H}} \tan^{-1} \bar{L} - \frac{(\bar{H}^2+1)^{1/2}}{\bar{L}} \tan^{-1} \frac{\bar{L}}{(\bar{H}^2+1)^{1/2}} \right.$$

$$\left. + \frac{1}{4} \ln \left[\left(\frac{(\bar{H}^2+\bar{L}^2)(1+\bar{L}^2)}{(\bar{H}^2+\bar{L}^2+1)\bar{L}^2} \right) \left(\frac{\bar{H}^2(\bar{H}^2+\bar{L}^2+1)}{(\bar{H}^2+\bar{L}^2)(\bar{H}^2+1)} \right)^{\bar{H}^2/\bar{L}^2} \left(\frac{(\bar{H}^2+\bar{L}^2+1)}{(\bar{L}^2+1)(\bar{H}^2+1)} \right)^{1/\bar{L}^2} \right] \right\},$$

(13.83)

$$F_{15} = \frac{1}{\pi \bar{L}} \left\{ \bar{L} \tan^{-1} \frac{1}{\bar{L}} + \bar{H} \tan^{-1} \frac{1}{\bar{H}} - (\bar{H}^2+\bar{L}^2)^{1/2} \tan^{-1} \frac{1}{(\bar{H}^2+\bar{L}^2)^{1/2}} \right.$$

$$\left. + \frac{1}{4} \ln \left[\left(\frac{(1+\bar{L}^2)(1+\bar{H}^2)}{(\bar{H}^2+\bar{L}^2+1)} \right) \left(\frac{\bar{L}^2(\bar{H}^2+\bar{L}^2+1)}{(\bar{H}^2+\bar{L}^2)(\bar{L}^2+1)} \right)^{\bar{L}^2} \left(\frac{\bar{H}^2(\bar{H}^2+\bar{L}^2+1)}{(\bar{H}^2+1)(\bar{H}^2+\bar{L}^2)} \right)^{\bar{H}^2} \right] \right\},$$

(13.84)

$$F_{25} = \frac{\bar{H}}{\pi \bar{L}} \left\{ \frac{\bar{L}}{\bar{H}} \tan^{-1} \frac{\bar{H}}{\bar{L}} + \frac{1}{\bar{H}} \tan^{-1} \bar{H} - \frac{(1+\bar{L}^2)^{1/2}}{\bar{H}} \tan^{-1} \frac{\bar{H}}{(1+\bar{L}^2)^{1/2}} \right.$$

$$\left. + \frac{1}{4} \ln \left[\left(\frac{(\bar{L}^2+\bar{H}^2)(\bar{H}^2+1)}{(\bar{H}^2+\bar{L}^2+1)\bar{H}^2} \right) \left(\frac{\bar{L}^2(\bar{H}^2+\bar{L}^2+1)}{(\bar{H}^2+\bar{L}^2)(\bar{L}^2+1)} \right)^{\bar{L}^2/\bar{H}^2} \left(\frac{(\bar{H}^2+\bar{L}^2+1)}{(\bar{H}^2+1)(\bar{L}^2+1)} \right)^{1/\bar{H}^2} \right] \right\}.$$

(13.85)

Equations 13.81 to 13.85 give an explicit expression for the view factor $F_{s\text{-surr}}$. However, the expressions for view factors F_{14}, F_{15}, F_{24}, and F_{25} are too complex and calculations are too lengthy. A simpler approximate correlation for the channel view factor was developed by noting the limiting values of $F_{s\text{-surr}}$ as $\bar{H}$ and $\bar{L}$ approach zero and infinity [9];

$$\lim_{\bar{H} \to 0} F_{s\text{-surr}} = 1, \qquad (13.86)$$

$$\lim_{\bar{H} \to \infty} F_{s\text{-surr}} = 1 + \frac{1}{\bar{L}} - \frac{(1+\bar{L}^2)^{1/2}}{\bar{L}}, \qquad (13.87)$$

$$\lim_{\bar{L} \to 0} F_{s\text{-surr}} = 1, \qquad (13.88)$$

$$\lim_{\bar{L} \to \infty} F_{s\text{-surr}} = \frac{1}{1+2\bar{H}}, \qquad (13.89)$$

Radiation Heat Transfer

and solving for $f_1(\bar{H},\bar{L})$ and $f_2(\bar{H},\bar{L})$ such that

$$F_{s-surr} = f_1(\bar{H},\bar{L})\left[\lim_{\bar{L}\to\infty} F_{s-surr}\right] + f_2(\bar{H},\bar{L})\left[\lim_{\bar{H}\to\infty} F_{s-surr}\right] \tag{13.90}$$

meets all the above conditions. The result is [9]

$$F_{s-surr} = 1 - \frac{2\bar{H}[(1+\bar{L}^2)^{1/2} - 1]}{2\bar{H}\,\bar{L} + (1+\bar{L}^2)^{1/2} - 1}. \tag{13.91}$$

Figure 13.18 shows the percentage error associated with Equation 13.91 compared to the exact view factor calculated with Equation 13.81. It is seen that the approximate expression for the channel view factor gives values within 11% of the exact values. However, Figure 13.18 shows that the error for most practical cases in which $\bar{L} > 5$ and $\bar{H} > 5$ is less than 5%. Derivative of Equation 13.80 with respect to F_{s-surr} shows that

$$\frac{d\dot{Q}_{ch,r}}{\dot{Q}_{ch,r}} = \frac{dF_{s-surr}}{F_{s-surr}}\left[\frac{\varepsilon}{\varepsilon + (1-\varepsilon)F_{s-surr}}\right]. \tag{13.92}$$

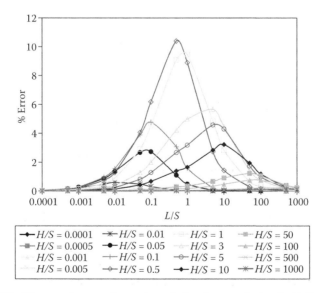

FIGURE 13.18 The error in approximate channel view factors calculated with Equation 13.91.

This shows that the maximum error in the calculated radiation heat transfer rate from a channel happens when the channel surface emissivity is equal to one and this error is exactly the same as the error associated with the channel view factor calculation.

Equation 13.77 assumes that the entire heat sink surface is at a uniform temperature T_s. A more accurate estimate of the radiation heat transfer rate from this heat sink is obtained by using the well-known fin equations to account for the temperature variation in the fins. First, a radiation heat transfer coefficient is defined as

$$h_r = \frac{\dot{Q}_{\text{total},r}}{A_{hs}(T_s - T_{\text{surr}})}, \qquad (13.93)$$

where $\dot{Q}_{\text{total},r}$ is given by Equation 13.77 and

$$A_{hs} = (N_f - 1)SL + N_f[t(L + 2H) + 2HL] + 2t_b(W + L) \qquad (13.94)$$

is the heat sink radiating surface area. Then, the radiation heat transfer rate from the heat sink is

$$\dot{Q}_{hs,r} = \eta_{hs} A_{hs} h_r (T_s - T_{\text{surr}}), \qquad (13.95)$$

where

$$\eta_{hs} = 1 - \frac{N_f A_f}{A_{hs}}(1 - \eta_f), \qquad (13.96)$$

$$A_f = 2(L + t)H + Lt, \qquad (13.97)$$

$$\eta_f = \frac{\tanh\left[(H + t/2)\sqrt{2h_r(L+t)/kLt}\right]}{(H + t/2)\sqrt{2h_r(L+t)/kLt}}. \qquad (13.98)$$

Equations 13.77 and 13.93 through 13.98 were used to calculate the radiation heat transfer rate for a plate-fin aluminum heat sink and the results were compared with the data measured in vacuum by Rea and West in Figure 13.19. The heat sink dimensions are $L = 127$ mm, $H = 38.3$ mm, and $W = 92.2$ mm. It has nine 3.3 mm-thick fins with the fin spacing of $S = 7.8$ mm. The base thickness is $t_b = 6.2$ mm. The analytical results were obtained using the exact view factor correlation, Equations 13.81 through 13.85, as well as the approximate view factor correlation, Equation 13.91. It is seen that both analytical methods predict the measured radiation heat transfer rate very well and there is basically no difference between the methods using the exact and the approximate view factor correlations.

Radiation Heat Transfer

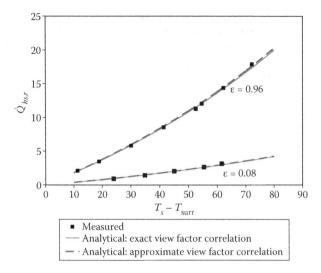

FIGURE 13.19 Measured and calculated radiation heat transfer rates from a plate-fin heat sink.

PROBLEMS

13.1 What is a gray surface? What is a diffuse surface? How do these surfaces differ from a blackbody?

13.2 An opaque surface does not reflect any radiation incident on it. What are its absorptivity and emissivity values?

13.3 A piece of glass reflects 20% and absorbs 40% of the radiation incident on its surface. What percentage of the incident radiation is transmitted through the glass?

13.4 The emissivity of an opaque surface is 0.35. What are its absorptivity and reflectivity?

13.5 If the view factor between surfaces 1 and 2, F_{12}, is 0.3, and the absorptivity of surface 2 is 0.8, what fraction of the radiation energy emitted by surface 1 is absorbed by surface 2?

13.6 Consider a 0.6 m × 0.45 m printed circuit board dissipating 200 W and sitting on the base of a 0.6 m × 0.45 m × 0.15 m enclosure as shown below. If the emissivity and temperature of the top and side surfaces are 30°C and 0.7, and the emissivity of the printed circuit board is 0.8, what is the temperature of the printed circuit board? Neglect any conduction and convection heat transfer.

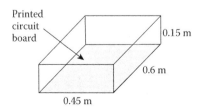

Two identical 0.8 m × 0.5 m plates are spaced 0.2 m apart from each other. One of the plates is at temperature $T_1 = 250°C$ and the other plate is at temperature $T_2 = 20°C$. The space between the plates is filled with still air. If the radiation heat transfer between these plates is as much as the conduction heat transfer through the air layer between them, determine the surface emissivity of the plates.

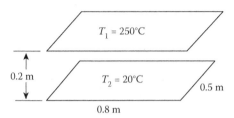

REFERENCES

1. Planck, M. 1959. *The theory of heat radiation.* New York: Dover Publication.
2. Modest, M. F. 1993. *Radiation heat transfer.* New York, NY: McGraw-Hill.
3. Incropera, F. P., D. P. DeWitt, T. L. Bergman, and A. S. Lavine. 2006. *Fundamentals of heat and mass transfer.* Sixth Edition. Hoboken, NJ: John Wiley & Sons, Inc.
4. Holman, J. P. 2010. *Heat transfer.* New York, NY: McGraw-Hill.
5. Cengel, Y. A. 2007. *Heat and mass transfer: A practical approach.* New York, NY: McGraw-Hill.
6. Rea, S. N., and West, S. E. 1976, June. Thermal radiation from finned heat sinks. *IEEE Transactions on Parts, Hybrids and Packaging* PHP-12 (2): 115–17.
7. Bilitzky, A. 1986. The effect of geometry on heat transfer by free convection from a fin array. MS Thesis, Department of Mechanical Engineering, Ben-Gurion University Of The Negev, Beer Sheva, Israel.
8. Ellison, N. G. 1979, December. Generalized computations of the gray body shape factor for thermal radiation from a rectangular U-channel. *IEEE Transactions on Components, Hybrids, and Manufacturing Technology* CHMT-2 (4): 517–22.
9. Shabany, Y. 2008, March 16-20. Radiation Heat Transfer from Plate-Fin Heat Sinks. Proceedings of 24th Annual IEEE Semiconductor Thermal Measurement and Management Symposium (SemiTherm 24), 133–37, San Jose, California.

14 Computer Simulations and Thermal Design

The three modes of heat transfer have been described in previous chapters and correlations were proposed to calculate these heat transfer rates for certain situations. For example, Fourier's law of heat conduction can be used to calculate conductive heat transfer if conduction is predominately in one direction or, as shown in Chapter 8, more complex analytical methods can be used to compute the temperature distribution and conduction heat transfer rate in two- and three-dimensional, steady-state or transient situations. Similarly, as shown in Chapters 9 through 12, Newton's law of cooling with appropriate correlations for convection heat transfer coefficient can be used to calculate either the convection heat transfer rate or the temperature difference between a surface and ambient if the other parameter is known. Finally, radiation heat transfer rate between a surface and its large or blackbody isothermal surrounding or between two surfaces can be calculated using the correlations given in Chapter 13 if the enclosure making the whole radiation domain is made of only two or three surfaces.

Many practical situations are not as simple as those mentioned above. Conduction heat transfer rate through a heat sink that is mounted on a smaller package is not one dimensional. Although, the concept of spreading resistance was introduced to account for this three-dimensional effect in a one-dimensional analysis, the underlying assumption of uniform heat flux on the fin side of the base is not completely accurate in most cases. Many electronic devices, such as notebook computers and power supplies, have complex shapes and are internally so congested that they may not be approximated as a flat plate or a duct. Besides, the assumption of uniform heat flux, uniform surface temperature, or even uniform ambient temperature may not be valid in many of these cases. For the radiation heat transfer, it is rare that we can assume the surrounding to be at constant temperature or only two or three radiating surfaces are involved. Therefore, radiation heat transfer calculations will involve calculating many view factors and solving a set of N simultaneous linear equations where N is the number of radiating surfaces. Again, similar to the convection heat transfer, the assumption of uniform surface temperature or uniform surface heat flux may not be accurate.

Computers have been extensively used to solve heat transfer problems that do not have analytical solutions. Computer-aided design (CAD) and computer-aided engineering (CAE) refer to the use of computer simulation techniques in a design process. This is a complex and extensive subject that can not be covered in this chapter and the interested reader is referred to many books written on the subject. It is also argued that, from a user's perspective, it is not necessary to know in detail how these techniques work; it is only important to know when to use these techniques in the design process and how to use them. However, the correct use of these techniques requires at least a basic understanding of how they work. Therefore, a brief introduction of the fundamentals of computer simulation techniques will be given for those who are unfamiliar with this subject. Then, the appropriate ways of using these techniques for thermal design of electronic equipment will be discussed.

14.1 HEAT TRANSFER AND FLUID FLOW EQUATIONS: A SUMMARY

Conduction heat flux at any point in a medium and in any direction n is described by Fourier's law of heat conduction in that direction,

$$\dot{q}_{\text{cond},n} = -k_n \frac{\partial T}{\partial n}. \tag{14.1}$$

Here, k_n is the thermal conductivity of the medium at that point and in the n direction. The temperature distribution and temperature gradient can be obtained by solving the heat conduction equation,

$$\frac{\partial}{\partial x}\left(k_x \frac{\partial T}{\partial x}\right) + \frac{\partial}{\partial y}\left(k_y \frac{\partial T}{\partial y}\right) + \frac{\partial}{\partial z}\left(k_z \frac{\partial T}{\partial z}\right) + \dot{g} = \rho C_p \frac{\partial T}{\partial t}. \tag{14.2}$$

Similarly, convection heat flux from a surface exposed to a moving fluid is

$$\dot{q}_{\text{conv},n} = -k_f \left.\frac{\partial T}{\partial n}\right|_{n=0}, \tag{14.3}$$

where k_f is the fluid thermal conductivity herein assumed to be the same for all directions, n is the direction normal to the surface, and $n = 0$ at the surface. If the fluid is moving, the temperature distribution in the fluid is obtained by solving the energy equation that, for an incompressible fluid with constant thermal conductivity, is

$$\rho C_p \left(\frac{\partial T}{\partial t} + U\frac{\partial T}{\partial x} + V\frac{\partial T}{\partial y} + W\frac{\partial T}{\partial z}\right) = k\left(\frac{\partial^2 T}{\partial x^2} + \frac{\partial^2 T}{\partial y^2} + \frac{\partial^2 T}{\partial z^2}\right) + \mu\Phi + \dot{g}, \tag{14.4}$$

$$\Phi = 2\left[\left(\frac{\partial U}{\partial x}\right)^2 + \left(\frac{\partial V}{\partial y}\right)^2 + \left(\frac{\partial W}{\partial z}\right)^2\right] + \left(\frac{\partial U}{\partial x} + \frac{\partial U}{\partial y}\right)^2 + \left(\frac{\partial W}{\partial y} + \frac{\partial V}{\partial z}\right)^2 + \left(\frac{\partial U}{\partial z} + \frac{\partial W}{\partial x}\right)^2. \tag{14.5}$$

Note that $\mu\Phi$ is the rate of frictional dissipation of flow kinetic energy into heat.

Computer Simulations and Thermal Design

U, V, and W are components of fluid velocity in x, y, and z directions and P is fluid pressure. They are obtained by solving fluid flow equations that, for an incompressible fluid with constant viscosity, are

$$\frac{\partial U}{\partial x} + \frac{\partial V}{\partial y} + \frac{\partial W}{\partial z} = 0, \tag{14.6}$$

$$\rho\left(\frac{\partial U}{\partial t} + U\frac{\partial U}{\partial x} + V\frac{\partial U}{\partial y} + W\frac{\partial U}{\partial z}\right) = -\frac{\partial P}{\partial x} + \mu\left(\frac{\partial^2 U}{\partial x^2} + \frac{\partial^2 U}{\partial y^2} + \frac{\partial^2 U}{\partial z^2}\right) + X, \tag{14.7}$$

$$\rho\left(\frac{\partial V}{\partial t} + U\frac{\partial V}{\partial x} + V\frac{\partial V}{\partial y} + W\frac{\partial V}{\partial z}\right) = -\frac{\partial P}{\partial y} + \mu\left(\frac{\partial^2 V}{\partial x^2} + \frac{\partial^2 V}{\partial y^2} + \frac{\partial^2 V}{\partial z^2}\right) + Y, \tag{14.8}$$

$$\rho\left(\frac{\partial W}{\partial t} + U\frac{\partial W}{\partial x} + V\frac{\partial W}{\partial y} + W\frac{\partial W}{\partial z}\right) = -\frac{\partial P}{\partial z} + \mu\left(\frac{\partial^2 W}{\partial x^2} + \frac{\partial^2 W}{\partial y^2} + \frac{\partial^2 W}{\partial z^2}\right) + Z. \tag{14.9}$$

X, Y, and Z in these equations are components of body forces, per unit volume of the fluid, in x, y, and z directions, respectively.

Analytical solutions of these equations for a few simple geometries and flows were given in Chapters 8 through 12. These solutions give temperature and heat flux as a continuous function of space and/or time. On the other hand, as will be explained in the next section, computer simulation tools give solutions of these equations at selected points in space and time. Therefore, it is very important to choose a set of points for which computer solutions are obtained at appropriate locations and times.

14.2 FUNDAMENTALS OF COMPUTER SIMULATION

Simple examples will be used here to explain how computers solve heat transfer equations.

14.2.1 Steady-State, One-Dimensional Heat Conduction

Consider the process that was followed to derive the one-dimensional heat conduction equation in Chapter 8, Section 8.1. We showed that the energy balance equation for a thin wall with thickness Δx was

$$\dot{Q}_x - \dot{Q}_{x+\Delta x} + \dot{g}A\Delta x = \rho C_p A \Delta x \frac{\Delta T}{\Delta t}. \tag{14.10}$$

Note that $A\Delta x$ is the volume of the thin wall. This equation is used to derive algebraic equations for temperatures at several points in a domain. These equations are solved by appropriate computer algorithms to obtain temperature distribution in the domain of interest.

The first step for obtaining a computer solution to this equation is to select the points or nodes for which this solution will be obtained. Figure 14.1 shows the cross section of a rectangular wall with seven nodes that are spaced from each other by distance $\Delta x = t/6$. It must be mentioned that neither Δx nor Δt has to be uniform across a domain and a uniform Δx is assumed here for simplicity only.

Some parameters such as temperature and pressure are defined at a point while other parameters such as heat transfer rate are defined at a surface. Therefore, it is necessary to define surfaces as well as points in a computational domain. The surfaces define volumes called control volumes. The points may be located at the center of each control volume, at the center of the faces of the control volume, or at the corners of the faces of the control volume. Figure 14.1 shows a case where the points are located at the center of one-dimensional control volumes that are strips with thickness Δx for points 2–6. These points are called internal points. The thickness of the control volumes is $\Delta x/2$ for point 1 and 7, which are boundary points. Note that since the thickness of the boundary strips is $\Delta x/2$, their volume is $A\Delta x/2$, and the energy balance equations for these boundary strips is

$$\dot{Q}_x - \dot{Q}_{x+\Delta x/2} + \dot{g}A\Delta x/2 = \rho C_p A\Delta x/2 \frac{\Delta T}{\Delta t}. \quad (14.11)$$

Consider a steady-state situation where the right-hand side of Equations 14.10 and 14.11 disappear and let's assume that thermal conductivity and volumetric heat generation are uniform across the domain. For an internal point, such as point 4,

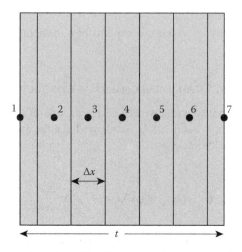

FIGURE 14.1 Cross section of a wall with seven nodes spaced uniformly.

Computer Simulations and Thermal Design

$\dot{Q}_x$ and $\dot{Q}_{x+\Delta x}$ can be approximated by the finite-difference form of the Fourier's law of conduction;

$$\dot{Q}_x = -kA\frac{T_4 - T_3}{\Delta x}, \tag{14.12}$$

$$\dot{Q}_{x+\Delta x} = -kA\frac{T_5 - T_4}{\Delta x}. \tag{14.13}$$

Substituting these into the finite-difference form of the steady-state, one-dimensional heat conduction equation, Equation 14.10, gives

$$T_5 - 2T_4 + T_3 + \frac{\dot{g}}{k}(\Delta x)^2 = 0. \tag{14.14}$$

This equation contains three unknown temperatures T_3, T_4 and T_5. Similar equations can be derived for other internal points. These equations have the general form:

$$T_{m+1} - 2T_m + T_{m-1} + \frac{\dot{g}}{k}(\Delta x)^2 = 0, \tag{14.15}$$

where m is an internal point with temperature T_m and $m-1$ and $m+1$ are its left and right side neighbor points, respectively.

The treatment of the boundary points is different and depends on the boundary conditions. If the left and right sides of the wall are at uniform temperature T_L and T_R, respectively, the equations for points 1 and 7 will have the following simple forms:

$$\begin{aligned} T_1 &= T_L \\ T_7 &= T_R. \end{aligned} \tag{14.16}$$

If there is a uniform heat flux $\dot{q}_L$ applied to the left side of this wall, $\dot{Q}_x = \dot{q}_L A$ for point 1 and the equation for this point is

$$T_2 - T_1 + \frac{\dot{q}_L}{k}\Delta x + \frac{\dot{g}}{2k}(\Delta x)^2 = 0. \tag{14.17}$$

Similarly, if the right side of this wall is exposed to convection heat transfer with a fluid at temperature T_∞ and heat transfer coefficient h, $\dot{Q}_{x+\Delta x/2} = hA(T_7 - T_\infty)$ for point 7 and the equation for this point is

$$T_6 - \left(1 + \frac{h\Delta x}{k}\right)T_7 + \frac{h\Delta x}{k}T_\infty + \frac{\dot{g}}{2k}(\Delta x)^2 = 0. \tag{14.18}$$

The equations for internal points 2 through 6 plus one equation for each boundary point 1 and 7 constitute a set of seven equations for seven temperatures T_1 to T_7. This set of equations can be written in the following matrix form

$$\begin{bmatrix} -a_{11} & a_{12} & 0 & 0 & 0 & 0 & 0 \\ 1 & -2 & 1 & 0 & 0 & 0 & 0 \\ 0 & 1 & -2 & 1 & 0 & 0 & 0 \\ 0 & 0 & 1 & -2 & 1 & 0 & 0 \\ 0 & 0 & 0 & 1 & -2 & 1 & 0 \\ 0 & 0 & 0 & 0 & 1 & -2 & 1 \\ 0 & 0 & 0 & 0 & 0 & a_{76} & -a_{77} \end{bmatrix} \begin{bmatrix} T_1 \\ T_2 \\ T_3 \\ T_4 \\ T_5 \\ T_6 \\ T_7 \end{bmatrix} = \begin{bmatrix} c_1 \\ -\dot{g}(\Delta x)^2/2k \\ -\dot{g}(\Delta x)^2/2k \\ -\dot{g}(\Delta x)^2/2k \\ -\dot{g}(\Delta x)^2/2k \\ -\dot{g}(\Delta x)^2/2k \\ c_7 \end{bmatrix}, \quad (14.19)$$

where constants a_{11}, a_{12}, a_{76}, a_{77}, c_1, and c_7 depend on the boundary conditions. For example, if the left side of the wall is at uniform temperature T_L, $a_{11} = 1$, $a_{12} = 0$ and $c_1 = -T_L$.

14.2.2 STEADY-STATE, TWO-DIMENSIONAL HEAT CONDUCTION

Consider the rectangular cross section of a long rod as shown in Figure 14.2. If boundary conditions on each side of the rod are uniform along the length of the rod, the heat transfer in the cross section of the rod will be two-dimensional. Let's locate 20 nodes in this domain on a 5×4 network and draw their corresponding control volumes as shown in Figure 14.2. Note that the nodes are indicated by two digit numbers such that the first digit shows the row and the second digit shows the column that a node belongs to. For simplicity we assume that the distances between the points are $\Delta x = L/4$ in x direction and $\Delta y = W/3$ in y direction. Note that Δx and Δy do not have to be the same. The energy balance equation for an internal node with a control volume whose dimensions are Δx, Δy and H is

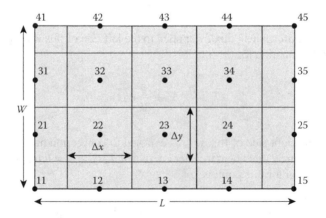

FIGURE 14.2 Cross section of a long rod with a 5×4 node network.

Computer Simulations and Thermal Design

$$\dot{Q}_x - \dot{Q}_{x+\Delta x} + \dot{Q}_y - \dot{Q}_{y+\Delta y} + \dot{g}H\Delta x\Delta y = \rho C_p H \Delta x \Delta y \frac{\Delta T}{\Delta t}, \quad (14.20)$$

where H is the length of the long rod normal to the xy plane.

Similar to the one-dimensional case, consider a steady-state situation where the right-hand side of the above equation disappears and let's assume that thermal conductivity and volumetric heat generation are uniform across the domain. For an internal point, such as point 22, $\dot{Q}_x$, $\dot{Q}_{x+\Delta x}$, $\dot{Q}_y$, and $\dot{Q}_{y+\Delta y}$ can be approximated by the finite difference from the Fourier's law of conduction:

$$\dot{Q}_x = -kH\Delta y \frac{T_{22}-T_{21}}{\Delta x}, \quad \dot{Q}_{x+\Delta x} = -kH\Delta y \frac{T_{23}-T_{22}}{\Delta x}, \quad (14.21)$$

$$\dot{Q}_y = -kH\Delta x \frac{T_{22}-T_{12}}{\Delta y}, \quad \dot{Q}_{y+\Delta y} = -kH\Delta x \frac{T_{32}-T_{22}}{\Delta y}. \quad (14.22)$$

Substituting these into the finite-difference form of the steady-state, two-dimensional heat conduction equation gives

$$\frac{1}{(\Delta x)^2}T_{21} + \frac{1}{(\Delta x)^2}T_{23} + \frac{1}{(\Delta y)^2}T_{12} + \frac{1}{(\Delta y)^2}T_{32} - \left[\frac{2}{(\Delta x)^2} + \frac{2}{(\Delta y)^2}\right]T_{22} + \frac{\dot{g}}{k} = 0.$$

$$(14.23)$$

Note that if $\Delta x = \Delta y$ and there is no heat generation:

$$T_{21} + T_{23} + T_{12} + T_{32} - 4T_{22} = 0, \quad (14.24)$$

which means that temperature at any internal node is the average of the temperatures of its neighboring nodes.

The equations for the boundary nodes are obtained using the appropriate forms of the finite-difference heat conduction equation for those nodes. These are

$$\dot{Q}_x - \dot{Q}_{x+\Delta x/2} + \dot{Q}_y - \dot{Q}_{y+\Delta y} + \dot{g}H\Delta x\Delta y/2 = \rho C_p H \Delta x \Delta y/2 \frac{\Delta T}{\Delta t}$$

for points 21, 31, 25, and 35,

$$\dot{Q}_x - \dot{Q}_{x+\Delta x} + \dot{Q}_y - \dot{Q}_{y+\Delta y/2} + \dot{g}H\Delta x\Delta y/2 = \rho C_p H \Delta x \Delta y/2 \frac{\Delta T}{\Delta t}$$

for points 12–14 and 42–44,

$$\dot{Q}_x - \dot{Q}_{x+\Delta x/2} + \dot{Q}_y - \dot{Q}_{y+\Delta y/2} + \dot{g}H\Delta x\Delta y/4 = \rho C_p H\Delta x\Delta y/4 \frac{\Delta T}{\Delta t}$$

for points 11, 15, 41, and 45.

Let's assume that the left-hand side of this rod is exposed to a fluid with temperature T_∞ and convection heat transfer coefficient h and the bottom side is exposed to a constant heat flux $\dot{q}_b$. Then, for point 21:

$$\dot{Q}_x = hH\Delta y(T_\infty - T_{21}), \qquad \dot{Q}_{x+\Delta x/2} = -kH\Delta y \frac{T_{22} - T_{21}}{\Delta x},$$

$$\dot{Q}_y = -kH \frac{\Delta x}{2} \frac{T_{21} - T_{11}}{\Delta y} \quad \text{and} \quad \dot{Q}_{y+\Delta y} = -kH \frac{\Delta x}{2} \frac{T_{31} - T_{21}}{\Delta y}.$$

(14.25)

Substituting these into the energy balance equation, and assuming steady-state condition, gives the following equation for node 21:

$$\frac{2}{(\Delta x)^2} T_{22} + \frac{1}{(\Delta y)^2} T_{11} + \frac{1}{(\Delta y)^2} T_{31}$$

$$- \left[\frac{2}{(\Delta x)^2} + \frac{2}{(\Delta y)^2} - \frac{2h}{k\Delta x} \right] T_{21} + \frac{\dot{g}}{k} + \frac{2h}{k\Delta x} T_\infty = 0.$$

(14.26)

For node 11:

$$\dot{Q}_x = hH\Delta y/2(T_\infty - T_{11}), \qquad \dot{Q}_{x+\Delta x/2} = -kH\Delta y/2 \frac{T_{12} - T_{11}}{\Delta x},$$

$$\dot{Q}_y = \dot{q}_b H\Delta x/2 \qquad \text{and} \qquad \dot{Q}_{y+\Delta y/2} = -kH \frac{\Delta x}{2} \frac{T_{21} - T_{11}}{\Delta y},$$

(14.27)

and the energy balance equation for this node in steady-state condition results in

$$\frac{1}{(\Delta x)^2} T_{12} + \frac{1}{(\Delta y)^2} T_{21}$$

$$- \left[\frac{1}{(\Delta x)^2} + \frac{1}{(\Delta y)^2} - \frac{h}{k\Delta x} \right] T_{11} + \frac{\dot{g}}{2k} + \frac{h}{k\Delta x} T_\infty + \frac{\dot{q}_b}{k\Delta y} = 0.$$

(14.28)

Similar procedures can be used to obtain equations for other boundary nodes. The result is a set of 20 equations for 20 unknown temperatures that has the following form:

Computer Simulations and Thermal Design

$$\begin{bmatrix} x & x & 0 & 0 & 0 & x & 0 & 0 & 0 & 0 & 0 & 0 & 0 & 0 & 0 & 0 & 0 & 0 & 0 \\ x & x & x & 0 & 0 & 0 & x & 0 & 0 & 0 & 0 & 0 & 0 & 0 & 0 & 0 & 0 & 0 & 0 \\ 0 & x & x & x & 0 & 0 & 0 & x & 0 & 0 & 0 & 0 & 0 & 0 & 0 & 0 & 0 & 0 & 0 \\ 0 & 0 & x & x & x & 0 & 0 & 0 & x & 0 & 0 & 0 & 0 & 0 & 0 & 0 & 0 & 0 & 0 \\ 0 & 0 & 0 & x & x & 0 & 0 & 0 & 0 & x & 0 & 0 & 0 & 0 & 0 & 0 & 0 & 0 & 0 \\ x & 0 & 0 & 0 & 0 & x & x & 0 & 0 & 0 & x & 0 & 0 & 0 & 0 & 0 & 0 & 0 & 0 \\ 0 & x & 0 & 0 & 0 & x & x & x & 0 & 0 & 0 & x & 0 & 0 & 0 & 0 & 0 & 0 & 0 \\ 0 & 0 & x & 0 & 0 & 0 & x & x & x & 0 & 0 & 0 & x & 0 & 0 & 0 & 0 & 0 & 0 \\ 0 & 0 & 0 & x & 0 & 0 & 0 & x & x & x & 0 & 0 & 0 & x & 0 & 0 & 0 & 0 & 0 \\ 0 & 0 & 0 & 0 & x & 0 & 0 & 0 & x & x & 0 & 0 & 0 & 0 & x & 0 & 0 & 0 & 0 \\ 0 & 0 & 0 & 0 & 0 & x & 0 & 0 & 0 & 0 & x & x & 0 & 0 & 0 & x & 0 & 0 & 0 \\ 0 & 0 & 0 & 0 & 0 & 0 & x & 0 & 0 & 0 & x & x & x & 0 & 0 & 0 & x & 0 & 0 \\ 0 & 0 & 0 & 0 & 0 & 0 & 0 & x & 0 & 0 & 0 & x & x & x & 0 & 0 & 0 & x & 0 \\ 0 & 0 & 0 & 0 & 0 & 0 & 0 & 0 & x & 0 & 0 & 0 & x & x & x & 0 & 0 & 0 & x \\ 0 & 0 & 0 & 0 & 0 & 0 & 0 & 0 & 0 & x & 0 & 0 & 0 & x & x & 0 & 0 & 0 & x \\ 0 & 0 & 0 & 0 & 0 & 0 & 0 & 0 & 0 & 0 & x & 0 & 0 & 0 & 0 & x & x & 0 & 0 \\ 0 & 0 & 0 & 0 & 0 & 0 & 0 & 0 & 0 & 0 & 0 & x & 0 & 0 & 0 & x & x & x & 0 \\ 0 & 0 & 0 & 0 & 0 & 0 & 0 & 0 & 0 & 0 & 0 & 0 & x & 0 & 0 & 0 & x & x & x \\ 0 & 0 & 0 & 0 & 0 & 0 & 0 & 0 & 0 & 0 & 0 & 0 & 0 & x & 0 & 0 & 0 & x & x \\ 0 & 0 & 0 & 0 & 0 & 0 & 0 & 0 & 0 & 0 & 0 & 0 & 0 & 0 & x & 0 & 0 & 0 & x \end{bmatrix} \begin{bmatrix} T_{11} \\ T_{12} \\ T_{13} \\ T_{14} \\ T_{15} \\ T_{21} \\ T_{22} \\ T_{23} \\ T_{24} \\ T_{25} \\ T_{31} \\ T_{32} \\ T_{33} \\ T_{34} \\ T_{35} \\ T_{41} \\ T_{42} \\ T_{43} \\ T_{44} \\ T_{45} \end{bmatrix} = \begin{bmatrix} c_{11} \\ c_{12} \\ c_{13} \\ c_{14} \\ c_{15} \\ c_{21} \\ c_{22} \\ c_{23} \\ c_{24} \\ c_{25} \\ c_{31} \\ c_{32} \\ c_{33} \\ c_{34} \\ c_{35} \\ c_{41} \\ c_{42} \\ c_{43} \\ c_{44} \\ c_{45} \end{bmatrix}.$$

(14.29)

Note that the letter x is used only to indicate the generally nonzero elements of the coefficient matrix. This does not mean that these elements are the same. It is seen that the nonzero elements of this matrix are located on five diagonal lines. This clear order helps in storing this matrix in computer memory and in the methods used to solve this matrix equation.

The extension of this analysis to three-dimensional cases is straightforward. Due to the additional dimension, the resulting coefficient matrix will have seven nonzero diagonal lines.

14.2.3 Transient Heat Conduction

Let's consider the two-dimensional domain of Figure 14.2 in a transient condition. The energy equation for an internal node is

$$\dot{Q}_x - \dot{Q}_{x+\Delta x} + \dot{Q}_y - \dot{Q}_{y+\Delta y} + \dot{g} H \Delta x \Delta y = \rho C_p H \Delta x \Delta y \frac{\Delta T}{\Delta t}. \quad (14.30)$$

The time-dependent temperature change on the right-hand side of this equation can be expressed as

$$\frac{\Delta T}{\Delta t} = \frac{T(t+\Delta t) - T(t)}{\Delta t} = \frac{T^{n+1} - T^n}{\Delta t}, \qquad (14.31)$$

where T^n and T^{n+1} refer to temperatures at times t and $t + \Delta t$. The left-hand side of Equation 14.30 can be expressed as in the last section. However, the temperatures have to be at time t, $t + \Delta t$ or a combination of both as described below.

Explicit Method: If the temperatures on the left side of the energy balance equation are at time t, the equation for node 22 will have the following form:

$$\frac{1}{(\Delta x)^2} T_{21}^n + \frac{1}{(\Delta x)^2} T_{23}^n + \frac{1}{(\Delta y)^2} T_{12}^n + \frac{1}{(\Delta y)^2} T_{32}^n - \left[\frac{2}{(\Delta x)^2} + \frac{2}{(\Delta y)^2}\right] T_{22}^n + \frac{\dot{g}}{k}$$
$$= \frac{\rho C_p}{k} \frac{T_{22}^{n+1} - T_{22}^n}{\Delta t}. \qquad (14.32)$$

Similar equations are derived for other nodes. Temperatures at time t (i.e., T^n) are known when temperatures at time $t + \Delta t$ (i.e., T^{n+1}) are calculated. Therefore, the above equation is an explicit equation to calculate T_{22} at time $t + \Delta t$ knowing temperatures at the same and the neighboring nodes at time t. That is called an **explicit method**.

The explicit method is very simple and straightforward. However, it does not work for every arbitrary value of Δx and Δt and may give physically impossible values that may oscillate in sign and increase in value indefinitely. The method is said to be unstable in these situations. To demonstrate this, let's consider the simpler case of the one-dimensional heat conduction with no heat generation. The energy balance equation for an internal node m is

$$\frac{1}{(\Delta x)^2}\left(T_{m+1}^n - 2T_m^n + T_{m-1}^n\right) = \frac{\rho C_p}{k} \frac{T_m^{n+1} - T_m^n}{\Delta t}. \qquad (14.33)$$

The **Fourier number** is defined by

$$\text{Fo} = \frac{\alpha \Delta t}{(\Delta x)^2}, \qquad (14.34)$$

Where $\alpha = k/\rho C_p$ is the fluid thermal diffusivity. Equation 14.33 can be written as

$$T_m^{n+1} = \text{Fo}\left(T_{m+1}^n + T_{m-1}^n\right) + (1 - 2\text{Fo})T_m^n. \qquad (14.35)$$

Computer Simulations and Thermal Design

The **stability criteria** for this equation is

$$1 - 2\text{Fo} \geq 0 \implies \text{Fo} \leq 0.5. \tag{14.36}$$

To demonstrate this, consider a plane wall with an initial temperature of 50°C whose left and right side temperatures are suddenly changed to 200°C. Let's use only three nodes; one interior node and one at each side as shown in Figure 14.3. The $T_1 = T_3 = 200$°C and T_2 has to be calculated as a function of time. It is obvious that T_2 will increase from 50°C to the equilibrium value of 200°C and can not go higher than that. Figure 14.3 shows that

- If Fo > 1, T_2 oscillates back and forth around the equilibrium value of 200°C and gets farther from it as iteration continues.
- If Fo = 1, the value of T_2 oscillates between 50°C and 350°C indefinitely.
- If Fo = 0.5, the correct value of $T_2 = 200$°C is obtained in one iteration.
- If 0.5 < Fo < 1, the value of T_2 will finally converge toward the correct value of 200°C but exceeds that value in some intermediate iteration.
- For any value of Fo < 0.5, T_2 will always remain less than 200°C and approaches this value as n increases. Note that, smaller Fo numbers represent smaller time steps. Therefore, more time steps are required to reach the equilibrium value of T_2.

Note that the stability criteria have to be met for all the points in a domain including the boundary points. The most stringent stability criterion in a domain dictates the largest time step that can be used for that particular problem.

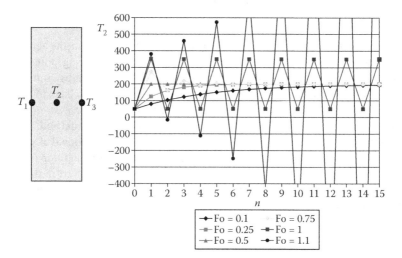

FIGURE 14.3 A thin wall and calculated temperatures at its center at different time steps as a function of the Fourier number.

The stability criteria for the two-dimensional cases are obtained by requiring the coefficient of the T_{mn}^n to be positive for all the nodes. For example, the stability criterion of the equation for node T_{22} is

$$1 - 2\text{Fo}_x - 2\text{Fo}_y \geq 0, \quad (14.37)$$

where $\text{Fo}_x = \alpha \Delta t/(\Delta x)^2$ and $\text{Fo}_y = \alpha \Delta t/(\Delta y)^2$.

Although explicit methods are very simple, their stability criteria may enforce either very small time steps and long computation times or large spatial steps and spatially coarse results.

Implicit Method: If the temperatures on the left side of the energy balance equation are at time $t + \Delta t$, the equation for node 22 will have the following form:

$$\frac{1}{(\Delta x)^2} T_{21}^{n+1} + \frac{1}{(\Delta x)^2} T_{23}^{n+1} + \frac{1}{(\Delta y)^2} T_{12}^{n+1} + \frac{1}{(\Delta y)^2} T_{32}^{n+1}$$

$$- \left[\frac{2}{(\Delta x)^2} + \frac{2}{(\Delta y)^2} \right] T_{22}^{n+1} + \frac{\dot{g}}{k} = \frac{\rho C_p}{k} \frac{T_{22}^{n+1} - T_{22}^n}{\Delta t}. \quad (14.38)$$

Using the notations $\text{Fo}_x = \alpha \Delta t/(\Delta x)^2$ and $\text{Fo}_y = \alpha \Delta t/(\Delta y)^2$, this equation can be written as

$$\text{Fo}_x T_{21}^{n+1} + \text{Fo}_x T_{23}^{n+1} + \text{Fo}_y T_{12}^{n+1} + \text{Fo}_y T_{32}^{n+1}$$

$$- \left[1 + 2\text{Fo}_x + 2\text{Fo}_y \right] T_{22}^{n+1} + \frac{\dot{g}}{k} + T_{22}^n = 0. \quad (14.39)$$

Note that the temperature at node 22, at time $t + \Delta t$ depends on the temperature of this point at time t, as well as temperatures of its neighboring points at time $t + \Delta t$. Thus, it can not be found without knowing those temperatures. Similar equations are derived for other nodes in the domain. These equations have to be solved simultaneously to obtain temperatures at all the nodes at time $t + \Delta t$, given those temperatures at time t. Such an approach is called an **implicit method**. Note that each time step in the implicit method requires the same computational effort as that for a steady-state problem. However, implicit methods are unconditionally stable. Therefore, larger time steps can be used compared to those allowed in the explicit methods and the total computation time to reach a certain physical time may be less with an implicit method.

14.2.4 FLUID FLOW AND ENERGY EQUATIONS

Computer simulations of heat transfer in a moving fluid are more complex than simulating pure conduction in a medium. The continuity equation, three momentum equations and energy equation have to be solved to calculate temperatures and heat

Computer Simulations and Thermal Design

transfer rates in a moving fluid. The problem becomes more complicated due to the lack of an independent equation for the pressure whose gradient appears in the momentum equations. A brief description of how these equations are solved for an incompressible fluid with constant properties is given here and the interested reader is referred to dedicated detailed texts for further explanations [1,2].

The continuity equation, the momentum equations, and the energy equation for incompressible fluids with constant properties are

$$\frac{\partial U}{\partial x} + \frac{\partial V}{\partial y} + \frac{\partial W}{\partial z} = 0, \quad (14.40)$$

$$\frac{\partial U}{\partial t} + \frac{\partial U^2}{\partial x} + \frac{\partial VU}{\partial y} + \frac{\partial WU}{\partial z}$$
$$= -\frac{1}{\rho}\frac{\partial P}{\partial x} + \frac{\partial}{\partial x}\left(\nu\frac{\partial U}{\partial x}\right) + \frac{\partial}{\partial y}\left(\nu\frac{\partial U}{\partial y}\right) + \frac{\partial}{\partial z}\left(\nu\frac{\partial U}{\partial z}\right) + \frac{X}{\rho}, \quad (14.41)$$

$$\frac{\partial V}{\partial t} + \frac{\partial UV}{\partial x} + \frac{\partial V^2}{\partial y} + \frac{\partial WV}{\partial z}$$
$$= -\frac{1}{\rho}\frac{\partial P}{\partial y} + \frac{\partial}{\partial x}\left(\nu\frac{\partial V}{\partial x}\right) + \frac{\partial}{\partial y}\left(\nu\frac{\partial V}{\partial y}\right) + \frac{\partial}{\partial z}\left(\nu\frac{\partial V}{\partial z}\right) + \frac{Y}{\rho}, \quad (14.42)$$

$$\frac{\partial W}{\partial t} + \frac{\partial UW}{\partial x} + \frac{\partial VW}{\partial y} + \frac{\partial W^2}{\partial z}$$
$$= -\frac{1}{\rho}\frac{\partial P}{\partial z} + \frac{\partial}{\partial x}\left(\nu\frac{\partial W}{\partial x}\right) + \frac{\partial}{\partial y}\left(\nu\frac{\partial W}{\partial y}\right) + \frac{\partial}{\partial z}\left(\nu\frac{\partial W}{\partial z}\right) + \frac{Z}{\rho}, \quad (14.43)$$

$$\frac{\partial T}{\partial t} + \frac{\partial UT}{\partial x} + \frac{\partial VT}{\partial y} + \frac{\partial WT}{\partial z}$$
$$= \frac{\partial}{\partial x}\left(\alpha\frac{\partial T}{\partial x}\right) + \frac{\partial}{\partial y}\left(\alpha\frac{\partial T}{\partial y}\right) + \frac{\partial}{\partial z}\left(\alpha\frac{\partial T}{\partial z}\right) + \frac{\mu\Phi + \dot{g}}{\rho C_p}. \quad (14.44)$$

These equations can be expressed in the following general form:

$$\frac{\partial \phi}{\partial t} + \frac{\partial U\phi}{\partial x} + \frac{\partial V\phi}{\partial y} + \frac{\partial W\phi}{\partial z} = \frac{\partial}{\partial x}\left(\Gamma_\phi\frac{\partial \phi}{\partial x}\right) + \frac{\partial}{\partial y}\left(\Gamma_\phi\frac{\partial \phi}{\partial y}\right) + \frac{\partial}{\partial z}\left(\Gamma_\phi\frac{\partial \phi}{\partial z}\right) + S_\phi, \quad (14.45)$$

where the general dependent variable ϕ, the diffusion coefficient Γ_ϕ, and the source term S_ϕ are defined in Table 14.1.

TABLE 14.1
Values of the General Variable ϕ, the Diffusion Coefficient Γ_ϕ, and the Source Term S_ϕ

Equation	ϕ	Γ_ϕ	S_ϕ
Continuity	1	0	0
x-Momentum	U	ν	$-\dfrac{1}{\rho}\dfrac{\partial P}{\partial x}+\dfrac{X}{\rho}$
y-Momentum	V	ν	$-\dfrac{1}{\rho}\dfrac{\partial P}{\partial y}+\dfrac{Y}{\rho}$
z-Momentum	W	ν	$-\dfrac{1}{\rho}\dfrac{\partial P}{\partial z}+\dfrac{Z}{\rho}$
Energy	T	α	$\dfrac{\mu\phi+\dot{g}}{\rho C_p}$

Equation 14.45 is integrated over individual control volumes around each node in a domain, and over the time interval Δt, to obtain the corresponding finite-volume form of the equation. Let's consider the two-dimensional case for illustration as the extension to three dimensions is straightforward. The control volume in this case has dimensions Δx and Δy. The integration of the transport equation for ϕ over this control volume and over the time interval Δt is described below.

1. The transient term, $\partial\phi/\partial t$, is considered to be constant over the control volume. Therefore:

$$\int_x^{x+\Delta x}\int_y^{y+\Delta y}\int_t^{t+\Delta t}\frac{\partial\phi}{\partial t}dxdydt=\left(\phi^{n+1}-\phi^n\right)\Delta x\Delta y, \qquad (14.46)$$

where ϕ^n and ϕ^{n+1} represent values of ϕ at times t and $t+\Delta t$, respectively.

2. The convection terms in the x and y directions are considered to be constant along the y and x directions, respectively, and during the time interval Δt. The implicit method is used, which means that the values of all variables are those at time $t+\Delta t$:

$$\int_x^{x+\Delta x}\int_y^{y+\Delta y}\int_t^{t+\Delta t}\frac{\partial U\phi}{\partial x}dxdydt=\left(\left(U\phi\right)_{x+\Delta x}^{n+1}-\left(U\phi\right)_x^{n+1}\right)\Delta y\Delta t \qquad (14.47)$$

$$\int_x^{x+\Delta x}\int_y^{y+\Delta y}\int_t^{t+\Delta t}\frac{\partial V\phi}{\partial y}dxdydt=\left(\left(V\phi\right)_{y+\Delta y}^{n+1}-\left(V\phi\right)_y^{n+1}\right)\Delta x\Delta t. \qquad (14.48)$$

Computer Simulations and Thermal Design

U and V are the x and y components of the fluid velocity at the boundaries of a control volume. A so-called upwind scheme is used to determine the value of ϕ at a boundary of a control volume. This means that, for example, if velocity at a boundary is from left to right, ϕ at that boundary will be equal to ϕ at the first node on the left side of the boundary.

3. The diffusion terms in the x and y directions are considered to be constant along the y and x directions, respectively, and during the time interval Δt. The implicit method is used, which means that the values of all parameters are those at time $t + \Delta t$:

$$\int_x^{x+\Delta x} \int_y^{y+\Delta y} \int_t^{t+\Delta t} \frac{\partial}{\partial x}\left(\Gamma_\phi \frac{\partial \phi}{\partial x}\right) dx\, dy\, dt \\ = \left(\left(\Gamma_\phi \frac{\partial \phi}{\partial x}\right)_{x+\Delta x}^{n+1} - \left(\Gamma_\phi \frac{\partial \phi}{\partial x}\right)_x^{n+1}\right) \Delta y \Delta t, \quad (14.49)$$

$$\int_x^{x+\Delta x} \int_y^{y+\Delta y} \int_t^{t+\Delta t} \frac{\partial}{\partial y}\left(\Gamma_\phi \frac{\partial \phi}{\partial y}\right) dx\, dy\, dt \\ = \left(\left(\Gamma_\phi \frac{\partial \phi}{\partial y}\right)_{y+\Delta y}^{n+1} - \left(\Gamma_\phi \frac{\partial \phi}{\partial y}\right)_y^{n+1}\right) \Delta x \Delta t. \quad (14.50)$$

Piecewise linear profiles are assumed for the variations of ϕ between two adjacent nodes and therefore derivatives of ϕ with respect to x and y are equal to slopes of such lines. The value of the diffusion coefficient, Γ_ϕ, is a suitable average of the values of this parameter at the neighboring points.

4. The source term is simply integrated by assuming an average value, $\overline{S_\phi}$, over the control volume and during the time interval Δt:

$$\int_x^{x+\Delta x} \int_y^{y+\Delta y} \int_t^{t+\Delta t} S_\phi\, dx\, dy\, dt = \overline{S_\phi} \Delta x \Delta y \Delta t. \quad (14.51)$$

Figure 14.4 shows a two-dimensional computational network. The nodes for U and V velocities are indicated by small arrows in the x and y directions, respectively, and the nodes for all other variables including temperature, pressure, and fluid properties are indicated by small circles. This is called a staggered node arrangement and results in separate control volumes for different variables as shown in Figure 14.4 [1].

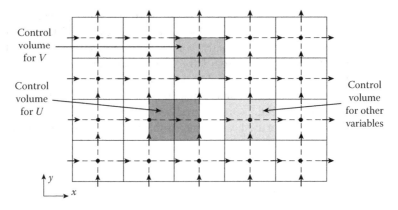

FIGURE 14.4 A two-dimensional computational domain with staggered nodes and control volumes.

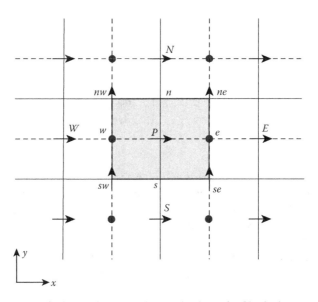

FIGURE 14.5 A typical two-dimensional control volume for U velocity.

Let's examine the use of this method by obtaining the finite-difference form of the x-momentum equation (i.e., $\phi = U$). Figure 14.5 shows a typical two-dimensional control volume for U velocity. Let's assign the letter P for the node located at the center of this control-volume (i.e., the node for which we will obtain the finite-difference equation) and letters W, E, N, and S for the nodes on the immediate left, right, top, and bottom of this node, respectively. Let's also assign small letters w, e, n, and s to the centers of the left, right, top, and bottom sides of the control volume, and nw, ne, sw, and se for the corners of the control volume as shown in Figure 14.5.

Computer Simulations and Thermal Design

Finite-difference forms of different terms of the x-momentum equation are obtained as described below.

- Transient term:

$$\left(\phi_P^{n+1} - \phi_P^n\right)\Delta x \Delta y = \left(U_P^{n+1} - U_P^n\right)\Delta x \Delta y \tag{14.52}$$

- Convection term (x direction): Assume that the function $\max(A,B)$ returns the maximum value of A and B. Then:

$$\left((U\phi)_{x+\Delta x}^{n+1} - (U\phi)_x^{n+1}\right)\Delta y \Delta t = \left[\max(U_e^{n+1},0)U_P^{n+1} - \max(-U_e^{n+1},0)U_E^{n+1}\right.$$

$$\left. - \max(U_w^{n+1},0)U_W^{n+1} + \max(-U_w^{n+1},0)U_P^{n+1}\right]\Delta y \Delta t. \tag{14.53}$$

Note that $U_{x+\Delta x} = U_e$ and $U_x = U_w$ are appropriate weighted averages of U_P and U_E, and U_P and U_W, respectively. If the boundaries are located midway between the nodes, $U_e = 0.5(U_P + U_E)$ and $U_w = 0.5(U_P + U_W)$. However, an upwind value was assigned to the value of $\phi = U$ at each boundary.

- Convection term (y direction):

$$\left((V\phi)_{y+\Delta y}^{n+1} - (V\phi)_y^{n+1}\right)\Delta x \Delta t = \left[\max(V_n^{n+1},0)U_P^{n+1} - \max(-V_n^{n+1},0)U_N^{n+1}\right.$$

$$\left. - \max(V_s^{n+1},0)U_S^{n+1} + \max(-V_s^{n+1},0)U_P^{n+1}\right]\Delta x \Delta t. \tag{14.54}$$

Note again that $V_{y+\Delta y} = V_n$ and $V_y = V_s$ are appropriate weighted averages of V_{nw} and V_{ne}, and V_{sw} and V_{se}, respectively. If the boundaries are located midway between the nodes, $V_n = 0.5(V_{nw} + V_{ne})$ and $V_s = 0.5(V_{sw} + V_{se})$. Similar to the x direction, an upwind scheme was used for $\phi = U$.

- Diffusion term (x direction):

$$\left(\left(\Gamma_\phi \frac{\partial \phi}{\partial x}\right)_{x+\Delta x}^{n+1} - \left(\Gamma_\phi \frac{\partial \phi}{\partial x}\right)_x^{n+1}\right)\Delta y \Delta t = \left(\left(v\frac{\partial U}{\partial x}\right)_e^{n+1} - \left(v\frac{\partial U}{\partial x}\right)_w^{n+1}\right)\Delta y \Delta t$$

$$= \left(v_e \frac{U_E^{n+1} - U_P^{n+1}}{\Delta x} - v_w \frac{U_P^{n+1} - U_W^{n+1}}{\Delta x}\right)\Delta y \Delta t. \tag{14.55}$$

Note that we assumed Δx to be constant on the left and right sides of point P. The extension to a nonuniform node distribution is straightforward. Since fluid properties such as viscosity, v, are given at nodes indicated by circles in Figure 14.5, no interpolations are required to find v_e and v_w.

- Diffusion term (y direction):

$$\left(\left(\Gamma_\phi \frac{\partial \phi}{\partial y}\right)^{n+1}_{y+\Delta y} - \left(\Gamma_\phi \frac{\partial \phi}{\partial y}\right)^{n+1}_y\right)\Delta x \Delta t = \left(\left(v\frac{\partial U}{\partial y}\right)^{n+1}_n - \left(v\frac{\partial U}{\partial y}\right)^{n+1}_s\right)\Delta x \Delta t$$

$$= \left(v_n \frac{U^{n+1}_N - U^{n+1}_P}{\Delta y} - v_s \frac{U^{n+1}_P - U^{n+1}_S}{\Delta y}\right)\Delta x \Delta t.$$

(14.56)

Again, we assume Δy is constant on the top and bottom sides of point P. Values of v_n and v_s are calculated by taking the average of v at their respective four neighboring nodes.

- Source term: The source term includes a pressure gradient term and a body force term:

$$\overline{S_\phi} \Delta x \Delta y \Delta t = \left(-\frac{1}{\rho}\frac{\partial P}{\partial x} + \frac{X}{\rho}\right)\Delta x \Delta y \Delta t = \left(-\frac{1}{\rho_P}\frac{P_e - P_w}{\Delta x} + \frac{X}{\rho_P}\right)\Delta x \Delta y \Delta t$$

$$= \frac{1}{\rho_P}\left(\frac{P_w - P_e}{\Delta x} + \overline{X}\right)\Delta x \Delta y \Delta t.$$

(14.57)

Note that a piecewise linear profile was assumed for the variation of pressure in a control volume and therefore the average value of pressure gradient is equal to the slope of this line. The density at node P is equal to the weighted average of the densities at nodes e and w. If node P is at the center of the control volume, $\rho_P = 0.5(\rho_w + \rho_e)$. $\overline{X}$ is a suitable average of the body force per unit volume in the control volume.

A finite-volume form of the x-momentum equation is obtained by combining all these terms:

$$\left(U^{n+1}_P - U^n_P\right)\Delta x \Delta y$$

$$+ \left[\max(U^{n+1}_e, 0)U^{n+1}_P - \max(-U^{n+1}_e, 0)U^{n+1}_E - \max(U^{n+1}_w, 0)U^{n+1}_W + \max(-U^{n+1}_w, 0)U^{n+1}_P\right]\Delta y \Delta t$$

$$+ \left[\max(V^{n+1}_n, 0)U^{n+1}_P - \max(-V^{n+1}_n, 0)U^{n+1}_N - \max(V^{n+1}_s, 0)U^{n+1}_S + \max(-V^{n+1}_s, 0)U^{n+1}_P\right]\Delta x \Delta t$$

$$= \left(v_e \frac{U^{n+1}_E - U^{n+1}_P}{\Delta x} - v_w \frac{U^{n+1}_P - U^{n+1}_W}{\Delta x}\right)\Delta y \Delta t + \left(v_n \frac{U^{n+1}_N - U^{n+1}_P}{\Delta y} - v_s \frac{U^{n+1}_P - U^{n+1}_S}{\Delta y}\right)\Delta x \Delta t$$

$$+ \frac{1}{\rho_P}\left(\frac{P_w - P_e}{\Delta x} + \overline{X}\right)\Delta x \Delta y \Delta t.$$

Computer Simulations and Thermal Design

This equation can be written in the following form:

$$a_P U_P^{n+1} + a_E U_E^{n+1} + a_W U_W^{n+1} + a_N U_N^{n+1} + a_S U_S^{n+1} = c, \quad (14.58)$$

where

$$a_E = -\max(-U_e^{n+1}, 0)\Delta y - \frac{v_e \Delta y}{\Delta x},$$

$$a_W = -\max(U_w^{n+1}, 0)\Delta y - \frac{v_w \Delta y}{\Delta x},$$

$$a_N = -\max(-V_n^{n+1}, 0)\Delta x - \frac{v_n \Delta x}{\Delta y},$$

$$a_S = -\max(V_s^{n+1}, 0)\Delta x - \frac{v_s \Delta x}{\Delta y},$$

$$\quad (14.59)$$

$$a_P = \frac{\Delta x \Delta y}{\Delta t} + \max(U_e^{n+1}, 0)\Delta y + \max(-U_w^{n+1}, 0)\Delta y + \max(V_n^{n+1}, 0)\Delta x$$

$$+ \max(-V_s^{n+1}, 0)\Delta x + \frac{(v_e + v_w)\Delta y}{\Delta x} + \frac{(v_n + v_s)\Delta x}{\Delta y},$$

$$c = U_P^n \frac{\Delta x \Delta y}{\Delta t} + \frac{1}{\rho_P}\left(\frac{P_w - P_e}{\Delta x} + \bar{X}\right)\Delta x \Delta y.$$

Similar equations can be derived for all the U and V nodes in the computational domain and the set of the equations have to be solved simultaneously to obtain U and V velocities at those points.

The finite-difference equation for the x-momentum equation looks very similar to the finite-difference equation for the two-dimensional heat conduction equation. However, there are two important differences:

1. Some of the coefficients are functions of U. This will require an iterative procedure in which the current velocity distribution is used to calculate those coefficients and a new velocity distribution is calculated. Then, the newly calculated velocities are used to update those coefficients. This process continues until the velocity change from one iteration to the next becomes very small.
2. The major difference between the finite-difference equations for x- momentum and heat conduction equations is that pressure appears in the x-momentum equation while there is no explicit transport equation for obtaining pressure. The pressure field is indirectly specified via the continuity equation; if the correct pressure field is substituted into the momentum equations, the resulting velocity field satisfies the continuity equation. This observation is used to obtain an equation for pressure from the continuity equation as described below.

Let's assume that the solution of the x-momentum equation with prescribed pressure field, P^*, is U^*. Then, U^* satisfies the finite-difference equation

$$a_P U_P^{*n+1} + a_E U_E^{*n+1} + a_W U_W^{*n+1} + a_N U_N^{*n+1} + a_S U_S^{*n+1} = C + \frac{\Delta y}{\rho_P}(P_w^* - P_e^*). \quad (14.60)$$

The correct pressure and x-component velocity are P and U. Therefore

$$a_P U_P^{n+1} + a_E U_E^{n+1} + a_W U_W^{n+1} + a_N U_N^{n+1} + a_S U_S^{n+1} = C + \frac{\Delta y}{\rho_P}(P_w - P_e). \quad (14.61)$$

If the pressure and velocity corrections P' and U' are defined as

$$P = P^* + P',$$
$$U = U^* + U', \quad (14.62)$$

a finite-difference equation for U' is obtained by subtracting the equation for U^* from that for U:

$$a_P U'^{n+1}_P + a_E U'^{n+1}_E + a_W U'^{n+1}_W + a_N U'^{n+1}_N + a_S U'^{n+1}_S = \frac{\Delta y}{\rho_P}(P'_w - P'_e). \quad (14.63)$$

This equation is usually simplified by assuming that velocity corrections at neighboring nodes have no effect on the velocity of the node under consideration [1]. This is not strictly true. However, as will be seen shortly, pressure and velocity corrections diminish as the solution procedure converges to the true pressure and velocity fields and this assumption does not cause an error. This assumption results in

$$U'^{n+1}_P = \frac{\Delta y}{\rho_P a_P}(P'_w - P'_e), \quad (14.64)$$

and

$$U^{n+1}_P = U_P^{*n+1} + \frac{\Delta y}{\rho_P a_P}(P'_w - P'_e). \quad (14.65)$$

This is called the velocity correction formula. Similar equations can be derived for U and V velocities at other nodes.

A finite-difference form for the continuity equation is obtained by integrating this equation over a control volume similar to the one used for temperature and pressure. This control volume is staggered with respect to the control volume used

Computer Simulations and Thermal Design

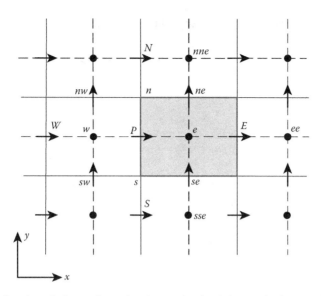

FIGURE 14.6 A typical two-dimensional control volume for continuity equation.

for the x-component of velocity and is shown in Figure 14.6. The integration of the continuity over this control volume gives

$$(U_E^{n+1} - U_P^{n+1})\Delta y + (V_{ne}^{n+1} - V_{se}^{n+1})\Delta x = 0, \tag{14.66}$$

where U_P^{n+1} is given above and

$$U_E^{n+1} = U_E^{*n+1} + \frac{\Delta y}{\rho_E a_E}(P'_e - P'_{ee}), \tag{14.67}$$

$$V_{ne}^{n+1} = V_{ne}^{*n+1} + \frac{\Delta x}{\rho_{ne} a_{ne}}(P'_e - P'_{nne}), \tag{14.68}$$

$$V_{se}^{n+1} = V_{se}^{*n+1} + \frac{\Delta x}{\rho_{se} a_{se}}(P'_{sse} - P'_e). \tag{14.69}$$

Substituting these in the finite-difference form of the continuity equation, Equation 14.66, gives

$$(U_E^{*n+1} - U_P^{*n+1})\Delta y + (V_{ne}^{*n+1} - V_{se}^{*n+1})\Delta x + \frac{(\Delta y)^2}{\rho_E a_E}(P'_e - P'_{ee})$$
$$- \frac{(\Delta y)^2}{\rho_P a_P}(P'_w - P'_e) + \frac{(\Delta x)^2}{\rho_{ne} a_{ne}}(P'_e - P'_{nne}) - \frac{(\Delta x)^2}{\rho_{se} a_{se}}(P'_{sse} - P'_e) = 0. \tag{14.70}$$

This equation can be written in the following compact form:

$$a_e P'_e + a_{ee} P'_{ee} + a_w P'_w + a_{nne} P'_{nne} + a_{sse} P'_{sse} = b, \quad (14.71)$$

where

$$a_{ee} = -\frac{(\Delta y)^2}{\rho_E a_E},$$

$$a_w = -\frac{(\Delta y)^2}{\rho_P a_P}.$$

$$a_{nne} = -\frac{(\Delta x)^2}{\rho_{ne} a_{ne}},$$

$$a_{sse} = -\frac{(\Delta x)^2}{\rho_{se} a_{se}},$$

$$a_p = -a_{ee} - a_w - a_{nne} - a_{sse},$$

$$b = -(U_E^{*n+1} - U_P^{*n+1})\Delta y - (V_{ne}^{*n+1} - V_{se}^{*n+1})\Delta x. \quad (14.72)$$

This is a finite-difference equation for the pressure correction P'. Note that the source term is the negative of the finite-difference form of the continuity equation. If the source term, b, is zero everywhere in the computational domain, the velocity fields U^* and V^* satisfy the continuity equation, and therefore are the same as U and V. On the other hand, if $b = 0$ everywhere in the domain, the pressure corrections would all be zero and $P = P^*$ everywhere in the domain.

The procedure described above for the calculation of the flow field is called SIMPLE, which stands for Semi-Implicit Method for Pressure-Linked Equations [1]. The sequence of operations will be given here for reference:

1. Guess the pressure field P^*
2. Solve the momentum equations to obtain U^*, V^* (and in the three-dimensional cases W^*)
3. Solve the equation for pressure correction P'
4. Calculate $P = P^* + P'$
5. Calculate U, V (and in the three-dimensional cases W) from their starred values and pressure correction P' using the velocity-correction formulas similar to those given above
6. Solve the equations for other ϕ's such as temperature if they affect the flow field through fluid properties

7. Treat the pressure field P as the new guessed pressure P^* and return to step 2. Repeat the whole procedure until the guessed pressure P^* is the same as the calculated pressure P

14.3 TURBULENT FLOWS

All turbulent flows are three-dimensional, highly fluctuating and unsteady, and include various features that are characterized by a large range of length and time scales. For example, consider turbulent flow inside a pipe. The net flow is obviously one-dimensional and along the pipe direction. However, as shown in the famous nineteenth-century experiment by Osborne Reynolds, there are three-dimensional and unsteady velocity fluctuations that cause the spreading of the die shown in Figure 14.7 [3]. The largest length scales are on the order of the pipe diameter and the length of the flow under consideration. The largest time scale is on the order of the time it takes for the flow to pass through the length of the pipe under consideration. On the other hand, the smallest time and length scales are on the order of the period and the size of the turbulent fluctuations that are also called turbulent eddies.

The method described in Section 14.2.4 for solving flow and energy equations are applicable to both laminar and turbulent flows. This method, when used for turbulent flows, is called Direct Numerical Simulation (DNS). However, because of the large difference between smallest and largest length and time scales, any computational domain has to be larger than the largest scale while the control volumes have to be smaller than the smallest scales. This will require a domain with many nodes and many small time steps and consequently will require a significant amount of computational time. The great progress in computer power and efficient numerical algorithms over the last few decades have made DNS possible for some flows.

But, even with the rapid improvement in the computational capabilities of computers, DNS is still limited to simple flows and geometries and can not be used for more sophisticated engineering applications. Methods based on statistical **turbulence modeling** are used in most of these cases. Turbulence models are based on the observations that turbulent flows have orderly large-scale structures despite the small-scale fluctuations and apparent randomness. Therefore, a general variable ϕ can be written as the sum of an average value, $\bar{\phi}$, and a fluctuation about that average value, ϕ';

FIGURE 14.7 Repetitions of Reynolds experiment shows the die injected at the center of a pipe with turbulent flow is spread throughout the cross section of the pipe by turbulent velocity fluctuations. (Courtesy of Johannesen and Lowe.) (From Van Dyke, M. *An Album of Fluid Motion*. Stanford, CA: The Parabolic Press, 1982.)

$$\phi = \bar{\phi} + \phi'. \tag{14.73}$$

For a flow with a steady large scale structures, the average value is defined as

$$\bar{\phi} = \lim_{T \to \infty} \frac{1}{T} \int_0^T \phi \, dt. \tag{14.74}$$

T is the averaging time interval and must be large compared to the typical time scale of the fluctuations. If *T* is large enough, $\bar{\phi}$ will not depend on the time at which the averaging is taken.

For flows with unsteady large scale structures, the time averaging can not be used. Instead, the ensemble averaging is defined as

$$\bar{\phi} = \lim_{N \to \infty} \frac{1}{N} \sum_{n=1}^{N} \phi, \tag{14.75}$$

where *N* is the number of observations of ϕ that makes the ensemble (an imagined set of flows in which all controllable variables are identical). *N* must be large enough to eliminate the effects of fluctuations.

Using either of the above averaging schemes, velocity, temperature, and pressure in a turbulent flow can be expressed as

$$U = \bar{U} + U' \quad V = \bar{V} + V' \quad W = \bar{W} + W' \quad P = \bar{P} + P' \quad T = \bar{T} + T'. \tag{14.76}$$

Substituting these into the momentum and temperature equations and taking the average of those equations results in the following set of equations for the averaged velocities, pressure, and temperature:

$$\frac{\partial \bar{U}}{\partial x} + \frac{\partial \bar{V}}{\partial y} + \frac{\partial \bar{W}}{\partial z} = 0, \tag{14.77}$$

$$\frac{\partial \bar{U}}{\partial t} + \bar{U}\frac{\partial \bar{U}}{\partial x} + \bar{V}\frac{\partial \bar{U}}{\partial y} + \bar{W}\frac{\partial \bar{U}}{\partial z}$$
$$= -\frac{1}{\rho}\frac{\partial \bar{P}}{\partial x} + \nu\left(\frac{\partial^2 \bar{U}}{\partial x^2} + \frac{\partial^2 \bar{U}}{\partial y^2} + \frac{\partial^2 \bar{U}}{\partial z^2}\right) - \frac{\overline{\partial U'^2}}{\partial x} - \frac{\overline{\partial U'V'}}{\partial y} - \frac{\overline{\partial U'W'}}{\partial z} + \frac{\bar{X}}{\rho}, \tag{14.78}$$

$$\frac{\partial \bar{V}}{\partial t} + \bar{U}\frac{\partial \bar{V}}{\partial x} + \bar{V}\frac{\partial \bar{V}}{\partial y} + \bar{W}\frac{\partial \bar{V}}{\partial z}$$

$$= -\frac{1}{\rho}\frac{\partial \bar{P}}{\partial y} + \nu\left(\frac{\partial^2 \bar{V}}{\partial x^2} + \frac{\partial^2 \bar{V}}{\partial y^2} + \frac{\partial^2 \bar{V}}{\partial z^2}\right) - \frac{\partial \overline{U'V'}}{\partial x} - \frac{\partial \overline{V'^2}}{\partial y} - \frac{\partial \overline{V'W'}}{\partial z} + \frac{\bar{Y}}{\rho}, \quad (14.79)$$

$$\frac{\partial \bar{W}}{\partial t} + \bar{U}\frac{\partial \bar{W}}{\partial x} + \bar{V}\frac{\partial \bar{W}}{\partial y} + \bar{W}\frac{\partial \bar{W}}{\partial z}$$

$$= -\frac{1}{\rho}\frac{\partial \bar{P}}{\partial z} + \nu\left(\frac{\partial^2 \bar{W}}{\partial x^2} + \frac{\partial^2 \bar{W}}{\partial y^2} + \frac{\partial^2 \bar{W}}{\partial z^2}\right) - \frac{\partial \overline{U'W'}}{\partial x} - \frac{\partial \overline{V'W'}}{\partial y} - \frac{\partial \overline{W'^2}}{\partial z} + \frac{\bar{Z}}{\rho}, \quad (14.80)$$

$$\frac{\partial \bar{T}}{\partial t} + \bar{U}\frac{\partial \bar{T}}{\partial x} + \bar{V}\frac{\partial \bar{T}}{\partial y} + \bar{W}\frac{\partial \bar{T}}{\partial z}$$

$$= \alpha\left(\frac{\partial^2 \bar{T}}{\partial x^2} + \frac{\partial^2 \bar{T}}{\partial y^2} + \frac{\partial^2 \bar{T}}{\partial z^2}\right) - \frac{\partial \overline{U'T'}}{\partial x} - \frac{\partial \overline{V'T'}}{\partial y} - \frac{\partial \overline{W'T'}}{\partial z} + \frac{\nu\Phi}{C_p} + \frac{\bar{g}}{\rho C_p}. \quad (14.81)$$

These equations include nine new variables. These are the averages of the products of the fluctuating components of velocities and temperature; the terms involving velocity fluctuations only are called **turbulent momentum fluxes** or **Reynolds stresses** and those involving velocity and temperature fluctuations are known as **turbulent heat fluxes**. Since no new equation has been introduced, the result is five equations and 14 unknowns. This is called the **closure problem**. **Turbulence models** have been developed to approximate these new models as functions of the averaged velocities and temperature. Most of these models are semiempirical, meaning that they are based on a mix of experimental observations and theoretical arguments.

Many turbulence models are based on an analogy between laminar diffusion of momentum and heat and their corresponding turbulent fluxes. Consider a slightly different form of the *x*-momentum equation, Equation 14.78,

$$\frac{\partial \bar{U}}{\partial t} + \bar{U}\frac{\partial \bar{U}}{\partial x} + \bar{V}\frac{\partial \bar{U}}{\partial y} + \bar{W}\frac{\partial \bar{U}}{\partial z}$$

$$= -\frac{1}{\rho}\frac{\partial \bar{P}}{\partial x} + \frac{\partial}{\partial x}\left(\nu\frac{\partial \bar{U}}{\partial x} - \overline{U'^2}\right) + \frac{\partial}{\partial y}\left(\nu\frac{\partial \bar{U}}{\partial y} - \overline{U'V'}\right) + \frac{\partial}{\partial z}\left(\nu\frac{\partial \bar{U}}{\partial z} - \overline{U'W'}\right) + \frac{\bar{X}}{\rho}. \quad (14.82)$$

Note that the turbulent momentum fluxes and viscous shears appear together in the same terms on the right side of this equation. On the other hand, as viscous shears are the results of molecular diffusion of momentum, Reynolds stresses are the consequences of transfer of fluctuating momentums by velocity fluctuations. This analogy suggest defining **turbulent** or **eddy viscosity**, ν_t, such that

$$-\overline{U'^2} = 2\nu_t \frac{\partial \overline{U}}{\partial x} - \frac{2}{3}k, \qquad (14.83)$$

$$-\overline{V'^2} = 2\nu_t \frac{\partial \overline{V}}{\partial y} - \frac{2}{3}k, \qquad (14.84)$$

$$-\overline{W'^2} = 2\nu_t \frac{\partial \overline{W}}{\partial z} - \frac{2}{3}k, \qquad (14.85)$$

$$-\overline{U'V'} = \nu_t \left(\frac{\partial \overline{U}}{\partial x} + \frac{\partial \overline{V}}{\partial y} \right), \qquad (14.86)$$

$$-\overline{U'W'} = \nu_t \left(\frac{\partial \overline{U}}{\partial z} + \frac{\partial \overline{W}}{\partial z} \right), \qquad (14.87)$$

$$-\overline{V'W'} = \nu_t \left(\frac{\partial \overline{V}}{\partial z} + \frac{\partial \overline{W}}{\partial y} \right), \qquad (14.88)$$

where

$$k = (\overline{U'^2} + \overline{V'^2} + \overline{W'^2})/2 \qquad (14.89)$$

is the turbulent kinetic energy, and defining turbulent diffusivity, α_t, or turbulent Prandtl number, $Pr_t = \nu_t/\alpha_t$, such that

$$-\overline{U'T'} = \alpha_t \frac{\partial \overline{T}}{\partial x} = \frac{\nu_t}{Pr_t} \frac{\partial \overline{T}}{\partial x}, \qquad (14.90)$$

$$-\overline{V'T'} = \alpha_t \frac{\partial \overline{T}}{\partial y} = \frac{\nu_t}{Pr_t} \frac{\partial \overline{T}}{\partial y}, \qquad (14.91)$$

$$-\overline{W'T'} = \alpha_t \frac{\partial \overline{T}}{\partial z} = \frac{\nu_t}{Pr_t} \frac{\partial \overline{T}}{\partial z}. \qquad (14.92)$$

It is seen that the problem of determining turbulent shear stresses and turbulent heat fluxes was reduced to the problem of determining the turbulent viscosity and turbulent diffusivity (or Prandtl number). Some of the most commonly used models for determining these variables are described below [4,5].

Computer Simulations and Thermal Design

- **Zero-Equation Model**

  The simplest model for turbulent viscosity sets it equal to the product of a length scale and a velocity scale. The length scale l, which has yet to be determined, is a measure of the distance the effects of turbulent mixing penetrates into the flow and therefore is called the **mixing length**. The velocity scale is defined as the product of the mixing length and the modulus of the mean strain rate tensor, defined as

  $$S = \sqrt{2\overline{S_{ij}}\,\overline{S_{ij}}}. \tag{14.93}$$

  The mean strain rate tensor is defined by

  $$\overline{S_{ij}} = \frac{1}{2}\left(\frac{\partial \overline{U_i}}{\partial x_j} + \frac{\partial \overline{U_j}}{\partial x_i}\right). \tag{14.94}$$

  Turbulent viscosity is therefore equal to

  $$\nu_t = l^2 S. \tag{14.95}$$

  Since no new differential equation was used to determine turbulent viscosity, this model is called **zero-equation** or **mixing-length model**. If d is the distance from the nearest surface, the mixing length is specified by the functional form:

  $$l = \max(\kappa d, 0.09 d_{\max}), \tag{14.96}$$

  where $\kappa = 0.41$ is the von Karman constant.

  Note that the turbulent viscosity for two-dimensional boundary layers with boundary layer thickness δ is given by

  $$\nu_t = l^2 \left|\frac{\partial \overline{U}}{\partial y}\right|, \tag{14.97}$$

  where $l = \max(\kappa y, 0.09\delta)$.

  A value of $Pt_t = 0.85$–0.9 gives good results for turbulent heat flux in boundary layers, pipe flows, and other similar flows.

- **One-Equation Models**

  There are two problems with the mixing-length model. (1) Turbulent viscosity will be zero where the modulus of the mean strain rate tensor is zero. However, it has been experimentally shown that, turbulent viscosity at the center of a plane channel is very high [4]. This discrepancy is not of much direct importance to computing the mean velocity. However, for a channel flow that is heated on one side and cooled on the other side, the mixing-length assumption results in an erroneous temperature distribution. (2) The other main defect of the mixing-length model is that it does not account for upstream or downstream effects on turbulent viscosity. It is determined by a length scale and mean strain rate at the point of interest only. However, convection and diffusion from upstream or downstream locations will influence the value of turbulent viscosity at a point.

One proposal to overcome these shortcomings is to use the square root of turbulent kinetic energy, k, as the velocity scale and to define turbulent viscosity as

$$v_t = C_\mu L \sqrt{k}, \tag{14.98}$$

where L is a length scale. An exact transport equation for k can be derived. However, this equation contains other unknown turbulent variables that need to be modeled before the equation is used to obtain turbulent kinetic energy. The modeled form of the transport equation for turbulent kinetic energy is [4,5]

$$\frac{\partial k}{\partial t} + \bar{U}\frac{\partial k}{\partial x} + \bar{V}\frac{\partial k}{\partial y} + \bar{W}\frac{\partial k}{\partial z} = \frac{\partial}{\partial x}\left(v + \frac{v_t}{\sigma_k}\frac{\partial k}{\partial x}\right)$$
$$+ \frac{\partial}{\partial y}\left(v + \frac{v_t}{\sigma_k}\frac{\partial k}{\partial y}\right) + \frac{\partial}{\partial z}\left(v + \frac{v_t}{\sigma_k}\frac{\partial k}{\partial z}\right) + P_k - C_D \frac{k^{3/2}}{l}. \tag{14.99}$$

The last two terms on the right hand side of this equation represent the rate of production and dissipation of the turbulent kinetic energy. Production term is equal to $v_t(\partial \bar{U}/\partial y)^2$ for a turbulent boundary layer but is a more complex term in a general three-dimensional flow. σ_k is the Prandtl number for turbulent kinetic energy. Values of $C_\mu = 0.09$, $C_D = 0.08$ and $\sigma_k = 1.0$ are recommended. Length scale for a boundary layer was given by $L = \max(0.2y, 0.09\delta)$. However, there is no precise algebraic prescription for length scale in a general flow and that makes this model of limited use.

In another category of one-equation turbulence models, a transport equation is derived for the turbulent viscosity itself. This transport equation has the following general form:

$$\frac{\partial v_t}{\partial t} + \bar{U}\frac{\partial v_t}{\partial x} + \bar{V}\frac{\partial v_t}{\partial y} + \bar{W}\frac{\partial v_t}{\partial z} = \frac{\partial}{\partial x}\left(v + \frac{v_t}{\sigma_v}\frac{\partial v_t}{\partial x}\right)$$
$$+ \frac{\partial}{\partial y}\left(v + \frac{v_t}{\sigma_v}\frac{\partial v_t}{\partial y}\right) + \frac{\partial}{\partial z}\left(v + \frac{v_t}{\sigma_v}\frac{\partial v_t}{\partial z}\right) + P_v - D_v, \tag{14.100}$$

where σ_v is called the Prandtl number for turbulent viscosity and P_v and D_v are the production and dissipation of the turbulent viscosity, respectively.

- **Two-Equation Models**

    There are two major problems with one-equation models. (1) There is no general algebraic equation for length scale. (2) Convection and diffusion effects from upstream and downstream of a point on the turbulent length scale at that point are not accounted for by an algebraic expression. This made one-equation models only marginally superior or almost the same as mixing-length models.

    One proposal for improving turbulence models was to develop a transport equation for length scale, or another turbulent quantity $z = k^m l^n$, and use that with the transport equation for turbulent kinetic energy to determine l and k and therefore turbulent viscosity v_t. The most common choices for z were turbulent energy dissipation rate, $\varepsilon = k^{3/2}/l$, and specific dissipation rate, $\omega = k^{1/2}l$ [4,5].

    The standard form of the k-ε model includes the following transport equations for the turbulent kinetic energy k and the turbulent energy dissipation rate ε:

$$\frac{\partial k}{\partial t} + \bar{U}\frac{\partial k}{\partial x} + \bar{V}\frac{\partial k}{\partial y} + \bar{W}\frac{\partial k}{\partial z} = \frac{\partial}{\partial x}\left(v + \frac{v_t}{\sigma_k}\frac{\partial k}{\partial x}\right)$$
$$+ \frac{\partial}{\partial y}\left(v + \frac{v_t}{\sigma_k}\frac{\partial k}{\partial y}\right) + \frac{\partial}{\partial z}\left(v + \frac{v_t}{\sigma_k}\frac{\partial k}{\partial z}\right) + P_k - \varepsilon, \quad (14.101)$$

$$\frac{\partial \varepsilon}{\partial t} + \bar{U}\frac{\partial \varepsilon}{\partial x} + \bar{V}\frac{\partial \varepsilon}{\partial y} + \bar{W}\frac{\partial \varepsilon}{\partial z} = \frac{\partial}{\partial x}\left(v + \frac{v_t}{\sigma_\varepsilon}\frac{\partial \varepsilon}{\partial x}\right)$$
$$+ \frac{\partial}{\partial y}\left(v + \frac{v_t}{\sigma_\varepsilon}\frac{\partial \varepsilon}{\partial y}\right) + \frac{\partial}{\partial z}\left(v + \frac{v_t}{\sigma_\varepsilon}\frac{\partial \varepsilon}{\partial z}\right) + C_{\varepsilon 1}\frac{\varepsilon}{k}P_k - C_{\varepsilon 2}\frac{\varepsilon^2}{k}. \quad (14.102)$$

P_k is the production of the turbulent kinetic energy that is usually modeled as

$$P_k = v_t S^2, \quad (14.103)$$

where S is the modulus of the mean strain rate tensor. The turbulent viscosity is given by

$$v_t = C_\mu \frac{k^2}{\varepsilon}. \quad (14.104)$$

This model contains five constants and the most commonly used values for those are

$$C_\mu = 0.09, \quad C_{\varepsilon 1} = 1.44, \quad C_{\varepsilon 2} = 1.92, \quad \sigma_k = 1.0, \quad \sigma_\varepsilon = 1.3. \quad (14.105)$$

The standard form of the k-ω model includes the following transport equations for the turbulent kinetic energy k and the specific dissipation rate ω:

$$\frac{\partial k}{\partial t} + \bar{U}\frac{\partial k}{\partial x} + \bar{V}\frac{\partial k}{\partial y} + \bar{W}\frac{\partial k}{\partial z} = \frac{\partial}{\partial x}\left(\nu + \frac{\nu_t}{\sigma_k}\frac{\partial k}{\partial x}\right)$$
$$+ \frac{\partial}{\partial y}\left(\nu + \frac{\nu_t}{\sigma_k}\frac{\partial k}{\partial y}\right) + \frac{\partial}{\partial z}\left(\nu + \frac{\nu_t}{\sigma_k}\frac{\partial k}{\partial z}\right) + P_k - \varepsilon, \qquad (14.106)$$

$$\frac{\partial \omega}{\partial t} + \bar{U}\frac{\partial \omega}{\partial x} + \bar{V}\frac{\partial \omega}{\partial y} + \bar{W}\frac{\partial \omega}{\partial z} = \frac{\partial}{\partial x}\left(\nu + \frac{\nu_t}{\sigma_\omega}\frac{\partial \omega}{\partial x}\right)$$
$$+ \frac{\partial}{\partial y}\left(\nu + \frac{\nu_t}{\sigma_\omega}\frac{\partial \omega}{\partial y}\right) + \frac{\partial}{\partial z}\left(\nu + \frac{\nu_t}{\sigma_\omega}\frac{\partial \omega}{\partial z}\right) + C_{\omega 1}\frac{\omega}{k}P_k - C_{\omega 2}\frac{\omega}{k}\varepsilon. \qquad (14.107)$$

Here, the turbulent energy dissipation rate is

$$\varepsilon = C_\varepsilon \omega k, \qquad (14.108)$$

and the turbulent viscosity is given by

$$\nu_t = \frac{k}{\omega}. \qquad (14.109)$$

This model contains five constants with the following commonly used values:

$$C_\varepsilon = 0.09, \quad C_{\varepsilon 1} = 5/9, \quad C_{\varepsilon 2} = 3/40, \quad \sigma_k = 2.0, \quad \sigma_\varepsilon = 2.0. \qquad (14.110)$$

- **Reynolds Stress Models**
  The main deficiency of the turbulence models based on the turbulent viscosity is that the simple relationship between Reynolds stresses and mean strain rate may not be valid in some situations. This means that turbulent viscosity may not be a scalar quantity. Both measurements and DNS have shown that turbulent viscosity is a tensor quantity. Reynolds stress models are based on the transport equations for Reynolds stresses and turbulent heat fluxes itself. These equations can be derived from the momentum and temperature equations but are very complex and contain new unknown turbulent quantities that need to be modeled.

14.4 SOLUTION OF FINITE-DIFFERENCE EQUATIONS

Using a concise matrix notation, the set of finite-difference equations derived in Section 14.2 can be written as

$$[A][T] = [C]. \qquad (14.111)$$

This is the general form of the matrix equation that is solved by computer simulation tools. If N is the number of nodes in the domain, the **coefficient matrix** $[A]$ is an $N \times N$ matrix, and $[T]$ and $[C]$ are vectors with N elements.

If $[A]^{-1}$ is the inverse of matrix $[A]$ such that $[A]^{-1}[A] = [I]$ where $[I]$ is the unity matrix, the solution of this matrix equation will be

$$[T] = [A]^{-1}[C]. \qquad (14.112)$$

However, obtaining the inverse of a matrix is not a trivial task. There are many numerical techniques such as Gauss–Seidel and incomplete LU decomposition that are used to approximate the inverse of a matrix. These methods will not be described here and the interested readers are referred to appropriate texts on numerical linear algebra [6].

14.5 COMMERCIAL THERMAL SIMULATION TOOLS

Developing algorithms and computer codes that can be used for thermal simulation of systems such as electronic devices and equipment is a tremendous and time-consuming task. Almost none of the engineers who use computer simulation in their thermal design process have either the expertise or the time to develop their own computer codes. Instead they use commercial thermal simulation tools from companies whose main business is to develop and sell such software. Most of these tools are interactive and object-oriented. They are offered in various degrees of complexities and capabilities and may be tailored toward a group of thermal problems such as component-level, board-level, system-level, rack-level, or data-center-level thermal designs.

Commercial thermal simulation tools can be classified into two main categories of conduction-only and full computational fluid dynamics tools. Conduction-only simulation tools solve the heat conduction equation in three-dimensional geometries such as single electronic devices or heat sinks. They do not solve the flow equations and account for the convection and radiation heat transfers by using the available analytical correlations to set oppropriate boundary conditions. These tools are very easy to setup and solutions are obtained in seconds or minutes rather than hours and days. However, their accuracy is partially dependent on the accuracy of the boundary conditions. On the other hand, full computational fluid dynamics tools calculate the flow and temperature fields in and around an electronic device or system. They solve for conduction, convection, and radiation heat transfer rates simultaneously and there is no need to specify any of these as a boundary condition. These simulation tools are more complex than the conduction-only tools, require more expertise in the user, need more inputs, and take longer to obtain solutions. However, if used appropriately, calculated convection and radiation heat transfer rates will be more accurate than the boundary conditions used with the conduction-only tools.

Any thermal simulation process starts with creating a thermal model that includes the geometry being simulated. Then, the model is divided into computational elements, otherwise called control volumes or meshes, and flow and/or temperature equations are solved at each of those elements as described in Sections 14.2, 14.3 and 14.4. Finally, the results are reviewed and presented in tabular and/or graphical formats. Computer algorithms and programs to perform each of these steps are different from each other and have been developed by different individuals, research groups, and companies. However, most companies that offer these thermal simulation tools combine the capabilities that are necessary to perform all these steps into one interactive and user-friendly environment. The user performs these steps sequentially and usually does not need to transfer data from one tool to another to perform the next step. This will increase the speed and efficiency of thermal simulation and will eliminate any error due to mistakes in data transfer. However, a possible drawback of the easy-to-use simulation tools is that a user may rely too much on the tool's default options for performing an analysis instead of using more relevant analysis options for the problem under analysis. It is important to recognize that every problem is special and requires special attention. Each of the steps of a thermal simulation process will be explained in detail, and appropriate recommendations to improve the accuracy and speed of the thermal simulation will be given here.

14.5.1 Creating the Thermal Model

The first step in any thermal simulation analysis is to build a thermal model. A wrong perception that is sometimes seen in the design community is that a thermal model is simply a mechanical three-dimensional file imported into the thermal simulation tool. Even if this file import is that easy, it only creates the geometry of the thermal model. In addition to creating the geometry, a thermal model requires material properties, power dissipation data, environmental conditions, flow type, and so forth, each of which will be explained in this section.

- **Creating Model Geometry**
  There are basically three ways to create the model geometry in a thermal simulation tool. Most commercial thermal simulation tools allow the users to create different geometrical objects inside the tool itself. Therefore, an expert user will be able to create approximate model geometry for a package, a board or a full system by a combination of those individual geometrical objects. This is usually the only method to create model geometry if the design is still in a conceptual or feasibility stage. The reason for calling it "approximate model geometry" is that thermal model geometries are, and shall be, much simpler than the corresponding mechanical models. For example, in a board-level or system-level thermal model, many fine features such as surface curvatures and small components such as nuts, bolts, small resistors, capacitors, inductors, and so forth are not included. This intentional and necessary omission has a minor impact on the accuracy of the results and results in a significant improvement in the efficiency of the

Computer Simulations and Thermal Design

thermal model. Figure 14.8 shows thermal models of two electronic systems created inside the thermal simulation tool.

The second method for creating model geometry in a thermal simulation tool is to import a mechanical CAD file into the tool and to use surfaces and lines from that CAD file to generate three-dimensional objects in the thermal model. The advantage of this method is that the locations and the dimensions of different objects are correctly transferred from the mechanical file and do not need to be measured or approximated. However, this method can only be used in a late stage when mechanical design is relatively mature and mechanical CAD file is available.

The last method for creating model geometry in a thermal simulation tool is to import a CAD file into the tool such that the complete three-dimensional objects are imported. Similar to the method mentioned above, it can only be used in later stages of design, when mechanical design is relatively mature. There are two potential drawbacks for this method. If the original CAD file is not simplified for thermal analysis purpose, too many fine details are imported into the thermal simulation tool and need to be removed to make thermal analysis feasible. Also, some mechanical and electrical CAD tools add the keep-out areas around a component to its own volume and therefore, the dimensions of those components have to be corrected after the file is imported. Figure 14.9 shows an imported CAD file with many fine details and the simpler less complex thermal model built from it.

- **Material and Components Properties**
  Geometrical objects created during the previous phase represent different parts of a system such as printed circuit board, resistors, capacitors, transistors, integrated circuits, connectors, heat sinks, fans, internal or external shields, casings, and so on. These objects have to be identified by their thermal and/or fluid properties. Some of them, such as cases, shields, resistors, and heat sinks, are relatively easy to model by their geometry and material properties. Material properties that are needed for a thermal simulation are thermal conductivity, density, and specific

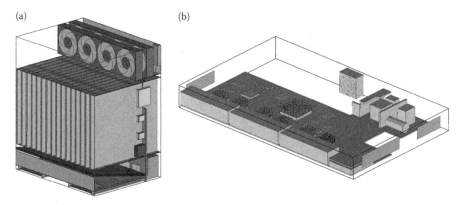

FIGURE 14.8 (See color insert following page 240.) Thermal models of two electronic systems created inside a thermal simulation tool.

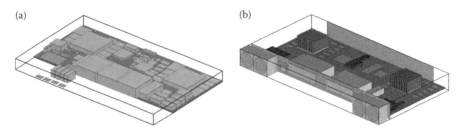

FIGURE 14.9 (See color insert following page 240.) (a) An imported CAD file and (b) the thermal model built based on that.

heat. Density and specific heat are necessary inputs for transient thermal simulations only but thermal conductivity is always needed. Since the accuracy of the results is partially dependent on the accuracy of input data, enough care should be given to obtain correct material properties for the model.

Other objects such as printed circuit boards, transistors, integrated circuits, fans, and so forth are very complex entities. Each of these can be individually modeled and their thermal behaviors can be accurately estimated if detailed mechanical and material properties, and enough computational power, are available. In fact many component manufacturers use thermal simulation tools for thermal design and characterization of their products, and many fans are designed and optimized using computational fluid dynamics tools. However, when a thermal simulation tool is used for a board-level or system-level thermal analysis, many components are involved and detailed modeling of all those components is not practical. In these cases, the so called compact models are used to represent fluid and thermal properties of those components. These compact models are simple representations of complex components and mimic their important fluid and thermal properties. The term "important" shall be emphasized here, as no compact model is able to mimic all fluid and thermal properties of an object. Therefore, the right compact model to choose is the one that represents critical fluid and thermal properties of the object it is representing, as far as the objectives of the thermal simulation are concerned.

Printed circuit boards are complex multilayer structures in which layers of high thermal conductivity copper (trace, ground, and power layers) are sandwiched between layers of dielectric and low thermal conductivity glass epoxy (see Figure 5.25). Recent developments in thermal simulation tools allow the users to import board files from electrical CAD tools into the thermal simulation tools. These files contain detailed information about the shape, thickness and location of the copper layers in printed circuit boards. This information is used to calculate the local values of thermal conductivities of the printed circuit board. A printed circuit board in this case is modeled as a single object whose thermal conductivity is a function of space coordinates. This method gives a relatively accurate representation of the printed circuit board but it can not be used in early design stages, when board stack up is not determined yet, and it can not be used to investigate thermal effects of changes in the board stack up

Computer Simulations and Thermal Design 385

and the copper content. Both of these are important as the main objectives of thermal engineers are guiding the product toward a thermally feasible design and providing recommendations to the design team on how to improve the thermal performance. A few commonly used models for printed circuit boards will be described below.

The copper content in trace layers is relatively low and the traces are usually dissociated from each other as shown in Figure 14.10. On the other hand, power and ground layers are solid copper layers that cover most of the board area. Therefore, trace layers may be neglected for simplicity, and a printed circuit board may be modeled as a composite layer of dielectric, power, and ground layers. Note that printed circuit boards are about 1 to 5 mm thick and may contain 5 to 10 power and ground layers that may be as thin as 15 microns each. Therefore, even this simplified model requires many computational elements and makes the system-level thermal models too complex. The next simpler alternative, which is available in some commercial simulation tools, models power and ground layers as objects with zero thickness but treats them as thick objects in their internal solver algorithms. This will reduce the number of computational elements significantly, without compromising the accuracy. The simplest model for a printed circuit board treats them as single objects with orthotropic thermal conductivities calculated as described in Chapter 5, Section 5.8. This model requires the least number of computational elements but may result in significant errors, as discussed in that section. A compromise between modeling all the power and ground layers explicitly and modeling a printed circuit board as a single orthotropic object is to model top and bottom copper layers, if they exist, explicitly and the rest of the board as a single orthotropic object. This produces reasonably accurate results, as the most important copper layers are those that are right under the heat dissipating components.

Printed circuit boards may include thermal vias, as shown in Figure 14.11, to transfer heat through the board, as the power and ground layers spread the heat mainly in the plane of the board. These thermal vias improve thermal conductivity of the board in the normal direction significantly and have to

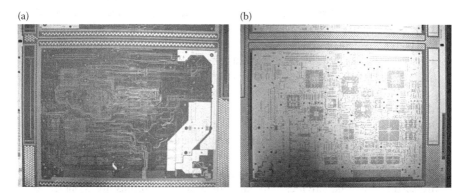

FIGURE 14.10 (See color insert following page 240.) (a) A trace or signal layer and (b) a power layer.

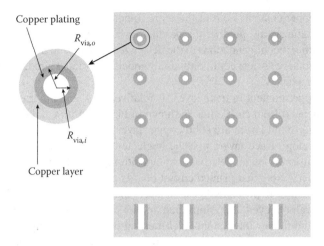

FIGURE 14.11 A PCB with thermal vias.

be included in the thermal model to obtain more accurate results. The thermal vias can be modeled in detail in component-level thermal simulations, but appropriate compact models are needed in board-level and system-level thermal simulations. The simplest compact model for a PCB with thermal via is a one-dimensional model described in Chapter 5, Section 5.8 and a more accurate model was described in Chapter 7, Section 7.3.

Many different compact models for transistors and integrated circuits have been used by thermal engineers. The simplest models represent a component with a single solid block with assigned thermal properties and volumetric power dissipation. The obvious advantage of this modeling technique is its simplicity. However, it is too simple to represent the heat flow paths inside the component, or to meaningfully predict either its case or its junction temperature, in most situations. It can be shown that if a high conductivity value is assigned to the block representing the component and if all the heat generated by the component is dissipated through the bottom or the top of the component, this model gives meaningful prediction for the bottom or the top temperature of the component, respectively. The junction temperature of the component, if required, can be obtained analytically using the predicted top or bottom temperature and the junction-to-case or junction-to-board thermal resistance, respectively.

A more complex compact model of a transistor or an integrated circuit uses a combination of solid blocks with appropriate thermal properties to represent the die, substrate, die attach, underfill, flip-chip solder bumps, plastic overmold or metal heat spreader, solder balls, and so forth, separately. This compact model is still simple compared to the actual geometry of the component. For example, solder balls and the air gap between those and flip-chip solder bumps and the underfill between those are modeled by single blocks with appropriate orthotropic thermal conductivities. Similarly, multilayer substrates are modeled as single blocks with orthotropic thermal conductivities as described in Chapter 5, Section 5.8. The advantage of these compact models is that it can predict the heat flow paths inside the component and results in meaningful predictions for top, bottom, and junction temperatures of the component. Although these compact models increase the simulation time, this is not a real issue with today's computing capability. The main issue is that the dimensional specifications and material properties needed to build these compact models are not normally available. If these models are built with estimated or guessed, rather than exact dimensions and material properties, the error that may be introduced by the incorrect inputs offsets any presumed accuracy that would have been gained.

A widely used compact model for transistors and other microelectronic packages is the network model. This model is geometrically as simple as the single-block model mentioned earlier. However, it is capable of predicting the correct heat flow paths inside the component. A network model is made of multiple surfaces that collectively surround an empty volume representing the component itself. The relationship between those surfaces and a virtual internal die is defined by a thermal resistance network. The simplest network model is a two-resistance model that represents a component by three points (junction or die, top and bottom) and two thermal resistances (junction-to-case and junction-to-board thermal resistances) as shown in Figure 14.12. The sides are treated as adiabatic surfaces. The thermal resistances needed to create a two-resistance network model are usually provided by component vendors. However, some vendors report junction-to-case and junction-to-air thermal resistances only. Junction-to-board thermal resistance of those components can be estimated by building a thermal model for the conditions in which their junction-to-air thermal resistances were measured. The component is modeled as a two-resistance network with the specified junction-to-case thermal resistance and a guessed value for its junction-to-board thermal resistance. This model is used for iterating on the junction-to-board thermal resistance value until it predicts the reported junction-to-air thermal resistance.

The two-resistance model assumes that four vertical sides of the component shown in Figure 14.12 are adiabatic and do not contribute in heat transfer from the device. Their temperatures are assumed to be the average of the top and bottom surface temperatures. This is, of course, not true and may result

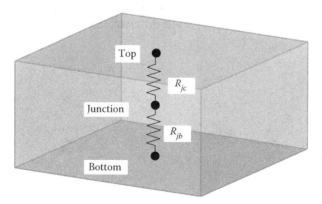

FIGURE 14.12   A two-resistance network model.

in unacceptable errors in simulation results and an odd-looking temperature distribution for the component. The star network model that is obtained by adding junction-to-side thermal resistances to the two-resistance network model and the shunt network model that connects all sides of a component to each other and to its junction, through appropriate thermal resistances, were proposed to improve the network model for a package (see Figure 14.13). However, the main difficulty with using the star or shunt network model is that the thermal resistance data they need is usually not provided by the component vendor.

Fans and blowers are other types of complex objects that require compact models for board-level and system-level thermal simulations. Most thermal simulation tools use fan curves to model fans and blowers as planar or volumetric pressure sources. The flow rate a fan or blower generates in a system is obtained by a balance between the pressure loss in the system and the magnitude of the pressure source generated by the simulation tool. A few simple guidelines in modeling fans and blowers will improve the accuracy of thermal models. (1) Thermal engineers are encouraged to measure the fan curve in conditions similar to the final intended use, rather than using the standard fan curve provided by the fan vendor. A fan curve is usually measured while there is no obstruction close to its inlet or exhaust and fan flows into the fan from all directions uniformly. However, airflow blockages are usually present close to the inlet and/or exhaust of a fan mounted in a system. On the other hand, in some electronic systems, air can enter the fan from certain directions only. The presence of a blockage near the inlet or exhaust of a fan will restrict the airflow direction and alter the fan curve. Using the original fan curve may result in erroneous predictions in these cases. If the fan curve is measured in a situation similar to its final use, and then is used in the thermal simulation, the inaccuracy associated with an erroneous fan curve will be eliminated or significantly reduced. (2) The fan curves are measured at standard air pressure and temperature and a few fan voltages or speeds. It was shown in Chapter 11, Section 11.7.2 that fan curves change with air density and fan speed. If the final intended use of a fan is at a speed different from the standard speed at which the fan curve is reported, or the environmental conditions are different from standard pressure and

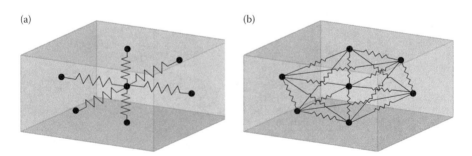

FIGURE 14.13 (a) A star network model and (b) a shunt network model.

temperature in which the fan curve is measured, the fan curve has to be corrected for use in the thermal simulation tool. (3) If multiple fans are put close to each other in series, parallel, or mixed arrangement, they may affect each other's performance. This is especially important when two or more fans are used in series and there is not much air gap between those fans. Therefore, the fan curve measured for each fan individually may not represent its performance when operating near other fans. A reverse procedure in which the performance of the whole fan arrangement is measured first and the individual fan performances are obtained from this measurement is recommended in these situations.

It has been shown that if an axial fan is used to blow air into a system, the resulting flow will be unsteady and the air jet takes the form of a hollow, conical vortex, which interacts in a complex manner with the electronics inside the system [7]. This is in contrast to the case where a fan is located at the exhaust of the system and creates a uniform and steady airflow. Thermal simulations based on fan curves predict total airflow rate and the velocity distribution for the case with a fan at the exhaust reasonably well. However, their predicted velocity fields can be significantly wrong for the case with a fan at the inlet of a system. Some simulation tools allow the users to specify the flow direction and swirl at the exhaust of a fan. However, both of these flow characteristics are too complex to be identified in advance by a user. Although thermal models based on a fan curve are very useful for thermal design of electronic systems, thermal engineers need to be aware of the capabilities and limitations of these models.

- **Power Dissipation**

    Temperature of an active component is a function of its own power dissipation, $\dot{Q}_{comp}$, power dissipation of other components around it, the heat flow path from the component to the outside environment, and the ambient temperature, T_{amb}. If the heat flow path and the effects of power dissipations by other components in the system are represented by a thermal resistance between the component and the outside ambient, $R_{comp\text{-}amb}$, the component's temperature will be

$$T_{comp} = T_{amb} + \dot{Q}_{comp} \cdot R_{comp-amb}.$$

Note that there is a direct relationship between power dissipation of a component and the difference between its temperature and ambient temperature. Even if the thermal resistance between a component and its ambient is predicted accurately, any error in the power dissipation of a component results in the same percentage error in its temperature rise from ambient. This shows the importance of having accurate power dissipation data for the thermal simulations. On the other hand, these data can not be accurately determined until the system is fully designed and tested. This should not prevent a thermal engineer from using simulation to design an appropriate thermal solution for the system. The thermal design can be updated and optimized as power dissipation data are updated.

Note that the word "component" is used as a general term. For example, it can refer to a power transistor inside a phone charger, the entire board inside the charger, or even the whole charger itself. The above equation shows that if the power dissipation of a component is not known, its temperature can not be determined even if the power dissipations of all the other components around it are known. For example, if the total power dissipation of the board inside a charger is known but power dissipations of the individual components on that board are not known, the best a thermal simulation tool can do is to predict the average board temperature and the average external surface temperature of that charger.

Some electronic equipment may be exposed to heat flux coming from other sources. For example, equipment operating outdoors in a sunny day are exposed to the solar and atmospheric radiations as described in Chapter 13, Section 13.4. These additional heat fluxes will affect the thermal performance of such equipment and have to be considered in thermal simulations. The incident solar radiation can be accounted for by assigning a uniform heat flux over the exposed surfaces and the atmospheric radiation is modeled as described under the next heading.

- **Environmental Conditions**
Electronic equipment are designed to operate in extreme environmental conditions for their specific market. For example, personal computer manufacturers design and build their products to be used on a cold day in San Francisco, California as well as a hot day in Denver, Colorado. On the other hand, a personal computer designed for military application has to operate in more severe environmental conditions than a similar computer designed for the consumer market. The environmental conditions are specified by the type of cooling fluid, its temperature, and its flow and thermal properties. Air is the cooling fluid in most electronic applications and its properties are functions of pressure and temperature. Higher temperature and lower pressure air has lower densities and is less efficient in cooling electronics. Therefore, it is important to set the air properties in the thermal simulation tool to those corresponding to the highest design temperature and the lowest design pressure (highest design altitude). For example, if the operating temperature and altitude ranges of an equipment are 0–35°C and 0–3000 m, respectively, air temperature of 35°C and air properties at 35°C temperature and 3000 m altitude have to be used in the thermal simulation of that equipment.

Another environmental condition, which is especially important if the equipment dissipates its heat partially through radiation from its external surfaces, is the temperature of the surfaces surrounding that equipment. These are the wall, floor, and ceiling temperatures for equipment operating

Computer Simulations and Thermal Design 391

indoors and sky, ground, and building temperatures for equipment operating outdoors. The atmospheric radiation described in Chapter 13, Section 13.4 can be accounted for in thermal simulation if the temperatures of the computational surfaces representing the sky are set to the appropriate sky temperature.

- **Flow Type**
  The flow inside and around the electronic system under thermal simulation may be laminar, turbulent, or mixed. However, none of the commercial thermal simulation tools is able to determine whether a flow is locally laminar, turbulent, or in transition. Therefore, the user needs to specify the flow type in advance as a global parameter. A good rule of thumb for electronic systems is that if the flow is forced convection, it will most likely be turbulent and if it is natural convection, it will most likely be laminar. If the flow is assumed to be turbulent, a turbulence model that will be used to calculate turbulent parameters has to be specified. Some simulation tools allow the user to specify different flow types in different regions of a model. This is useful in cases where flow may change from laminar to turbulent or vice versa. For example, flow between the fins of a plate-fin heat sink can be laminar even if the flow upstream and downstream of the heat sink is turbulent. In these cases, the heat sink is treated as a separate region where flow is assigned to be laminar.

  If natural convection heat transfer is important in the system under simulation, direction and magnitude of the acceleration of gravity have to be specified. If radiation heat transfer plays a role in cooling the electronic system, it must be considered in thermal simulation as well. Generally, radiation heat transfer is negligible compared to convection heat transfer for forced convection systems but is comparable to convection heat transfer for natural convection systems.

  The flow field and thermal behavior of a system may be steady or transient. The time steps in a transient thermal simulation have to be small enough to capture the changes with time but large enough such that the entire simulation is done in a reasonable time. The best strategy is to use a variable time step where small time steps are used whenever a time-dependent change happens in the system and large time steps are used whenever the conditions under which the system is operating are constant with time. For example, transient thermal simulation of a die in a transistor, from the time that it is powered on, has to start with time steps on the order of milliseconds and ends with time steps on the order of seconds or even minutes.

14.5.2 Creating the Mesh

The next step in a thermal simulation process is to divide the computational domain into many small volumes called the mesh. This step has been integrated in most of the current thermal simulation tools. Computational volumes that make a mesh may have different shapes. A Cartesian mesh, which uses rectangular prisms or cuboids, as shown in Figure 14.14a and b, will modify non-Cartesian geometries such as cylinders and inclined surfaces. A more complex mesh uses a combination of triangular and general quadrilateral prisms as shown in Figure 14.14c and d to create a so-called body-fitted mesh. Note that even a body-fitted mesh may modify geometries but the modifications are minor and become less distinguishable as the mesh is refined.

Many users tend to accept the default mesh that is created by thermal simulation tools, and to start the thermal simulation as soon as they create the model. However, dimensions, shapes, and quantity of the mesh have significant impacts on the accuracy of the results as well as on the time it takes to obtain the results. Therefore, the time that is spent to create a good mesh will be justified by more accurate results, if not compensated by the time that is saved to obtain those results. Figure 14.15 shows the default mesh generated by a thermal simulation software and a mesh generated after user defined mesh control parameters were used. The user-controlled mesh has a few advantages as will become obvious below.

It is usually mentioned that the accuracy of thermal simulation results increases if smaller computational volumes (i.e., a finer mesh) is used. However, the fine mesh is only needed in the regions where significant changes in velocity or temperature are present. These include near surface regions where boundary layers are present, immediately upstream and downstream of an object, and at intake and exhaust of a system where flow may contract or expand. On the other hand, there is no need to have fine mesh around symmetry lines or outside the boundary layers where flow variables do not change. A nonuniform mesh in which dimensions of the computational volumes change gradually shall be used to generate a fine mesh where it is needed and a coarse mesh where it can be accepted. The user-controlled mesh in Figure 14.15b has small elements near surfaces and larger elements away from them while the default mesh in Figure 14.15a has rather arbitrary distribution of mesh in this domain.

The detailed geometrical specifications of each individual computational volume may affect the accuracy of the thermal simulation. Two of the most important geometrical features that need to be studied in every mesh are the maximum aspect ratio and the maximum angle in a computation volume. Aspect ratio is defined

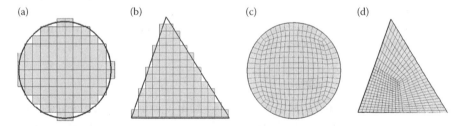

FIGURE 14.14 Cartesian (a and b) and body-fitted (c and d) meshes for a cylindrical and a triangular object.

Computer Simulations and Thermal Design

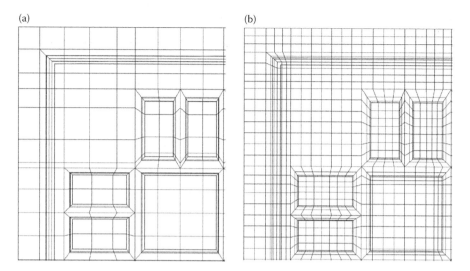

FIGURE 14.15 (a) A software default mesh and (b) a user-controlled mesh for the same model.

as the ratio of the dimension of one side to the other. High aspect ratio volumes, also called thin or long volumes, shall be avoided especially if changes in flow parameters along the long side of the volume are important. A common cause for the creation of thin volumes is small misalignments between the faces of different objects in a mechanical model that has been imported into the thermal simulation tool. These small misalignments do not affect the flow and temperature distributions and shall be corrected to obtain a better mesh and smaller mesh count. Figure 14.16 shows the top left corner of the meshes shown in Figure 14.15. It is seen that the high aspect ratio elements in the default mesh had disappeared in the user-controlled mesh.

The default mesh creation algorithm in thermal simulation tools may create elements that have internal angles much larger than 90° as shown in Figure 14.17a. Large angles in elements of a mesh prevent or delay the convergence of the solution to the continuity equation. These elements shall be modified to reduce the angle close to 90° as shown in Figure 14.17b.

14.5.3 Solving Flow and Temperature Equations

Now that the thermal model has been created and it is divided into many small computational volumes, finite-difference forms of the flow and temperature equations can be solved in each of those computational volumes. A few parameters have to be set before the "solve" button is pressed.

The solution of the finite-difference flow and temperature equations is an iterative procedure. The results obtained in each iteration, or step, contain some errors. If the model is built and meshed properly, this error will be reduced, although not necessarily monotonically, as the iterations proceed. If the errors are small enough,

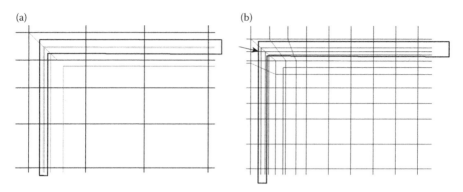

FIGURE 14.16 High aspect ratio meshes (a) in the default mesh are eliminated by (b) user controlled mesh parameters.

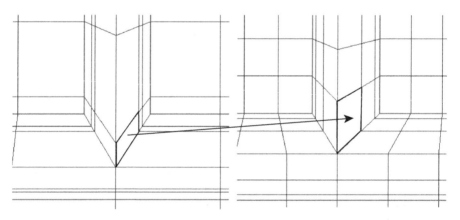

FIGURE 14.17 Computational volumes with angles larger than 90° shall be modified to reduce those angles close to 90°.

further iterations do not change flow and temperature fields significantly. The solution is said to be converged. However, the magnitude of an acceptable error may not be known in advance. A user shall specify the maximum number of iterations as well as an estimate for the acceptable error. The solution will end if either the error is below the set acceptable value or the maximum number of iterations is reached. If the error is below the set value, a second solution shall be obtained with a smaller acceptable error. If the maximum number of iterations is reached before the error goes below the set value, a second solution shall be obtained by further iterations. If the results from the two solutions are close enough, the solution is said to be converged. Otherwise, further iterations are needed to obtain a converged solution.

Computer simulations are done with single or double precisions. Single precision calculations use less memory and take less time per iteration while double precision calculations are more accurate and may converge in fewer iterations. Most

Computer Simulations and Thermal Design

electronic cooling thermal simulations can be done in single precision without compromising the accuracy. However, in cases where very small computational volumes are involved, only temperature equation is solved, or when strongly anisotropic properties are present, a converged single precision solution may not give accurate results.

If transient thermal simulation is required, the user has to specify the time steps as well as the number of iterations and acceptable errors for each time step. Since storing the solution results in every time step may require a huge amount of storage capacity, the user shall also specify the times for which the solution results shall be stored.

Some thermal simulation tools allow the user to monitor values of velocity, temperature, and pressure at specified points as the solution process continues. This is a very useful capability and shall be used as often as possible to judge whether a solution is converging or not. For example, if velocities at a few different locations and temperatures of a few critical components approach certain values and stay at those values as the solution process continues, the solution is practically converged even if the set error limit is not met yet. Figure 14.18 shows plots of solution errors for continuity, momentum, and temperature equations and values of velocity and temperature at four points in the computational domain. Although the solution errors continuously decrease, the velocity and temperature in the monitoring locations do not reach their converged values until approximately 150 iterations.

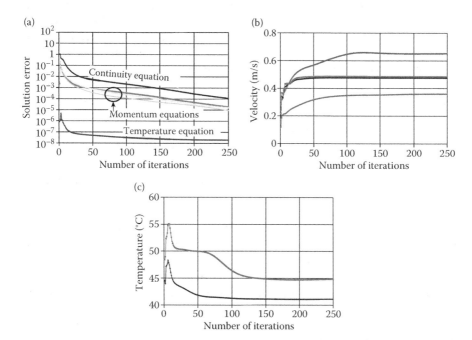

FIGURE 14.18 Plots of (a) solution error, (b) calculated velocity, and (c) calculated temperature at a few points in the computational domain.

14.5.4 Review the Results

It is very important that the results of a thermal simulation are reviewed to make sure they make sense before making any design decisions based on those results. Some may argue that thermal simulation tools have been tested and benchmarked against similar thermal models before being released to market and one should trust the results they produce. However, this argument is only partially true. There is no doubt that most of these tools have been aggressively tested and are capable of predicting flow and temperature fields with reasonable accuracies. However, the accuracy of any simulation result strongly correlates with the validity of the modeling assumptions and boundary conditions, and the accuracy of the input data including ambient conditions and material properties. All these inputs are provided by the user and incorrect assumptions or unintended mistakes in entering these data into the tool will cause erroneous results.

The principles of conservation of mass and energy govern the performance of every physical system. Therefore, simulations of open systems have to satisfy these two conservation principles, and simulations for closed systems have to satisfy the conservation of energy. The conservation of mass can be checked by calculating the difference between mass flow rates entering a system and mass flow rates leaving that system. The conservation of energy can be verified by calculating the difference between the sum of the conduction, convection, and radiation heat transfer rates from the external surfaces of the system and the total heat generation inside the system. If the difference between inlet and exhaust mass flow rates is significant, either the solution is not converged or there are mass leaks in the thermal model. The leaks can be found by plotting velocity vectors on the external surfaces of the system and looking for nonzero velocities in regions other than intended intake and exhaust areas. If the difference between the heat transfer rate from external surfaces of a system and the heat generated inside the system is significant, either the solution is not converged or some internal heat generating objects are not meshed properly, or there are mass leaks in the thermal model, or radiation heat transfer between different surfaces inside the system are not accounted for properly.

Velocity, temperature, and pressure vary smoothly inside and around a system and simulations shall preserve this continuity. This can be checked by determining maximum calculated velocities and maximum and minimum calculated temperatures throughout the computational domain. If either of these global maximum or minimum values are significantly different from the rest of the computational domain, or if they happen only in a small isolated region of the computational domain, they are usually the result of bad mesh elements in that part of the domain. The user should check the mesh elements in that region, create new mesh with good quality mesh elements there, and run the simulation again.

Temperatures of active devices can be estimated by their power dissipations and thermal properties as explained in Chapter 6. A rough agreement between the estimated and calculated temperatures of those active components is a testimony to the accuracy of the thermal model. It is also useful to check temperature gradients on both active and passive devices such as heat sinks. Temperature gradients inside an active device can be estimated by their junction-to-case and junction-to-board thermal resistances and power dissipations to the case and board, respectively, which can be obtained from simulation results. Common mistakes in modeling active components are specifying power dissipations on the wrong side of a device, incorrect resistance network assignments, and overriding the active surface of a device by the surface of another object. These will result in erroneous temperature values and temperature gradients inside the device.

Another common modeling mistake is assigning incorrect thermal conductivity for passive objects such as heat sink, heat pipe, and system case. These mistakes can be captured by noticing unexpected temperature gradients on those objects. For example, heat sinks are made of high conductivity materials such as aluminum and copper and temperature gradients in fins should be relatively low unless fins are long and thin or heat transfer coefficient is very high. Therefore, a predicted high temperature gradient for fins of a heat sink may be due to a low thermal conductivity assigned for that heat sink. For example, thermal conductivity of extruded aluminum is about 200 W/m°C. If a user mistakenly enters 20 instead of 200, the results will be significantly different.

14.5.5 Presenting the Results

Thermal engineers use thermal simulation to study thermal behavior of electronic systems and to come up with design variations that optimize their thermal performance. They work in a multidisciplinary team and rely on other engineers to implement changes they recommend for the system. However, some of the recommended changes may not be easy to implement, or may alter the configurations preferred by mechanical, electrical, power, and industrial design engineer or even sale and marketing managers. Appropriate presentation of thermal simulation results is a critical step in thermal engineer's pursuit of appropriate thermal design. The thermal engineer should present the results of the thermal simulation with the goal of educating the rest of the design team on the thermal design challenges for the system, and specific thermal consequences of any design change. For example, if cooling a CPU is the most important aspect of thermal design for a specific system, a thermal engineer shall indicate that to the team and explains the reasons why that is a challenge. Then, that CPU temperature shall be calculated and reported for any recommended design change.

The best way to present the results of a thermal simulation is through a report. This report should be brief, while conveying the main points to the design team. It shall include modeling assumptions that come from or need to be verified by the rest of the design team, power dissipation and thermal specifications of major active components, one or more plots of the thermal model, a few plots showing relevant results, a summary of the calculated parameters of interest such as airflow rates, and temperatures of critical components, conclusions, and recommendations for improving thermal design if needed. Thermal engineers typically include only plots that carry specific messages to the design team. For example, no one is interested in seeing plots of pressure distribution or, most of the time, air velocity in a system. The design team needs to know whether all components are within safe temperature range or not. Figure 14.19 shows predicted air velocity and air temperature in a telecommunication chassis.

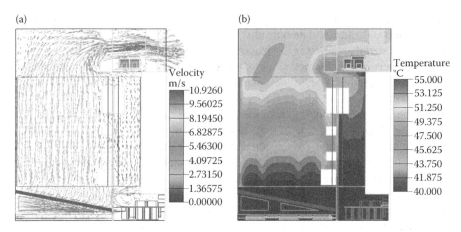

FIGURE 14.19 (See color insert following page 240.) Predicted air velocity and air temperature in a telecommunication system.

14.6 IMPORTANCE OF MODELING AND SIMULATION IN THERMAL DESIGN

Prototype building is costly and time consuming and design cycles that involve building and testing multiple prototypes until the final product is released are long and expensive. On the other hand, the accuracy of fluid flow and thermal modeling and simulation techniques has been significantly improved in the last few decades. In parallel, with faster and less expensive computers, computation power became cheaper and easily accessible. As a result, thermal simulation became a faster and cheaper alternative to the traditional slower and more expensive design by experiment that required building and testing multiple prototypes. Appropriate thermal modeling of electronic systems, a long time before their first prototype is built, allows the design team to evaluate design variations, to rule out those that do not meet thermal requirements, and to select one or more design options that do meet thermal requirements, as well as the requirements from the rest of the design team. In cases where other design or marketing requirements, such as a small size or compact electrical layout, makes cooling the electronics inside the system a challenging task, thermal simulations can be used either to come up with innovative cooling techniques, or to prove necessary changes in size or component layout are needed for cooling to become feasible.

Thermal engineers need to specify their objectives for their simulations before building the thermal models. For example, if the objective of thermal simulation is to understand the airflow pattern inside telecommunication equipment, it will not be necessary to model all the active components on every board inside this equipment. On the other hand, if the objective is to design cooling solutions for active components of a board inside this equipment, the entire equipment including other boards inside it shall not be modeled. The effects of the rest of the system on this board can be modeled by applying appropriate boundary conditions on the computational domain that surrounds this board. Building the thermal models based on the main objectives of the thermal simulation avoids creating unnecessarily complex models and results in faster simulation results and shorter design time.

There is a tendency in the design community to equate "thermal simulation" and "thermal design." As important and necessary as thermal simulation is for thermal design, it is important to distinguish between the two. Thermal design is a process in which different design tools and techniques are used to design a thermally feasible system. Thermal simulation is a design tool that may be used during the thermal design process. Thermal engineers may also decide to use other tools such as analytical calculations and experimental measurements, instead of relying on thermal simulations. In fact, it is strongly recommended that thermal engineers use quicker and simpler analytical methods in earlier design phase and perform experimental measurements to verify or correct thermal simulation predictions as soon as those measurements can be done.

REFERENCES

1. Patankar, Suhas V. 1980. *Numerical heat transfer and fluid flow.* New York, NY: Hemisphere Publishing Corporation.
2. Ferziger, J. H., and M. Peric. 1996. *Computational methods for fluid dynamics.* Berlin, Germany, Springer Publishing.
3. Van Dyke, M. 1982. *An album of fluid motion.* Stanford, CA: The Parabolic Press.
4. Launder, B. E., and D. B. Spalding. 1972. *Lectures in mathematical models of turbulence.* London, England, Academic Press.
5. Kays, W., M. Crawford, and B. Weigand. 2005. *Convective heat and mass transfer.* New York, NY: McGraw-Hill.
6. Jennings, A., and J. J. McKeown. 1992. *Matrix computation.* Chichester, England, John Wiley & Sons, Inc.
7. Grimes, R., M. Davis, J. Punch, T. Dalton, and R. Cole. 2001. Modeling Electronic Cooling Axial Fan Flows, *ASME Journal of Electronic Packaging*, Vol. 123, pp. 112–119.
8. Fox, R. W., A. T. McDonald, and P. J. Pritchad. 2006. *Introduction to fluid mechanics.* Hoboken, NJ: John Wiley & Sons, Inc.
9. Munson, B. R., D. F. Young, T. H. Okiishi, and W. W. Huebsch. 2009. *Fundamentals of fluid mechanics.* Hoboken, NJ: John Wiley & Sons, Inc.

15 Experimental Techniques and Thermal Design

As mentioned in Chapter 1, experimental measurements play important roles in a thermal design process. They are used to provide input to thermal modeling (pre-analysis tests), to either verify that a design meets the requirements or to identify areas of concern (Engineering Verification Tests (EVT) and Design Verification Tests (DVT)), and to certify the compatibility of a product with the applicable industry standards and requirements (certification tests). The extent of experimental measurements and the parameters that are measured depend on the product and how it is cooled. Airflow rate through an air-cooled unit and air velocity and air temperature in such systems, fan or pump performance curve, component's case or junction temperature, heat sink temperature, exterior surface temperature of a unit, and noise level are among the parameters that may be measured during a thermal design process. Some of the techniques used to measure these parameters are described in this chapter.

15.1 FLOW RATE MEASUREMENT TECHNIQUES

The most widely used technique for measuring airflow rate through electronic systems is based on the Bernoulli equation. It states that the sum of the pressure, P, and the kinetic energy per unit volume, $\rho V^2/2$, of a fluid is constant if the flow is frictionless and there is no heat transfer to or from the flow [1]. Consider a nozzle whose cross sectional area reduces from A_{n1} to A_{n2} as shown in Figure 15.1. The Bernoulli equation can be written as

$$P_{n1} + \frac{1}{2}\rho_{n1}V_{n1}^2 = P_{n2} + \frac{1}{2}\rho_{n2}V_{n2}^2. \tag{15.1}$$

The mass balance between cross sections 1 and 2 gives

$$\dot{m}_{n1} = \dot{m}_{n2} \Rightarrow \rho_{n1}A_{n1}V_{n1} = \rho_{n2}A_{n2}V_{n2}. \tag{15.2}$$

If pressure and temperature variations from section 1 to 2 are small such that the air density can be treated as constant (i.e., $\rho_{n1} = \rho_{n2} = \rho$), Equations 15.1 and 15.2 can be simultaneously solved to give

$$V_{n2} = \sqrt{\frac{2(P_{n1} - P_{n2})}{\rho[1 - (A_{n2}/A_{n1})^2]}}. \tag{15.3}$$

Therefore, the volumetric airflow rate is

$$\dot{V}_{n2} = A_{n2}V_{n2} = A_{n2}\sqrt{\frac{2(P_{n1}-P_{n2})}{\rho[1-(A_{n2}/A_{n1})^2]}}. \tag{15.4}$$

It is seen that the airflow rate through the nozzle shown in Figure 15.1 can be calculated by measuring the pressure drop from section 1 to 2. Equation 15.4 gives the airflow rate through an ideal frictionless and adiabatic nozzle. The airflow rate through a real nozzle is calculated by adding a correction factor C:

$$\dot{V}_{n2} = A_{n2}V_{n2} = CA_{n2}\sqrt{\frac{2(P_{n1}-P_{n2})}{\rho[1-(A_{n2}/A_{n1})^2]}}, \tag{15.5}$$

which is a function of the nozzle geometry, flow speed, and fluid properties (Reynolds number). Detailed dimensional specifications of nozzles used for airflow measurements and their corresponding correction factors are given by the American Society of Mechanical Engineers, the Air Movement and Control Association, and the American National Standards and Institutes [2].

The equipment that is built based on the Bernoulli equation, and is used to measure airflow rate through a system, as well as system impedance curve and fan curve, is called a flow bench or airflow chamber [2]. A schematic of a flow bench is shown in Figure 15.2. The main components of a flow bench are a variable speed (flow rate) blower, two plenums, flow straighteners in each plenum, a blast gate, a plate with a

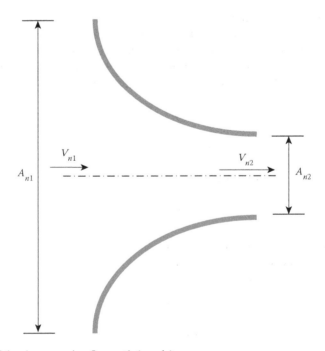

FIGURE 15.1 A converging flow path (nozzle).

Experimental Techniques and Thermal Design

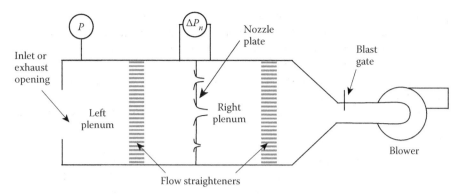

FIGURE 15.2 Schematic of a flow bench.

FIGURE 15.3 (**See color insert following page 240.**) A commercial flow bench. (Courtesy of Airflow Measurement Systems.)

series of nozzles between the two plenums, and a series of pressure taps and pressure manifolds connected to those taps. Although the nozzle plate has a few nozzles with different dimensions, not all of those have to be open at the same time. Most flow benches are designed such that either intake or exhaust of the blower can be attached to the flow bench based on the desired flow direction. In either case, the nozzles have to be located such that the air moves in the direction that air flow contraction happens. The flow straighteners are used to create a uniform flow into those nozzles. The blast gate is used to control the flow rate through the flow bench. Figure 15.3 shows a commercial flow bench.

The two sets of pressure taps on the two sides of the nozzle plate are connected to two pressure manifolds. The pressure manifolds record the average values of the pressures upstream and downstream the nozzles, P_{n1} and P_{n2}. These two manifolds are connected to two sides of a manometer to record the pressure difference across the nozzles; $\Delta P_n = P_{n1} - P_{n2}$. Another pressure manifold is connected to a set of pressure taps on the other side of the plenum that is not attached to the blower and records the average value of the air pressure there, P. This manifold is connected to one side of a manometer whose other side is sensing the atmospheric pressure to record the gauge pressure; $\Delta P = P - P_{atm}$.

Consider a device under test (DUT) such as a desktop or laptop computer, a projector, a server, a network router or switch, a power supply, or any other similar system. The airflow rate through the inlet(s) or the exhaust(s) of these systems can be measured following the steps outlined below.

1. Connect either the inlet(s) or the exhaust(s) of the DUT to the left plenum of the flow bench shown in Figure 15.2 in order to measure airflow rate through those inlet(s) or exhaust(s), respectively. If the total inlet or exhaust airflow rate through the DUT has to be measured, connect all its inlets or exhausts to the flow bench. If a system is leak free, the total inlet and exhaust airflow rates will be equal. However, all mechanical systems may have some air leaks and measuring both the total inlet and the total exhaust airflow rates is recommended to quantify these leaks.
2. Connect either the inlet or the exhaust of the blower to the right plenum of the flow bench as shown in Figure 15.2 such that the airflow generated inside the flow bench, by the DUT and the blower, are in the same direction.
3. Select an appropriate nozzle size, based on a guess of the amount of airflow rate and the corresponding pressure drop through that nozzle at that airflow rate, and unplug that nozzle. Note that the pressure drop through a nozzle increases with airflow rate. This pressure drop should be within the measurable range of the manometer if that nozzle is to be selected to measure the airflow rate. If the airflow rate is so high such that the pressure drop through the largest nozzle is higher than the maximum pressure that can be measured by the manometer, two or more nozzles are unplugged.
4. Align the nozzles such that they contract in the same direction as the airflow in the flow bench.
5. Turn on the fans in the DUT. This will generate either a negative or a positive pressure difference ΔP depending on whether the inlet(s) or the exhaust(s) of the DUT are connected to the flow bench, respectively. Note that the fans inside the DUT have to overcome the impedance of the DUT as well as the impedance of the flow bench including the open nozzles. That is the reason for a negative ΔP when DUT pulls air from the flow bench and a positive ΔP when DUT pushes air into the flow bench.
6. A DUT will not be connected to a flow bench during its normal operation and shall not need to overcome its impedance. Therefore, the blower of the flow bench has to be used to balance the negative or positive ΔP mentioned above. This blower is turned on and its speed is increased until $\Delta P = 0$.
7. ΔP_n is measured and the correlations between ΔP_n and nozzle airflow rate is used to calculate the airflow rate through that nozzle, which is the same as the airflow rate through the inlet(s) or exhaust(s) of the DUT connected to the flow bench.

Experimental Techniques and Thermal Design

A widely used technique to measure the flow rate of liquids is based on the drag force a moving fluid imposes on an object. Consider an object exposed to a moving fluid with average velocity U_m. The drag force applied to this object is given by [1]

$$F_D = \frac{1}{2} C_D A_F \rho U_m^2, \qquad (15.6)$$

where C_D is the drag coefficient, A_F is the frontal area of the object, and ρ is the fluid density. Note that C_D is a function of the geometry of the object as well as the fluid properties and the flow velocity (Reynolds number).

The schematic of a rotameter, which is a flow meter based on drag force, is shown in Figure 15.4a. It is a tapered vertical tube with a floating object inside it. The flow enters the rotameter at the bottom of the tube and causes the floating object to move upward. Let's assume that the object stays afloat at some height y. A force balance on the object gives

$$F_D + \rho \mathcal{V}_b g = \rho_b \mathcal{V}_b g, \qquad (15.7)$$

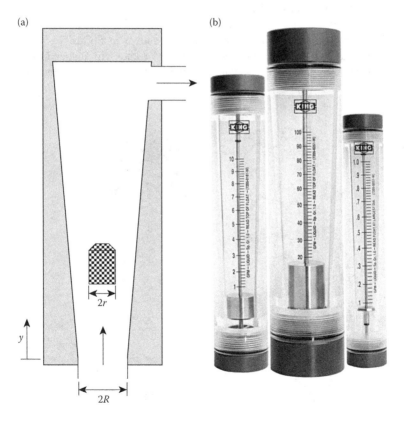

FIGURE 15.4 (a) Schematic of a rotameter and (b) three commercial rotameters. (Courtesy of King Instruments.)

where ρ_b and $\mathcal{V}_b$ are density and volume of the floating object, respectively. The second term on the left side of Equation 15.7 is the buoyancy force applied on the object. Substituting for F_D from Equation 15.6 and solving for U_m gives

$$U_m = \sqrt{\frac{2\mathcal{V}_b g}{C_D A_F}\left(\frac{\rho_b}{\rho} - 1\right)}. \tag{15.8}$$

The values of C_D is determined such that U_m in Equation 15.6 is the average fluid velocity in the annular space between the object and the tube. If A is the annular flow cross section area around the floating object, the flow rate is

$$\dot{\mathcal{V}} = A U_m = A\sqrt{\frac{2\mathcal{V}_b g}{C_D A_F}\left(\frac{\rho_b}{\rho} - 1\right)}. \tag{15.9}$$

The annular cross section area in a rotameter is usually related to the tube inlet radius, R, the radius of the floating object, r, and the distance from the flow entrance, y, by

$$A = \pi\left[(R + \alpha y)^2 - r^2\right], \tag{15.10}$$

where α is a measure of the tube taper.

Another popular device for measuring liquid or gas flow rate is the turbine flow meter. Two commercial turbine flow meters are shown in Figure 15.5. The fluid impinges on the blades of a turbine and forces it to rotate with a speed that is proportional to the flow rate. Most turbine flow meters have flow straighteners at their

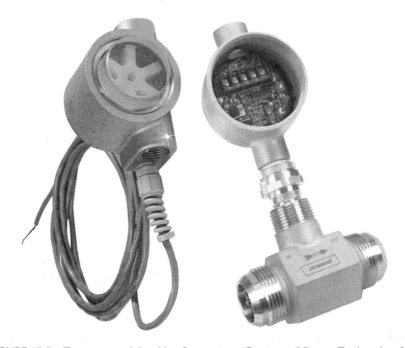

FIGURE 15.5 Two commercial turbine flow meters. (Courtesy of Omega Engineering, Inc.)

Experimental Techniques and Thermal Design 407

inlets to reduce turbulent fluctuations and to provide relatively uniform flows to the blades. The mechanical rotation of the turbine is converted into an electrical signal whose frequency is proportional to the flow rate. This frequency is measured and converted to the fluid flow rate.

15.2 SYSTEM IMPEDANCE MEASUREMENT

A plot of pressure drop versus airflow rate through a system is called the system impedance curve. A flow bench can be used to obtain the system impedance curve of a DUT as described below:

1. Determine the minimum airflow rate for which the DUT impedance has to be measured and unplug the smallest nozzle that allows that airflow rate to be measured.
2. If the fans in the DUT are close to its inlets and push air into the DUT (i.e., the DUT is a push system), connect the inlets of the DUT to the left plenum of the flow bench shown in Figure 15.2. Otherwise (i.e., if the DUT is a pull system), connect the exhausts of the DUT to that plenum.
3. Connect either the intake or the exhaust of the blower to the right plenum of the flow bench such that the flow bench blows air toward the push system or draws air from a pull system.
4. Align the nozzles such that they contract in the same direction as the airflow in the flow bench.
5. Run the blower of the flow bench and change its speed until ΔP_n corresponding to the desired airflow rate is obtained.
6. Record ΔP and ΔP_n and the size of the unplugged nozzle. ΔP_n and the nozzle size give the airflow rate through the DUT and ΔP gives the pressure drop through it. This is the first point on the system impedance curve of the DUT (i.e., ΔP_1 and $\dot{V}_1$).
7. Increase the airflow rate through the DUT by increasing the blower speed and measure the airflow rates $\dot{V}_i$ at a few different ΔP_i values. If necessary, larger nozzle sizes have to be used to measure higher airflow rates.
8. A plot of the pressure drops versus airflow rates is called the system impedance curve. It was mentioned in Chapter 11, section 11.7.2 that most system impedance curves can be fit in the form:

$$\Delta P = C_1 \dot{V} + C_2 \dot{V}^n, \tag{15.11}$$

where n is a value close to 2. This form can be used to find a curve fit to the measured data if an analytical expression for the system impedance curve is needed. Figure 15.6 shows the measured points and a curve fit for the impedance curve of a typical DUT.

Note that the fans inside the DUT shall be off during the system impedance measurement. However, if those fans are left inside the DUT, they will create additional

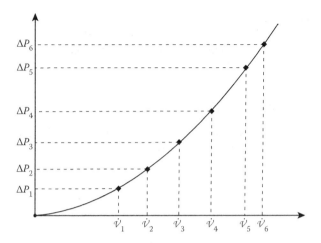

FIGURE 15.6 System impedance curve of a DUT.

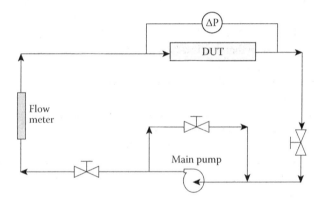

FIGURE 15.7 A test setup for measuring the impedance of a liquid cooling device.

flow impedance and if they are removed out of the DUT, the airflow path and therefore the system impedance will be altered. One may measure the system impedance curves with fans inside and with fans outside the DUT and consider those as the envelopes that contain the actual system impedance curve. Another solution is to measure the system impedance curve while the fan frames, without the fan blades, are kept inside the DUT.

The impedance of a liquid cooling device such as a cold plate can be measured using a test setup similar to the one shown in Figure 15.7. The main pump and the flow meter in the test setup play the same roles as the blower and the nozzle in the flow bench. The flow rate through the device under test is controlled by the valve that is parallel with the main pump, as well as the two valves upstream of the flow meter and downstream of the DUT. The pressure drop through the DUT is measured by a differential pressure sensor connected to the inlet and exhaust of the DUT. A plot of ΔP versus flow rate gives the impedance curve of the DUT.

15.3 FAN AND PUMP CURVE MEASUREMENTS

Flow benches are also used to measure the performance curve of a single fan or fan trays that consists of a set of series and/or parallel fans. The procedure is as described below:

1. Connect the fan or the fan tray under test to the left plenum of the flow bench shown in Figure 15.2.
2. Align the nozzles such that they contract in the same direction as the airflow generated by the fan.
3. Connect the appropriate side of the blower to the right plenum of the flow bench such that the airflow generated by the blower is in the same direction as the airflow generated by the fan or the fan tray under test.
4. Plug all nozzles and close the blast gate such that no air can pass through the flow bench and make sure there is no leak.
5. Turn on the fan, set the voltage at the desired value, and measure pressure difference ΔP. Since there is no airflow through the flow bench, this pressure is the maximum fan pressure at zero airflow rate.
6. Unplug the smallest nozzle and open the blast gate. This will cause a drop in ΔP. Record both ΔP and ΔP_n. The measured ΔP and the calculated airflow rate using the measured ΔP_n gives another point of the fan curve (i.e. ΔP_1 and $\dot{V}_1$).
7. Turn on the blower of the flow bench and increase its speed to reduce ΔP further and record both ΔP and ΔP_n. This gives another point on the fan curve (i.e. ΔP_2 and $\dot{V}_2$).
8. Continue increasing the speed of the blower until either ΔP_n is outside the range to be measured by the manometer, in which case the small nozzle is plugged and the next size nozzle is unplugged, or $\Delta P = 0$ in which case the maximum airflow rate the fan can blow, at free flow or zero pressure condition, is achieved. Record both ΔP and ΔP_n at several steps between maximum ΔP and $\Delta P = 0$. These give points $\Delta P_3, \Delta P_4, ..., 0$ vs. $\dot{V}_3, \dot{V}_3, ... \dot{V}_{max}$.
9. Plot measured pressure rises ΔP_i versus airflow rates $\dot{V}_i$ as shown in Figure 15.8. This is called the fan performance curve.

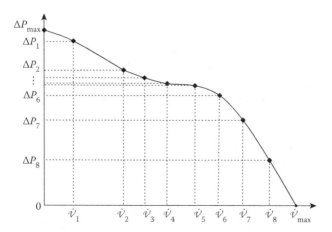

FIGURE 15.8 A measured fan performance curve.

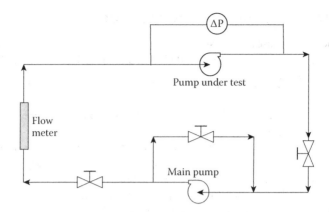

FIGURE 15.9 A test setup to measure performance curve of a pump.

The performance curve of a pump can be measured using a test setup similar to the one shown in Figure 15.9. The flow rate through the pump is measured by the flow meter and the pressure rise through the pump is measured by the differential pressure sensor connected to the inlet and exhaust of the pump. The flow rate through the pump under test, and therefore the pressure rise through the pump, can be varied by the valve that is parallel with the main pump as well as the two valves upstream of the flow meter and downstream of the pump under test.

15.4 VELOCITY MEASUREMENT METHODS

A simple method to measure fluid velocity is based on the Bernoulli equation, which states that the sum of the static pressure and the kinetic energy per unit volume of a fluid is constant if the flow is frictionless and adiabatic:

$$P + \frac{1}{2}\rho V^2 = P_t = \text{constant}. \tag{15.12}$$

P_t is called the total or stagnation pressure of the fluid and is equal to its static pressure where the flow comes to rest. Pitot-static tube is a device that measures fluid velocity by measuring its total and static pressure at the same time [3]. It is made of two concentric tubes that are connected to two sides of a manometer as shown in Figure 15.10. The inner tube is open in the front of the probe, and, therefore, the side of the manometer that is connected to it senses the total pressure of the fluid. There are small holes around the periphery of the outer tube, at least eight outer diameters away from the front of the probe, that sense the static pressure of the fluid. The difference between the total and static pressure of the fluid is measured by the manometer and is converted to the fluid velocity. Pitot-static tubes have simple constructions, are highly accurate, and relatively cheap. However, they are prone to error if not well aligned with the flow direction, and may alter the flow. Pitot-static tubes are widely

Experimental Techniques and Thermal Design

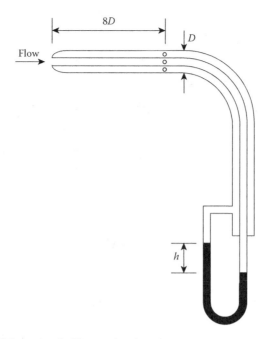

FIGURE 15.10 Schematic of a Pitot-static tube.

used for measuring aircraft air speed but used only for the calibration of the other velocity measurement probes that are used in the electronic equipment.

A device that is often used to measure velocity in electronic equipment is hot-wire anemometer. It is usually made of a platinum or tungsten wire that is about 5–10 micrometers in diameter and about 1 mm in length. If such a wire is heated electrically and exposed to a flow, the heat transfer rate from it is given by

$$\dot{Q} = (a + bU^{0.5})(T_w - T_\infty), \quad (15.13)$$

where T_w and T_∞ are the wire and fluid temperatures, U is the fluid velocity, and a and b are constants obtained from the calibration of the device. The heat transfer rate from the wire is equal to the heat generated by the electrical current through the wire:

$$\dot{Q} = I^2 R_w = I^2 R_0 \left[1 + \alpha(T_w - T_0)\right]. \quad (15.14)$$

I is the current through the wire and R_w is the electrical resistance of the wire at temperature T_w. It is assumed that the electrical resistance of the wire changes linearly with temperature and α is the temperature coefficient of resistance.

The hot wire is connected to a bridge circuit as shown in Figure 15.11. The current through the wire is determined by measuring the voltage drop across the standard resistance R_s and the wire resistance is determined from the bridge circuit [3]:

$$R_w + R_s = \frac{R_1 R_3}{R_2}. \quad (15.15)$$

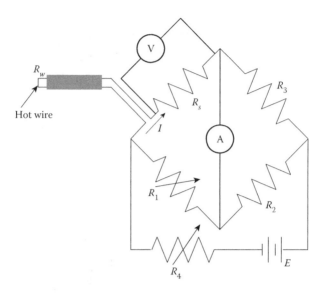

FIGURE 15.11 Schematic of a bridge circuit for a hot-wire anemometer device.

Having known I and R_w, the heat transfer rate and T_w are determined from Equation 15.14 and the fluid velocity is determined from Equation 15.13 if fluid temperature is measured independently.

The main advantages of hot-wire anemometers are that they can be built as small and low intrusion probes and have a very fast response. They are suitable for measuring both steady-state and transient velocities including measurement of turbulent velocity fluctuations. However, their responses depend on the angle the flow makes with the axis of the wire. The best result is usually obtained if the flow is perpendicular to the axis of the wire. A cross-wire anemometer is made of two or three wires that are perpendicular to each other and allows the simultaneous measurement of different components of the velocity at a point. Disadvantages of hot-wires are that they are fragile, accumulate dust and particulates, and are hard to clean. Hot-film anemometers, which are small insulating cylinders coated with a thin metallic film, such as a quartz rod with a platinum film, are more robust and less susceptible to fouling and easier to clean. Hot-film anemometers are used where hot-wires easily break such as in water flow measurements.

Hot-wire anemometers are commercially available as individual sensors that can be connected to a common data acquisition unit as shown in Figure 15.12 or as stand-alone probes with their own portable data loggers as shown in Figure 15.13. An important fact that should be considered when deciding which hot-wire anemometry device to use, is the amount of flow blockage created by the device compared to the entire flow passage area. Most electronic systems are very compact and flow passages are relatively narrow. The probe section of most handheld hot-wire anemometer devices is big enough to either block or significantly alter the flow through these flow passages. This must be kept in mind when interpreting velocity measurements taken by these probes.

Experimental Techniques and Thermal Design 413

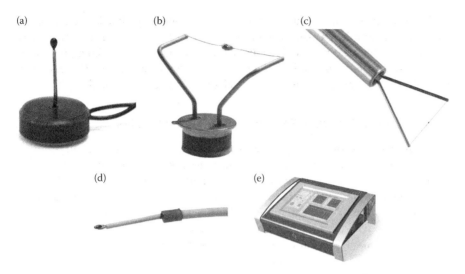

FIGURE 15.12 (**See color insert following page 240.**) Examples of hot-wire probes (a, b, c, and d) and data acquisition unit (e) (Courtesy of Advanced Thermal Solutions, Inc.).

FIGURE 15.13 A portable handheld hot-wire anemometer. (Courtesy of Omega Engineering, Inc.)

Another type of sensor that is commonly used to measure air velocity uses glass-encapsulated thermistors instead of wires or films [4]. Since the thermistors are glass encapsulated, they are more rugged and easier to handle. The working principle of these sensors is the same as the hot-wire and hot-film anemometers. However, they normally measure both air temperature and air velocity at the same time.

15.5 TEMPERATURE MEASUREMENT TECHNIQUES

The most widely known temperature measurement device is the mercury-in-glass thermometer. A relatively large amount of mercury is kept inside a bulb at the bottom of the thermometer. The mercury expands when heated and rises in a capillary tube attached to the bulb. The height of the mercury in a capillary tube is an indication of the temperature of the medium the bulb is exposed to. These thermometers are widely used to measure room air temperature and body temperature. However, the use of mercury has been diminished due to the environmental concerns and alcohol and alternative metals such as galinstan have been used instead.

Electrical resistance of many metals such as copper, aluminum, nickel, tungsten, and platinum varies with temperature. The variation of resistance with temperature is linear if the temperature change is small;

$$R = R_0[1 + \alpha(T - T_0)]. \qquad (15.16)$$

For a wider temperature range, the resistance of the material is expressed by a quadratic relation;

$$R = R_0(1 + a(T - T_0) + b(T - T_0)^2). \qquad (15.17)$$

R_0 is the electrical resistance of the material at temperature T_0, α is the linear temperature coefficient of resistance, and a and b are experimentally determined constants specific to each material. Resistance temperature detectors (RTDs) are resistors whose electrical resistance is measured and converted to the temperature they are exposed to using either Equation 15.16 or 15.17.

Thermistors are semiconductor devices whose resistance is inversely proportional to temperature. The variation of the resistance of a thermistor with temperature is given by an exponential relation such as

$$R = R_0 \exp\left[\beta\left(\frac{1}{T} - \frac{1}{T_0}\right)\right], \qquad (15.18)$$

where R_0 is the resistance at temperature T_0 and β is an experimentally determined constant. Thermistors are very sensitive devices and are accurate within 0.01°C.

The most widely used sensor to measure temperature in electronic cooling applications is the thermocouple. It is a circuit formed by two dissimilar metals that are connected electrically in series as shown in Figure 15.14. The connection points are called junctions. Thermocouples work based on the Seebeck effect, which was discovered by the German physicist Thomas J. Seebeck in 1821. It states that if the

Experimental Techniques and Thermal Design

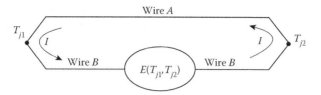

FIGURE 15.14 Illustration of Seebeck effect.

two junctions of a thermocouple are kept at different temperatures T_{j1} and T_{j2}, an electromotive force (emf) or voltage, $E(T_{j1}, T_{j2})$, arises that is a function of those temperatures (see Figure 15.14). Common metal pairs are copper versus constantan (type T), iron versus constantan (type J), chromel versus alumel (type K), chromel versus constantan (type E), and platinum versus 90/10 platinum rhodium (type S). Different types of thermocouples are used for different temperature ranges and have different tolerances. Although type T thermocouples have lower tolerances than type K thermocouples, both are widely used to measure temperatures in electronic equipment.

The Seebeck effect gives the temperature difference between two junctions as a function of the voltage created between those junctions. Therefore, if temperature at a reference junction is known, temperature at the other junction can be determined. A common method to calibrate thermocouples is to keep the reference junction at 0°C by inserting it into an ice bath and to measure voltage as a function of the temperature at the other junction; $E(T) = E(T, 0)$. However, the reference junction of a thermocouple is at a temperature other than 0°C in most applications. The law of intermediate temperatures states that if a thermocouple generates the voltage $E(T_1, T_2)$ when operated between temperatures T_1 and T_2 and the voltage $E(T_2, T_3)$ when operated between temperatures T_2 and T_3, then it will generate $E(T_1, T_3) = E(T_1, T_2) + E(T_2, T_3)$ when operated between temperatures T_1 and T_3. The calibration curve, the law of intermediate temperatures, and measurement of the voltage generated by the thermocouple and the temperature of the reference junction by a thermistor in thermal contact with that junction are used to calculate the temperature on the other junction of the thermocouple. For example, let's assume that the measured temperature at the reference junction, and the voltage generated by the thermocouple are T_{j1} and E_{tc}, respectively. The calibration curve gives the thermocouple voltage between 0°C and T_{j1} as $E(T_{j1})$, and the law of intermediate temperatures give the thermocouple voltage between 0°C and the other junction temperature T_{j2} as $E(T_{j2}) = E_{tc} + E(T_{j1})$. Having calculated $E(T_{j2})$, T_{j2} can be found from the calibration curve. Modern thermocouple meters have built in reference temperatures and automatically convert the thermocouple voltage into a temperature.

Another well-known method for measuring temperature of an object is radiation thermometry. It was shown in Chapter 13 that the total thermal radiation emitted from an object at temperature T is

$$E(T) = \varepsilon(T)\sigma T^4, \tag{15.19}$$

where $\varepsilon(T)$ is the total emissivity of the object at temperature T. Equation 15.19 shows that if thermal radiation emitted from a surface is measured and its total emissivity is known, its temperature can be calculated. However, the accuracy of the calculated temperature depends on the accuracy with which the emissivity of the surface is known. Equation 15.19 shows that the error in calculated temperature is correlated to the error in surface emissivity by

$$\frac{\Delta T}{T} = -\frac{1}{4}\frac{\Delta \varepsilon}{\varepsilon}. \tag{15.20}$$

This equation shows that the absolute error in the calculated temperature, ΔT, for a given absolute error in the assumed surface emissivity, $\Delta \varepsilon$, is smaller for surfaces with higher emissivity values.

Radiation coming from a surface is not all emitted from that surface. It also includes the reflected part of the irradiation that is incident on the surface; the sum of the emitted and reflected radiations from a surface was named radiosity in Chapter 13. Measuring radiosity instead of the emitted radiation will result in an additional error in the calculated temperature. Both this and the concern regarding the error in the value of surface emissivity can be addressed by adding a thin layer of paint over the surfaces whose temperatures are measured. Many paints have emissivity values close to one and reflectivity values close to zero, for the thermal radiation, and therefore minimize both errors. Figure 15.15 shows an aluminum plate covered with paints and tapes of various colors at different locations. It is seen that all those provide accurate measurements of the temperature regardless of the color.

Figure 15.16 shows a spot or point measuring radiation thermometer and a full thermal imaging camera that simultaneously records temperatures over an area and creates a two-dimensional temperature map similar to the one shown in Figure 15.17. Both are very useful in quick diagnostic measurements where the goal is to find the hot components on a board or the hottest areas on the exterior surfaces of an electronic product such as a cell phone.

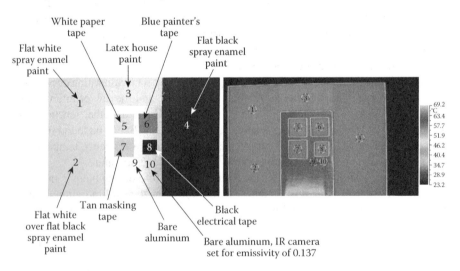

FIGURE 15.15 (See color insert following page 240.) An aluminum plate covered with different paints and tapes and the corresponding infrared temperature measurement. (Courtesy of Dave Dlouhy and Doug Twedt, Rockwell Collins.)

Experimental Techniques and Thermal Design 417

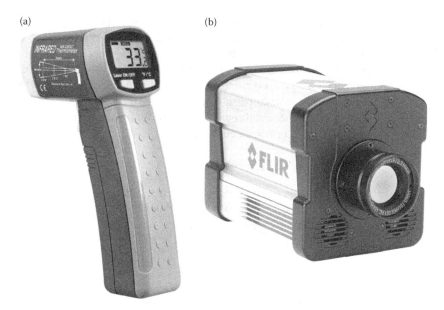

FIGURE 15.16 (**See color insert following page 240.**) (a) A handheld spot radiation thermometer (Courtesy of Omega Engineering, Inc.) and (b) a thermal imaging camera. (Courtesy of FLIR systems.)

FIGURE 15.17 (**See color insert following page 240.**) A two-dimensional temperature map obtained by a thermal imaging camera. (Courtesy of FLIR Systems.)

15.6 ACOUSTIC NOISE MEASUREMENTS

Sound waves are pressure waves. Therefore, their strength is characterized either by the pressure at a specific distance from the sound source or by the total energy or power of the source. Sound pressure is referred to the root mean square value of the instantaneous pressures created by a sound wave during a time interval. Sound pressure is what causes our ears to hear the sound. It is affected by the medium the sound travels through, and therefore diminishes at large distances from the source. It is measured with the unit Pascal (Pa). The sound pressure at the threshold of hearing is about 20 µPa and the hearing damage due to a long-term exposure happens at a sound pressure of about 0.356 Pa. Sound power is the total power emitted from a sound source in all directions. It is a property of the sound source and does not depend on the medium the sound travels through or the distance from the sound source. Sound power is measured in watts (W).

Sound pressure level (*SPL* or L_p) is defined as

$$SPL = 20\log_{10}\left(\frac{P}{P_{ref}}\right), \tag{15.21}$$

where $P_{ref} = 20$ µPa is the reference sound pressure and *SPL* is measured with the unit decibels (dB). Note that $SPL = 0$ at the threshold of hearing and $SPL = 85$ when hearing damage due to a long-term exposure may happen.

Sound power level (*SWL* or L_w) is defined as

$$SWL = 10\log_{10}\left(\frac{W}{W_{ref}}\right), \tag{15.22}$$

where $W_{ref} = 10^{-12}W$ is the minimum sound power that can be heard at 28 cm away from the sound source. Sound power level is measured with the unit decibels (dB).

Sound measurements are performed in anechoic or acoustic chambers or rooms, such as the one shown in Figure 15.18, which create an environment with a very low background noise. The sound pressure is measured by special microphones or sound level meters at a distance from the sound source that is specified by the applicable standards [5]. The sound pressure should be measured on different sides of a product to find the maximum noise emitted from it.

A common method to calculate sound power level uses measured sound pressures at a few locations on an area that surrounds the DUT [5]. If the average sound pressure level of *N* different measured values SPL_i is

FIGURE 15.18 (See color insert following page 240.) The interior of an anechoic chamber. (Courtesy of Eckel Industries, Inc.)

Experimental Techniques and Thermal Design

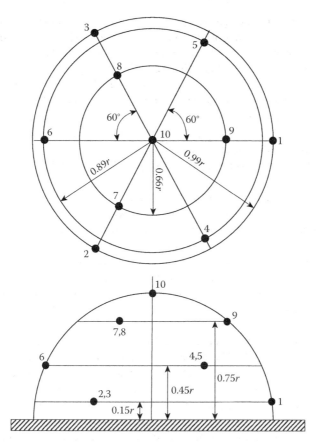

FIGURE 15.19 Locations of the microphones in a typical test setup for measuring sound power level. Each of the ten microphones cover the same area on the hemisphere surrounding the device under test.

$$\overline{SPL} = 10\log_{10}\left(\frac{1}{N}\sum_{i=1}^{N}10^{0.1SPL_i}\right), \quad (15.23)$$

the sound power level is calculated by

$$SWL = \overline{SPL} + 10\log_{10} S, \quad (15.24)$$

where S is the total area of the surrounding surface in m². Figure 15.19 shows the locations of the microphones in a typical test setup used for measuring sound power level.

15.7 IMPORTANCE OF EXPERIMENTAL MEASUREMENTS IN THERMAL DESIGN

Experimental measurements are used to verify that a prototype or product meets design requirements. They are also used to calibrate a thermal model that in turn is used for the thermal design of a product. A prototype can sometimes be easily altered or modified to mimic minor product redesigns. These modified prototypes can be used to investigate the effects of the proposed design changes on the thermal performance of the product. This is thermal design through experiments and, as long as design modifications are done cost effectively and fast, is preferred to the thermal design through modeling and analysis. For example, consider a telecom chassis whose prototype was built and the airflow rate through the chassis was measured and found to be less than required. The thermal engineer decides to look at other fans with similar dimensions to get the required airflow. The fans in the chassis can easily be replaced with other fans and airflow tests will determine whether the airflow rate with those fans is enough.

Although experimental measurements play an important role in the thermal design process, it is important to distinguish between "thermal design" and "thermal measurement." Thermal design is a process in which appropriate heat transfer techniques are used to appropriately cool the electronics inside a system. On the other hand, thermal measurement is a tool that is used to verify that the system is cooled appropriately. Therefore, most thermal design work shall be done before a system is ready for thermal measurement. However, airflow and temperature tests are sometimes used to validate or to calibrate a thermal model that is used in a thermal design process.

The objectives of experimental measurements have to be clearly identified in advance and a test plan shall be prepared to achieve those goals. This test plan shall specify the conditions under which the measurements will be performed, the parameters that have to be measured, the locations those parameters will be measured at, the target values for those parameters if applicable and known, and how the test results are interpreted or used. Having a test plan eliminates, or significantly reduces, the possibility of measuring too many or too few parameters, at too many or too few locations, and reduces measurement time and cost.

All experimental measurements shall be recorded in a report. The report shall be concise but include the objectives of the experiments, pictorial description of the test setup, list of parameters and locations those parameters are measured at, tabulated or graphical data, clear interpretations of the data, and possible recommendations to improve thermal performance of the unit under test. The report shall be self-explanatory such that its message is clearly conveyed.

REFERENCES

1. Fox, R. W., A. T. McDonald, and P. J. Pritchad. 2006. Introduction to fluid mechanics. John Wiley & Sons, Inc.
2. Air Movement and Control Association. 2000. Laboratory Methods of Testing Fans for Aerodynamic Performance Rating, ANSI/AMCA Standard 210–99.
3. Holman, J. P. 2000. Experimental Methods for Engineers. McGraw Hill.
4. Raouf, I. Air Velocity Measurement Using Thermistors, http://www.degreec.com/downloads/Air_Velocity_Measurement_Using_Thermistors.pdf
5. ISO3745. 2003. Acoustics - Determination of Sound Power Levels of Noise Sources Using Sound Pressure - Precision Methods for Anechoic and Hemi-Anechoic Rooms.

16 Advanced Cooling Technologies

The heat transfer mechanisms and simple techniques for cooling electronic equipment were described in previous chapters. Although a continuous rise in the power dissipation of electronics, accompanied by a reduction in their dimensions, has placed stringent conditions on the required cooling solutions, these simple techniques are still used to cool most electronic devices. However, there have been situations where the cooling requirements can not be satisfied through the conventional techniques such as simple air-cooled heat sinks. The methods described in this chapter have been used to address those situations. Although many of these techniques have been used widely and enough to be considered as conventional methods, we classify all of them as advanced cooling techniques as opposed to the simpler and cheaper air-based cooling techniques.

16.1 HEAT PIPES

Heat pipes are passive devices that operate based on the fact that saturated liquids evaporate by absorbing heat from a higher temperature, and saturated vapors condense by releasing heat to a lower temperature. Heat pipe was first introduced by Gaugler [1] at General Motors Corporation. However, his idea was not used by General Motors. Grover [2] reinvented heat pipes independently at Los Alamos National Laboratory and built several prototypes that used water or sodium as the working fluid. He introduced the name heat pipe for his invention and described it as "equivalent to a material having a thermal conductivity greatly exceeding any known metal."

A heat pipe is a sealed pipe whose internal surfaces are covered with a capillary structure or wick as shown in Figure 16.1. It is partially evacuated and filled with a small amount of a working fluid before being sealed. The operation of a heat pipe can be illustrated by dividing it into the evaporator, adiabatic, and condenser sections as shown in Figure 16.1. The pressure inside a heat pipe is equal to the saturation pressure of its working fluid. The heat that is applied to the evaporator section vaporizes the working fluid and increases the pressure there. The higher pressure pushes the vapor through the center of the pipe to the condenser section where it releases its latent heat to a colder temperature and condenses to liquid. The liquid is absorbed by the wick and pumped back to the evaporator section by the capillary force in the wick. Therefore, a heat pipe can continuously transport heat from its evaporator to its condenser section as long as the capillary force is large enough to pump the condensed liquid back to the evaporator section.

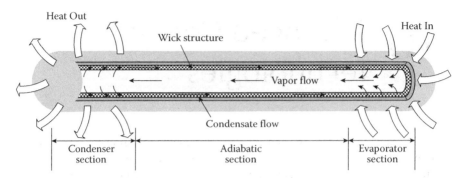

FIGURE 16.1 Schematic of a heat pipe.

Heat pipes are designed with different working fluids, pipe materials, wick materials, and structures. The working fluid is selected based on the intended operating temperature range of the heat pipe. For example, nitrogen is suitable for the temperature range of 70–103 K, silver is used for the temperature range of 2073–2573 K, ammonia is used for the temperature range of 213–373 K, while water is used for the temperature range of 303–473 K [3]. Since most electronic cooling applications require the temperature range of 25–150°C, water heat pipes are the most commonly used ones. The pipe and wick materials are chosen such that they are chemically compatible with each other and with the working fluid. For example, most water heat pipes that are used in electronic applications are made of copper, silver heat pipes are made of tungsten, and ammonia heat pipes are made of aluminum or stainless steel [3].

The wick structure in a heat pipe is needed to return the condensed fluid to the evaporator section. Three important properties of the wicks used in heat pipes are their pore radius, permeability, and thermal conductivity. The pore radius should be small enough to create the capillary pressure difference needed to pump the liquid from the condenser to the evaporator. Permeability is inversely related to the wick resistance to the liquid flow and should be large such that the liquid pressure drop through the wick is low. Finally, larger values of thermal conductivity results in smaller temperature drops across the wick. It is noted that high thermal conductivity and permeability, and low pore radius are somewhat contradictory properties for a wick. Therefore, compromises need to be made between these competing properties to obtain an optimal wick design for a heat pipe.

The three most commonly used wick structures for the heat pipes used in electronic cooling applications are the screen, sintered, and groove wicks. The screen wick is made of a metal fabric that is wrapped around a mandrel, inserted into the heat pipe, and is held there by the tension of the wrapped screen after the mandrel is removed. The capillary pressure is determined by the size of the rectangular pores of the screen and the permeability is determined by the looseness of the wraps that create annular gaps for the condensate flow. Sintered metal wicks are manufactured by packing tiny metal particles, in powder form, between the inner wall of the heat pipe and a mandrel. This assembly is then heated until the metal particles are sintered to each other and to the inner wall of the heat pipe. The mandrel is then removed leaving an open central portion in the heat pipe. Groove wicks are formed

Advanced Cooling Technologies

by extruding the grooves into the inner wall of the heat pipe. Rectangular, triangular, trapezoidal, and circular grooves are the used although trapezoidal grooves are the most common. Since the size of the grooves is large compared to the screen or sintered pore, capillary pumping pressure of the grooved heat pipes is very small. However, their permeability and wick thermal conductivity are very high.

The heat transfer capability of a heat pipe is limited by several factors that must be addressed when designing a heat pipe for a specific application. These heat transfer limits will be described in the following sections [3].

16.1.1 Capillary Limit

Consider a typical heat pipe in which the condenser section is located below the evaporator section as shown in Figure 16.2. The capillary pumping pressure, ΔP_c, is the only driving force to circulate the working fluid in this heat pipe. On the other hand, the pressure losses along the vapor–liquid path are the liquid pressure loss from the condenser to the evaporator, ΔP_l, the vapor pressure loss from the evaporator to the condenser, ΔP_v, and the gravitational pressure drop, ΔP_g. The proper operation of a heat pipe requires that:

$$\Delta P_c \geq \Delta P_l + \Delta P_v + \Delta P_g. \tag{16.1}$$

The pressure drop in the liquid path is given by

$$\Delta P_l = \frac{\mu_l L_{\text{eff}} \dot{Q}}{\rho_l K A_l h_{fg}}, \tag{16.2}$$

where $\dot{Q}$ is the heat transfer rate to the heat pipe, μ_l and ρ_l are the dynamic viscosity and density of the liquid working fluid, h_{fg} is the latent heat of the working fluid, K is

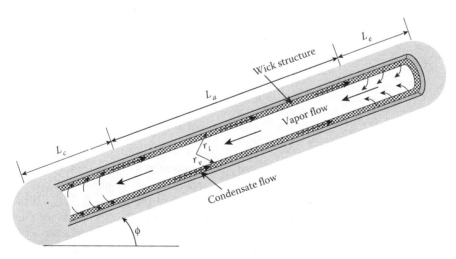

FIGURE 16.2 A typical orientation for heat pipe where it makes an angle with the horizontal direction.

the permeability of the wick, A_l is the wick cross section area, and the effective length of the heat pipe, L_{eff}, is related to the evaporator, condenser, and adiabatic lengths by

$$L_{\text{eff}} = L_a + 0.5(L_e + L_c). \quad (16.3)$$

The vapor pressure drop is the sum of the inertial pressure drop to bring the vapor from the wick to the center of the pipe and to give it the initial push in the axial direction, and the frictional pressure drop at radius r_v where the vapor contacts the wick surface;

$$\Delta P_v = \frac{\dot{Q}^2}{\rho_v A_v^2 h_{fg}^2} + \frac{8\mu_v L_{\text{eff}} \dot{Q}}{\rho_v \pi\, r_v^4 h_{fg}}. \quad (16.4)$$

Here, μ_v and ρ_v are dynamic viscosity and density of the vapor working fluid and $A_v = \pi r_v^2$ is the vapor flow area. The first term is usually neglected compared to the second term.

The gravitational pressure drop is given by

$$\Delta P_g = \rho_l g L_{\text{eff}} \sin\phi, \quad (16.5)$$

where g is the acceleration of gravity and ϕ is the angle between the heat pipe and the horizontal direction. Note that $\sin\phi > 0$ if $\phi > 0$ and gravity works against the liquid flow while $\sin\phi < 0$ if $\phi < 0$ and gravity helps the flow of the liquid working fluid.

The maximum capillary pressure is given by

$$\Delta P_{c,\max} = \frac{2\sigma}{r_c}, \quad (16.6)$$

where σ is the liquid surface tension and r_c is the wick pore radius.

Inserting Equations 16.2, 16.4, 16.5, and 16.6 into Equation 16.1, neglecting the first term in Equation 16.4, and solving for $\dot{Q}$ gives the capillary limit of a heat pipe:

$$\dot{Q} \leq \frac{2\sigma/r_c - \rho_l g L_{\text{eff}} \sin\phi}{\dfrac{\mu_l L_{\text{eff}}}{\rho_l K A_l h_{fg}} + \dfrac{8\mu_v L_{\text{eff}}}{\rho_v \pi r_v^4 h_{fg}}}. \quad (16.7)$$

16.1.2 Boiling Limit

The boiling mechanism around a heated surface submerged in a stagnant volume of liquid, and the associated heat flux from it, depends on the temperature difference between the surface and the saturation temperature of the liquid, $T_w - T_{\text{sat}}$. If the temperature difference is very small, heat transfer rate from the surface to the fluid is due to natural convection only. As the temperature difference increases, vapor bubbles are formed at the surface by absorbing the latent heat from that surface. This is called the nucleate boiling regime where very high heat fluxes are obtained with relatively small temperature difference. At higher temperature differences, bubbles

Advanced Cooling Technologies

become so large and their numbers grow so fast that they create a continuous vapor film near the surface. This layer prevents the liquid from direct contact with the surface and reduces the heat transfer rate from the surface. The maximum heat flux that is reached before the heat transfer rate is decreased is called the critical heat flux. As the temperature difference is further increased, radiation through the vapor film becomes important and the heat transfer rate starts to increase again.

Although heat transfer from a wick surface is more complex than that from a flat surface, similarities exist between the two. If temperature difference between the wick surface and the saturation temperature of the working fluid is very low, no boiling happens within the wick. The wick will always be filled with liquid and vaporization takes place at the liquid–vapor interface. As heat flux is increased at the same low temperature difference, the evaporation at the liquid–vapor interface is intensified and the amount of liquid in the evaporator section is reduced. If the capillary pressure is not able to pump enough liquid back to the evaporator section, the heat pipe burns out due to the capillary limit. However, if the temperature difference across the wick is increased, nucleate boiling takes place within the wick. Nucleate boiling is a favorable heat transfer mechanism unless the bubbles are so large that they can not escape from the wick and block the return of the liquid from the condenser. As the temperature difference across the wick is increased, a large number of bubbles coalesce together and form a layer of vapor that prevents liquid from reaching the evaporator. The heat pipe wall temperature will increase rapidly and the heat pipe may burn out. This is the boiling heat transfer limit for a heat pipe.

The critical temperature difference and the corresponding boiling heat transfer limit are given by [3]

$$\Delta T_{crit} = T_w - T_v = \frac{2\sigma T_v}{h_{fg}\rho_v}\left(\frac{1}{R_b} - \frac{1}{R_{men}}\right) \qquad (16.8)$$

and

$$\dot{Q}_b = \frac{2\pi L_e k_w \Delta T_{crit}}{\ln(r_i/r_v)}. \qquad (16.9)$$

Here, T_w and T_v are the heat pipe wall and vapor temperatures at the evaporator section, L_e is the length of the evaporator section, k_w is the effective thermal conductivity of the wick, and r_i and r_v are the inner radius of the pipe wall and the vapor space radius, respectively. R_b is the effective radius of the bubble, and is approximated by

$$R_b = \sqrt{\frac{2\sigma T_{sat} k_l (\rho_l - \rho_v)}{h_{fg} \dot{q}_r \rho_l \rho_v}}, \qquad (16.10)$$

where $\dot{q}_r$ is the radial heat flux, k_l is the thermal conductivity of liquid working fluid and T_{sat} is the saturation temperature of the working fluid. R_{men} is the radius of the liquid-vapor meniscus, which is taken to be the same as the effective pore radius of the wick, r_c.

16.1.3 Sonic Limit

The maximum velocity of a compressible fluid passing through a converging–diverging nozzle is equal to the speed of sound in that fluid, and happens at the throat of the nozzle. The vapor flow in the evaporator of a heat pipe is similar to the flow in a converging–diverging nozzle. The vapor velocity increases along the evaporator and reaches a maximum, at the end of the evaporator section, that can not be higher than the local speed of sound. This flow condition is called the sonic limit and the associated heat transfer limit is given by [3]:

$$\dot{Q}_s = \frac{\rho_0 c_0 h_{fg} A_v}{\sqrt{2(\gamma+1)}}, \qquad (16.11)$$

Where ρ_0 and c_0 are the vapor density and the speed of sound at the evaporator and $\gamma = C_p/C_v$ is the ratio of the vapor specific heats. Note that the speed of sound is given by

$$c_0 = \sqrt{\lambda R_v T_0}, \qquad (16.12)$$

where T_0 is the vapor temperature at the evaporator and R_v is the gas constant for the vapor.

The sonic limit is usually associated with the liquid–metal heat pipes and occurs during start up or steady operation if the heat transfer coefficient at the condenser is very high. Note that the sonic limit does not represent a serious failure. It represents a limit for the heat transport from the evaporator to the condenser. The only way to increase this limit is to increase the evaporator temperature as can be seen from Equations 16.11 and 16.12.

16.1.4 Entrainment Limit

Since the vapor and liquid move in opposite directions, a shear force exists at the liquid–vapor interface. If the vapor velocity is very high, droplets of liquid are entrained into the vapor and transported to the condenser section rather than flowing to the evaporator section. Excessive entrainment of liquid into the vapor results in the dry out of the evaporator. The heat transfer rate at which this happens is called the entrainment limit and is given by [3]:

$$\dot{Q}_{ent} = A_v h_{fg} \left(\frac{\sigma \rho_v}{2 R_{h,w}} \right)^{1/2}, \qquad (16.13)$$

where $R_{h,w}$ is the hydraulic radius of the wick pore. For screen wicks $R_{h,w} = 0.5\ W$, for axial grooved $R_{h,w} = W$, and for packed sphere wicks $R_{h,w} = 0.205\ D$ where W is the wire spacing for screen wick and the groove width for axial grooves and D is the sphere diameter for packed sphere wicks.

16.1.5 OTHER HEAT PIPE PERFORMANCE LIMITS

Other heat pipe performance limitations that are less common in heat pipes used for electronic cooling applications are the viscous limit, the condenser limit, the frozen start-up limit, and the continuum flow limit. These performance limits are briefly described in this section.

At low operating temperatures, viscous forces are dominant in the vapor flow from the evaporator to the condenser and cause a steep pressure drop in the vapor. The vapor pressure at the condenser end may reach its lowest value. The vapor transport and therefore the heat transfer capacity of the heat pipe will be limited under this condition. This heat transfer limit is called the viscous or the vapor pressure limit and is give by [3]:

$$\dot{Q}_{vis} = \frac{\pi r_v^4 h_{fg} \rho_0 P_0}{16 \mu_0 L_{eff}}. \quad (16.14)$$

If the heat sink that is connected to the condenser section of a heat pipe is not able to dissipate the maximum heat transport capability of the heat pipe, the performance of the heat pipe is limited by the condenser. This is called the condenser heat transfer limit and can be given by

$$\dot{Q}_c = \frac{T_c - T_\infty}{R_c}, \quad (16.15)$$

where T_c and T_∞ are the heat pipe wall temperature at the condenser and the temperature of the cooling fluid around the condenser, and R_c is the total thermal resistance from the condenser wall to the cooling fluid.

The frozen start-up limit happens mainly in the heat pipes designed for operation at high temperatures. The working fluid of these heat pipes is in solid state at normal ambient temperature. If heat is applied to the evaporator section and the temperature exceeds the melting temperature of the working fluid, the working fluid starts to liquefy and evaporation begins to take place at the fluid–vapor interface. The vapor moves toward the condenser section. However, if the temperature in the adiabatic and condenser sections of the heat pipe has not reached the melting temperature of the working fluid, the vapor solidifies. This may deplete the working fluid in the evaporator section and cause the dry out of the evaporator.

The vapor flow in conventional heat pipes is in the continuum state. However, for microheat pipes and for heat pipes operating at very low temperatures, the vapor flow in the heat pipe may be in the free molecular or rarefied condition. The heat pipe becomes a very ineffective heat transfer device in this condition. The continuum criterion is expressed in terms of the Knudsen number:

$$Kn = \frac{\lambda}{2r_v} > 0.01, \quad (16.16)$$

where λ is the mean free path of the vapor molecules.

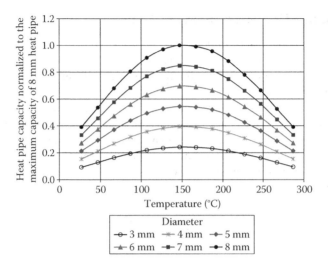

FIGURE 16.3 Effect of the pipe diameter on the performance of typical horizontal heat pipes with water working fluid.

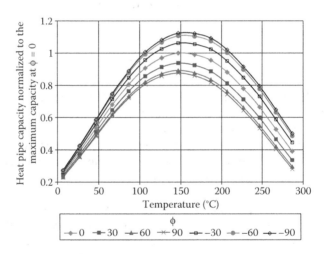

FIGURE 16.4 Effect of gravity on the performance of a typical water-based heat pipe ($\phi > 0$ is for operation against gravity and $\phi < 0$ is for gravity assisted operation.

Figure 16.3 through 16.5 show how the performance of a typical heat pipe change with operating temperature, angle with respect to the direction of gravity, and the heat pipe length and diameter.

16.1.6 Heat Pipe Applications in Electronic Cooling

Heat pipes are extensively used in electronic cooling applications to either transfer heat from where it is originally generated to where it can be dissipated, or to spread a hot spot and to reduce the spreading thermal resistance. There is probably no

Advanced Cooling Technologies

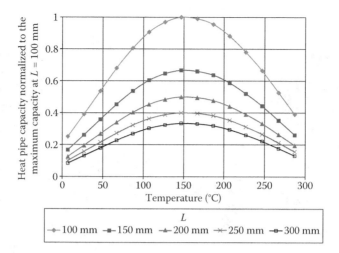

FIGURE 16.5 Effect of length on the performance of a typical heat pipe with water working fluid.

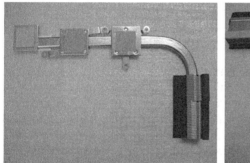 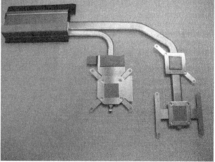

FIGURE 16.6 (See color insert following page 240.) Examples of notebook cooling solutions with heat pipes. (Courtesy of Auras Technology Co. Ltd.)

notebook computer that is built currently and does not have one or more heat pipes in its cooling solution. The reason is that the main heat generating devices such as CPU and GPU are located in places where there is very limited space for any heat sink or fan. Heat pipe based cooling solutions similar to those shown in Figure 16.6 are used to transfer heat from those heat generating components to somewhere close to the periphery of the notebook where heat sink and fan can be placed. Similar situations exist in applications where low noise limit, low maintenance, and high reliability requirements rule out the use of fans. As natural convection heat transfer is less effective than forced convection heat transfer, larger heat sinks are required in these cases. Heat pipes are used to transfer heat to one or multiple heat sinks such that their combined performance meets the cooling requirement.

Many CPU coolers for desktop computers, such as those shown in Figure 16.7, are built with one or more heat pipes. The heat pipes in this case are used to improve the

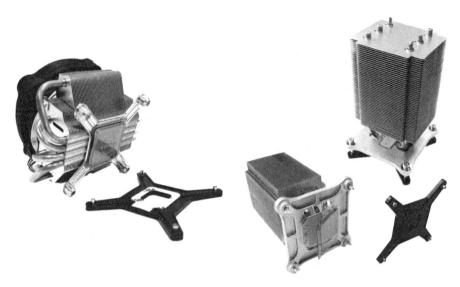

FIGURE 16.7 (See color insert following page 240.) Examples of desktop CPU coolers with heat pipes. (Courtesy of Auras Technology Co. Ltd.)

heat conductivity of the heat sink base (to reduce the spreading resistance) and/or to efficiently transfer heat to the fins or areas of the fins that are farther from the heat sink base and therefore to improve the heat sink efficiency.

Two commonly used heat sink materials are aluminum and copper. The thermal conductivity of copper is about two times that of aluminum but it is heavier and more expensive. Heat pipe is practically equivalent to a material with very high thermal conductivity along its length. It has been experimentally verified that an aluminum heat sink base with appropriately embedded heat pipes has an effective thermal conductivity that is very close to or even higher than a copper base while being significantly lighter. Embedding heat pipes in aluminum or even copper base, similar to those shown in Figure 16.8, is a common practice in electronics cooling to spread a hot spot and to reduce the spreading thermal resistance of the heat sink base.

Heat pipes are usually bent and flattened in order to fit within the specific geometrical restrictions of the intended application. These operations shall be done carefully such that the vapor space is not blocked. However, every bend and flattening has an effect on the performance of a heat pipe. A good practice is to keep the bend radius at least three times the heat pipe diameter, to minimize the number of bends, and to avoid the excessive flattening of the heat pipes. Figures 16.9 and 16.10 show how flattening and bending affect the performance of a typical heat pipe [4].

16.1.7 Heat Pipe Selection and Modeling

Selecting a heat pipe for a particular situation is not an easy task. Thermal engineers have to meet their cooling requirements while keeping the design as simple and as inexpensive as possible. The methods by which heat is transferred to the evaporator section of the heat pipe and the design of the condenser heat sink have to be

Advanced Cooling Technologies 431

FIGURE 16.8 (See color insert following page 240.) Examples of heat sink base with embedded heat pipes. (Courtesy of Aavid Thermalloy.)

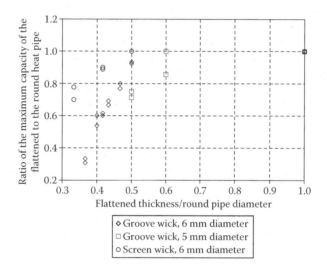

FIGURE 16.9 Effect of flattening on the performance of a heat pipe with water working fluid.

considered as the heat pipe is selected. The following steps are recommended in order to select a heat pipe:

1. Determine the heat dissipation from the heat source.
2. If the heat pipe is used to transport heat from one to another location in a system, identify location of the heat source (heat pipe evaporator) and possible locations for the heat sink (heat pipe condenser). It is always better to have a shorter heat pipe, to minimize operation against gravity, to avoid flattening, and to minimize number and angle of bends.

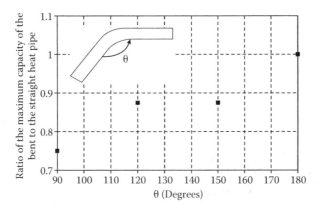

FIGURE 16.10 Effect of bend angle on the performance of a 6 mm-diameter heat pipe, flattened to 2.5 mm thickness, with water working fluid.

3. Identify the smallest diameter heat pipe that can comfortably transfer the heat generated in the source to the condenser section. The heat capacity of a straight heat pipe can be obtained from the heat pipe supplier and the effect of length, flattening, and bending can be estimated using Figures 16.5, 16.9, and 16.10 or similar data for that heat pipe if available.
4. Design the interface between the heat source and the evaporator section of the heat pipe. This is usually a copper block with the evaporator section of the heat pipe embedded in or laid on it and then soldered to it.
5. Design the heat sink for the condenser section and select an appropriate fan or blower if necessary.
6. Create an analytical or computational model for the entire cooling module including heat source, heat pipe, heat sink, and fan and calculate heat source temperature.

Thermal performance of a heat pipe is sometimes specified by its thermal resistance, which is defined by the temperature difference between the evaporator and the condenser divided by the heat transfer through the heat pipe:

$$R_{hp} = \frac{T_e - T_c}{\dot{Q}}. \qquad (16.17)$$

An analytical model for the thermal resistance of a heat pipe is obtained by tracking the heat through the heat pipe as shown in Figure 16.11. A small amount of the heat that enters the heat pipe at the evaporator section is conducted axially through the wall of the heat pipe to the condenser section, and the rest goes into the wick. Most of the heat that is transferred to the wick is conducted radially to the liquid-vapor interface where it is absorbed by the vaporizing liquid as latent heat, and the rest is conducted axially through the wick thickness. The vapor transfers the heat to the condenser section where it is transferred to the wick by condensation and conducted through the wick and heat pipe wall to the condenser.

Advanced Cooling Technologies

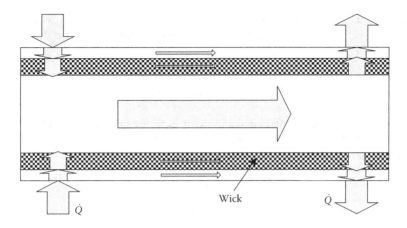

FIGURE 16.11 Heat flow path in a typical heat pipe.

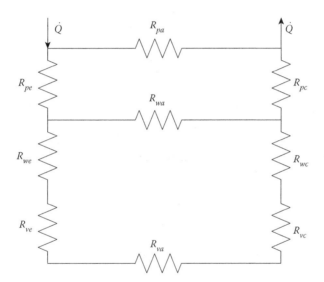

FIGURE 16.12 Thermal resistance network for a typical heat pipe.

Figure 16.12 shows a resistance network for the heat pipe shown in Figure 16.11. R_{pe} and R_{pc} are the radial conduction thermal resistances of the heat pipe wall at the evaporator and condenser sections, respectively, and are given by

$$R_{pe} = \frac{\ln(r_o/r_i)}{2\pi L_e k_p}, \tag{16.18}$$

$$R_{pc} = \frac{\ln(r_o/r_i)}{2\pi L_c k_p}, \tag{16.19}$$

where r_o and r_i are the outer and inner radii of the heat pipe wall and k_p is its thermal conductivity. R_{pa} is the axial thermal resistance of the heat pipe wall at the adiabatic section and is given by

$$R_{pa} = \frac{L_a}{\pi k_p \left(r_o^2 - r_i^2\right)}. \tag{16.20}$$

Similarly, if k_w is the thermal conductivity of the wick material, the radial thermal resistances of the wick at the evaporator and condenser sections and its axial thermal resistance at the adiabatic section are given by

$$R_{we} = \frac{\ln(r_i/r_v)}{2\pi L_e k_w}, \tag{16.21}$$

$$R_{wc} = \frac{\ln(r_i/r_v)}{2\pi L_c k_w}, \tag{16.22}$$

$$R_{wa} = \frac{L_a}{\pi k_w \left(r_i^2 - r_v^2\right)}. \tag{16.23}$$

R_{ve} and R_{vc} are the thermal resistances corresponding to the vapor–liquid surfaces at the evaporator and the condenser, respectively. They are usually very small and can be neglected [5]. R_{va} is an equivalent thermal resistance for the axial heat transfer at the core of the heat pipe due to the transfer of vapor from the evaporator to the condenser section. It is defined as

$$R_{va} = \frac{T_e - T_c}{\dot{Q}} = \frac{\Delta T_v}{\dot{Q}}. \tag{16.24}$$

Clapeyron equation gives the slope of the $P - T$ curve as

$$\frac{\Delta P}{\Delta T} = \frac{h_{fg}}{T\Delta v}, \tag{16.25}$$

where $\Delta v = v_v - v_l$ is the change in the specific volume of the fluid due to the phase change [6]. If we approximate the vapor as an ideal gas and neglect the volume of the liquid compared to the volume of the vapor, $\Delta v = RT_v/P_v$. This results in

$$\Delta T_v = \frac{RT_v^2}{h_{fg}P_v}\Delta P_v. \tag{16.26}$$

Neglecting the first term in Equation 16.4 and substituting it into 16.26 give

$$\Delta T_v = \frac{8\mu_v L_{\text{eff}} T_v}{\pi \rho_v^2 r_v^4 h_{fg}^2}\dot{Q} \tag{16.27}$$

FIGURE 16.13 A computational model for a heat pipe.

and

$$R_{va} = \frac{8\mu_v L_{eff} T_v}{\pi \rho_v^2 r_v^4 h_{fg}^2}. \tag{16.28}$$

A computational model for a heat pipe can be built using the sandwich structure shown in Figure 16.13 where a central vapor core, with an effective thermal conductivity $k_{v,eff}$, is surrounded by a wick structure and a solid wall. The effective thermal conductivity of the vapor core is obtained from its thermal resistance given by Equation 16.28 assuming that it can be treated as a conduction thermal resistance:

$$k_{v,eff} = \frac{L_{eff}}{R_{va} \pi r_v^2} = \frac{\rho_v^2 r_v^2 h_{fg}^2}{8\mu_v T_v}. \tag{16.29}$$

In most cases, similar results can be obtained with a simpler model where the heat pipe is modeled with a square shape as long as the contact areas with heat source and heat sink, the cross sectional area of the vapor core, and the thicknesses of the wick and pipe wall are preserved.

16.1.8 THERMOSYPHONS, LOOP HEAT PIPES, AND VAPOR CHAMBERS

Thermosyphons are basically wickless heat pipes that rely on gravity for returning the condensate from the condenser to the evaporator section. The cross section of a conventional thermosyphon is shown in Figure 16.14. Similar to a heat pipe, it is divided into three sections. Heat is transferred to the thermosyphon at the bottom, where there is a liquid pool, and vaporizes the working fluid. The vapor rises and passes through the adiabatic section to the condenser section where it condenses and transfers its latent heat. The condensate returns to the evaporator section by the gravitational force.

The heat transport limitations of thermosyphons are flooding, entrainment, dry out, and boiling limits. Flooding and entrainment limits are due to the viscous force between the liquid current from the condenser to the evaporator and the faster moving vapor current in the opposite direction. If this force is large enough to retard the return of the liquid to the evaporator, the thermosyphon is said to have reached its flooding limit. If the viscous force is larger than the surface

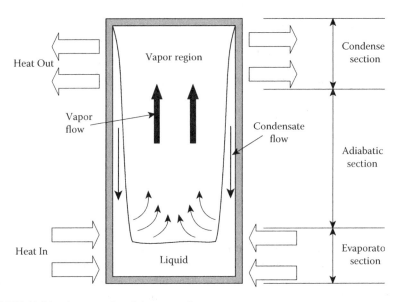

FIGURE 16.14 A conventional thermosyphon.

tension force, liquid droplets are entrained into the vapor region of the evaporator and the thermosyphon would be at its entrainment limit. Both the flooding and the entrainment limits result in the liquid accumulation in the condenser section and possible dry out of the evaporator section. Dry out limit for a thermosyphon is reached when the entire fluid is circulating throughout the thermosyphon, either as a falling condensate film or in the vapor flow, and no pool of liquid is present at the evaporator section. Dry out starts at the bottom of the evaporator section and increases in size as heat input to the evaporator section is increased. The boiling limit in a thermosyphon is reached when a large heat flux results in the creation of a vapor film at the bottom of the evaporator and blocks the liquid working fluid from contact with the wall.

Thermosyphons with capillary wicks have been built to decrease the vapor-liquid interaction, which causes the flooding and entrainment limits, to improve the circumferential distribution of liquid and to allow the use of thermosyphons in a tilted configuration, to promote nucleate boiling sites in the evaporator, and to enhance condensation heat transfer in the condenser section [3].

Heat pipes are very effective heat transfer devices as long as their length is short enough such that the capillary force can drive the liquid from the condenser to the evaporator section. The heat transfer capability of a conventional heat pipe will decrease if its length is increased or it works against the gravity as shown in Figures 16.4 and 16.5. Loop heat pipes have been developed to operate at long distances and at any orientation. Figure 16.15 shows the schematic of a typical loop heat pipe. It consists of an evaporator and a condenser section similar to a conventional heat pipe. However, liquid and vapor flows have their separate lines. Therefore, the shear force caused by the countercurrent liquid and vapor flows, that resulted in the flooding and entrainment limits in heat pipes and thermosyphons, is

Advanced Cooling Technologies

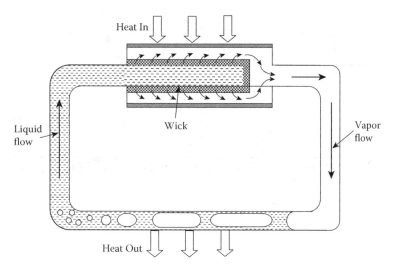

FIGURE 16.15 A loop heat pipe.

not present in loop heat pipes. The only force against the flow of vapor and liquid is a small shear force between the fluid and the pipe surface. Another important difference between the conventional heat pipe and loop heat pipe is that the wick structure in a loop heat pipe is needed only in the evaporator section to promote nucleate boiling sites.

Heat pipes and loop heat pipes are one-dimensional heat transfer devices that transport heat very effectively from the evaporator to the condenser. A vapor chamber is a flat, thin, and two-dimensional heat pipe similar to that shown in Figure 16.16. Let's assume that heat is applied to a small region at the center of the bottom plate of the vapor chamber. This heat vaporizes the working fluid and the vapor moves away from the center of the vapor chamber in all directions. As the vapor reaches the cooler areas of the vapor chamber including its top plate, it will condense and be absorbed by the wick structure through which it is transported back to the heat source area. Although the wick structure on the inner side of the top plate helps to improve the distribution of the liquid and increase the capillary pumping pressure, most vapor chambers are built with no wick structure on the top plate.

Vapor chambers are used as heat spreaders with very high thermal conductivity. The most common configuration is the one in which a heat source is attached to a small area on one side and a heat sink covers the entire other side of the vapor chamber as shown in Figure 16.17a. An equivalent resistance network for this configuration is shown in Figure 16.17b. R_{cb} and R_{ct} are the thermal resistances of the bottom and top plates respectively, R_w is the thermal resistance of the wick on the inner side of the bottom plate, and R_v is the thermal resistance of the vapor core. The first three thermal resistances can be obtained if the thermal conductivity, area and thickness of the plates, and the wick structure, and dimensions of the heat source, to account for the spreading thermal resistance, are known. Thermal resistance of the vapor core can be defined as

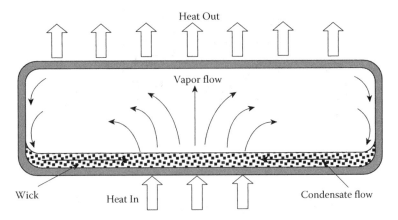

FIGURE 16.16 Cross section of a vapor chamber.

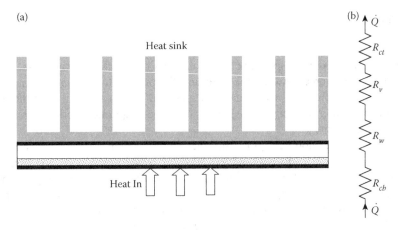

FIGURE 16.17 (a) A typical application of vapor chamber as a heat spreader and (b) its equivalent thermal resistance network.

$$R_v = \frac{t_v}{k_{v,\text{eff}} A}, \qquad (16.30)$$

where t_v, $k_{v,\text{eff}}$ and A are thickness, effective thermal conductivity, and area of the vapor core, respectively. The effective thermal conductivity of the vapor core can be calculated by [5]

$$k_{v,\text{eff}} = \frac{h_{fg}^2 \rho_v^2 t_v^2}{12\mu_v T_v}. \qquad (16.31)$$

A computational model for a vapor chamber can be built using four solid blocks representing the bottom plate, the wick structure, the vapor core, and the top plate as shown in Figure 16.18.

Advanced Cooling Technologies

FIGURE 16.18 A computational model for a vapor chamber.

16.2 LIQUID COOLING

Many liquids have higher density, specific heat, and thermal conductivity than gases and therefore are more effective in transferring heat. Natural convection along a vertical flat plate and flow inside a circular pipe will be used to demonstrate the superiority of liquid cooling compared to air cooling. Let's consider two vertical plates whose heights are 400 mm and are kept at 100°C. Let's assume that one of these plates is exposed to air at 25°C while the other is immersed in a water bath at 25°C. Using the correlations given in Chapter 12, it can be shown that the heat flux from the plate immersed in water is more than 200 times the heat flux from the plate exposed to air. Next, let's consider two pipes whose internal diameters are 10 mm and are kept at 100°C. Airflow with a mean velocity of 3.87 m/s and water flow with a mean velocity of 0.09 m/s creates laminar flows inside these two pipes with a Reynolds number of about 2000. Using the correlations given in Chapter 11, it can be shown that the convection heat transfer coefficient in the fully developed region of the pipe with water flow is more than 20 times that in the pipe with air flow.

Liquid cooling systems are classified into direct or indirect and single-phase or two-phase cooling systems. Figure 16.19 shows the schematic of one direct or immersion cooling system. Since the liquid is in direct contact with the electronics, it must be a dielectric liquid such as fluorocarbon-based liquids FC-72 and FC-77. A pump is used to push the liquid through the electronic component of the system being cooled. The liquid absorbs the heat from those components and leaves the system at a higher temperature. Then, it is cooled in an external heat exchanger before returning to the system. The heat transfer from the electronic components to the cooling fluid can be through natural or forced convection, and the fluid may remain in liquid phase or go through a phase change by boiling on the surfaces of the hot components.

The most widely known direct or immersion cooling system is the Cray-2 supercomputer made by Cray Research in 1985 and shown in Figure 16.20. The packaging of the electronics inside this supercomputer was very tight as shown in Figure 16.21. The FC-77 was used as the cooling fluid and a chilled water heat exchange was used to cool the warm fluid coming out of the supercomputer.

The cooling fluid in indirect liquid cooling systems does not come in contact with heat sources. Instead, it passes through the channels or pipes of a so-called cold plate that is in contact with those heat sources. The main components of an indirect liquid cooling loop are pump, cold plate, heat exchanger, and expansion

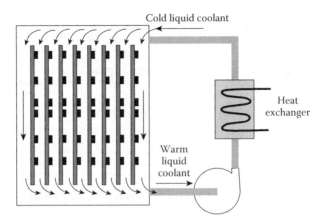

FIGURE 16.19 A direct or immersion liquid cooling system.

FIGURE 16.20 (See color insert following page 240.) Liquid-cooled Cray-2 supercomputer. (http://en.wikipedia.org/wiki/cray-2)

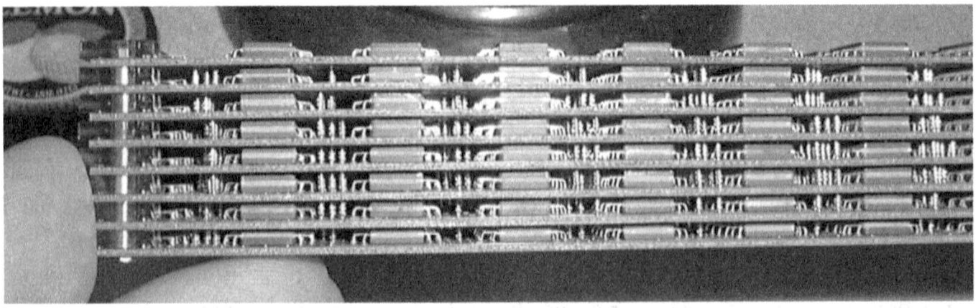

FIGURE 16.21 (See color insert following page 240.) A typical module of Cray-2. (http://en.wikipedia.org/wiki/cray-2)

Advanced Cooling Technologies

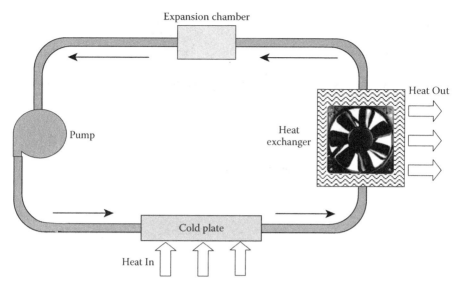

FIGURE 16.22 An indirect liquid cooling system.

tank as shown in Figure 16.22. The pump pushes the liquid into the cold plate where it absorbs heat from the heat sources the cold plate is attached to. The warm liquid exits the cold plate and enters a heat exchanger where it cools down and is ready to be pumped to the cold plate again. Indirect liquid cooling systems are closed systems filled with a certain volume of the liquid. However, the volume of liquid changes with temperature and while expansion may apply excessive pressure to the piping and other components of the system, contraction may create bubbles that are not suitable for the operation of the pump. An expansion tank is located upstream of the pump to allow minor expansion of the liquid and to compensate for the liquid contraction.

Cold plates have been built in different sizes, shapes, and complexities. They are sometimes classified as macroscale, mesoscale, and microscale cold plates based on the hydraulic diameter of the fluid passages being larger than 1 mm, between 0.1 to 1 mm and less than 0.1 mm, respectively. Figure 16.23 shows tubed cold plates that consist of copper or stainless steel tubes pressed into a channeled aluminum extrusion. These relatively simple cold plates are suitable for low-to-medium heat transfer rates. Applications include lasers, power electronics, RF generators and transmitters, semiconductor processing equipment, and uninterruptible power supplies.

The Nusselt number in the fully developed region of a laminar single-phase internal flow is constant. Since $Nu = hD_h/k$, the convection heat transfer coefficient will increase if dimensions of the flow passage and therefore its hydraulic diameter is reduced. This guided the cold plate designers to using multiple smaller flow channels instead of one large one. This design will also offer more surface area for heat transfer. However, smaller flow passages and more surface area usually translate to higher pressure drop and the need for a stronger pump. The main objective

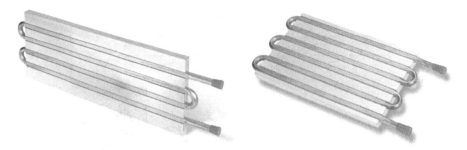

FIGURE 16.23 (See color insert following page 240.) Examples of tubed cold plates. (Courtesy of Lytron Inc.)

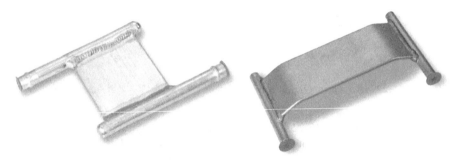

FIGURE 16.24 (See color insert following page 240.) Two flat tube cold plates. (Courtesy of Lytron Inc.)

in the cold plate design is to optimize its heat transfer capacity per unit pumping power requirement. Figure 16.24 shows two compact and lightweight cold plates that consist of header tubes, which are welded to the microchannel fins. These cold plates are about 3 mm thick and are ideal for cooling small, but high power density, electronics.

Meso and microscale cold plates with plate or pin fin structure can be built by precision machining, metal injection molding, and skiving. They can also be built using porous structures, such as those made by metal screen or foam or powder, as the heat transfer surface. These techniques can produce cold plates with very large heat transfer area to volume ratio and allow the miniaturization of cold plates. Cold plates with meso and microscale plate fins are called microchannels and are the most commonly researched and manufactured cold plates [7].

Thermal performance of a cold plate is characterized by its thermal resistance that is defined by

$$R_{cp} = \frac{T_{cp,\max} - T_{\text{ref}}}{\dot{Q}}, \qquad (16.32)$$

where $T_{cp,\max}$ is the maximum temperature of the cold plate where the heat flux is applied, $\dot{Q}$ is the heat transfer rate to the cold plate and T_{ref} is a reference temperature.

Advanced Cooling Technologies

The reference temperature is either the temperature of the liquid entering the cold plate or the average liquid temperature inside the cold plate. Another important characteristic of a cold plate is the resistance or impedance it shows against the liquid flow. As flow rate through a cold plate increases, its thermal resistance will decrease while the pressure drop through it will increase. Figure 16.25 shows thermal resistance and impedance curves of a cold plate. It is seen that the thermal resistance curve flattens while the impedance curve becomes steeper as flow rate increases. It is seen that the thermal resistance is reduced from 0.034°C/W to 0.03°C/W (i.e. about 11% reduction) if the flow rate is increased from 0.0025 m^3/min to 0.004 m^3/min. The same change in flow rate causes the pressure drop to increase from 45 kPa to 100 kPa (i.e. about 122% rise). The useful operating range of a cold plate is determined by the available pumping power of the pump that determines the flow rate.

Conventional centrifugal and reciprocating pumps are the most commonly used pumps in liquid cooling systems. These and a few other pump types have been described is Chapter 11, Section 11.7.4. As cold plates shrink in size, the need for smaller pumps that allow thermal engineers to fit a complete liquid cooling loop into small form factor electronics grows proportionally. Figure 16.26 shows two small-scale centrifugal pumps for the liquid cooling loops that are used in electronic

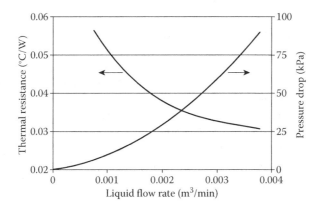

FIGURE 16.25 Thermal resistance and impedance curve of a cold plate.

FIGURE 16.26 Two centrifugal pumps used in small form-factor liquid cooling loops. (Courtesy of ITT Corp)

FIGURE 16.27 (See color insert following page 240.) A fin-and-tube heat exchanger used in indirect liquid cooling systems. (Courtesy of Lytron Inc.)

equipment. A detailed review of the recent advances in microscale pumping technologies was given by Iverson and Garimella [8].

Heat exchangers used in liquid cooling loops are mainly fin-and-tube, air-to-liquid heat exchangers similar to that shown in Figure 16.27. One or more fans blow the cooling air into the fins of these heat exchangers. They are characterized by their large heat transfer area to volume ratios and, with the appropriately designed inlet and exhaust manifolds, they provide parallel and low impedance paths for the liquid flow.

The liquid used in single-phase liquid cooled loops must have high thermal conductivity, high specific heat, low viscosity, high dielectric strength, low freezing point and high boiling point, and be low cost, chemically inert, stable, and nontoxic. The exact composition of the liquids used in most commercially available liquid cooling loops is proprietary. However, most of them are based on deionized and distilled water mixed with some form of antifreeze such as ethylene glycol or propylene glycol and a small amount of additives to prevent corrosion and to lubricate the pump. Another important factor to consider when designing a liquid cooling loop is the compatibility between the liquid and the solid materials that the cold plate, heat exchanger, pump, and piping are made of. If the surface and liquid are not compatible, galvanic corrosion will create leaks in the flow passages, or will cause blockages of those passages by the corroded materials. Copper and stainless steel are compatible with water-glycol solutions as well as fluorocarbon-based dielectric fluids used in direct liquid cooling loops. Stainless steel is also compatible with deionized water. Aluminum is compatible with water-glycol solutions and fluorocarbon-based fluids but not with water and deionized water. Table 16.1 gives material properties of common fluids used in direct and indirect liquid cooling loops.

Advanced Cooling Technologies

TABLE 16.1
Material Properties of Common Liquid Cooling Fluids at 25°C and Atmospheric Pressure

Material	Density kg/m³	Specific Heat J/kg°C	Thermal Conductivity W/m°C	Dynamic Viscosity N.s/m²	Freezing Point °C	Boiling Point °C
Water	997	4180	0.607	0.891×10^{-3}	0	100
Ethylene Glycol Water Solution						
20/80	1023	3817	0.57		−8	102.2
40/60	1052	3470	0.45		−25	104.4
50/50	1064	3283	0.39		−37	107.2
Deionized Water						
FC-72	1680	1100	0.057	0.64×10^{-3}		56
FC-77	1780	1100	0.063	1.3×10^{-3}		97

16.3 THERMOELECTRIC COOLERS

A thermoelectric cooler (TEC) is a solid-state refrigeration device that operates based on the Peltier effect discovered by the French physicist Jean C. A. Peltier in 1834. A thermocouple is a circuit formed by two dissimilar metals that are connected electrically in series as shown in Figure 16.28. The connection points are called junctions. The Peltier effect states that if a current passes through a thermocouple, a temperature difference will be created between its two junctions. As a result, heat is absorbed at one junction and rejected at the other junction of the thermocouple, as shown in Figure 16.28. If I is the current flow through a thermocouple made of metals A and B, the heat transfer rate due to the Peltier effect is

$$\dot{Q} = \pi_{AB} I, \tag{16.33}$$

where π_{AB} is called the Peltier coefficient between materials A and B [9].

The Peltier effect is the inverse of the Seebeck effect, which is the basis for temperature measurement using thermocouples. It states that if two junctions of a thermocouple are at different temperatures, a voltage difference is developed that is proportional to the temperature difference between those junctions. If dT and dV are the temperature and voltage differences between the two junctions of a thermocouple made of materials A and B, the Seebeck effect is described by

$$dV = S_{AB} dT, \tag{16.34}$$

where $S_{AB} = S_A - S_B$ is the Seebeck coefficient between materials A and B, and S_A and S_B are the Seebeck coefficients of materials A and B, respectively [9]. The Seebeck coefficient in metals is low but it is moderate in some semiconductors. It may be

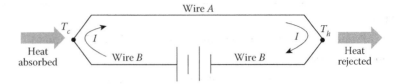

FIGURE 16.28 Illustration of the Peltier effect. The direction of heat flow is for the indicated direction of current and $S_{AB} > 0$.

either positive or negative. Both the Seebeck and the Peltier coefficients are temperature dependent and the relationship between them is given by

$$\pi_{AB} = S_{AB} T \qquad (16.35)$$

where T is the absolute temperature [9]. Using Equation 16.35 in Equation 16.33 gives

$$\dot{Q} = S_{AB} IT \qquad (16.36)$$

as the Peltier heat transfer to a thermocouple junction at absolute temperature T due to the current I in the thermocouple. Note that if $S_{AB} = S_A - S_B > 0$, heat is absorbed at the junction that current flows from material B to material A and is rejected at the other junction.

Thermoelectric coolers consist of pairs of p-type and n-type semiconductor pellets called couples. The couples are connected electrically in series and thermally in parallel, and are sandwiched between two thermally conducting and electrically insulating plates as shown in Figure 16.29. The Peltier effect can be explained by considering that electrons need additional energy to leave the p-type pellets and to enter the n-type pellets. They absorb this energy from a heat source at a cold junction. However, they loose this additional energy to a heat sink when they enter the hot junction. This creates a net heat flux from the cold to the hot junction of the TEC couple.

Consider a thermoelectric module made of N couples as shown in Figure 16.29 and assume that temperatures at the cold and hot junctions of this couple are T_c and T_h, respectively. The net rate of heat transfer to the cold side of this couple is the sum of the Peltier heat transfer and the heat transfer rate due to the heat conduction through the couple [10],

$$\dot{Q}_c = N \left(S_{AB} IT_c - 2kA_p \frac{dT}{dx}\bigg|_{x=0} \right). \qquad (16.37)$$

Here k and A_p are the thermal conductivity and the cross section area of the pellet, respectively. Similarly, the net rate of heat transfer from the hot side of this couple is the sum of the Peltier heat transfer and the heat transfer rate due to the heat conduction through the couple:

$$\dot{Q}_h = N \left(S_{AB} IT_h - 2kA_p \frac{dT}{dx}\bigg|_{x=L} \right). \qquad (16.38)$$

Advanced Cooling Technologies

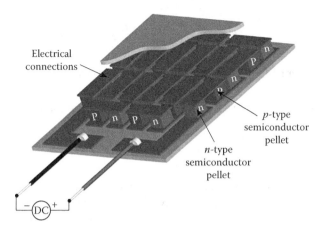

FIGURE 16.29 (See color insert following page 240.) A thermoelectric module. (Courtesy of Marc Hodes, Tuft University.)

Hodes [10] solved the one-dimensional heat conduction equation, with boundary conditions $T(x = 0) = T_c$ and $T(x = L) = T_h$, for one couple and derived the following expressions for the net rates of heat transfer to the cold and from the hot side of a thermocouple module using Equations 16.37 and 16.38,

$$\dot{Q}_c = N\left[S_{AB}\, IT_c - \frac{T_h - T_c}{R_t} - \frac{1}{2}I^2R\right], \quad (16.39)$$

$$\dot{Q}_h = N\left[S_{AB}\, IT_h - \frac{T_h - T_c}{R_t} + \frac{1}{2}I^2R\right]. \quad (16.40)$$

Here, $R = 2\rho L/A_p$, where ρ is the electrical resistivity, and $R_t = L/2kA_p$ are the electrical and conduction resistances of one couple, respectively.

The power input to the thermoelectric module is equal to the difference between the net heat transfer rates at the hot and cold sides,

$$P = \dot{Q}_h - \dot{Q}_c = N[S_{AB}\, I(T_h - T_c) + I^2R]. \quad (16.41)$$

Since $P = VI$, the voltage difference across the whole module is

$$V = N[S_{AB}(T_h - T_c) + IR]. \quad (16.42)$$

Equation 16.39 shows that the Peltier cooling varies linearly with the current while Joule heating is proportional to the current squared. The value of the current I at which the maximum cooling is achieved is obtained by taking the derivative of Equation 16.39 with respect to I and setting it to zero. This gives $I_{max} = S_{AB}T_c/R$. Inserting this into Equation 16.39 gives the maximum cooling capacity of a TEC,

$$\dot{Q}_{c,\max} = N\left[\frac{S_{AB}^2 T_c^2}{2R} - \frac{T_h - T_c}{R_t}\right]. \quad (16.43)$$

As the temperature difference between the hot and cold sides increases, the maximum cooling capacity of the TEC decreases. Setting $\dot{Q}_{c,\max} = 0$ in Equation 16.43 gives the maximum temperature difference at which the TEC has any cooling capacity:

$$\Delta T_{\max} = (T_h - T_c)_{\max} = \frac{R_t S_{AB}^2 T_c^2}{2R}. \quad (16.44)$$

Equation 16.39 shows that thermoelectric materials with higher Seebeck coefficient, S_{AB}, and lower electrical resistivity, ρ, and thermal conductivity, k, will have higher cooling capacity. The figure of merit for a TEC is defined as

$$Z = \frac{S_{AB}^2}{\rho k}, \quad (16.45)$$

and is used to compare different thermoelectric materials. Typical materials for TEC couples are bismuth telluride (Bi_2Te_3), Silicon-Germanium (SiGe), and Silicon (Si).

The coefficient of performance (COP) of a TEC is defined as

$$\text{COP} = \frac{\dot{Q}_c}{P} = \frac{N}{VI}\left[S_{AB} IT_c - \frac{T_h - T_c}{R_t} - \frac{1}{2}I^2 R\right]. \quad (16.46)$$

The optimum current that results in the maximum COP is

$$I_{opt} = \frac{S_{AB}(T_h - T_c)}{R[\sqrt{1 + Z(T_h + T_c)/2} - 1]}, \quad (16.47)$$

and the maximum COP is [9]

$$\text{COP}_{\max} = \frac{T_c[\sqrt{1 + Z(T_h + T_c)/2} - T_h/T_c]}{(T_h - T_c)[\sqrt{1 + Z(T_h + T_c)/2} + 1]}. \quad (16.48)$$

More detailed analyses shows that both the pellet height and the current shall be optimized in order to obtain the optimum COP of a TEC [11]. Heat transfer rate between the hot and cold junctions is proportional to the current and the direction of transfer is reversed if the direction of current is reversed. Therefore, TECs can be used for cooling, heating, and temperature control. TECs are used in electronic equipment when a device has to be cooled below ambient temperature or the heat sink alone can not cool a component below the required temperature. On the other hand, component such as an optical electronic or a medical sample must be kept within certain temperature range, heat sink alone can not cool a component below the required temperature. As the direction and amount of the net heat transfer from a TEC junction depends on the direction and amount of the current, they can be used for the precise temperature control in these situations.

Advanced Cooling Technologies

16.4 ELECTROHYDRODYNAMIC FLOW

An electrically charged fluid can be moved in the presence of a strong enough electric field. This flow is known as ionic wind, corona wind, and electro-aerodynamic or electrohydrodynamic flow. The concept has been known for over a century but has recently been considered for the electronic cooling applications. Consider air between a sharp electrode, such as a metal wire, and a blunt electrode, such as a metal plate, as shown in Figure 16.30. If a large voltage difference is applied between these two electrodes, an intense electric field is generated near the sharp electrode that ionizes some of the air molecules around it. The charged air molecules are pulled through the electric field from the sharp electrode, called the corona electrode, toward the blunt electrode, called the collector electrode. As the charged air molecules move toward the collector electrode, they collide with other neutral air molecules, transfer momentum to them, and create a bulk airflow toward the collector electrode.

If the voltage difference is applied in the presence of a bulk forced convection flow, such as the flow between the fins of a heat sink or the flow over a plate, the ionic wind distorts the boundary layer as shown in Figure 16.31 and therefore increases the local heat transfer rate. An increase of more than 200% was observed in the local convection heat transfer coefficient of the low-speed laminar flows due to the ionic wind and it was observed that the heat transfer coefficient varied linearly with the fourth root of the corona current [12].

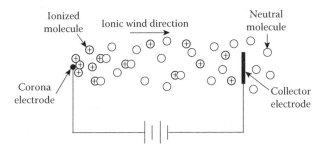

FIGURE 16.30 Schematic of an electrohydrodynamic flow generation mechanism.

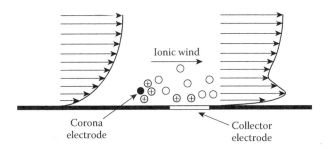

FIGURE 16.31 Effect of the ionic wind on a boundary layer.

Electrohydrodynamic fans, that operate based on the ionic wind principle, offer a solid-state and silent alternative to the conventional fans for use in small electronic devices such as notebook computers. In addition to their silent operation, these fans can be made in flexible form factors to fit around the other components in the system. Recent research on these fans concentrated on the miniaturization, reducing their operating voltage and the reliability [13, 14].

16.5 SYNTHETIC JET

A type of flow that could significantly enhance the air cooling envelope without the use of fans is the synthetic jet. Synthetic jets are created by the periodic suction and ejection of a fluid through an orifice, due to the periodic motion of a diaphragm, as shown in Figure 16.32. As the diaphragm pushes the air out of the orifice during the ejection phase, a pair of vortices and a jet are created and convected away from the orifice. By the time the suction phase begins, the vortex flow has moved far enough from the orifice that it can not be entrained into the cavity. Instead, the quiescent air from the vicinity of the orifice is sucked in through the cavity. This will produce a flow with a finite momentum while the net mass flow through the orifice is zero [15]. The far field characteristics of synthetic jets, such as their rate of lateral spreading and the decay of their centerline velocity, are similar to the conventional turbulent jets.

The periodic motion of the diaphragm is produced by piezoelectric, electromagnetic, electrostatic, and combustion driven pistons. The most commonly used actuators are piezoelectric or electromagnetic based. While piezoelectric diaphragms have weight and power consumption advantage, electromagnetic actuators have better noise and reliability performance [16].

Conventional jet ejectors create the primary jet by ducting the net mass flow as a continuous jet into the entry region of a channel. The low pressure created by the primary jet results in the entrainment of the quiescent ambient fluid thus creating an increase in the overall flow rate through the channel. The use of synthetic jets as the primary jet is an attractive option since the only required input to the primary jet is electrical and no plumbing or flow sources are needed. Similar to the conventional jet ejectors, a low-pressure region is created at the inlet region of the channel during the ejection phase resulting in the entrainment of the ambient fluid

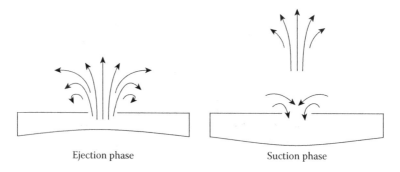

Ejection phase   Suction phase

FIGURE 16.32 Formation of a synthetic jet.

Advanced Cooling Technologies

as shown in Figure 16.33a. However, larger secondary flow entrainment is created during the suction phase of the synthetic jet as shown in Figure 16.33b. The fluid that is sucked in during the suction phase will be pushed out during the next ejection phase [17].

Mahalingam et al. [17] reported the airflow and thermal properties of a synthetic jet ejector. Figure 16.34 shows the airflow rate created by a synthetic jet ejector through a channel as a function of the channel width. It is seen that the airflow rate increases monotonically although it tends to an asymptotic level for channel widths larger than 25 mm. The Nusselt number for a synthetic jet ejector and a fully developed turbulent channel flow are compared with each other in Figure 16.35. The Reynolds number is based on the measured centerline velocity and the hydraulic diameter of the channel. It is seen that the Nusselt numbers of the synthetic jet ejector can be as high as six times those in the corresponding fully developed turbulent channel flows.

Because of their ability to direct airflow along the heated surfaces in confined environments and to induce small-scale mixing, synthetic jets are suited for cooling applications at the package and heat sink levels. Also, the ability to accurately control the directionality and location of a synthetic jet enables efficient localized heat removal from hot spots. Some of the advantages of synthetic jets to the

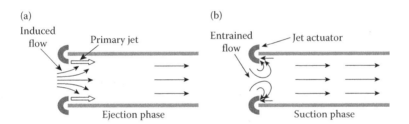

FIGURE 16.33 Principle of operation of a synthetic jet ejector.

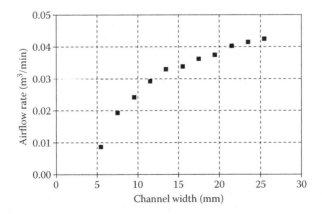

FIGURE 16.34 Induced volumetric flow rate by a synthetic jet ejector.

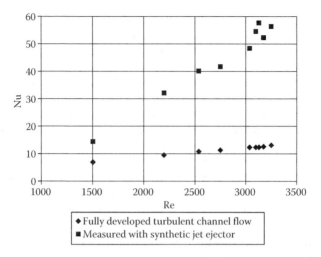

FIGURE 16.35 Nusselt number for a synthetic jet ejector and fully developed turbulent channel flow.

conventional fans are lower noise, lower power consumption, flexible form factor, higher reliability, easier miniaturization, and less fouling problem [18].

REFERENCES

1. Gaugler, R. 1942. Heat transfer devices. US Patent 2,350,348, filed Dec. 21,.1942, and issued June 6, 1944.
2. Grover, M. G. 1963. Evaporation-condensation heat transfer device. US Patent 3,229,759, filed Dec. 2, 1963, and issued Jan. 18, 1966.
3. Faghri, A. 1995. *Heat pipe science and technology.* Washington, DC: Taylor & Francis.
4. Private Communication. 2003 Chaun Choung Technology.
5. Prasher, R. S. 2003. A simplified conduction based modeling scheme for design sensitivity study of thermal solution utilizing heat pipe and vapor chamber technology. *Journal of Electronic Packaging* 125 (3): 378–85.
6. Moran, M. J., and H. N. Shapiro. 2003. *Fundamentals of engineering thermodynamics.* Fifth Edition. Hoboken, NJ: John Wiley & Sons, Inc.
7. Sobhan, C. B., and Garimella S. V. 2001. A comparative analysis of studies on heat transfer and fluid flow in microchannels. *Microscale Thermophysical Engineering* 5 (4): 293–311.
8. Iverson, B. D., and S. V. Garimella. 2008. Recent advances in microscale pumping technologies: A review and evaluation. *Microfluidics and Nanofluidics* 5:145–74.
9. Rowe, D. M. 2006. General principles and basic considerations. In *Thermoelectrics handbook macro to nano,* ed. D. M. Row. pp.1–13, Boca Raton, FL: Taylor & Francis.
10. Hodes, M. 2005. On one-dimensional analysis of thermoelectric modules (TEM). *IEEE Transactions on Components and Packaging Technologies* 28 (2): 218–29.
11. Hodes, M. 2007. Optimal Pellet Geometries for Thermoelectric Refrigeration. IEEE *Transaction on Components and Packaging Technologies* 30 (1): 50–58.
12. Go, D. B., R. A. Maturana, T. S. Fisher, and S. V. Garimella. 2008. Enhancement of external forced convection by ionic wind. *International Journal of Heat and Mass Transfer* 51:6047–53.

13. Jewell-Larsen, N. E., H. Ran, Y. Zhang, M. Schwiebert, K. A. Honer, and A. V. Mamishev. 2009, March 15–19. Electrohydrodynamic (EHD) cooled laptop. Proceedings of 24th Annual IEEE Semiconductor Thermal Measurement and Management Symposium (SemiTherm 25), San Jose, California.
14. Schlitz, D., and V. Singhal. 2008, March 16–20. An electro-aerodynamic solid-state fan and cooling system. Proceedings of 24th Annual IEEE Semiconductor Thermal Measurement and Management Symposium (SemiTherm 24), pp. 47–50, San Jose, California.
15. Smith, B. L., and A. Glezer. 1998. The formation and evolution of synthetic Jets. *Physics of Fluids* 10 (2): 2281–97.
16. Mahalingam, R., S. Heffington, L. Jones, and R. Williams. 2007. Synthetic jets for forced air cooling of electronics. *Electronics Cooling* 13 (2): 12–18.
17. Mahalingam, R., N. Rumigny, and A. Glezer. 2004. Thermal management using synthetic jet ejectors. *IEEE Transactions on Components and Packaging Technologies* 27 (3): 439–44.
18. Lasance, C. J. M., and R. M. Aarts. 2008. Synthetic jet cooling, Part I: Overview of heat transfer and acoustics. Proceedings of 24th Annual IEEE Semiconductor Thermal Measurement and M Management Symposium (SemiTherm 24), San Jose, California.

Appendix: Tables of Material Properties

LIST OF TABLES

TABLE A.1	Properties of Air at 1 atm (101.33kPa) Pressure	456
TABLE A.2	Properties of Carbon Dioxide at 1 atm (101.33kPa) Pressure	457
TABLE A.3	Properties of Carbon Monoxide at 1 atm (101.33kPa) Pressure	458
TABLE A.4	Properties of Hydrogen at 1 atm (101.33kPa) Pressure	459
TABLE A.5	Properties of Nitrogen at 1 atm (101.33kPa) Pressure	460
TABLE A.6	Properties of Oxygen at 1 atm (101.33kPa) Pressure	461
TABLE A.7	Properties of Methane at 1 atm (101.33kPa) Pressure	462
TABLE A.8	Properties of Ethane at 1 atm (101.33kPa) Pressure	463
TABLE A.9	Properties of Propane at 1 atm (101.33kPa) Pressure	464
TABLE A.10	Properties of n-Butane at 1 atm (101.33kPa) Pressure	465
TABLE A.11	Properties of Saturated Water/Steam	466
TABLE A.12	Properties of Saturated Liquid and Vapor Ammonia	468
TABLE A.13	Properties of Solid Materials Commonly used in Electronic Packaging.	470

TABLE A.1
Properties of Air at 1 atm (101.33kPa) Pressure

T °C	ρ kg/m³	C_p J/kg°C	$\mu \times 10^6$ N·s/m²	$\nu \times 10^6$ m²/s	k W/m°C	$\alpha \times 10^6$ m²/s	Pr
−20	1.3947	1004.44	16.23	11.63	0.0225	16.08	0.7236
−10	1.3417	1004.71	16.73	12.47	0.0233	17.28	0.7216
0	1.2926	1005.06	17.23	13.33	0.0241	18.52	0.7198
10	1.2469	1005.50	17.72	14.21	0.0248	19.79	0.7181
20	1.2044	1006.02	18.21	15.12	0.0256	21.10	0.7165
30	1.1647	1006.62	18.68	16.04	0.0263	22.43	0.7150
40	1.1275	1007.30	19.15	16.98	0.0270	23.80	0.7137
50	1.0926	1008.06	19.61	17.95	0.0277	25.19	0.7124
60	1.0598	1008.88	20.06	18.93	0.0285	26.62	0.7112
70	1.0289	1009.78	20.51	19.93	0.0292	28.07	0.7101
80	0.9998	1010.75	20.95	20.95	0.0299	29.55	0.7091
90	0.9722	1011.79	21.38	21.99	0.0305	31.06	0.7081
100	0.9462	1012.89	21.81	23.05	0.0312	32.59	0.7073
110	0.9215	1014.05	22.23	24.13	0.0319	34.15	0.7064
120	0.8980	1015.28	22.65	25.22	0.0326	35.74	0.7057
130	0.8758	1016.56	23.06	26.33	0.0333	37.35	0.7050
140	0.8546	1017.90	23.47	27.46	0.0339	38.99	0.7044
150	0.8344	1019.30	23.87	28.61	0.0346	40.65	0.7038
160	0.8151	1020.74	24.27	29.77	0.0352	42.34	0.7033
170	0.7967	1022.25	24.66	30.96	0.0359	44.05	0.7028
180	0.7791	1023.80	25.05	32.15	0.0365	45.78	0.7024
190	0.7623	1025.39	25.44	33.37	0.0372	47.53	0.7020
200	0.7462	1027.04	25.82	34.60	0.0378	49.31	0.7016
210	0.7308	1028.72	26.19	35.84	0.0384	51.11	0.7013
220	0.7159	1030.45	26.57	37.11	0.0391	52.93	0.7010
230	0.7017	1032.23	26.94	38.39	0.0397	54.78	0.7008
240	0.6880	1034.04	27.30	39.68	0.0403	56.64	0.7005
250	0.6749	1035.88	27.66	40.99	0.0409	58.52	0.7003
260	0.6622	1037.77	28.02	42.31	0.0415	60.43	0.7001
270	0.6500	1039.68	28.37	43.65	0.0421	62.36	0.7000
280	0.6383	1041.63	28.72	45.00	0.0428	64.30	0.6998
290	0.6269	1043.61	29.07	46.36	0.0434	66.27	0.6997
300	0.6160	1045.62	29.41	47.74	0.0440	68.25	0.6995

Source: ρ is calculated using the ideal gas law, C_p, μ and k are calculated using the correlations given by Irvine, T. F. Jr., and P. E. Liley. *Steam and gas tables with computer equations.* Orlando, FL: Academic Press. 1984. With permission, $\nu = \mu/\rho$, $\alpha = k/\rho C_p$ and Pr = ν/α.

TABLE A.2
Properties of Carbon Dioxide at 1 atm (101.33kPa) Pressure

T °C	ρ kg/m³	C_p J/kg°C	$\mu \times 10^6$ N·s/m²	$\nu \times 10^6$ m²/s	k W/m°C	$\alpha \times 10^6$ m²/s	Pr
−50	2.4039	760.57	11.37	4.73	0.0110	6.04	0.7833
−40	2.3008	772.21	11.86	5.15	0.0117	6.60	0.7809
−30	2.2061	783.66	12.34	5.59	0.0124	7.19	0.7784
−20	2.1190	794.90	12.82	6.05	0.0131	7.80	0.7758
−10	2.0385	805.94	13.30	6.52	0.0139	8.44	0.7732
0	1.9638	816.79	13.77	7.01	0.0146	9.10	0.7706
10	1.8945	827.44	14.24	7.52	0.0153	9.79	0.7680
20	1.8299	837.89	14.70	8.04	0.0161	10.50	0.7655
30	1.7695	848.15	15.17	8.57	0.0169	11.23	0.7632
40	1.7130	858.21	15.62	9.12	0.0176	11.99	0.7609
50	1.6600	868.09	16.07	9.68	0.0184	12.76	0.7587
60	1.6101	877.77	16.52	10.26	0.0192	13.56	0.7567
70	1.5632	887.26	16.97	10.85	0.0199	14.38	0.7548
80	1.5190	896.57	17.41	11.46	0.0207	15.22	0.7530
90	1.4771	905.69	17.84	12.08	0.0215	16.08	0.7514
100	1.4375	914.63	18.28	12.71	0.0223	16.96	0.7498
110	1.4000	923.39	18.71	13.36	0.0231	17.85	0.7483
120	1.3644	931.97	19.13	14.02	0.0239	18.77	0.7470
130	1.3306	940.38	19.55	14.70	0.0247	19.71	0.7457
140	1.2984	948.62	19.97	15.38	0.0254	20.66	0.7445
150	1.2677	956.69	20.39	16.08	0.0262	21.63	0.7434
160	1.2384	964.60	20.80	16.79	0.0270	22.62	0.7423
170	1.2105	972.34	21.21	17.52	0.0278	23.63	0.7413
180	1.1838	979.92	21.61	18.26	0.0286	24.66	0.7404
190	1.1582	987.35	22.01	19.01	0.0294	25.70	0.7394
200	1.1337	994.62	22.41	19.77	0.0302	26.77	0.7385
210	1.1103	1001.75	22.81	20.54	0.0310	27.85	0.7377
220	1.0877	1008.73	23.20	21.33	0.0318	28.95	0.7368
230	1.0661	1015.56	23.59	22.12	0.0326	30.06	0.7359
240	1.0454	1022.26	23.97	22.93	0.0333	31.20	0.7350
250	1.0254	1028.81	24.35	23.75	0.0341	32.36	0.7340
260	1.0061	1035.24	24.73	24.58	0.0349	33.53	0.7331
270	0.9876	1041.53	25.11	25.42	0.0357	34.73	0.7320
280	0.9698	1047.70	25.48	26.28	0.0365	35.95	0.7309
290	0.9525	1053.74	25.85	27.14	0.0373	37.19	0.7298
300	0.9359	1059.66	26.22	28.02	0.0381	38.46	0.7285

Source: ρ is calculated using the ideal gas law, C_p, μ and k are calculated using the correlations given by Irvine, T. F. Jr., and P. E. Liley. *Steam and gas tables with computer equations.* Orlando, FL: Academic Press. 1984. With permission, $\nu = \mu/\rho$, $\alpha = k/\rho C_p$ and $Pr = \nu/\alpha$.

TABLE A.3
Properties of Carbon Monoxide at 1 atm (101.33kPa) Pressure

T °C	ρ kg/m³	C_p J/kg°C	$\mu \times 10^6$ N·s/m²	$\nu \times 10^6$ m²/s	k W/m°C	$\alpha \times 10^6$ m²/s	Pr
−20	1.3486	1039.80	15.52	11.51	0.0216	15.44	0.7455
−10	1.2974	1039.79	16.02	12.35	0.0224	16.62	0.7429
0	1.2499	1039.84	16.51	13.21	0.0232	17.84	0.7404
10	1.2058	1039.95	17.00	14.10	0.0240	19.10	0.7380
20	1.1646	1040.14	17.47	15.00	0.0247	20.39	0.7358
30	1.1262	1040.41	17.94	15.93	0.0254	21.72	0.7337
40	1.0902	1040.76	18.40	16.88	0.0262	23.07	0.7317
50	1.0565	1041.19	18.86	17.85	0.0269	24.46	0.7298
60	1.0248	1041.71	19.31	18.84	0.0276	25.88	0.7281
70	0.9949	1042.32	19.75	19.85	0.0283	27.33	0.7264
80	0.9668	1043.02	20.19	20.88	0.0290	28.80	0.7249
90	0.9401	1043.82	20.62	21.93	0.0297	30.31	0.7236
100	0.9149	1044.70	21.04	23.00	0.0304	31.84	0.7223
110	0.8911	1045.69	21.46	24.08	0.0311	33.39	0.7212
120	0.8684	1046.76	21.87	25.18	0.0318	34.97	0.7201
130	0.8469	1047.93	22.28	26.31	0.0325	36.57	0.7192
140	0.8264	1049.20	22.68	27.44	0.0331	38.20	0.7184
150	0.8068	1050.55	23.08	28.60	0.0338	39.85	0.7177
160	0.7882	1051.99	23.47	29.77	0.0344	41.52	0.7172
170	0.7704	1053.52	23.85	30.96	0.0351	43.20	0.7167
180	0.7534	1055.14	24.24	32.17	0.0357	44.91	0.7163
190	0.7371	1056.84	24.61	33.39	0.0363	46.63	0.7160
200	0.7216	1058.62	24.99	34.63	0.0370	48.38	0.7158
210	0.7066	1060.48	25.35	35.88	0.0376	50.14	0.7157
220	0.6923	1062.42	25.72	37.15	0.0382	51.91	0.7156
230	0.6785	1064.42	26.08	38.43	0.0388	53.70	0.7157
240	0.6653	1066.50	26.44	39.73	0.0394	55.51	0.7158
250	0.6526	1068.63	26.79	41.05	0.0400	57.33	0.7160
260	0.6404	1070.83	27.14	42.38	0.0406	59.16	0.7163
270	0.6286	1073.09	27.48	43.72	0.0412	61.01	0.7166
280	0.6172	1075.40	27.82	45.08	0.0417	62.87	0.7170
290	0.6062	1077.77	28.16	46.45	0.0423	64.75	0.7175
300	0.5957	1080.17	28.50	47.84	0.0429	66.63	0.7180

Source: ρ is calculated using the ideal gas law, C_p, μ and k are calculated using the correlations given by Irvine, T. F. Jr., and P. E. Liley. *Steam and gas tables with computer equations.* Orlando, FL: Academic Press. 1984. With permission, $\nu = \mu/\rho$, $\alpha = k/\rho C_p$ and $Pr = \nu/\alpha$.

TABLE A.4
Properties of Hydrogen at 1 atm (101.33kPa) Pressure

T °C	ρ kg/m³	C_p J/kg°C	$\mu \times 10^6$ N·s/m²	$\nu \times 10^6$ m²/s	k W/m°C	$\alpha \times 10^6$ m²/s	Pr
−20	0.9706	14054.80	7.97	8.21	0.1580	11.58	0.7085
−10	0.9337	14116.90	8.18	8.76	0.1632	12.38	0.7074
0	0.8995	14169.33	8.39	9.33	0.1683	13.21	0.7064
10	0.8678	14213.34	8.60	9.91	0.1733	14.05	0.7055
20	0.8382	14250.17	8.81	10.51	0.1781	14.91	0.7048
30	0.8105	14281.14	9.01	11.12	0.1828	15.79	0.7041
40	0.7846	14307.50	9.21	11.74	0.1873	16.69	0.7037
50	0.7604	14330.49	9.41	12.38	0.1918	17.60	0.7035
60	0.7375	14351.25	9.61	13.03	0.1960	18.52	0.7036
70	0.7160	14370.75	9.81	13.69	0.2002	19.45	0.7039
80	0.6958	14389.80	10.00	14.37	0.2042	20.39	0.7046
90	0.6766	14408.89	10.19	15.06	0.2081	21.35	0.7057
100	0.6585	14428.20	10.38	15.77	0.2119	22.30	0.7070
110	0.6413	14447.47	10.57	16.49	0.2156	23.27	0.7086
120	0.6250	14465.92	10.77	17.22	0.2192	24.25	0.7104
130	0.6095	14482.19	10.96	17.98	0.2227	25.24	0.7123
140	0.5947	14494.19	11.15	18.74	0.2262	26.24	0.7142
150	0.5807	14499.02	11.34	19.52	0.2297	27.28	0.7158
160	0.5673	14494.70	11.53	20.32	0.2331	28.35	0.7169
170	0.5545	14494.70	11.72	21.13	0.2365	29.42	0.7182
180	0.5422	14494.70	11.91	21.96	0.2399	30.52	0.7194
190	0.5305	14494.70	12.09	22.79	0.2433	31.64	0.7203
200	0.5193	14494.70	12.27	23.63	0.2468	32.79	0.7208
210	0.5086	14494.70	12.45	24.48	0.2503	33.96	0.7207
220	0.4982	14496.10	12.62	25.32	0.2540	35.17	0.7200
230	0.4883	14497.90	12.82	26.26	0.2577	36.40	0.7214
240	0.4788	14500.06	13.00	27.14	0.2611	37.60	0.7218
250	0.4697	14502.59	13.17	28.04	0.2645	38.83	0.7221
260	0.4609	14505.47	13.34	28.95	0.2679	40.08	0.7224
270	0.4524	14508.71	13.52	29.88	0.2713	41.34	0.7227
280	0.4442	14512.31	13.69	30.81	0.2748	42.62	0.7229
290	0.4363	14516.25	13.86	31.76	0.2782	43.93	0.7231
300	0.4287	14520.54	14.03	32.72	0.2816	45.24	0.7232

Source: ρ is calculated using the ideal gas law, C_p, μ and k are calculated using the correlations given by Irvine, T. F. Jr., and P. E. Liley. *Steam and gas tables with computer equations*. Orlando, FL: Academic Press. 1984. With permission, $\nu = \mu/\rho$, $\alpha = k/\rho C_p$ and Pr = ν/α.

TABLE A.5
Properties of Nitrogen at 1 atm (101.33kPa) Pressure

T °C	ρ kg/m³	C_p J/kg°C	$\mu \times 10^6$ N·s/m²	$\nu \times 10^6$ m²/s	k W/m°C	$\alpha \times 10^6$ m²/s	Pr
10	1.2058	1039.16	17.02	14.12	0.0247	19.70	0.7167
20	1.1646	1039.08	17.50	15.02	0.0254	21.00	0.7155
30	1.1262	1039.10	17.96	15.95	0.0261	22.32	0.7145
40	1.0902	1039.21	18.42	16.90	0.0268	23.68	0.7135
50	1.0565	1039.41	18.87	17.86	0.0275	25.06	0.7127
60	1.0248	1039.70	19.32	18.85	0.0282	26.48	0.7120
70	0.9949	1040.08	19.76	19.86	0.0289	27.92	0.7113
80	0.9668	1040.55	20.19	20.88	0.0296	29.38	0.7107
90	0.9401	1041.10	20.62	21.93	0.0302	30.88	0.7102
100	0.9149	1041.74	21.04	22.99	0.0309	32.39	0.7098
110	0.8911	1042.46	21.45	24.08	0.0315	33.94	0.7095
120	0.8684	1043.25	21.86	25.18	0.0322	35.50	0.7092
130	0.8469	1044.12	22.27	26.30	0.0328	37.09	0.7090
140	0.8264	1045.07	22.67	27.43	0.0334	38.70	0.7088
150	0.8068	1046.10	23.06	28.58	0.0340	40.33	0.7087
160	0.7882	1047.19	23.45	29.75	0.0347	41.99	0.7086
170	0.7704	1048.36	23.84	30.94	0.0353	43.66	0.7086
180	0.7534	1049.60	24.22	32.14	0.0359	45.36	0.7087
190	0.7371	1050.90	24.59	33.36	0.0365	47.07	0.7088
200	0.7216	1052.27	24.96	34.59	0.0371	48.80	0.7089
210	0.7066	1053.70	25.33	35.84	0.0376	50.55	0.7091
220	0.6923	1055.20	25.69	37.11	0.0382	52.32	0.7093
230	0.6785	1056.75	26.05	38.39	0.0388	54.10	0.7096
240	0.6653	1058.37	26.40	39.68	0.0394	55.90	0.7099
250	0.6526	1060.04	26.75	40.99	0.0399	57.72	0.7102
260	0.6404	1061.76	27.10	42.31	0.0405	59.55	0.7106
270	0.6286	1063.54	27.44	43.65	0.0410	61.40	0.7110
280	0.6172	1065.37	27.78	45.00	0.0416	63.26	0.7114
290	0.6062	1067.26	28.11	46.37	0.0421	65.14	0.7119
300	0.5957	1069.19	28.44	47.75	0.0427	67.03	0.7124
310	0.5855	1071.16	28.77	49.14	0.0432	68.93	0.7129
320	0.5756	1073.18	29.10	50.55	0.0438	70.85	0.7135
330	0.5660	1075.25	29.42	51.97	0.0443	72.78	0.7141

Source: ρ is calculated using the ideal gas law, C_p, μ and k are calculated using the correlations given by Irvine, T. F. Jr., and P. E. Liley. *Steam and gas tables with computer equations.* Orlando, FL: Academic Press. 1984. With permission, $\nu = \mu/\rho$, $\alpha = k/\rho C_p$ and Pr = ν/α.

TABLE A.6
Properties of Oxygen at 1 atm (101.33kPa) Pressure

T °C	ρ kg/m³	C_p J/kg°C	$\mu \times 10^6$ N·s/m²	$\nu \times 10^6$ m²/s	k W/m°C	$\alpha \times 10^6$ m²/s	Pr
−20	1.5407	910.93	17.94	11.64	0.0227	16.20	0.7187
−10	1.4822	912.31	18.53	12.50	0.0236	17.43	0.7175
0	1.4279	913.81	19.11	13.39	0.0244	18.69	0.7164
10	1.3775	915.43	19.69	14.29	0.0252	19.98	0.7154
20	1.3305	917.16	20.26	15.23	0.0260	21.31	0.7146
30	1.2866	918.99	20.82	16.18	0.0268	22.67	0.7139
40	1.2455	920.94	21.38	17.16	0.0276	24.06	0.7133
50	1.2070	922.98	21.92	18.16	0.0284	25.48	0.7128
60	1.1707	925.11	22.46	19.19	0.0292	26.93	0.7124
70	1.1366	927.34	22.99	20.23	0.0299	28.41	0.7121
80	1.1044	929.65	23.52	21.30	0.0307	29.91	0.7119
90	1.0740	932.05	24.04	22.38	0.0315	31.44	0.7118
100	1.0452	934.53	24.55	23.49	0.0322	33.00	0.7118
110	1.0180	937.07	25.06	24.62	0.0330	34.58	0.7119
120	0.9921	939.69	25.56	25.76	0.0337	36.18	0.7120
130	0.9675	942.38	26.05	26.93	0.0345	37.81	0.7123
140	0.9440	945.12	26.54	28.11	0.0352	39.46	0.7125
150	0.9217	947.93	27.02	29.32	0.0359	41.13	0.7129
160	0.9005	950.78	27.50	30.54	0.0367	42.82	0.7133
170	0.8801	953.68	27.97	31.78	0.0374	44.53	0.7138
180	0.8607	956.63	28.44	33.04	0.0381	46.26	0.7143
190	0.8421	959.62	28.90	34.32	0.0388	48.01	0.7148
200	0.8243	962.64	29.35	35.61	0.0395	49.78	0.7154
210	0.8073	965.70	29.80	36.92	0.0402	51.56	0.7160
220	0.7909	968.78	30.25	38.25	0.0409	53.37	0.7167
230	0.7752	971.88	30.69	39.59	0.0416	55.19	0.7174
240	0.7601	975.00	31.13	40.95	0.0423	57.03	0.7181
250	0.7455	978.14	31.56	42.33	0.0429	58.88	0.7189
260	0.7316	981.29	31.99	43.73	0.0436	60.76	0.7197
270	0.7181	984.44	32.41	45.13	0.0443	62.65	0.7205
280	0.7051	987.59	32.83	46.56	0.0450	64.55	0.7213
290	0.6926	990.74	33.24	48.00	0.0456	66.48	0.7221
300	0.6805	993.89	33.66	49.46	0.0463	68.41	0.7229

Source: ρ is calculated using the ideal gas law, C_p, μ and k are calculated using the correlations given by Irvine, T. F. Jr., and P. E. Liley. *Steam and gas tables with computer equations.* Orlando, FL: Academic Press. 1984. With permission, $\nu = \mu/\rho$, $\alpha = k/\rho C_p$ and Pr = ν/α.

TABLE A.7
Properties of Methane at 1 atm (101.33kPa) Pressure

T °C	ρ kg/m³	C_p J/kg°C	$\mu \times 10^6$ N·s/m²	$\nu \times 10^6$ m²/s	k W/m°C	$\alpha \times 10^6$ m²/s	Pr
10	0.6906	2185.66	10.62	15.38	0.0316	20.96	0.7337
20	0.6670	2211.83	10.95	16.42	0.0330	22.34	0.7349
30	0.6450	2238.81	11.28	17.48	0.0343	23.76	0.7358
40	0.6244	2266.59	11.60	18.58	0.0357	25.22	0.7364
50	0.6051	2295.12	11.92	19.70	0.0371	26.73	0.7367
60	0.5869	2324.37	12.23	20.84	0.0386	28.29	0.7368
70	0.5698	2354.32	12.54	22.01	0.0401	29.88	0.7367
80	0.5537	2384.93	12.85	23.21	0.0416	31.52	0.7365
90	0.5385	2416.18	13.16	24.43	0.0432	33.19	0.7361
100	0.5240	2448.02	13.46	25.68	0.0448	34.91	0.7356
110	0.5104	2480.43	13.75	26.95	0.0464	36.66	0.7351
120	0.4974	2513.38	14.05	28.25	0.0481	38.45	0.7346
130	0.4850	2546.83	14.34	29.56	0.0497	40.27	0.7341
140	0.4733	2580.76	14.63	30.90	0.0515	42.13	0.7336
150	0.4621	2615.14	14.91	32.26	0.0532	44.01	0.7331
160	0.4514	2649.92	15.19	33.65	0.0549	45.92	0.7327
170	0.4413	2685.08	15.47	35.05	0.0567	47.86	0.7323
180	0.4315	2720.60	15.74	36.48	0.0585	49.82	0.7321
190	0.4222	2756.43	16.01	37.92	0.0603	51.81	0.7319
200	0.4133	2792.55	16.28	39.39	0.0621	53.82	0.7319
210	0.4047	2828.92	16.54	40.87	0.0639	55.84	0.7320
220	0.3965	2865.51	16.80	42.38	0.0658	57.88	0.7321
230	0.3886	2902.30	17.06	43.90	0.0676	59.94	0.7324
240	0.3811	2939.25	17.32	45.45	0.0695	62.02	0.7328
250	0.3738	2976.33	17.57	47.01	0.0713	64.11	0.7333
260	0.3668	3013.50	17.82	48.59	0.0732	66.21	0.7339
270	0.3600	3050.74	18.07	50.19	0.0750	68.32	0.7346
280	0.3535	3088.02	18.32	51.81	0.0769	70.45	0.7354
290	0.3472	3125.30	18.56	53.45	0.0788	72.59	0.7363
300	0.3412	3162.55	18.80	55.10	0.0806	74.74	0.7373
310	0.3353	3199.75	19.04	56.78	0.0825	76.90	0.7384
320	0.3297	3236.85	19.27	58.47	0.0844	79.07	0.7395
330	0.3242	3273.84	19.51	60.18	0.0862	81.25	0.7406

Source: ρ is calculated using the ideal gas law, C_p, μ and k are calculated using the correlations given by Irvine, T. F. Jr., and P. E. Liley. *Steam and gas tables with computer equations.* Orlando, FL: Academic Press. 1984. With permission. $\nu = \mu/\rho$, $\alpha = k/\rho C_p$ and Pr = ν/α.

TABLE A.8
Properties of Ethane at 1 atm (101.33kPa) Pressure

T °C	ρ kg/m³	C_p J/kg°C	$\mu \times 10^6$ N·s/m²	$\nu \times 10^6$ m²/s	k W/m°C	$\alpha \times 10^6$ m²/s	Pr
10	1.2943	1690.51	8.96	6.92	0.0197	8.98	0.7710
20	1.2501	1733.06	9.26	7.41	0.0209	9.64	0.7678
30	1.2089	1775.61	9.55	7.90	0.0222	10.33	0.7649
40	1.1703	1818.12	9.84	8.41	0.0235	11.03	0.7623
50	1.1341	1860.59	10.13	8.93	0.0248	11.75	0.7599
60	1.1000	1903.00	10.41	9.46	0.0261	12.49	0.7577
70	1.0680	1945.34	10.69	10.01	0.0275	13.25	0.7555
80	1.0377	1987.61	10.97	10.57	0.0289	14.03	0.7535
90	1.0092	2029.78	11.25	11.15	0.0304	14.83	0.7517
100	0.9821	2071.84	11.52	11.73	0.0318	15.65	0.7499
110	0.9565	2113.78	11.80	12.33	0.0333	16.48	0.7483
120	0.9321	2155.59	12.07	12.95	0.0348	17.33	0.7469
130	0.9090	2197.26	12.34	13.57	0.0364	18.20	0.7456
140	0.8870	2238.77	12.60	14.21	0.0379	19.09	0.7444
150	0.8661	2280.11	12.87	14.86	0.0395	19.98	0.7434
160	0.8461	2321.26	13.13	15.52	0.0410	20.90	0.7425
170	0.8270	2362.22	13.39	16.19	0.0426	21.82	0.7419
180	0.8087	2402.97	13.64	16.87	0.0442	22.76	0.7413
190	0.7913	2443.49	13.90	17.57	0.0458	23.71	0.7410
200	0.7745	2483.78	14.15	18.27	0.0475	24.67	0.7408
210	0.7585	2523.83	14.40	18.99	0.0491	25.64	0.7407
220	0.7431	2563.61	14.65	19.72	0.0507	26.62	0.7408
230	0.7284	2603.12	14.90	20.46	0.0523	27.61	0.7409
240	0.7142	2642.35	15.14	21.20	0.0540	28.61	0.7413
250	0.7005	2681.28	15.39	21.96	0.0556	29.62	0.7417
260	0.6874	2719.89	15.63	22.73	0.0573	30.63	0.7422
270	0.6747	2758.18	15.87	23.52	0.0589	31.66	0.7427
280	0.6625	2796.14	16.10	24.31	0.0606	32.70	0.7433
290	0.6508	2833.74	16.34	25.11	0.0622	33.75	0.7440
300	0.6394	2870.98	16.57	25.92	0.0639	34.81	0.7446
310	0.6284	2907.85	16.80	26.74	0.0656	35.88	0.7453
320	0.6178	2944.33	17.03	27.57	0.0672	36.96	0.7459
330	0.6076	2980.40	17.26	28.41	0.0689	38.06	0.7465

Source: ρ is calculated using the ideal gas law, C_p, μ and k are calculated using the correlations given by Irvine, T. F. Jr., and P. E. Liley. *Steam and gas tables with computer equations.* Orlando, FL: Academic Press. 1984. With permission, $\nu = \mu/\rho$, $\alpha = k/\rho C_p$ and Pr = ν/α.

TABLE A.9
Properties of Propane at 1 atm (101.33kPa) Pressure

T °C	ρ kg/m³	C_p J/kg°C	$\mu \times 10^6$ N·s/m²	$\nu \times 10^6$ m²/s	k W/m°C	$\alpha \times 10^6$ m²/s	Pr
10	1.8985	1598.50	7.81	4.11	0.0164	5.39	0.7627
20	1.8337	1647.01	8.07	4.40	0.0174	5.78	0.7624
30	1.7732	1695.15	8.34	4.70	0.0185	6.17	0.7628
40	1.7166	1742.89	8.60	5.01	0.0196	6.56	0.7640
50	1.6635	1790.25	8.87	5.33	0.0207	6.96	0.7656
60	1.6136	1837.21	9.13	5.66	0.0219	7.37	0.7676
70	1.5665	1883.77	9.39	5.99	0.0230	7.79	0.7699
80	1.5222	1929.93	9.65	6.34	0.0241	8.21	0.7725
90	1.4803	1975.69	9.91	6.69	0.0253	8.63	0.7752
100	1.4406	2021.03	10.16	7.06	0.0264	9.07	0.7781
110	1.4030	2065.96	10.42	7.43	0.0276	9.51	0.7811
120	1.3673	2110.47	10.67	7.81	0.0287	9.95	0.7842
130	1.3334	2154.55	10.93	8.19	0.0299	10.41	0.7873
140	1.3011	2198.21	11.18	8.59	0.0311	10.87	0.7904
150	1.2704	2241.44	11.43	8.99	0.0323	11.33	0.7936
160	1.2410	2284.23	11.67	9.41	0.0335	11.81	0.7966
170	1.2130	2326.58	11.92	9.83	0.0347	12.29	0.7997
180	1.1863	2368.48	12.17	10.26	0.0359	12.78	0.8027
190	1.1607	2409.94	12.41	10.69	0.0371	13.27	0.8056
200	1.1361	2450.94	12.65	11.14	0.0384	13.78	0.8085
210	1.1126	2491.49	12.89	11.59	0.0396	14.29	0.8112
220	1.0901	2531.58	13.13	12.05	0.0409	14.80	0.8139
230	1.0684	2571.20	13.37	12.52	0.0421	15.33	0.8165
240	1.0476	2610.35	13.61	12.99	0.0434	15.86	0.8189
250	1.0275	2649.03	13.84	13.47	0.0447	16.40	0.8213
260	1.0083	2687.22	14.08	13.96	0.0459	16.95	0.8235
270	0.9897	2724.94	14.31	14.46	0.0472	17.51	0.8257
280	0.9718	2762.17	14.54	14.96	0.0485	18.08	0.8277
290	0.9546	2798.91	14.77	15.48	0.0498	18.66	0.8295
300	0.9379	2835.16	15.00	15.99	0.0512	19.24	0.8313
310	0.9218	2870.90	15.23	16.52	0.0525	19.83	0.8329
320	0.9063	2906.15	15.45	17.05	0.0538	20.44	0.8344

Source: ρ is calculated using the ideal gas law, C_p, μ and k are calculated using the correlations given by Irvine, T. F. Jr., and P. E. Liley. *Steam and gas tables with computer equations.* Orlando, FL: Academic Press. 1984. With permission, $\nu = \mu/\rho$, $\alpha = k/\rho C_p$ and Pr = ν/α.

TABLE A.10
Properties of *n*-Butane at 1 atm (101.33 kPa) Pressure

T °C	ρ kg/m³	C_p J/kg°C	$\mu \times 10^6$ N·s/m²	$\nu \times 10^6$ m²/s	k W/m°C	$\alpha \times 10^6$ m²/s	Pr
10	2.5026	1622.98	7.16	2.86	0.0144	3.55	0.8056
20	2.4172	1668.79	7.41	3.06	0.0154	3.81	0.8051
30	2.3375	1714.31	7.65	3.27	0.0163	4.07	0.8047
40	2.2628	1759.54	7.89	3.49	0.0173	4.34	0.8043
50	2.1928	1804.48	8.13	3.71	0.0183	4.61	0.8041
60	2.1270	1849.10	8.37	3.94	0.0193	4.90	0.8039
70	2.0650	1893.42	8.61	4.17	0.0203	5.19	0.8040
80	2.0065	1937.41	8.85	4.41	0.0213	5.48	0.8042
90	1.9513	1981.08	9.09	4.66	0.0224	5.79	0.8046
100	1.8990	2024.42	9.33	4.91	0.0234	6.10	0.8051
110	1.8494	2067.42	9.56	5.17	0.0245	6.42	0.8059
120	1.8024	2110.07	9.80	5.44	0.0256	6.74	0.8068
130	1.7577	2152.37	10.03	5.71	0.0267	7.07	0.8079
140	1.7151	2194.31	10.27	5.99	0.0278	7.40	0.8092
150	1.6746	2235.88	10.50	6.27	0.0290	7.74	0.8107
160	1.6359	2277.07	10.73	6.56	0.0301	8.08	0.8124
170	1.5990	2317.89	10.97	6.86	0.0312	8.42	0.8142
180	1.5637	2358.32	11.20	7.16	0.0324	8.77	0.8163
190	1.5300	2398.35	11.43	7.47	0.0335	9.13	0.8185
200	1.4976	2437.99	11.66	7.79	0.0346	9.48	0.8209
210	1.4666	2477.21	11.89	8.11	0.0358	9.84	0.8236
220	1.4369	2516.02	12.12	8.43	0.0369	10.21	0.8264

Source: ρ is calculated using the ideal gas law, C_p, μ and k are calculated using the correlations given by Irvine, T. F. Jr., and P. E. Liley. *Steam and gas tables with computer equations.* Orlando, FL: Academic Press. 1984. With permission. $\nu = \mu/\rho$, $\alpha = k/\rho C_p$ and $Pr = \nu/\alpha$.

TABLE A.11
Properties of Saturated Water/Steam

T °C	P kPa	ρ kg/m³ Liquid	ρ kg/m³ Vapor	h_{fg} kJ/kg	C_p J/kg°C Liquid	C_p J/kg°C Vapor	C_p/C_v Vapor	$\mu \times 10^6$ N·s/m² Liquid	$\mu \times 10^6$ N·s/m² Vapor	k W/m°C Liquid	k W/m°C Vapor	Pr Liquid	Pr Vapor	$\sigma \times 10^3$ N/m
0.01	0.61	999.8	0.0049	2501	4220	1884	1.329	1791.2	9.22	0.5610	0.0171	13.4739	1.0176	75.65
10	1.23	999.7	0.0094	2477	4196	1895	1.328	1306.0	9.46	0.5800	0.0176	9.4482	1.0174	74.22
20	2.34	998.2	0.0173	2454	4184	1906	1.327	1001.6	9.73	0.5984	0.0182	7.0032	1.0173	72.74
30	4.25	995.6	0.0304	2430	4180	1918	1.327	797.4	10.01	0.6155	0.0189	5.4153	1.0164	71.19
40	7.38	992.2	0.0512	2406	4180	1931	1.327	653.0	10.31	0.6306	0.0196	4.3285	1.0157	69.60
50	12.35	988.0	0.0831	2382	4182	1947	1.328	546.8	10.62	0.6436	0.0204	3.5530	1.0156	67.94
60	19.95	983.2	0.1304	2358	4185	1965	1.328	466.4	10.93	0.6543	0.0212	2.9832	1.0136	66.24
70	31.20	977.7	0.1984	2333	4190	1986	1.330	403.9	11.26	0.6631	0.0221	2.5522	1.0132	64.48
80	47.41	971.8	0.2937	2308	4197	2012	1.332	354.3	11.59	0.6700	0.0230	2.2194	1.0134	62.67
90	70.18	965.3	0.4239	2282	4205	2043	1.334	314.4	11.93	0.6753	0.0240	1.9577	1.0147	60.82
99.97	101.33	958.4	0.5977	2256	4216	2080	1.337	281.8	12.27	0.6791	0.0251	1.7495	1.0172	58.92
100	101.42	958.3	0.5982	2256	4216	2080	1.337	281.7	12.27	0.6791	0.0251	1.7489	1.0168	58.91
110	143.38	950.9	0.8269	2230	4228	2124	1.341	254.7	12.61	0.6817	0.0262	1.5797	1.0207	56.96
120	198.67	943.1	1.1221	2202	4244	2177	1.346	232.1	12.96	0.6832	0.0275	1.4418	1.0271	54.97
130	270.28	934.8	1.4970	2174	4261	2239	1.352	212.9	13.30	0.6837	0.0288	1.3268	1.0354	52.93
140	361.54	926.1	1.9668	2144	4283	2311	1.359	196.5	13.65	0.6833	0.0301	1.2317	1.0466	50.86
150	476.16	917.0	2.5481	2114	4307	2394	1.368	182.5	13.99	0.6820	0.0316	1.1525	1.0599	48.74
160	618.23	907.4	3.2597	2082	4335	2488	1.379	170.2	14.34	0.6800	0.0331	1.0850	1.0769	46.59
170	792.19	897.5	4.1222	2049	4368	2594	1.392	159.6	14.68	0.6770	0.0348	1.0297	1.0958	44.41
180	1002.80	887.0	5.1589	2014	4405	2713	1.407	150.1	15.03	0.6733	0.0365	0.9820	1.1187	42.19

Appendix: Tables of Material Properties

190	1255.20	876.1	6.3955	1978	4447	2844	1.425	141.8	15.37	0.6688	0.0382	0.9429	1.1431	39.95
200	1554.90	864.7	7.8610	1940	4496	2990	1.447	134.3	15.71	0.6633	0.0401	0.9103	1.1711	37.67
210	1907.70	852.7	9.5886	1900	4551	3150	1.472	127.6	16.06	0.6570	0.0421	0.8839	1.2019	35.38
220	2319.60	840.2	11.6158	1857	4615	3329	1.501	121.5	16.41	0.6497	0.0442	0.8630	1.2368	33.07
230	2797.10	827.1	13.9860	1813	4688	3528	1.536	116.0	16.76	0.6413	0.0464	0.8480	1.2749	30.74
240	3346.90	813.4	16.7504	1765	4772	3754	1.578	110.9	17.12	0.6318	0.0487	0.8376	1.3189	28.39
250	3976.20	798.9	19.9681	1715	4870	4011	1.627	106.1	17.49	0.6212	0.0513	0.8318	1.3686	26.04
260	4692.30	783.6	23.7135	1662	4986	4308	1.686	101.7	17.88	0.6092	0.0540	0.8324	1.4256	23.69
270	5503.00	767.5	28.0741	1604	5123	4656	1.757	97.5	18.28	0.5959	0.0571	0.8382	1.4903	21.34
280	6416.60	750.3	33.1675	1543	5289	5073	1.845	93.5	18.70	0.5811	0.0606	0.8510	1.5652	18.99
290	7441.80	731.9	39.1389	1477	5493	5582	1.954	89.7	19.15	0.5650	0.0647	0.8721	1.6519	16.66
300	8587.90	712.1	46.1681	1405	5750	6220	2.094	85.9	19.65	0.5474	0.0697	0.9023	1.7548	14.36
310	9865.10	690.7	54.5554	1326	6085	7045	2.277	82.2	20.21	0.5287	0.0758	0.9461	1.8774	12.09
320	11284.30	667.1	64.6412	1238	6537	8159	2.528	78.4	20.85	0.5092	0.0839	1.0065	2.0274	9.86
330	12858.10	640.8	77.0416	1140	7186	9753	2.889	74.5	21.61	0.4891	0.0949	1.0946	2.2200	7.70
340	14600.70	610.7	92.7644	1027	8210	12240	3.450	70.4	22.55	0.4685	0.1109	1.2337	2.4886	5.63
350	16529.40	574.7	113.6364	893	10120	16690	4.460	65.9	23.82	0.4474	0.1360	1.4906	2.9243	3.67
360	18666.00	527.6	143.8849	720	15000	27360	6.830	60.3	25.72	0.4257	0.1815	2.1247	3.8769	1.88
370	21043.60	451.4	202.0202	444	45160	96600	21.150	52.1	29.68	0.4250	0.3238	5.5361	8.8534	0.39
373.95	22064.00	322.0	321.5434	0	∞	∞	∞	—	—	∞	∞	—	—	0.00

Source: Adapted from American Society of Heating, Refrigerating and Air Conditioning Engineers. *ASHRAE Handbook of Fundamentals*. Atlanta, GA: ASHRAE. 2009. With permission.

TABLE A.12
Properties of Saturated Liquid and Vapor Ammonia

T °C	P kPa	ρ kg/m³ Liquid	ρ kg/m³ Vapor	h_{fg} kJ/kg	C_p J/kg°C Liquid	C_p J/kg°C Vapor	C_p/C_v Vapor	$\mu \times 10^6$ N·s/m² Liquid	$\mu \times 10^6$ N·s/m² Vapor	k W/m°C Liquid	k W/m°C Vapor	Pr Liquid	Pr Vapor	$\sigma \times 10^3$ °C
−77.65	6.09	732.9	0.0641	1484	4202	2063	1.325	559.6	6.84	0.8190	0.0196	2.8711	0.7185	62.26
−70	10.94	724.7	0.1110	1466	4245	2086	1.327	475.0	7.03	0.7921	0.0197	2.5456	0.7433	59.10
−60	21.89	713.6	0.2125	1442	4303	2125	1.330	391.3	7.30	0.7570	0.0199	2.2243	0.7783	55.05
−50	40.84	702.1	0.3806	1416	4360	2178	1.335	328.9	7.57	0.7223	0.0202	1.9853	0.8146	51.11
−40	71.69	690.2	0.6438	1389	4414	2244	1.342	281.2	7.86	0.6881	0.0206	1.8038	0.8545	47.26
−33.33	101.33	682.0	0.8895	1370	4448	2297	1.348	255.5	8.05	0.6657	0.0210	1.7072	0.8818	44.75
−30	119.43	677.8	1.0374	1360	4465	2326	1.351	244.1	8.15	0.6546	0.0212	1.6650	0.8963	43.52
−20	190.08	665.1	1.6033	1329	4514	2425	1.363	214.4	8.45	0.6220	0.0218	1.5560	0.9413	39.88
−10	290.71	652.1	2.3906	1297	4564	2542	1.378	190.2	8.75	0.5901	0.0225	1.4711	0.9886	36.34
0	429.38	638.6	3.4566	1262	4617	2680	1.398	170.1	9.06	0.5592	0.0234	1.4044	1.0390	32.91
10	615.05	624.6	4.8678	1226	4676	2841	1.422	153.0	9.36	0.5291	0.0244	1.3522	1.0912	29.59
20	857.48	610.2	6.7024	1186	4745	3030	1.453	138.3	9.68	0.4999	0.0255	1.3127	1.1493	26.38
30	1167.20	595.2	9.0531	1144	4828	3250	1.492	125.5	10.00	0.4714	0.0269	1.2854	1.2104	23.28
40	1555.40	579.4	12.0337	1099	4932	3510	1.541	114.0	10.33	0.4435	0.0284	1.2678	1.2776	20.29
50	2034.00	562.9	15.7853	1050	5064	3823	1.605	103.8	10.67	0.4163	0.0302	1.2627	1.3525	17.43
55	2311.10	554.2	18.0050	1024	5143	4005	1.643	99.0	10.86	0.4029	0.0312	1.2637	1.3958	16.04

Appendix: Tables of Material Properties

60	2615.60	545.2	20.4918	997	5235	4208	1.687	94.5	11.05	0.3896	0.0323	1.2698	1.4414	14.69
70	3313.50	526.3	26.4061	939	5465	4699	1.799	85.9	11.47	0.3632	0.0348	1.2925	1.5488	12.08
80	4142.00	505.7	33.8868	874	5784	5355	1.955	78.0	11.95	0.3371	0.0380	1.3383	1.6840	9.61
90	5116.70	482.8	43.4783	801	6250	6291	2.187	70.5	12.55	0.3110	0.0422	1.4168	1.8691	7.30
100	6255.30	456.6	56.1167	716	6991	7762	2.562	63.2	13.32	0.2848	0.0484	1.5514	2.1379	5.15
110	7578.30	425.6	73.5294	613	8360	10460	3.260	56.0	14.42	0.2581	0.0583	1.8139	2.5859	3.20
120	9112.50	385.5	100.1001	480	11940	17210	5.040	48.3	16.21	0.2312	0.0784	2.4944	3.5583	1.50
130	10897.70	312.3	156.7398	247	54210	76490	20.660	37.3	20.63	0.2219	0.1604	9.1124	9.8384	0.18
132.25	11333.00	225.0	225.2252	0	8000	8000	8.000	—	—	0.0080	0.0080			0.00

Source: Adapted from American Society of Heating, Refrigerating and Air Conditioning Engineers. *ASHRAE Handbook of Fundamentals.* Atlanta, GA: ASHRAE. 2009. With permission.

TABLE A.13
Properties of Solid Materials Commonly used in Electronic Packaging

Material	ρ kg/m³	C_p J/kg°C	k W/m°C
Alumina (Al2O3)	3970	765	15–33
Aluminum (Al)	2700	900	237
Aluminum Nitride (AlN)	3260	745	82–320
Beryllia (BeO)	3000	1047–2093	150–300
BT			0.2
Copper (Cu)	9000	385	403
Diamond	3510	471–509	2000–2300
Epoxy			0.19–0.35
Gallium Arsenide (GaAs)	5316	322	44–58
Germanium (Ge)	5340	322–335	64
Gold(Au)	1930	129	319
Lead-Tin Alloy 95Pb/5Sn Solder			35.5
Lead-Tin Alloy 37Pb/63Sn Solder			50.6
Kovar	8360	439	15.5–17.0
Polyimide			0.18–0.35
Silicon (Si)	2330	702–712	124–148
Silicon Carbide (SiC)	3210	670	283
Silicon Oxide (SiO$_2$)	2190–2320		1.5
Tellurium (Te)	6230	197	2.0–3.4

Source: Adapted from Pecht, M., R., Agarwal, F. P., McCluskey, T. J., Dishongh, S., Javadpour, and R., Mahajan. *Electronic packaging materials and their properties*. Boca Raton, FL: CRC Press. 1999. With permission.

REFERENCES

1. Irvine, T. F. Jr., and P. E. Liley. 1984. *Steam and gas tables with computer equations.* Orlando, FL: Academic Press.
2. American Society of Heating, Refrigerating and Air Conditioning Engineers. 2009. *ASHRAE handbook of fundamentals.* Atlanta, GA: ASHRAE.
3. Pecht, M., R. Agarwal, F. P. McCluskey, T. J. Dishongh, S. Javadpour, and R. Mahajan. 1999. *Electronic packaging materials and their properties*. Boca Raton, FL: CRC Press.

Index

A

Acoustic noise
    level of telecommunication chassis, 18
    measurements
        anechoic chamber, interior, 418
        sound measurements, 418
        sound pressure level (SPL), 418–419
        sound waves, 417
Adiabatic fin tip
    boundary conditions for, 132
    effectiveness, 142
    efficiency, 144
    solution of equation, 132
    thermal resistance, 140
    total heat transfer rate from, 133
Air properties, 456
Aluminum
    as fin materials, 142
    as heat sink materials, 430
    thermal conductivity, 55–56
Approximate model geometry, 382–383
Architectural thermal design, 16–17; *see also* Thermal design
Average friction coefficient, 225
Axial thermal resistance, 434

B

Backward curved impeller in radial fans, 272–273
Ball grid array (BGA) package, 107
    bottom sides of, 108
Bare-die package, 10
Bernoulli equation, 401
    velocity measurement methods, 410
BGA, *see* Ball grid array (BGA) package
Biot number, 203
Blackbody radiation
    defined, 324
    Planck distribution, 325
    and real body radiation, comparison with, 324
    spectral emissive power, 324–325
    spectral intensity, 324
    Stefan-Boltzmann law, 325
    total emissive power of, 325
    Wein's displacement law, 325
Board-to-air thermal resistance of package, 117
Bonded-fin heat sink, 160–161
Boundary and initial conditions for heat conduction
    constants, 177
    convection, 181–182
    general, 185
    heat flux, 179–180
    interface, 186–187
    radiation, 183–184
    temperature, 178
Boundary layer equations
    observations for, 217
    two-dimensional, 217–218
Boussinesq approximation, 290, 292–293
Bulk fluid mean temperature, 269
Buoyancy force
    Archimedes principle, 287
    fluids
        density of, 288–289
        forces applied on object in, 288
        submerged object in, 288
    and natural convection flows, 287
    net vertical force, 287
    volumetric thermal expansion coefficient
        Boussinesq approximation, 290
        for ideal gas, 289

C

CAD, *see* Computer-aided design (CAD)
CAE, *see* Computer-aided engineering (CAE)
Carbon dioxide properties, 457
Carbon monoxide properties, 458
Case-to-air thermal resistance, 115–116
Ceramic quad flat pack (CQFP), 119
Ceramic small outline package (CSOP), 104
Circular cylinder, flows across
    characteristics of, 239
    critical Reynolds number, 239
    drag force and coefficient, 242
    forward stagnation point, 239
    laminar flow regimes, 240
    Nusselt number equation constants, 243
    Reynolds number, 239
    total convection heat transfer rate, 242
    turbulent flow, 241
    variation of drag coefficient with Reynolds number, 242
Circular fin
    efficiency of, 147–148
    with nonuniform cross sections, 147

Clapeyron equation, 434
Closed system, 30
Coefficient of performance (COP), 448
Coefficient of thermal expansion (CTE), 9
    packaging materials for, 11
Cold plates, 441
    thermal performance of, 442
    thermal resistance and impedance
        curves, 443
    tubed cold plates, examples of, 442
Collector electrode, 449
Commercial thermal simulation tools, 381
    computer algorithms and programs, 382
    creating model geometry, 382–383
        electronic systems, thermal
            models of, 383
        imported CAD file and thermal
            model, 384
        methods, 382–383
    environmental conditions
        air, 390
        heat, 390–391
    flow and temperature equations
        plots of, 395
        results obtained in, 393–394
        single precision calculations use,
            394–395
        transient thermal simulation, 395
    flow type
        field and thermal behavior, 391
        heat transfer, 391
        turbulence model use, 391
    material and components properties, 383
        compact models, 384
        density and specific heat, 384
        printed circuit boards, 384–388
    mesh creating
        aspect ratio, 392–394
        body-fitted mesh, 392
        Cartesian and non-Cartesian geometries,
            391–392
        computational volumes, 391–392, 394
        software default and user-controlled
            mesh for, 393
    power dissipation
        data, 389–390
        thermal design, 390
    results, presentation
        air velocity and air temperature in, 398
        report, 397
    results, review
        common modeling mistake, 397
        conservation of mass and energy,
            principles, 396
        temperature gradients inside, 396–397
        velocity, temperature, and pressure, 396
Compact models, 384
Composite layer, thermal resistance network, 74
Compressed spring and potential energy, 22
Computer-aided design (CAD)
    design process, 351–352
    use of, 352
Computer-aided engineering (CAE)
    design process, 351–352
    use of, 352
Computer simulation tools
    fluid flow and energy equations
        continuity, momentum and energy
            equation for, 363
        control volume in, 364–365
        convection term, 367
        diffusion coefficient, values of, 364
        diffusion term, 367–368
        finite-difference forms of, 367
        general variable, values of, 364
        momentum equation, finite-volume
            form of, 368–370
        piecewise linear profiles, 365
        SIMPLE, 372–373
        source term, 364, 368–370
        staggered node arrangement and
            results in, 365–366
        temperatures and heat transfer
            rates in, 363
        transient term, 367
        two-dimensional control volume
            for, 366
        velocity correction formula, 370
    Fourier's law of conduction, 357
    nodes
        finite-difference heat conduction
            equation for, 357–358
    nonzero elements, matrix, 359
    one-dimensional heat conduction,
        353–354
        control volumes and internal points, 354
        cross section of wall with, 354
        temperature and pressure, 354
    steady-state situation, 354, 357–358
        boundary points, treatment of, 355
        Fourier's law of conduction, 355
        heat flux, 355
        set of equations, 356
    transient heat conduction
        energy equation for, 359
        explicit method, 360
        fluid thermal diffusivity, 360
        Fourier number, 360
        implicit method, 362
        stability criteria for, 361–362
        time-dependent temperature, 360
    two-dimensional heat conduction
        node network, cross section of, 356
Condenser heat transfer limit, 427

Index

Conduction heat transfer
  plane wall through, 53
    proportionality correlation, 54
  rate, 67
  vector in Cartesian coordinate system, 171
Conduction thermal resistance, 68
Constant temperature fin tip
  boundary conditions, 135
  solution of equation, 136
  total heat transfer rate, 136–137
Continuity equation; *see also* Mass conservation equation
  compact form, 372
  finite-difference form for, 370–371
  two-dimensional control volume for, 371
Convection
  heat transfer, 57
    coefficient, 58
    equivalent resistance for, 70
    simplified correlations in air, 58–59
  and radiation from fin tip
    boundary conditions for, 133
    effectiveness, 142
    efficiency, 144
    solution of equation, 133–134
    thermal resistance, 141
    total heat transfer rate from, 134
  term, 367
  thermal resistance, 69
Cooling electronic equipment, techniques
  electrohydrodynamic flow
    boundary layer, ionic wind effect of, 449
    corona and collector electrode, 449
    fans, 450
    flow generation mechanism, schematic of, 449
  heat pipes
    boiling mechanism, 424–425
    capillary pumping pressure, 423–424
    electronic cooling applications, 428–430
    entrainment limit, 426
    and loop heat pipe, 436–437
    performance limitations, 427–428
    pressure, 421
    schematic of, 422
    selection and modeling, 430–435
    sonic limit, 426
    thermosyphon, 435
    vapor chambers, 437–438
    water-based heat pipe performance, gravity effect of, 428
    and wick structure, 422–423
    working fluid, 421–423
  liquid cooling system
    aluminum, 444
    centrifugal and reciprocating pumps, 443
    cold plates, 441
    copper and stainless steel, 444
    direct or immersion, schematic of, 440
    fin-and-tube heat exchanger used in, 444
    indirect liquid cooling loop, components of, 439–440
    indirect liquid cooling system, 441
    liquid-cooled Cray-2 supercomputer, 440
    liquids used, exact composition of, 444
    material properties of, 445
    meso and microscale cold plates, 442
    module of Cray-2, 440
    Nusselt number in, 441–442
    pump use, 439, 441
    reference temperature, 443
    tubed cold plates, 442
  synthetic jets
    airflow and thermal properties, 451
    conventional jet ejectors, 450–451
    diaphragm, periodic motion of, 450
    formation of, 450
    induced volumetric flow rate, 451
    Nusselt number for, 451–452
    operation, principle of, 451
  thermoelectric cooler (TEC)
    coefficient of performance (COP), 448
    couples, 446
    heat transfer rate, 448
    junctions, 445
    maximum cooling capacity of, 448
    Peltier coefficient, 445–447
    Seebeck coefficient between, 445–446
    thermoelectric module, 446–447
COP, *see* Coefficient of performance (COP)
Copper
  as fin materials, 142
  as heat sink materials, 430
  heat slug, 105
  thermal conductivity, 55–56, 91–92
Corona electrode, 449
Corrosion failures and temperature, 12
CQFP, *see* Ceramic quad flat pack (CQFP)
Critical Reynolds number, 221
Cross-cut extrusion heat sinks, 160
Cross-wire anemometer, 412
Crystalline solids thermal conductivity, 55–56
CSOP, *see* Ceramic Small Outline Package (CSOP)
CTE, *see* Coefficient of thermal expansion (CTE)
Cylindrical pin fin heat sink, 150
  average Nusselt numbers for, 245
  contraction and expansion of air at inlet and exhaust, 246
  convection thermal resistance, 245
  friction factor, 246
  Reynolds number and flow in, 244
  total pressure drop, 246

D

Design verification test (DVT), 18, 401
Desktop computer, 103
Device under test (DUT), 404, 407
    impedance curve of, 408
Diamond thermal conductivity, 54–56
Diaphragm pump
    in suction cycle
        schematic of, 280
Die
    casting, 160
    die-cast heat sink, 160–161
    fracture, 10–12
    inner lead frame, 104
    junction, 110
    and substrate adhesion fatigue, 12
Diffuse emitter, 323–324
Diffusion
    coefficient values, 364
    equation, 176–177
    term, 367–368
Dimensionless group
    critical Reynolds number, 221
    Eckert number, 222
    Nusselt number, 223
    Prandtl number, 222
    Reynolds number
        inertia force/volume, 221
        viscous force/volume, 221
Direct numerical simulation (DNS), 373
Displacement of object, 19
Dittus-Boelter equation, 269
DNS, *see* Direct numerical simulation (DNS)
DUT, *see* Device under test (DUT)
DVD player, 103
DVT, *see* Design verification test (DVT)
Dynamic pumps, 279
Dynamic viscosity of fluid, 213

E

Eckert number, 222
Effective normal thermal conductivity, 93
Electrical failures and temperature, 12–13
Electrical overstress, 13
Electrical resistor
    conduction heat transfer rate, 67–68
    equivalent thermal resistance, 68
    Ohm's law and, 67
    potential difference between, 67–68
Electric foil heaters, 179
Electrohydrodynamic flow
    boundary layer, ionic wind effect of, 449
    corona and collector electrode, 449
    fans, 450
    flow generation mechanism, schematic of, 449

Electromagnetic waves
    emission, 319
    spectrum, 319–320
    thermal radiation, 320
Electromigration, 13
Electronic bonding energy, 23
Electronic equipment, 2–3
Electronics
    heat transfer importance in, 13–14
    MTBF of electronic system, 13–14
Emissivity defined, 326
Enclosures in natural convection
    boundary layer regime, 306, 308
        particle path and velocity vectors, 308
        temperature contours in, 308
    conduction limit, 305
        particle path and velocity vectors, 306
        temperature contours in, 306
    heat transfer
        mechanism, 306
        rate, 308
    Newton's law of cooling, 305
    Nusselt number, 305–306, 308
        correlation for, 309–310
    shallow limit, 305
        particle path and velocity vectors, 307
        and temperature contours, 307
    tall limit, 305
        particle path and velocity vectors, 307
        temperature contours, 307
    vertical boundary layers form, 308
    with vertical wall temperatures, 304
Energy
    balance equation, 29
        control mass, 31–36
        control volume, 36–38
        rate form, 30–31
        uses of, 33
        work/heat transfer, 33
    defined, 20
    flows in and out of control volume, 39
    per unit mass of system, 24
    system
        changes during time internal, 30
        entering and leaving, 30
        steady-state, 30
    transfer
        and heat transfer, 24
        per unit time, 25
    and work, relationship between, 19–20
Engineering systems, 30
    engineering verification test (EVT), 401
        design, 18
        phase, 17
Engine oil thermal conductivity, 55
Enthalpy per unit mass, 27
Entrainment limit, 426

Index

Ethane properties, 463
EVT, *see* Engineering verification test (EVT)
Exposed pad TQFP with die facing up/down cross section, 105
External flows, 209
External forced convection flow
   problem solution and calculations
      drag force, 247
      fluid properties, 246–247
      friction coefficient and Nusselt number, 247
      heat transfer rate and heat flux, 247–248
      Reynolds number based, 247
      unknown temperature, 247–248
External pump, 280
Extrusion heat sinks, 160–161

F

Fans
   based systems
      airflow rate and air velocity measurements, 17
   blowers
      characteristic curves of, 274
      examples of, 273
   curve, 388–389
   curve and system impedance curve
      altitude, effect, 275
      characteristic curves of, 274
      laws, 274–275
      observations, 275
      system impedance curve, 275–276
   mixed flow fans, 273
      characteristic curves of, 274
   and pump curve measurements
      performance curve, 409
      procedure, 409
      pump, test setup, 410
   radial fans, 272–273
   selection
      airflow rate, 277–278
      convection heat transfer coefficient, 276
      curve, 277
      impedance curves, 277–279
      pressure drop, 276
   tube-axial fan, 272
      characteristic curves of, 274
   vane-axial fan, 272
      characteristic curves of, 274
Fatigue failure, 8
FCBGA, *see* Flip-chip ball grid array (FCBGA)
FCPGA, *see* Flip-chip pin grid array (FCPGA)
Film temperature, 231

Fin
   adiabatic fin tip
      boundary conditions for, 132
      effectiveness, 142
      efficiency, 144
      solution of equation, 132
      thermal resistance, 140
      total heat transfer rate from, 133
   aluminum and copper as materials, 142
   circular fin, efficiency of, 147–148
      with nonuniform cross sections, 147
   conduction heat transfer, 128
   constant temperature fin tip
      boundary conditions, 135
      solution of equation, 136
      total heat transfer rate, 136–137
   convection and radiation from fin tip
      boundary conditions for, 133
      effectiveness, 142
      efficiency, 144
      solution of equation, 133–134
      thermal resistance, 141
      total heat transfer rate from, 134
   with corrected length
      temperature distribution and total heat transfer rate, 135
   effectiveness, 141–142
   efficiency, 143
   energy balance equation
      heat conduction and generation, 129
      steady-state condition and, 129
      total exposed area, 129
   equation, 127
      boundary conditions and, 130
      with constant thermal conductivity, cross section and perimeter, 130
   heat
      flow path, 128
      transfer rates, 128, 141
   infinitely long fin and
      boundary conditions for, 130
      effectiveness, 142
      efficiency, 144
      Fourier's law of heat conduction, 132
      heat transfer coefficient, 131
      solution of equation, 131
      temperature variation in, 131
      thermal resistance of, 140
   relationship between thermal resistance, effectiveness and efficiency, 144
   thermal conductivity of, 142–143
   triangular fin, efficiency of, 147
      with constant cross section, 148
      with nonuniform cross sections, 147
   with variable cross sections
      and constant thermal conductivity, 146

Finite-difference equations
    assumption results, 370
    momentum and two-dimensional heat conduction equation
        differences, 369–370
    solution, 380
        coefficient matrix, 381
        Gauss–Seidel decomposition and, 381
        matrix equation, general form, 381
    velocity correction formula, 370
First law of thermodynamics
    energy balance equation, 29
First-level interconnect, 91
First-order Bessel functions, 147
Fixed mass system, 30
Flat plate
    flow over
        average friction coefficient and average Nusselt number, 230, 236
        boundary layer equations, 227–228
        with constant temperature, 234
        convection heat transfer coefficient, 229
        friction coefficient, 229, 235
        normalized temperature, 228
        normalized velocity and temperature profiles, 229
        Nusselt number, 234–235
        Reynolds number, 235
        with uniform surface heat flux, 237
        velocity components, 228
    velocity boundary layer over, 213
Flip-chip ball grid array (FCBGA), 109, 112
    heat flow paths in, 115
Flip-chip metallic package, 10
Flip-chip pin grid array (FCPGA)
    cross section of, 107, 109
Flow rate measurement techniques
    Bernoulli equation, 401
        airflow chamber, 402
    blast gate use, 403
    commercial flow benches, 403
    device under test (DUT), 404
    drag force, 405
    flow bench, schematic of, 403
    mass balance between, 401
    nozzle plates, 401
        converging flow path, 402
        pressure taps, two sets, 403–404
    rotameter
        annular cross section area in, 406
        schematic of, 405
    turbine flow meter
        commercial, 406
        mechanical rotation, 407
    volumetric airflow rate, 402
Flows
    across circular cylinder
        characteristics of, 239
        critical Reynolds number, 239
        drag force and coefficient, 242
        forward stagnation point, 239
        laminar flow regimes, 240
        Nusselt number equation constants, 243
        Reynolds number, 239
        total convection heat transfer rate, 242
        turbulent flow, 241
        variation of drag coefficient with Reynolds number, 242
    external and internal flows, 209
    forced and natural convection flows, 210
    laminar and turbulent flows, 210–212
    over flat plate
        laminar flow with constant temperature, 227–231
        turbulent flow with uniform temperature, 234–236
        with uniform surface heat flux, 237
    steady-state and transient flows, 212
Fluid flow
    entry length and fully developed flow
        boundary layers form, 257–258
        fluid shell, thickness of, 258
        fluid velocity, 257
        hydrodynamic in pipe, 258
        inside pipe, 259
        for laminar flow, hydrodynamic entry length, 258
        thermal entry length of pipe, 258
        thermally fully developed region, 258
        turbulent flow, hydrodynamic entry length, 258
    inside pipe, 255–256
    laminar and turbulent pipe flows
        hydraulic diameter, 257
        Reynolds number for, 257
    mean velocity and mean temperature
        fluid temperature, 256
        mean fluid temperature, 257
        mean fluid velocity, 255
        total energy transfer rate, 256
        total mass flow rate, 255
    pipe, hydraulic diameter
        Reynolds number, 257
    pumping power and convection heat transfer
        average convection heat transfer coefficient, 262–263
        friction factor, 259
        local convection heat flux in, 260–262
        Newton's law of cooling for, 260, 263
        pressure drop in fully developed flow, 259
        small control volume of, 260
        total convection heat transfer rate, 260

Index

temperature profiles and convection heat transfer correlations
  bulk fluid mean temperature, 269
  convection heat transfer coefficient, 268–269
  Dittus-Boelter equation, 269
  fluid mean temperature, 268
  Nusselt number for, 268–270
  uniform surface heat flux, 267
velocity profiles and friction factor correlations, 263
  circular pipe, 265
  friction factor and Nusselt number for, 266
  fully developed turbulent flow, 267
  mean velocity for, 265
  parabolic velocity profile with, 265
  parallel plates, cross section of internal flow between, 264
  pumping power required, 265
Fluid mean temperature, 268
Folded fin heat sinks, 162–163
Force and total work done, 19
Forced convection, 57
  flows, 210
Forced convection internal flows
  entry length and fully developed flow
    boundary layers form, 257–258
    fluid shell, thickness of, 258
    fluid velocity, 257
    hydrodynamic in pipe, 258
    inside pipe, 259
    for laminar flow, hydrodynamic entry length, 258
    thermal entry length of pipe, 258
    thermally fully developed region, 258
    turbulent flow, hydrodynamic entry length, 258
  fans
    blowers, examples of, 273
    curve and system impedance curve, 274–276
    mixed flow fans, 273
    propeller fan, drawback, 271–273
    radial fans, 272
    radial fans, backward curved impeller, 273
    selection, 276–278
    tube-axial fan, 272
    vane-axial fan, 272
  laminar and turbulent pipe flows
    hydraulic diameter, 257
    Reynolds number for, 257
  mean velocity and mean temperature
    fluid temperature, 256
    mean fluid temperature, 257
    mean fluid velocity, 255

    total energy transfer rate, 256
    total mass flow rate, 255
  plate-fin heat sinks
    apparent friction factor, 282
    channel Reynolds number, 281–282
    convection heat transfer coefficient, 281
    pressure drop, 282
    schematic of, 281
  pumping power and convection heat transfer
    average convection heat transfer coefficient, 262–263
    friction factor, 259
    local convection heat flux in, 260–262
    Newton's law of cooling for, 260, 263
    pressure drop in fully developed flow, 259
    small control volume of, 260
    total convection heat transfer rate, 260
  pumps
    dynamic pumps, 279
    electroosmotic pumps and electromagnetic, 280–281
    positive displacement pumps, examples of, 279
  temperature profiles and convection heat transfer correlations
    bulk fluid mean temperature, 269
    convection heat transfer coefficient, 268–269
    Dittus-Boelter equation, 269
    fluid mean temperature, 268
    Nusselt number for, 268–270
    uniform surface heat flux, 267
  velocity profiles and friction factor correlations, 263
    circular pipe, laminar fully developed velocity profile, 265
    friction factor and Nusselt number for, 266
    fully developed turbulent flow, friction factor, 267
    mean velocity for, 265
    parabolic velocity profile with, 265
    parallel plates, cross section of internal flow between, 264
    pumping power required, 265
Fourier number, 360
Fourier's law, 67
  of conduction, 169
    Cartesian coordinate system and, 169–170
    local convection heat flux, calculation, 261
  of heat conduction, 54, 351
Fully developed laminar internal flows, 265
  friction factor and Nusselt number for, 266

G

GaAs FET, see Gallium arsenide field effect transistors (GaAs FET)
Gallium arsenide field effect transistors (GaAs FET), 94
Gamma rays, 319
Gases
  compressed in cylinder, 22
  conduction heat transfer, 53
  gas constant, 27
Gauss–Seidel decomposition and, 381
General form of heat conduction equation for isotropic medium, 176
General lumped system analysis, 198; see also Lumped systems
  heat balance equation, 199
  solution of equation, 200
  steady-state energy balance equation, 200
  with various boundary conditions, 199
General resistance network
  equivalent thermal resistance, 79–81
  total heat transfer rate, 80–81
General variable values, 364
Glass encapsulated thermistors, 414
Gold thermal conductivity, 54
Grashof numbers, 293
  net buoyancy force per unit volume, 294
  viscous force per unit volume, 294
Gravitational potential energy, 20–21
Gravitational pressure drop, 424
Gray surface, 327
Groove wicks, 422–423

H

Heat
  convection, 57
    coefficient, 58
    simplified correlations in air, 58–59
  radiation, 61–62
    coefficient, 63–64
  sink
    aspect ratio, 161–162
    base conduction thermal resistance, 153
    caloric thermal resistance, 154
    convection-radiation thermal resistance, 151
    cylindrical pin fin, 150
    effective heat transfer coefficient, 153
    effectiveness, 152
    efficiency, 153
    manufacturing processes, 160–164
    rectangular pin fin, 150
    rectangular plate fin, 150
    spreading resistance, 153
    thermal resistance of unfinned area of heat sink base, 151
    total fin area for, 152
    total heat transfer rate, 152
    total thermal resistance, 154
  slug, 105
  transfer, 53
    flux, 25
    from human body, 58
    importance in electronics, 13–14
    per unit time, 25
    through composite layer, 72
  transfer and fluid flow equations
    computer simulation tools, 353
    fluid flow equations, 353
    fluid pressure, 353
    Fourier's law of heat conduction, 352
    thermal conductivity fluid, 352
Heat conduction, 53
  boundary and initial conditions
    constants, 177
    convection, 181–182
    general, 185
    heat flux, 179–180
    interface, 186–187
    radiation, 183–184
    temperature, 178
  equation, 171
  one-dimensional heat conduction equation
    for plane wall, 171
    equations, 174–175
    heat balance equation, 172–173
    rate form, 172
    total heat transfers in and out of medium, 172
  in three dimensions
    finite-difference form, 176
    Fourier's law, 176
    general form of heat conduction equation, 176
    heat balance equation, 175
    Laplace equation, 177
    Poisson equation, 177
    steady-state heat conduction with, 177
    transient heat conduction with, 176
Heat pipes
  boiling mechanism, 424
    heat transfer from wick surface, 425
    temperature difference and, 425
  capillary pumping pressure
    gravitational pressure drop, 424
    orientation for, 423
    vapor pressure drop, 424
  electronic cooling applications, 428
    aluminum and copper, 430
    desktop CPU coolers as examples, 430

Index

embedded, heat sink base with, 431
notebook cooling as example, 429
entrainment limit, 426
loop heat pipe, 436
   schematic of, 437
performance
   bend angle effect of, 432
   flattening effect of, 431
performance limitations
   condenser heat transfer limit, 427
   frozen start-up limit, 427
   vapor flow in, 427
   vapor pressure, 427
pressure, 421
schematic of, 422
selection and modeling, 430–431
   computational model for, 435
   heat flow path in, 433
   thermal resistance network for, 432–434
   wick material, thermal conductivity of, 434
sonic limit, 426
thermosyphon
   with capillary wicks, 436
   dry out limit for, 436
   heat transport limitations of, 435–436
vapor chambers, 437
   computational model for, 438–439
   cross section and application of, 438
   thermal conductivity of, 438
   thermal resistance of, 437–438
water-based heat pipe performance
   gravity, effect of, 428
   and wick structure
      groove wicks, 422–423
      pore radius, 421
      screen wick, 422
      sintered metal wicks, 422
   working fluid, 421–423
      performance, diameter effect on, 428
High temperature and microelectronic device failures, 7
Hot-film anemometers, 412
Hot-wire anemometers, 412
   advantages and disadvantages, 412
   bridge circuit
      schematic of, 412
   examples and portable handheld, 413
   heat transfer rate, 411
Hydraulic diameter, pipe
   Reynolds number, 257
Hydrogen properties, 459

I

ICs, *see* Integrated circuits (ICs)
Ideal capacitor, 2
Ideal gas law, 27

Ideal product design cycle, 16
Inclined and horizontal plates
   natural convection
      average Nusselt number, 301
      buoyancy force, 300–301
      convection heat transfer, 301–302
      flow, 300–302
      and Rayleigh numbers, 301
Incompressible fluid
   energy equation with constant thermal conductivity, 216–217
   mass conservation equation for, 216
   momentum equations with constant dynamic viscosity, 216
Infinitely long fin
   boundary conditions for, 130
   effectiveness, 142
   efficiency, 144
   Fourier's law of heat conduction, 132
   heat transfer coefficient, 131
   solution of equation, 131
   temperature variation in, 131
   thermal resistance of, 140
In-line pin-fin heat sink, 244
Integrated circuits (ICs), 2
Intel's processors and Moore's Law, 3
Interface of two materials with perfect contact, 186
Internal energy per unit mass, 24
Internal flows, 209
   inside desktop computer and server, 210
Internal gear pumps
   schematic of, 280
International Technology Roadmap for Semiconductors (ITRS), 3
Ionic contamination failures, 13
Isothermal enclosure
   amount of incident radiation energy absorbed, 331
   object with temperature, 332
ITRS, *see* International Technology Roadmap for Semiconductors (ITRS)

J

Joule's law, 2
J-Shaped leads, 106
Junction, die part, 110
Junction-to-air thermal resistance, 110; *see also* Microelectronic packages
   JEDEC Solid State Technology Association enclosure, 111
   standards and publications for, 110
   test section for, 112
   measured temperatures and power dissipation, 111

Index

natural convection, 111
  variation with die size, 120
  variation with number of leads/pins, 119
  package, board and ambient conditions, 117
  power dissipation, 121
    measured temperatures and, 111
  test board and location package, 111–112
  variation with air velocity, 122
Junction-to-board thermal resistance, 113; *see also* Microelectronic packages
  thermal resistance measurement, setup for, 114
Junction-to-case thermal resistance, 112–113; *see also* Microelectronic packages
  power dissipation and, 121
Junction-to-top thermal characterization parameter, 114–115

K

Kinematic viscosity, 217–218
Kinetic energy, 20–21
Kirchhoff's law
  isothermal enclosure
    amount of incident radiation energy absorbed, 331
    object with temperature, 332
    rate of total radiation energy emitted, 331
Knudsen number, 427

L

Laminar and turbulent pipe flows
  hydraulic diameter, 257
  Reynolds number for, 257
Laminar flows, 210
  across circular cylinder
    regimes, 240
  on airfoil, 211
  hydrodynamic entry length, 258
  over circular cylinder, 211
  over flat plate, 211, 227
    normalized velocity and temperature profiles, 229
  thermal entry length of pipe, 258
Laminar forced convection air flow, 58
Laminar natural convection
  in vertical flat plate
    boundary layer equations for, 296
    normalized velocity and temperature profiles in, 297
    Nusselt number, 297–298
    Prandtl number, 297
    stream function with, 296–297
Laminar natural convection from vertical flat plate, 59
Laplace equation, 177

Large scale integrated circuits (LSICs), 2
Law of intermediate temperatures
  calibration curve, 415
14-Lead and 56-lead SOP, 104
240-Lead QFP and 128-lead LQFP with exposed top, 105
Liquid cooling system
  aluminum, 444
  centrifugal and reciprocating pumps, 443
  cold plates, 441
  copper and stainless steel, 444
  device
    pump and flow meter in, 408
    test setup for, 408
  direct or immersion, schematic of, 440
  fin-and-tube heat exchanger used in, 444
  indirect liquid
    cooling loop, components of, 439–440
    cooling system, 441
  liquid-cooled Cray-2 supercomputer, 440
  liquid-cooling modules, 13
  liquids used, exact composition of, 444
  material properties of, 445
  meso and microscale cold plates, 442
  module of Cray-2, 440
  Nusselt number in, 441–442
  pump use, 439, 441
  reference temperature, 443
  tubed cold plates, examples of, 442
Liquids, conduction heat transfer, 53
Lobe pump, 280
Local convection heat transfer coefficient, 215
Local friction coefficient, 214
  average friction coefficient, 225
  average Nusselt number, 226–227
  average of function, 225
  flow direction, 224
  functional form, 224
  local convection
    heat flux, 226
    heat transfer coefficient, 226
  Newton's law of cooling, 226
  normalized boundary layer energy equation, 225
  and Nusselt number, 223–224
  temperature gradient and, 225
  total convection heat transfer rate, 226
  total drag force, 225
  x-momentum equation, 224
Loop heat pipe, 436–437
Low profile quad flat pack (LQFP), 105
Low-profile rectangular surface-mount components, 104
LQFP, *see* Low profile quad flat pack (LQFP)
LSICs, *see* Large scale integrated circuits (LSICs)

Index   481

Lumped systems
  analysis validity
    Biot number, 203
    conduction heat flux, 203
    in steady-state condition, 202
    general lumped system analysis, 198, 201–202
    heat balance equation, 199
    solution of equation, 200
    steady-state energy balance equation, 200
    with various boundary conditions, 199
  simple lumped system analysis, 196
    heat balance equation, 197
    Newton's law of cooling, 197
    solution in terms of temperature variable, 198
    temperature variation with various time constants, 198
    time constant of, 198
    uniform temperature, 197

M

Macroscopic energy, 20
Manufacturing processes for heat sink
  bonded-fin heat sink technology, 160–161
  cold forming, 162
  die casting, 160
  extrusion, 160
  folded fin heat sink technology, 162
  skiving, 162–163
  swaging, 162–163
Mass conservation equation, 216
Mass flow at inlet and exhaust of control volume, 38
Materials expand and contract, temperature change, 8–9
Maximum allowable junction temperature of device, 110
Mean fluid
  temperature, 257
  velocity, 255
Mean strain rate tensor, 377
Mean time between failures (MTBF) of electronic system, 13–14
Mean velocity and mean temperature
  fluid temperature, 256
  mean fluid temperature, 257
  mean fluid velocity, 255
  total energy transfer rate, 256
  total mass flow rate, 255
Mechanical failures and temperature, 8–12
Mercury-in-glass thermometer, 414
Metallization and bond pads corrosion, 12
Metals thermal conductivity, 55–56

Methane
  properties of, 462
  thermal conductivity, 54–55
Microelectronic devices, 2
Microelectronic packages
  equivalent thermal resistance network for, 116
  packaging, 10
  parameters affecting thermal characteristics
    air velocity, 121
    board size and thermal conductivity, 122–123
    device power dissipation, 121
    die size, 119–120
    size and material, 119
  thermal resistance network, 115–117
  thermal specifications
    junction-to-air thermal resistance, 110–112
    junction-to-board thermal resistance, 113–114
    junction-to-case thermal resistance, 112–113
    package thermal characterization parameters, 114–115
    relationship between, 116
Microscopic energy, 20, 23
Microwaves, 319
Mixed convection, 313
  Nusselt number, 314
Mixed flow fans, 273
Modified radiation heat transfer coefficient, 64
Modified radiation thermal resistance coefficient, 70
Mold material, 104
Momentum diffusivity, *see* Kinematic viscosity
Momentum equations with constant dynamic viscosity, 216
Moore's law and Intel's processors, 3
MTBF, *see* Mean time between failures (MTBF) of electronic system
Multiple inlets and multiple exhausts
  control volume with, 39
  energy balance equation, 40–41
  energy balance for control mass, 40
  kinetic and potential energies, 41
  for steady-state system, 41–42
  work done on control mass, 40

N

National Technology Roadmap for Semiconductors (NTRS), 3
Natural convection velocity
  and temperature boundary layers, 290
  near vertical plate, 291
Natural/free convection, 57–58

from array of vertical plates
  convection heat flux from, 310, 312
  flow, 310
  heat flux, 312
  heat transfer coefficient, 312
  mounted on common base, 313
  Newton's law of cooling, 312
  Nusselt number, 312–313
  optimum plate spacing, 312–313
  outdoor electronic enclosure, 311
  printed circuit boards, 311
  uniform surface temperature, 310
  velocity profiles between, 311
from bottom side of horizontal flat plate, 59
in enclosures
  boundary layer regime, 306, 308
  conduction limit, 305–306
  with different vertical wall temperatures and, 304
  heat transfer mechanism, 306
  heat transfer rate, 308
  Newton's law of cooling for, 305
  Nusselt number, 305–306, 308–310
  shallow limit and particle path, 305, 307
  tall enclosure limit, 305
  tall limit and particle path, 307
  velocity vectors and temperature contours, 305, 307
  vertical boundary layers form, 308
flows, 210
  and buoyancy force and, 287
and horizontal cylinders
  convection heat transfer coefficient, 303–304
  Rayleigh and Nusselt numbers, 304
in inclined and horizontal plates
  average Nusselt number, 301
  buoyancy force, 300–301
  convection heat transfer, 301–302
  flow, 300–302
  and Rayleigh numbers, 301
from top side of horizontal flat plate, 59
vertical cylinders
  convection heat transfer coefficient, 303–304
  Rayleigh and Nusselt numbers, 304
n-butane properties, 465
Newton's law of cooling, 58, 63, 351
  forced convection internal flows
    pumping power and convection heat transfer, 260
  internal flow, 262
  for internal flow with uniform surface temperature, 263
Nitrogen properties, 460
Normalized boundary layer equations
  boundary layer equations, 220

coordinates and variables, 219–220
local convection heat transfer coefficient, 220
local friction coefficient, 220
steady-state boundary layer, 220
two-dimensional equations, 220
Normalized natural convection
  Grashof numbers, 293–294
Normalized natural convection boundary layer equations, 291
  Boussinesq approximation, 292–293
  convection heat transfer coefficient
    functional form, Nusselt number, 295–296
  Rayleigh number, 294
  vertical plate
    natural convection velocity profile, 292
No-slip boundary condition, 213
NTRS, see National Technology Roadmap for Semiconductors (NTRS)
NTRS/ITRS roadmaps
  for feature size, 3–4
  for high-performance processors power dissipation, 5–6
  for on-chip clock frequency, 4–5
  for power supply voltage, 5
  for transistor density, 3–4
Nuclear bonding energy, 23
Nucleate boiling, 425
Nusselt number
  circular cylinder, flows across
    equation constants, 243
  convection heat transfer coefficient, functional form, 295–296
  correlation for, 309–310
    enclosures natural convection, 309–310
  cylindrical pin fin heat sink
    average Nusselt numbers for, 245
  dimensionless group, 223
  equation constants
    for circular/noncircular cylinders, 243
  external forced convection flow
    problem solution and calculations, 247
  flat plate, flow over
    average friction coefficient and average Nusselt number, 236
  flows across circular cylinder
    equation constants, 243
  functional form of, 295
  laminar and turbulent natural convection
    in vertical flat plate, 297–298
  for laminar fully developed flow between
    surface heat flux, 268
    uniform surface temperature, 269
  and local friction coefficient, 223–224
    average Nusselt number, 226–227

Index

N (cont.)

natural convection
  from array of vertical plates, 312
  in enclosures, 305–306, 308
  in inclined and horizontal plates, 301
  vertical cylinders, 304
normalized natural convection boundary
    layer equations
  convection heat transfer coefficient,
      functional form, 295–296
  in shallow enclosure limit, 306, 308
  vertical and horizontal cylinders
    natural convection, 304
  in vertical enclosures
    boundary layer regime, correlation for,
        309–310

O

Observable energy, *see* Macroscopic energy
Ohm's law, 67
On-chip clock frequency, 4
One-dimensional heat conduction equation for
    plane wall, 171
  equations
    steady-state heat conduction, 174–175
    transient heat conduction with constant
        thermal conductivity, 174
  heat balance equation, 172–173
  rate form, 172
  for thin wall
    approach zero, 174
    Fourier law and, 174
    heat balance equation for, 173
    heat flow in and out, 173
    heat generation, 172
  total heat transfers in and out of medium, 172
One-dimensional steady-state heat
    conduction
  boundary conditions, 187–188
  conduction heat transfer rate, 189
  Fourier's law for, 187
  solution for equation, 190
  symmetric wall, 190
  temperature distribution, 187–188
  transfer, 67
One-equation models, 377
  channel flow and turbulent kinetic energy, 378
  transport equation, 378
Opaque object, 326
Oxygen properties, 461

P

Packages
  ball grid array (BGA) package, 107
    bottom sides of, 108
  bare-die package, 10
  ceramic small outline package (CSOP), 104
  flip-chip metallic package, 10
  packaging and
    cooling efficiency of, 110
    importance of, 103
    internal construction, 105
    thermal resistance network, 115–117
    types, 103–110
  pin grid array (PGA) packages, 106–107
  plastic ball grid array (PBGA) package, 112
    cross section of, 108
  plastic leaded chip carrier (PLCC) package
    cross section of, 106
    84-lead, 106
    mechanical and electrical construction
        of, 106
  plastic small outline package (PSOP), 104
  quad flat pack (QFP) packages, 104
    heat flow paths in, 115
    mechanical and electrical construction
        of, 106
    natural convection junction-to-air
        thermal resistance, variation of, 120
  shrink small outline package (SSOP), 104
  small outline package (SOP), 104
    I/O connections, 106
    mechanical and electrical construction
        of, 106
    soldering process, 107
  thin shrink small outline package
      (TSSOP), 104
  thin small outline package (TSOP), 104
  through-hole package, 110
  transistor outline (TO) package, 109–110
  wire-bonded plastic molded package, 10
Parallel thermal layers, 75
  heat transfer along, 76
  Ohm's law, 76
  steady-state condition, 76
PBGA, *see* Plastic ball grid array (PBGA)
    package
PCBs, *see* Printed circuit boards (PCBs)
Peclet number, 293
Peltier coefficient, 445–447
Piecewise linear profiles, 365, 368
Piezoelectric pumps, 279
Pin grid array (PGA) packages
  cross section, 106–107
  natural convection junction-to-air thermal
      resistance
    variation with die size, 120
  soldering process, 107
Pipes
  entry length and fully developed flow, 259
    boundary layers form, 257–258
    fluid shell, thickness of, 258
    fluid velocity, 257

hydrodynamic in pipe, 258
  for laminar flow, hydrodynamic entry
      length, 258
  thermal entry length of, 258
  thermally fully developed region, 258
  turbulent flow, hydrodynamic entry
      length, 258
laminar and turbulent pipe flows
  hydraulic diameter, 257
  Reynolds number for, 257
mean velocity and mean temperature
  fluid temperature, 256
  mean fluid temperature, 257
  mean fluid velocity, 255
  total energy transfer rate, 256
  total mass flow rate, 255
pumping power and convection heat transfer
  average convection heat transfer
      coefficient, 262–263
  friction factor, 259
  local convection heat flux in, 260–262
  Newton's law of cooling for, 260, 263
  pressure drop in fully developed flow, 259
  small control volume of, 260
  total convection heat transfer rate, 260
temperature profiles and convection heat
    transfer correlations
  bulk fluid mean temperature, 269
  convection heat transfer coefficient,
      268–269
  Dittus-Boelter equation, 269
  fluid mean temperature, 268
  Nusselt number for, 268–270
  uniform surface heat flux, 267
velocity profiles and friction factor
    correlations, 263
  circular pipe, laminar fully developed
      velocity profile, 265
  friction factor and Nusselt number for, 266
  fully developed turbulent flow, friction
      factor, 267
  mean velocity for, 265
  parabolic velocity profile with, 265
  parallel plates, cross section of internal
      flow between, 264
  pumping power required, 265
Pitot-static tube, 410
  schematic of, 411
Planck constant, 319
Planck distribution, 325
Plane wall
  thin plane wall
    with constant temperature of sides,
        194–195
    with convection on both sides, 196
    exposed to convection on both sides, 182
    exposed to radiation on both sides, 184
    with general boundary conditions on
        both sides, 186
    with uniform internal heat generation
        and convection on both sides, 189
    with uniform internal heat generation
        and fixed temperature surfaces, 188
    with two fixed temperature surfaces, 177–178
    with two specified heat flux surfaces, 180
Plastic ball grid array (PBGA) package, 112
  cross section of, 108
Plastic leaded chip carrier (PLCC) package
  cross section of, 106
  84-lead, 106
  mechanical and electrical construction
      of, 1060
Plastic packages cracking, 12
Plastic quad flat pack (PQFP), 104
  junction-to-air thermal resistance
    variation with air velocity, 122
    variation with size and thermal
        conductivity, 123
  junction-to-case thermal resistances
    variation with air velocity, 122
  natural convection junction-to-air thermal
      resistance
    variation with die size, 120
Plastic small outline package (PSOP), 104
Plate-fin heat sink, 154, 343–344
  channel view factor, 344–346
  error in, 347
  effective emittance, 344
  Stefan–Boltzmann constant, 344
PLCC, see Plastic leaded chip carrier (PLCC)
    package
Point/spot measuring radiation thermometer,
    416–417
Poisson equation, 177
Polymers thermal conductivity, 55–56
Potential energy, 20
Power, 25
PQFP, see Plastic Quad Flat Pack (PQFP)
Prandtl number, 222, 297, 376, 378
  laminar and turbulent natural convection
      in vertical flat plate, 297
  vertical flat plate
    laminar and turbulent natural
        convection, 297
Printed circuit boards (PCBs), 91
  axial fan use, 389
  compact model for, 386–387
  copper content in, 385
  cross section, 94
  effective normal thermal conductivity of,
      93, 96
  effective planar
    thermal conductivity, 94
    thermal resistance of, 93

Index

fans and blowers
  curves, 389
  flow rates, 388
  heat transfer path in, 92
  import board files from, 384–385
  junction temperature of, 386
  junction-to-board thermal resistance, 387
  modeling of, 94
  network model, 387
  orthotropic heat transfer properties, 92
  shunt network model, 388
  star network model, 388
  thermal conductivity of, 92
  thermal resistances, 387
  thermal vias, 95–96, 149, 385–386
  thickness of, 95
  with three internal copper layers, 92–93
  total glass epoxy and copper layer thicknesses, 92
  total thermal resistance of, 92
  trace/signal layer and power layer, 385
  two-resistance network model, 387–388
Process/production verification tests (PVT), 18
Propane properties, 464
Propeller fan, 271
  drawback, 272
PSOP, *see* Plastic small outline package (PSOP)
Pumping power and convection heat transfer
  average convection heat transfer coefficient, 262–263
  friction factor, 259
  local convection heat flux in, 260–262
  Newton's law of cooling for, 260, 263
  pressure drop in fully developed flow, 259
  small control volume of, 260
  total convection heat transfer rate, 260
PVT, *see* Process/production verification tests (PVT)

Q

QFP, *see* Quad flat pack (QFP) packages
Quad flat pack (QFP) packages, 104
  heat flow paths in, 115
  mechanical and electrical construction of, 106
  natural convection junction-to-air thermal resistance, variation of, 120

R

Radial fans, 272
  backward curved impeller, 273
Radial thermal resistances, 434
Radiation heat transfer, 61
  black bodies, between
    radiation exchange between, 340
    Stefan–Boltzmann law for, 340
  blackbody radiation
    defined, 324
    Planck distribution, 325
    and real body radiation, comparison with, 324
    spectral emissive power, 324–325
    spectral intensity, 324
    Stefan-Boltzmann law, 325
    total emissive power of, 325
    Wein's displacement law, 325
  coefficient, 63
    modified, 64
  electromagnetic waves
    emission, 319
    spectrum, 319–320
  equivalent resistance for, 70
  Kirchhoff's law
    isothermal enclosure, 331–332
    rate of total radiation energy emitted, 331
  net radiation heat transfer rate, 62
    coefficient, 63
    equation for, 63
  nonblack bodies, between
    space resistance to radiation, 341–342
    surface resistance to radiation, 341
  object and surrounding, between, 62
  plate-fin heat sink, 343–344
    channel view factor, 344–347
    coefficient, 348
    effective emittance, 344
    measured and calculated radiation heat transfer rates from, 349
    rate from heat sink, 348
    Stefan–Boltzmann constant, 344
  radiation intensity and emissive power, 320
    conical region, differential solid angle, 321–322
    diffuse emitter, 323–324
    solid angle, 322
    spectral intensity, 322–323
    spectral radiation flux, 323
    spherical coordinate system, coordinates of point and dimensions, 321
    surface element, 321
    surface radiation, 321
    total emissive power, 323
    total surface area of sphere, 321
  radiosity, defined, 336
  solar and atmospheric radiations
    effective sky temperature, 334–335
    net radiation heat flux, 335
    object in outdoor environment, 336
    room temperature solar absorptivity and surface emissivity, 335
    solar absorptivity, 333–334
    solar constant, 333
    solar radiation flux on, 333

spectral/Planck distribution of, 332
sun, apparent temperature, 332–333
surface absorptivity
  relationships between, 329
  spectral absorptivity, 328
  spectral-directional absorptivity, 328–329
  total absorptivity, 328
surface reflectivity
  spectral-directional reflectivity, 330
  spectral-hemispherical reflectivity, 329
  total-hemispherical reflectivity, 329
surfaces, radiation properties
  gray surface, 327
  surface emissivity, 326–328
  total emissive power and emissivity, 326
surface transmissivity
  spectral-directional transmissivity, 330
  spectral-hemispherical transmissivity, 330
  total-hemispherical transmissivity, defined, 330
view factor
  defined, 337
  reciprocity rule, 337
  for some three-dimensional geometries, 339
  for some two-dimensional geometries, 338
  summation rule, 337
  superposition rule, 338
  symmetry rule, 338–340
Radiation intensity and emissive power
  conical region, differential solid angle, 321–322
  diffuse emitter, 323–324
  solid angle, 322
  spectral intensity, 322–323
  spectral radiation flux, 323
  spherical coordinate system, coordinates of point and dimensions, 321
  surface element and radiation, 321
  total emissive power, 323
  total surface area of sphere, 321
Radiation thermal resistance, 70
Radiation thermometry, 415
Radio frequency (RF) transistors, 94
Radiosity defined, 336
Rayleigh numbers
  inclined and horizontal plates natural convection, 301
  normalized natural convection boundary layer equations, 294
Reciprocating pump, 279
  schematic of, 280
Reciprocity rule, 337
Rectangular pin fin heat sink, 150

Rectangular plate fin heat sink, 150
Reliability of system and MTBF, 14
Resistance network for two-layer composite with perfect and real interface, 84
Resistance temperature detectors (RTDs), 414
Resistors, heat/power dissipation in, 2
Reynolds number, 293
  forced convection internal flows laminar and turbulent pipe flows, 257
  inertia force/volume, 221
  pipe, hydraulic diameter, 257
  viscous force/volume, 221
Reynolds stress models, 380
Rotameter
  annular cross section area in, 406
  schematic of, 405
Rotary pumps, 279–280
Rotational energy, 23
RTD, *see* Resistance temperature detectors (RTDs)

S

Saturated ammonia thermal conductivity, 54–55
Saturated liquid and vapor ammonia properties of, 468–469
Saturated propane thermal conductivity, 54–55
Saturated water/steam properties, 466–467
Saturated water thermal conductivity, 55
Screen wick, 422
Second-level interconnect, 91
Seebeck coefficient, 445–446, 448
Seebeck effect, 414–415
Semiconductor Industry Association (SIA), 3
Semiconductor technology, 3–6
Semi-implicit method for pressure-linked equations (SIMPLE), 372–373
Series thermal layers, 72
  mathematical identity, 73
  qualitative temperature profile, 73
  resistance network for, 74
Shell-and-tube heat exchanger, 244
Shrink small outline package (SSOP), 104
SIA, *see* Semiconductor Industry Association (SIA)
Silicon thermal conductivity, 55–56
Simple compressible substance, 26
Simple lumped system analysis, 196; *see also* Lumped systems
  heat balance equation, 197
  Newton's law of cooling, 197
  solution in terms of temperature variable, 198
  temperature variation with various time constants, 198
  time constant of, 198
  uniform temperature, 197

Index

Sintered metal wicks, 422
Skived fin heat sinks, 163–164
Skiving processes, 162–163
Small outline integrated circuit (SOIC), 104
Small outline package (SOP), 104
   cross section, 104
   I/O connections, 106
   mechanical and electrical construction of, 106
   soldering process, 107
SOIC, *see* Small Outline Integrated Circuit (SOIC)
Solar and atmospheric radiations
   effective sky temperature, 334
   net radiation heat flux, 335
   object in outdoor environment, 336
   room temperature solar absorptivity and surface emissivity, 335
   solar absorptivity, 333–334
   solar constant, 333
   solar radiation flux on, 333
   spectral or Planck distribution of, 332
   sun, apparent temperature, 332–333
Solid angle defined, 322
Solid materials properties, 470
Sonic limit, 426
SOP, *see* Small outline package (SOP)
Sound power level (SWL), 418
Sound waves, 417
   sound pressure level (SPL), 418
      microphones, locations of, 419
Source term, 368–370
   densities, 368
   values of, 364
Specific heat
   at constant pressure and volume, 26
SPL, *see* Sound pressure level (SPL)
Spreading/constriction thermal resistance, 88
Spring force in Newton, 21
SSOP, *see* Shrink small outline package (SSOP)
Staggered pin-fin heat sink, 244
Steady-state flows, 212
Steady-state heat conduction
   one-dimensional
      boundary conditions, 187–188
      conduction heat transfer rate, 189
      Fourier's law for, 187
      solution for equation, 190
      symmetric wall, 190
      temperature distribution, 187–188
   two-dimensional
      solutions for equation, 192–193
      temperature boundary conditions, 192
      temperature distribution, 191–192
Steady-state heat transfer, 171
Steady-state system, 30

Stefan–Boltzmann law
   for blackbody emissive power, 340
   constant, 62–63, 325
Straight fin heat sink, 151
Stress corrosion in packages, 12
Summation rule, 337
Superposition rule, 338
Surface-mount package, 110
Surfaces
   absorptivity
      spectral absorptivity, defined, 328
      spectral-directional absorptivity, defined, 328–329
      total absorptivity, defined, 328
   contact area between, 83
   emissivity, 62–63
   radiation properties
      gray surface, 327
      spectral-directional emissivity, defined, 327
      spectral-hemispherical emissivity, 327
      surface emissivity, defined, 326
      total emissive power and emissivity, 326
      total-hemispherical emissivity, defined, 327–328
   reflectivity
      spectral-directional reflectivity, defined, 330
      spectral-hemispherical reflectivity, defined, 329
      total-hemispherical reflectivity, defined, 329
   transmissivity
      spectral-directional transmissivity, defined, 330
      spectral-hemispherical transmissivity, defined, 330
      total-hemispherical transmissivity, defined, 330
Swaged-fin heat sinks, 162
Swaging processes, 162–163
SWL, *see* Sound power level (SWL)
Symmetry rule, 338–340
Synthetic jets
   airflow and thermal properties of, 451
   conventional jet ejectors, 450–451
   diaphragm, periodic motion of, 450
   formation of, 450
   induced volumetric flow rate, 451
   Nusselt number for, 451–452
   operation, principle of, 451
System
   with clear boundaries, 33
   control mass and volumes, 30
      intersection of, 37
      mass balance equation, 37–38

mass flow at inlet and exhaust, 38
volumetric flow rate, 38
defined, 24
ways of, 33
energy
entering and leaving, 30
flows, in/out of, 33
during time internal, 30
equation of state, 25
internal energy per unit mass and pressure, 26
final and initial energies, 29
fixed amount of mass inside, 37
level cooling and test technologies, 6
steady-state, 30
system-level thermal design, 16
system-specific constant, 13
total energy of, 29
System impedance measurement
device under test (DUT), 407–408
impedance curve of, 408

T

TEC, see Thermoelectric cooler (TEC)
Temperature boundary layers
and natural convection velocity, 290
near vertical plate, 291
over surface, 214–215
Temperature-dependent failures, 6–7
corrosion failures, 12
electrical failures, 12–13
mechanical failures, 8–12
Temperature measurement techniques
aluminum plate covered with different paints and tapes, 416
law of intermediate temperatures
calibration curve, 415
mercury-in-glass thermometer, 414
metals, electrical resistance of, 414
point measuring radiation thermometer, 416
point/spot radiation thermometer, 416–417
radiation thermometry, 415–416
resistance temperature detectors (RTDs), 414
thermal imaging camera, 416
two-dimensional temperature map, 417
thermistors
resistance, variation of, 414
thermocouples
junctions, 414
metal pairs, 415
Seebeck effect, 414–415
Temperature profile in two layers with perfect and real interface, 83
Temperature profiles and convection heat transfer correlations
bulk fluid mean temperature, 269
convection heat transfer coefficient, 268–269
Dittus-Boelter equation, 269
fluid mean temperature, 268
Nusselt number for, 268–270
uniform surface heat flux, 267
Thermal boundary layer, 214
Thermal conductivity of material, 54
and electrical conductivities, 55–56
spectrum of, 55
Thermal contact conductance, 84
Thermal contact resistance, 82
Newton's law of cooling, 84
temperature change across interface, 84
temperature variation in, 83
thermal resistance networks, 84
Thermal design
experimental measurements, importance of, 419
prototype, 420
test plan and report, 420
modeling and simulation, importance
building thermal models, 399
design cooling solutions for, 398–399
process, 14
architectural thermal design, 16–17
concepts, 15
fans, dimensions and locations, 17
feasibility study, 16
high-level thermal designs, 16–17
phases of, 15
requirements, 15–16
system-level thermal design, 16
Thermal diffusivity of material, 176
Thermal imaging camera, 416
two-dimensional temperature map, 417
Thermal interface materials (TIM)
compliance/conformability, 86
compressed to, 85
dielectric strength, 87
ease of use and reworkability, 87
interface resistances, 85
long-term stability and reliability, 87
modulus of elasticity, 86
pressure and compression ratio, 87
thermal conductivity, 86
thermal resistance
with pressure, variation of, 86
with thickness, variation of, 87
thickness, 86
Thermally enhanced LQFP with die facing up/down cross section, 105
Thermal radiation defined, 320
Thermal resistance
approximation, 89
circular heat source with, 88
computational solutions, 89
concept of, 67

Index

convection, 89
heat sink, 88
source and ambient, 88
spreading, 88
Thermal runaway, 12–13
Thermal vias, 95
Thermistors resistance variation, 414
Thermocouples junctions, 414
Thermoelectric cooler (TEC)
    coefficient of performance (COP), 448
    couples, 446
    heat transfer rate, 448
    junctions, 445
    maximum cooling capacity of, 448
    Peltier coefficient, 445–447
    Seebeck coefficient between, 445–446, 448
    thermoelectric module, 446–447
Thermosyphon
    with capillary wicks, 436
    dry out limit for, 436
    heat transport limitations of, 435–436
Thickness of velocity boundary layer, 213–214
Thin quad flat pack (TQFP), 105
Thin rectangular plate/long rectangular rod
    with convection, adiabatic and constant temperature boundary condition, 193
    with four different sets of boundary conditions, 191
    with general temperature distribution on four sides, 193
Thin shrink small outline package (TSSOP), 104
Thin small outline package (TSOP), 104
Three dimensions heat conduction
    finite-difference form, 176
    Fourier's law, 176
    general form of heat conduction equation, 176
    heat balance equation, 175
    Laplace equation, 177
    Poisson equation, 177
    steady-state heat conduction with, 177
    transient heat conduction with, 176
Through-hole package, 110
TO, *see* Transistor outline (TO) package
Total absorptivity defined, 328
Total emissive power defined, 323, 326
Total-hemispherical reflectivity defined, 329
Total-hemispherical transmissivity defined, 330
Total internal energy, 23
Total irradiation defined, 326
TQFP, *see* Thin quad flat pack (TQFP)
Transient flows, 212
Transient heat conduction
    long wall in
        solution of heat conduction equation for, 196
    thin plane wall in
        temperature distribution, 194–195
Transient heat transfer, 170
Transient term, 367
Transistor outline (TO) package, 109–110
Transistors, heat dissipation in, 2
Translational energy, 23
Triangular fin
    efficiency of, 147
    with constant cross section, 148
    with nonuniform cross sections, 147
TSOP, *see* Thin small outline package (TSOP)
TSSOP, *see* Thin shrink small outline package (TSSOP)
Tube-axial fan, 272
Turbine flow meter
    mechanical rotation, 407
    two commercial, 406
Turbulent flows
    across circular cylinder, 241
    average value, 374
    closure problem, 375
    direct numerical simulation (DNS), 373
    eddy viscosity, 375–376
    fluid velocity in, 211
    hydrodynamic entry length, 258
    momentum and temperature equations, 374–375
    one-equation models, 377
        channel flow and turbulent kinetic energy, 378
        transport equation, 378
    over flat plate, 212, 227
        with uniform temperature, 234–236
    Prandtl number, 376
    Reynolds experiment, 375
        repetitions of, 373
    Reynolds stress models, 380
    time averaging, 374
    turbulent momentum fluxes, 375
    turbulent water jet, 212
    two-equation models
        dissipation rate, 379–380
        turbulent viscosity, 379
        values, 380
    zero-equation model
        and mean strain rate tensor, 377
        mixing length, 377
        turbulent viscosity for, 377
Turbulent forced convection air flow, 59
Turbulent kinetic energy
    transport equation for, 378
Turbulent natural convection
    in vertical flat plate
        boundary layer equations for, 296
        normalized velocity and temperature profiles in, 297

Nusselt number, 297–298
Prandtl number, 297
stream function with, 296–297
Turbulent natural convection from vertical flat plate, 59
Two-dimensional boundary layer equations, 217–218
Two-dimensional plate with internal heat generation, 194
Two-dimensional steady-state heat conduction
solutions for equation, 192–193
temperature boundary conditions, 192
temperature distribution, 191–192
Two-equation models
dissipation rate, 379–380
turbulent viscosity, 379
values, 380

U

Uniform heat flux surface, 179

V

Vacuum tubes, 2
Vane-axial fan, 272
Vapor chambers
computational model for, 438–439
cross section and application of, 438
thermal conductivity of, 438
thermal resistance of, 437–438
Vapor pressure
drop, 424
limit, 427
Vapor transfers, 432
Velocity
boundary layer, 213
over flat plate, 213
correction formula, 370
Velocity measurement methods
Bernoulli equation, 410
cross-wire anemometer, 412
glass encapsulated thermistors, 414
hot-film anemometers, 412
hot-wire anemometers, 411–412
advantages and disadvantages, 412
bridge circuit, schematic of, 412
examples of, 413
heat transfer rate, 411
pitot-static tube, 410
schematic of, 411
Velocity profiles and friction factor correlations, 263
circular pipe, laminar fully developed velocity profile, 265
friction factor and Nusselt number for, 266

fully developed turbulent flow, friction factor, 267
mean velocity for, 265
parabolic velocity profile with, 265
parallel plates, cross section of internal flow between, 264
pumping power required, 265
Vertical and horizontal cylinders
natural convection
convection heat transfer coefficient, 303–304
Rayleigh and Nusselt numbers, 304
Vertical flat plate
laminar and turbulent natural convection
boundary layer equations for, 296
convection heat flux, 298
normalized velocity and temperature profiles in, 297
Nusselt number, 297–298
Prandtl number, 297
stream function with, 296–297
Vertical plates, array
natural convection
heat flux, 312
heat transfer coefficient, 312
mounted on common base, 313
Nusselt number, 313
optimum plate spacing, 312–313
Vibrational energy, 23
View factor, radiation heat transfer
defined, 337
reciprocity rule, 337
for some three-dimensional geometries, 339
for some two-dimensional geometries, 338
summation rule, 337
superposition rule, 338
symmetry rule, 338–340
Viscous dissipation term, 218
Viscous force, 212
Viscous sheer stress, 212
local friction coefficient, 214
and velocity gradient at surface, 213
Volumetric thermal expansion coefficient
Boussinesq approximation, 290
for ideal gas, 289

W

Water, gravitational potential energy change in, 33
Water working fluid
heat pipe performance
bend angle effect of, 432
flattening effect of, 431

Index

Wein's displacement law, 325
Wet corrosion mechanism, 12
Wick material, thermal conductivity, 434
Wire-bonded plastic molded package, 10
Wire bond fatigue, 10
Wire fatigue, 10
Work
    defined, 19–20
    done on control mass, 40
    and energy
        relationship between, 19–20
        transfer, 24

X

X-rays, 319–320

Z

Zero-equation model
    and mean strain rate tensor, 377
    mixing length, 377
    turbulent viscosity for, 377
Zero-order Bessel functions, 147
Zero-velocity particles, 213